Traditional Herbal Medicine Research Methods

Traditional Herbal Medicine Research Methods

Identification, Analysis, Bioassay, and Pharmaceutical and Clinical Studies

Edited by

Willow J.H. Liu

WILEY

A John Wiley & Sons, Inc., Publication

Published by John Wiley & Sons, Inc., Hoboken, New Jersey
Published simultaneously in Canada

Library of Congress Cataloging-in-Publication Data:
Traditional herbal medicine research methods : identification, analysis, bioassay, and pharmaceutical and clinical studies/edited by Willow J.H. Liu.
p. ; cm.
Includes bibliographical references and index.
ISBN 978-0-470-14936-2 (cloth)
1. Herbs–Therapeutic use–Testing. 2. Materia medica, Vegetable–Research. I. Liu, Willow J. H. (Willow Jian Hua), 1963–
[DNLM: 1. Plant Extracts–analysis. 2. Plant Extracts–pharmacology. 3. Medicine, Chinese Traditional. 4. Phytotherapy–methods. 5. Plants, Medicinal–chemistry. QV 766 T763 2011]
RM666.H33T68 2011
615′.321072—dc22
2010023286

10 9 8 7 6 5 4 3 2 1

Dedication

This book is dedicated to all who are interested in traditional herbal medicines and willing to make their contribution to people's health with their efforts.

I first want to thank all of the professors in China, Germany, and the United States who have trained me or with whom I have worked; in chronological order, they are Professor Song-Song Yang, Xin-Sheng Yao, Rudolf Bauer, Koji Nakanishi, Norman R. Farnsworth, John M. Pezzuto, Judy L. Bolton, and Gail B. Mahady. Special acknowledgment is given to all of the authors of this book for contributing their time and knowledge, as well as to Dr. Hong-Jie Zhang for introducing some of them to me. Acknowledgment is also given to all of my previous colleagues and friends who have provided me with generous academic support.

And most of all, I want to thank my husband, Zhuo Chen, for his understanding and all unsung support to my career, and my daughter, Emily (Chen Chen), for her time on editing this book. They have been traveling with me to different countries and cities throughout the world, accompanying me in both body and soul. This book is also a special present to my lovely little son Derek, a precious gift sent from God while I was preparing this book.

Contents

Preface

After giving a presentation on the topic of "Modern Research on Traditional Herbal Medicine" in the American Chemical Society national meeting (September 2006, San Francisco), I received an e-mail from John Wiley & Sons asking if I would write a book with the same title.

Research on traditional herbal medicine involves botany, chemistry, biology, pharmacology, toxicology, clinical trials, and other disciplines. Chemical composition and biological or biochemical activities of many herbs have been studied by researchers in universities and pharmaceutical companies for purposes of investigation or new drug development. So far, there have been many books introducing functions or actions of herbs. Books on the chemistry of herbal medicines (often called phytochemistry), biochemistry, biology, and pharmacology of herbal medicines are also available. But there is no book giving a full description of all aspects of herbal research and development.

The purpose of research on traditional herbal medicine is not only for new drug development, but also for quality control and mechanism study of herbs. Unlike screening for new drug candidates simply using one or two bioassay tests, exploration of the mechanisms of traditional herbal medicines is much more complex. It requires close cooperation between scientists from many disciplines to unveil the secrets of the herbal kingdom. For maximal cooperation, pharmacologists, biologists, chemists, and clinical doctors need to have basic knowledge of the cooperating fields. To scientists who are very knowledgeable in modern science and have extensive research experience, their knowledge about the applications and theories of traditional medicines they are studying, for example, traditional Chinese medicine (TCM) and Ayurveda, might be limited. For this reason, their research designs for these herbs may simply copy those for new drug development. As a consequence, the results may not be accurate due to either inappropriate design of extract methods, insufficient experiment duration in animal study, or lower concentration of samples used for bioassay tests. This is why studies using the same assay for the same herb from different labs have often reported different results.

This book introduces the methodology of collection and identification of herbal materials, extraction and isolation of compounds from herbs, *in vitro* bioassay, *in vivo* animal test, toxicology, and clinical trial for herbal research. It is not written as a literature review. Instead, it introduces the basic content and methodology of each research field and the keys for the study of herbal medicine. The purpose of this book is to help scientists who are interested in the study of traditional herbal medicine gain a broader view of herbal medicine and knowledge about its research. I hope this book can be a bridge to provide scientists in different fields with basic information and knowledge about the progress of herbal study and to help them avoid unnecessary mistakes during their studies.

As for background information on my relationship with traditional herbal medicine, I received my B.S. in Chinese Herbal Medicine, M.S. in pharmacognosy, and Ph.D. in natural product chemistry in succession from Shenyang, China. My doctoral supervisor was Xinsheng Yao of Shenyang Pharmaceutical University, a well-known phytochemist and academician in China. I left China as a professor of phytochemistry at Liaoning University of Traditional Chinese Medicine and traveled to Germany as an AvH Research fellow. There I worked in the lab of Rudolf Bauer at Duesseldorf University, a world-renowned expert in plant medicine, in particular of Echinacea, and an aficionado of Chinese herbs. I focused on bioassay screening and standardization of herbal medicines. Later I moved to New York and worked in the lab of Koji Nakanishi at Columbia University, and then to Chicago, where I worked with Norman Farnsworth and John Pezzuto at the UIC/NIH Center for Botanical Dietary Supplement Research in Women's Health at the College of Pharmacy, University of Illinois at Chicago. The project there was the mechanism study of herbs for treating women's menopause symptoms, using *in vitro* bioassays and *in vivo* animal tests.

Unfortunately, I became afflicted with rheumatoid arthritis while I was in Chicago. My wrist gave me so much pain that I was too weak to even open a reagent bottle. Even while being treated with Western medicine, I was once paralyzed and could not get out of bed. For health reasons, I thus had to leave Chicago's harsh weather for California, and started to treat myself with Chinese herbs and acupuncture, in addition to treatment with Western medicine, while working in a research lab for pharmaceutical analysis. In my spare time, I taught Chinese herbology, TCM nutrition, as well as modern pharmacology and nutrition, at various schools of acupuncture. A few years later, I opened my own clinic of herbs and acupuncture. Since then, I have been treating patients with my combined knowledge on the functions of traditional herbs and their modern biological and pharmacological activities, meanwhile developing herbal products based on the efficacy of herbal formulas in clinical application. Research, teaching, plus clinical practice strongly consolidated my knowledge on both traditional and modern medicine, and helped me review TCM theories more deeply and from multiple perspectives.

My research on herbs in the past 25 years has told me that they work in a way that differs from modern drugs: the effect is not from one single compound in an herb, but is a synergetic result from many components working on many targets. And researchers should not be disappointed if their results show that the most bioactive compounds screened from an herbal extract in a bioassay are popular second metabolites in plants. Examples include flavonoids, fatty acids, or amines.

The successful treatment of a variety of diseases in my clinic with Chinese herbal formulas has reminded me of what I had first learned during college: the effective treatment with Chinese herbs is mostly based on formulas composed of several or more individual herbs, rather than single ones, and the formulation of a Chinese herbal prescription is guided by theories of TCM. But most scientists conducting research on traditional herbal medicine today are either unaware of or are ignoring this.

My research in China, Germany, and the United States has extended my knowledge of phytochemistry, analytical chemistry, biochemistry, biology, and pharmacology and experience with extraction, isolation, identification, and analysis of compounds in herbs and their bioassay screening and mechanism study with *in vitro* and *in vivo* tests. This is the reason that I boldly accepted the invitation to write this book. To make each chapter in this book more authoritative, I invited several experts from different fields in China to write some chapters. In the process of editing, necessary rewriting, rearrangement, additions, and clarification of contents were made with the agreement from authors of each chapter. However, due to limited space, it is impossible to cover all aspects or give detailed information in each chapter. I hope this book will work as a guideline for new scientists working with modern technologies and help them to explore more secrets in the treasury of traditional medicines.

Willow J.H. Liu

Contributors

Jia-Tao Feng, Dalian Institute of Chemical Physics, Chinese Academy of Sciences, Dalian, China

Zhi-jun Huang, The Third Xiangya Hospital in Central South University, Changsha, China

Yu Jin, East China University of Science and Technology, Shanghai, China

Yan-Xiong Ke, East China University of Science and Technology, Shanghai, China.

Chun-Li Li, Department of Pharmacology, Shenyang Pharmaceutical University, Shenyang, China

Ping Li, Department of Pharmacognosy, China Pharmaceutical University, Nanjing, China.

Xin-Miao Liang, Dalian Institute of Chemical Physics, Chinese Academy of Sciences, Dalian, China

Hong-Wei Liu, Institute of Microbiology, Chinese Academy of Sciences, Beijing, China

Hui-Juan Liu, Department of Pharmacognosy, China Pharmaceutical University, Nanjing, China

Willow J.H Liu, 21st Century Herbs & Health, Inc. CA, U.S.A.

Qi Luo, Changsha National Biomedical Industrial Base, Hunan Engineering Research Center of Botanical Extract, Liuyang, Hunan, China

Xuan Peng, Changsha National Biomedical Industrial Base, Hunan Engineering Research Center of Botanical Extract, Liuyang, Hunan, China

Man-Liang Tan, Changsha National Biomedical Industrial Base, Hunan Engineering Research Center of Botanical Extract, Liuyang, Hunan, China

Fang Wang, Department of Pharmacology, Shenyang Pharmaceutical University, Shenyang, China

Li-Hui Wang, Department of Pharmacology, Shenyang Pharmaceutical University, Shenyang, China

Chun-Fu Wu, Department of Pharmacology, Shenyang Pharmaceutical University, Shenyang, China

Guo-Ping Yang, The Third Xiangya Hospital in Central South University, Changsha, China

Jing-Yu Yang, Department of Pharmacology, Shenyang Pharmaceutical University, Shenyang, China

Ling Yi, R&D Department, School of Chinese Medicine, Hong Kong Baptist University, Kowloon, Hong Kong
Hong Yuan, The Third Xiangya Hospital in Central South University, Changsha, China
Jian-Guo Zeng, Hunan Agricultural University, Changsha, China

Abbreviations

AAS	atomic absorption spectrometry
AEs	adverse effects
AFLP	amplified fragment length polymorphism
AFS	atomic fluorescence spectrometry
AIC	Akaike Information Criterion
ALP	alkaline phosphatase
AMD	age-related macular degenerative disease
ANOVA	analysis of variance
AP	alkaline phosphatase
APCI	atmospheric pressure chemical ionization
API	atmospheric pressure ionization
AP-PCR	arbitrarily primed polymerase chain reaction
APPI	atmospheric pressure photoionization
ASTM	American Society for Testing and Materials
BMD	bone mineral density
CAG	coronary angiography
CAM	complementary and alternative medicine
CCCD	China Certification Committee for Drugs
CD	circular dichroism
CE	capillary electrophoresis
CHD	coronary heart disease
CHL	Chinese hamster lung cell
CI	chemical ionization
CID	collision-induced dissociation
CIOMS	Council for International Organizations of Medical Sciences
CNPIC	China National Pharmaceutical Industry Corporation Limited
CNS	central nervous system
COSY	chemical shift correlation spectroscopy
COX	cyclooxygenase
CP	cyclophosphamide
CPC	centrifugal partition chromatography
CQS	comprehensive quality systems
DAD	diode array detector
DCC	droplet countercurrent
DEPT	distortionless enhancement by polarization transfer
DMEM	Dulbecco's modified Eagle's medium
DMSO	dimethyl sulfoxide
DOPAC	3,4-dihydroxyphenylacetic acid

DPD	deoxypyridinoline
DPPH	2,2-diphenyl-1-picrylhydrazyl
E	enzyme
E_2	estradiol
ECD	electrochemical detector
ECG	electrocardiogram
ECL	enhanced chemiluminescence
EFPIA	European Federation of Pharmaceutical Industries Associations
EI	electron ionization
EIA	enzyme immunoassay
EIS	enzyme-inhibitor-substrate complex
ELISA	enzyme-linked immunosorbent assay
ELS	evaporative light scattering
ELSD	evaporative light scattering detector
EMEA	European Medicines Agency
ER	estrogenic receptor
ERB	Ethical Review Board
ERE	estrogen-responsive element
ERT	estrogen replacement therapy
ES	enzyme-substrate complex
ESI	electrospray ionization
EU	European Union
FAB	fast atom bombardment
FBS	fetal bovine serum
FC	flash chromatography
FCPC	fast centrifugal partition chromatography
FD	field desorption
FDA	Food and Drug Administration
FDCA	Federal Food, Drug, and Cosmetic Act
FI	field ionization
FLARE	fragment length associated repair enzyme
FOB	functional observatory battery
FT	Fourier transform
FT-ICR	Fourier transform ion cyclotron resonance
FTMS	Fourier transform mass spectrometry
GAP	good agriculture practice
GABA	γ-aminobutyric acid
GC	gas chromatography
GCP	good clinical practice
GE	gel electrophoresis
GEP	good extracting practice
GLP	good laboratory practice
GMP	good manufacturing practice
GOT	glutamate oxaloacetate transaminase
GPT	glutamate pyruvate transaminase

GSLS	Ginseng stem and leaf saponins
GTP	guanosine triphosphate
HBV	anti-hepatitis B virus
HHS	Department of Health and Human Services
5-HIAA	5-hydroxyindoleacetic acid
HILIC	hydrophilic interaction liquid chromatography
HMBC	heteronuclear multiple bond correlation
HMQC	heteronuclear multiple quantum coherence
HPLC	high-performance liquid chromatography
HRT	hormone replacement therapies
HSCC	high-speed countercurrent
HSQC	heteronuclear single quantum coherence
HT	serotonin
HTS	high-throughput screening
HVA	homovanillic acid
IBS	irritable bowel syndrome
ICH	International Conference on Harmonization of Technical Requirements for Registration of Pharmaceuticals for Human Use
ICP-MS	inductively coupled plasma mass spectroscopy
IEC	independent ethics committee
IND	investigational new drug
IR	infrared
IRB	institutional review board
ISSR	inter-simple sequence repeats
ITS	internally transcribed spacer
JPMA	Japan Pharmaceutical Manufacturers Association
LC	liquid chromatography
LD	lethal dose
LDL	low-density lipoprotein
LH	luteinizing hormone
LhRh	luteinizing hormone releasing hormone
LIT	linear ion trap
LOD	limit of detection
LOQ	limit of quantitation
LPH	lipotropic hormone
LPLC	low-pressure liquid chromatography
LS	light scattering
LSD	least significant difference
MAE	microwave-assisted extraction
MALDI	matrix-assisted laser desorption
MBC	metastatic breast cancer
MEM	minimum essential medium
MHLW	Ministry of Health, Labor, and Welfare
MOH	Ministry of Health
MPLC	medium-pressure liquid chromatography

MRM	multiple-reaction monitoring
MS	mass spectrum; mass spectrometer; mass spectrometry
MTD	maximum tolerated dose
MTS	3-(4,5-dimethylthiazol-2-yl)-5-(3-carboxymethoxyphenyl)-2-(4-sulfophenyl)-2H-tetrazolium
MTT	3-(4,5-dimethylthiazol-2-yl)-2,5-diphenyltetrazolium bromide
MW	molecular weight
NAD^+	nicotinamide adenine dinucleotide
NADH	reduced form of nicotinamide adenine dinucleotide
$NADP^+$	nicotinamide adenine dinucleotide phosphate
NADPH	reduced form of nicotinamide adenine dinucleotide phosphate
NCCAM	National Center for Complementary and Alternative Medicine
NCI	National Cancer Institute
NDA	new drug application
NEI	neuroendocrine-immune network
NIH	National Institutes of Health
NIR	near-infrared
NIRS	near infrared spectrometer; near infrared spectrometry
NLM	National Library of Medicine
NMR	nuclear magnetic resonance
NOAEL	no-observed-adverse-effect level
NOE	nuclear Overhauser effect
NOESY	nuclear Overhauser effect spectroscopy
NP–LC	normal phase liquid chromatography
ODS	octadecasilica
OHRP	Office for Human Research Protections
ORAC	oxygen radical absorbance capacity
ORD	optical rotatory dispersion
ORR	objective response rate
OVX	ovariectomized rat model
PB	particle beam
PBS	phosphate buffer saline
PC	paper chromatography
PCR	polymerase chain reaction
PD	pharmacodynamics
PE	phosphatidylethanolamine
PK	pharmacokinetics
PLE	pressurized liquid extraction
PMS	phenazine methosulfate
pQCT	peripheral quantitative computed tomography
PQR	product quality review
PR	progestin receptor
PRMA	Pharmaceutical Research and Manufacturers of America
PTLC	preparative thin layer chromatography
QA	quality assurance

QC	quality control
QOL	quality-of-life
QRM	quality risk management
QT	Q wave and T wave in ECG
QU	quality unit
RACE	rapid amplification of cDNA ends
RAPD	random amplified polymorphic DNA
RDA	retro-Diels–Alder
RFLP	restriction fragment length polymorphism
RI	refractive index
RP–LC	reversed phase liquid chromatography
RT-PCR	reverse transcriptase-polymerase chain reaction
SATCM	State Administration of Traditional Chinese Medicine
SAGE	serial analysis of gene expression
SCE	supercritical CO_2 extraction
SCGE	single cell gel electrophoresis
SDA	State Drug Administration
SEC	size exclusion chromatography
SFC	supercritical fluid chromatography
SFDA	State Food and Drug Administration
SFE	supercritical fluid extraction
SOP	standard operation procedure
SPE	solid phase extraction
SRM	selected-reaction monitoring
SSR	simple sequence repeats
SSRI	serotonin selective reuptake inhibitors
SWH	*Sambucus williamsii* HANCE extract
TCM	traditional Chinese medicine
TFA	trifluoroacetic acid
THF	tetrahydrofuran
TLC	thin layer chromatography
TMS	trimethylsyl
TOCSY	total correlation spectroscopy
TOF	time-of-flight
TQS	total quality systems
TSP	thermospray ionization
TTP	time to progression
UV	ultraviolet
UV-Vis	ultraviolet-visible spectrometry
VLC	vacuum liquid chromatography
WET	water suppression enhanced through T1 effects
WHO	World Health Organization
WMA	World Medical Assembly

Chapter 1

Introduction to Traditional Herbal Medicines and Their Study

Willow J.H. Liu

1.1 DEFINITION AND TRENDS OF TRADITIONAL HERBAL MEDICINES

According to the World Health Organization (WHO), traditional medicine refers to health practices, approaches, knowledge, and beliefs incorporating plant, animal, and mineral-based medicines, spiritual therapies, manual techniques, and exercises, applied singularly or in combination to treat, diagnose, and prevent illnesses or to maintain well-being. If the material being used is of plant origin, then it is called traditional herbal medicine.

Different types of traditional medicines are widely applied in Asia, Africa, and Latin America to meet primary health-care needs. Traditional medicine has maintained its popularity in most regions of the developing world. The application is also rapidly spreading in industrialized countries, where adaptations of traditional medicines are often termed "complementary" or "alternative." In the United States, the National Institutes of Health (NIH) uses the name complementary and alternative medicine (CAM) to cover health systems, practices, and products that are not considered part of conventional medicine. Worldwide, among all the different traditional medicine systems, traditional Chinese medicine (TCM) is currently the most popular, followed by Indian medicine. In Western terminology, the name "Oriental medicine" covers Chinese, Japanese, and Korean medicines preferred by immigrants from Korea, while "Asian medicine" is often used to include TCM, Indian (Ayurveda), and Tibetan medicine. Among all treatment methods in traditional medicine systems, medicinal herbs are the most widely applied.

Traditional Herbal Medicine Research Methods, Edited by Willow J.H. Liu

Medicine has been revolutionized in Europe by advances in chemistry, laboratory techniques, and equipment since Robert Koch discovered the transmission of disease by bacteria, followed by the discovery of antibiotics in the early 1900s. Thus, modern medicine is commonly called Western medicine even though there are also traditional medicines in Western countries. It is also called conventional medicine.

Webster's medical dictionary defines conventional medicine as medicine practiced by holders of medical doctor (M.D.) or doctor of osteopathy (D.O.) degrees and by their allied health professionals, such as physical therapists, psychologists, and registered nurses. Other terms for Western medicine or conventional medicine include allopathy and allopathic medicine, mainstream medicine, orthodox medicine, regular medicine, and biomedicine.

Although conventional medicine is the mainstream medicine in Western countries, application of traditional medicine, including herbal medicines, is growing worldwide for many reasons, in particular, the side effects or inefficacy of modern drugs. The following data are provided by the WHO.

- In Africa, up to 80% of the population uses traditional medicine for primary health care.
- In China, traditional herbal preparations account for 30–50% of the total medicinal consumption.
- In Europe, North America, and other industrialized regions, over 50% of the population has used complementary or alternative medicine at least once.
- In Germany, 90% of the population has used a natural remedy at some point in their lives.
- The global market for herbal medicines currently stands at over USD$60 billion annually, and is growing steadily.

Since the last century, scientists all over the world have studied herbal medicines from the fields of chemistry, biology, pharmacology, toxicology, and clinical trials. Recently, in addition to screening out new drug candidates, investigators also expect to explore the preventative and therapeutic mechanism of herbal medicines that play very important roles in most of the traditional medicine systems, such as TCM and Ayurveda medicine.

1.2 RESEARCH AND DEVELOPMENT OF HERBAL MEDICINES

The use of herbal medicines for treatment of diseases was documented several thousand years ago. As seen from journals, studies on herbal medicines have been encompassed under several different names, such as plant medicine, phytomedicine, pharmacognosy, and natural products. "Natural products" usually refer to products processed or derived from living organisms, including plants, animals, insects, microorganisms, and marine organisms.

Data from the WHO show that 25% of modern medicines are made from plants that were first used traditionally. Examples include atropine, morphine, quinine, ephedrine, warfarin, aspirin, digoxin, vincristine, taxol, and hyoscine.

Traditional medicine needs to be modernized in the twenty-first century. However, modernization of traditional medicine should not be simply Westernization. For herbal medicines, the purpose of a study is not only to screen out bioactive compounds from herbal extracts for new drug development, but also to standardize and control the quality of raw herbal materials and their products to ensure the safety and efficacy; and more importantly, to reveal their preventative and therapeutic mechanisms. So far, only a relatively small number of herbal medicines have been well studied from all of these aspects; these herbs include Echinacea, ginkgo, ginseng, and licorice.

To a large extent, the depth and progress of research on herbal medicines depend on the development of related technology and equipment, as well as the in-depth understanding of the human body and diseases. Mechanism study and functional evaluation of herbal medicine involve the fields of chemistry, biochemistry, biology, pharmacology, toxicology, and clinical study. Thus, organized and consistent teamwork is absolutely vital.

Researchers from different labs need to work closely together, discuss problems frequently, and analyze the results instantly. A scientist for extraction and isolation of herbal medicines in the chemistry lab should have enough knowledge of biology and pharmacology to provide an appropriate sample because an improperly extracted or isolated sample provided from his or her lab for biological and pharmacological study could lead to wrong results in the bioassay or animal test. The scientist in the bioassay or animal lab for screening or mechanism study of herbal medicines should make sure that the sample to be tested is correctly extracted, that the concentrations of tested samples are within a proper range, and that the design of the experiment is scientific enough to provide a true result. And to reach such a goal, an adequate understanding of the research target, the functions and indications, as well as clinic applications of the study herb is necessary. The following are several main aspects of herbal medicine research.

1.2.1 Extraction, Isolation, and Identification of Compounds in Herbal Medicines

All the substances in the universe, including plants, are composed of chemical compounds. To study herbal medicine, the major bioactive chemical components should be first known. Only after the biological compounds in herbs are correctly extracted, isolated, and identified can biochemical, biological, or pharmacological studies be performed scientifically.

Chemical studies of herbal medicines provide fundamental substances for further studies of biological and pharmacological activity. During the earlier decades of the 1800s, chemical studies in plants could only be performed on active compounds that were highly concentrated and isolated into a relatively pure form by techniques such as distillation or extraction with water, acid, base, or alcohol. Their structures were mainly determined by chemical degradation and proven by synthesis in an unambiguous manner. Scientists were unable to determine the stereochemistry of compounds.

The well-known example is the story of aspirin. According to records about willow leaves as an antipyretic treatment in Ebers papyrus, and following the same application of teas made from willow bark as an English herb, chemists and pharmacists successfully isolated salicin from the bark of the white willow, *Salix alba*, between 1825 and 1826. The compound responsible for the remedy was subsequently converted to salicylic acid via hydrolysis and oxidation, and proved as such a successful antipyretic (fever reducer) that it was actively manufactured and used worldwide. Due to severe gastrointestinal toxicity, salicylic acid was converted into acetylsalicylic acid via acetylation by scientists at Bayer. It was given its trade name of aspirin in 1899. Today, aspirin is still the most widely used analgesic and antipyretic drug in the world.

Since the 1950s, chromatography, including medium-pressure liquid chromatography (MPLC) and high-performance liquid chromatography (HPLC), and other methods such as supercritical fluid extraction (SFE), droplet countercurrent (DCC), and high-speed countercurrent (HSCC) have been popularly applied for isolation of natural products, while different types of spectral equipment such as infrared (IR), ultraviolet (UV), nuclear magnetic resonance (NMR), circular dichroism (CD), and mass spectrometer (MS), as well as MS coupled with gas chromatography (GC), have been commonly used for structure identification. Later on, LC-MS and LC-NMR also became available and gradually more popular in the last few decades. These advances have made the time for extraction, isolation, and identification of compounds from herbal medicines much shorter than that of a century ago. Modern extraction and isolation techniques, combined with all types of chromatography, are often guided by bioassays to isolate the active compounds. High-throughput screening with robots also dramatically lowers the screening times. Thus, structure-efficacy elucidation of newly isolated bioactive compounds is no longer a time-consuming and difficult process.

However, the process of finding new drug candidates from herbs for drug development is no longer as easy as the story of aspirin. The story of taxol is that of a difficult journey of a trace compound from a plant becoming a powerful new drug. Taxol is one of the most well-known diterpenes with a very complex steroid structure and anticancer activity. The extract of the bark of Pacific yew (*Taxus brevifolia*) was first found to be cytotoxic in a cellular assay in 1964. The active ingredient was isolated in 1966 with a very low amount, and the structure was published in 1971. By 1969, 28 kg of crude extract had been isolated from almost 1200 kg of bark, but yielded only 10 g of pure material. The research result showed that it acts to stabilize the mitotic apparatus in cells, causing them to act as normal cells rather than undergo rapid proliferation as they do in cancer. But it was not until the late 1980s that its value as an anticancer drug was confirmed.[1]

Current modern methods and techniques such as all kinds of chromatography and spectrometry, and their combined application make the extraction, isolation, and structure identification of bioactive compounds from herbs dramatically faster than half a century ago. Highly accurate analytical equipment, such as HPLC coupled with UV and/or MS and other detectors, makes the quality control and standardization of herbal products more reliable for pharmacological and clinical studies. Advanced biochemical and biological technologies, such as microarray, allow scientists to easily explore the mechanism study at the enzyme, receptor, and gene levels quantitatively using only small amounts of samples. These advanced technologies and their applications to herbal study will be introduced in the following chapters. With all these available high technologies, time for isolation and identification of compounds from herbs is becoming shorter and trace bioactive compounds are more easily obtained. With the popularity of various spectroscopy methods, identification of isolated compounds is becoming much easier than it was decades ago. Application of hyphenated LC-UV/MS and LC-NMR techniques greatly accelerates the systematic identification of compounds in an herbal extract.

To perform any herbal study, identification of the herbal materials used for study should never be neglected. Morphological, microscopic, physical, or chemical identification can all be applied to identify the raw materials. The availability of HPLC chromatogram or gene fingerprints makes identification of species highly accurate.

1.2.2 Bioassay Screening and Mechanism Study of Herbal Medicines

Scientists have spent over a hundred years trying to screen new drug candidates from herbal medicines. Recently, due to the rapid growth of products of herbal medicine or alternative medicine all over the world, their efficacy and safety have become more and more important. More attention has been drawn to the preventative and therapeutic mechanism study of herbal medicines. For both reasons, bioassay study on herbs is indispensible. Thanks to the advancement of biological technologies, more and more bioassays are available for mechanism study. The mechanism of many effective herbal medicines has been elucidated, such as the well-known ginkgo, Echinacea, red clover, black cohosh, ginseng, and many Chinese and other traditional herbs. Bioassays *in vitro* are usually followed by *in vivo* animal tests to further confirm the functional mechanism and understand the absorption, metabolism, and toxicity in living bodies.

Bioassay is commonly performed using enzymes, receptors, genes, cells, and sometimes tissues. In comparison to screening for new drug candidates of single compounds, screening herbal extracts or fractions is relatively difficult due to the solubility or complex composition in herbal samples. Compounds in an extract might interfere with each other, or more specifically, the activity of one compound might be masked by another in the mixture due to the adverse effect or toxicity of the latter. So, the bioassay result of an herbal extract should be carefully evaluated, particularly when a high-throughput method is applied, not only due to the mentioned interference, but also because of the dramatically varied concentrations of bioactive components in different samples prepared under the same conditions.

Mechanism study for herbal medicine does not necessarily use high-technology equipment. The most important thing is to select the right targets. Different enzymes, receptors, or genes should be tested for mechanism of an herbal extract. Assays at different levels should be applied to ensure the positive or negative research results. Evaluation of estrogenic activity of red clover and black cohosh extracts using different bioassays can be used as an example.[2]

In many cases, the corresponding bioactive components for the functional mechanism of herbal medicines are common or universally distributed compounds. Such results may disappoint researchers looking for new drug development, but they are very helpful to scientists who are dedicated to explaining the functions of herbs or willing to understand more about physiological functions of these common compounds in the human body. Examples include linolic acid, a cyclooxygenase (COX) inhibitor in *Angelica pubescens*[3] and an estrogenic agonist in *Vitex agnus-castus* L. (chaste berry),[4] and N_{ω}-methylserotonin, a serotonin agonist in black cohosh.[5]

1.2.3 Pharmacological and Toxicological Study of Herbal Medicines

Similar to modern pharmaceutical study, pharmacological study of herbal medicines include pharmacodynamic (PD) and pharmacokinetic (PK) aspects. Broadly, toxicology is also part of the pharmacology.

PD study of traditional herbal medicines is not always easy. Up to now, only the most popularly used herbs, a very small fraction of the total number used, have been well known with respect to pharmacological effects on animals. One reason is that herbs might treat diseases in a way different from known modern drugs. Black cohosh is one example. This herb has long been used in North America for menopause symptoms in women, but *in vivo* animal study indicated that its extract did not exhibit effects in ovariectomized Sprague–Dawley rats. Further study showed that instead of directly binding to estrogen receptors, extract of black cohosh was reported acting as a mixed competitive ligand and partial agonist of the serotonin and opiate receptor,[6,7] which indicates that this herb might treat menopause symptoms through regulation of the central nervous system.

Chinese scientists have done numerous pharmacological studies on Chinese herbs. Therapeutic mechanisms of the most commonly used Chinese herbs have been known by systematic PD studies.[8–10] However, there is another challenge in the pharmacological study of Chinese herbs; that is, in the vast majority of cases, the practitioners prescribe formulas that consist of several (sometimes over 20) herbal ingredients for the treatment. This makes the study difficult not only due to the complex analysis of chemical composition for quality control of the test samples, which is important to keep good reproducibility of the results, but also because of the complex theories of TCM behind the combination of different herbs, which will be mentioned in Chapter 10.

Many people mistakenly believe that herbal products are safe. Although most herbal medicines are relatively safe in comparison with modern drugs, results from toxicological studies show that this is not always true. To a large extent, the safety of herbs depends on dosage and period of administration. It is necessary to mention that purification of some herbal extracts may increase their toxicity. This is because, while the active components are concentrated, the concentration of toxic compounds may also be increased. Sometimes, the active components are toxic. In this case, while the therapeutic effect is enhanced, the toxicity is also increased. Examples include ephedra extract and herbal extracts from the Aristolochia family. Studies of aristolochic acid found in several herbs in Aristolochia family have shown its significant carcinogenic and mutagenic effects and poisoning of the kidney.[11–13] In TCM, processing of raw herbal materials with different methods, such as extended heating with steaming or boiling to decompose the chemical bonds of toxic ester or glycoside compounds in herbs, has been long applied to reduce the toxicity of Chinese herbs. Examples include aconitine in radix Aconiti and sennosides in rhubarb.

PK study of herbal medicines is so far mainly applied to herbs with known active compounds. The concentrations of these active index compounds in the blood, urine, and other body liquids or tissues after a certain period of administration are measured and compared by means of UV, MS, GC-MS, HPLC-MS, and other analytical methods to analyze the distribution of the compounds and change of concentrations with time. To herbs with unclear composition or whose concentration could not be monitored with analytical methods, their efficacies are measured and time-efficacy curves are drawn. In addition, PK–PD models are also applied to the study of herbal PK.

This book covers the PD and toxicology studies of herbal medicines, but not the PK. The reason is that the methods of sample collection for PK study of herbal medicine are the same as those for modern drugs. The analytical methods for absorbed and metabolized known compounds in herbs can refer to the qualitative and quantitative analysis of herbal medicines in Chapter 9. Keep in mind that the complex chemical composition of herbal preparations always makes the analysis relatively difficult.

In comparison with so many PD study reports of herbal medicines, only a few systematic PK studies for herbal preparations have been reported; one example is the PK of alkamides in Echinacea purpurea.[14] Progress of the PK study is covered in recent review articles.[15–17]

1.2.4 Chemical Standardization and Quality Control of Herbal Medicines

Substitute or counterfeit herbal materials are often found in the market. Even for the right species, the chemical composition and concentrations of bioactive compounds may vary dramatically with different collection seasons and regions as well as storage. Therefore, it is necessary to chemically standardize the herbal extracts or products for biological, pharmacological, and clinical studies.

The complex composition of herbal medicines makes the quality control of herbal products much more complicated. With the increase in knowledge about the bioactive and main compounds in most of the commonly used herbs and the popular application of various analytical instruments such as HPLC, equipped with UV, MS, and other detectors, fingerprint chromatograms are becoming powerful qualitative and quantitative methods for standardization of herbal medicines. Such standardization is not only necessary for quality control of final herbal products, but also important to guide the species collection and cultivation, as well as the optimization of the processing procedure.

1.2.5 Clinical Studies of Herbal Medicines

Anything that exists on the earth has a need for survival. Many traditional herbs have been used on human beings to prevent and treat diseases for hundreds or even thousands of years. The fact should be acknowledged that most of the herbs have been used by countless people. Take Chinese herbal medicine as an example. The efficacies, toxicities, therapeutic and toxic dosages, as well as cautions and contraindications of most herbs have been well recorded in many traditional Chinese herbal books. Although the terminologies used for diagnosis and treatment of diseases in traditional and modern medicines are different, researchers are encouraged to figure out the symptoms described in traditional terminologies for the application of traditional medicine and try to match them to that of modern diseases for scientific clinical trial.

A successful clinical trial depends on accurate scientific design. In comparison to the trial for a single chemical drug, that for an herbal product is more complicated due to the complex composition and difficult quality control of the components. The extract method, the concentrations of the main or bioactive compounds in the products (or the purity of the products), the number and criteria of patients selected, the route and dosage of the administration, the period of the trial, and the method to collect and process the data will all influence the results of the trial.

Unfortunately, many of the reported results of clinical studies on herbal medicine so far are not reliable due to more or less unscientific design. Quite often, the results of clinical trial for one herbal medicine obtained by different research groups vary significantly. A well-known example is St. John's Wort. Some reported this herb to have an effect on mild depression; others reported no such effect. Possible reasons have been mentioned in the above paragraph. A difference in any step of the experimental design will affect the result.

To obtain reliable clinical trial results for herbal medicines, double-blind experiments should be applied with enough patients selected, ideally using the standard of clinical trial for new drug development. Of course, budgetary constraints are often a hindrance to carrying out such trials.

1.3 COMMON MISTAKES SEEN IN RESEARCH ON TRADITIONAL HERBAL MEDICINES

Before starting research on herbal medicines, researchers should carefully search for literature that is related to the study. After reviewing the literature, they should develop a research plan by writing a detailed procedure design. The following common mistakes should be avoided.

1. Starting preparations of samples without identification of herbal materials.

For many reasons, substituted or adulterated herbal medicines are often seen in the markets. Sometimes they are not easily distinguished from the right material with the naked eye.

2. Starting biological or pharmacological experiments without chemical identification and standardization of samples.

I recall that one day an American friend showed me a bag containing an herbal product. The label on the bag said: No chemicals, all natural. This can lead to a popular misconception among consumers. But as scientists, we should know that chemicals are the fundamental substances of biological activities of herbal medicines, and nature is made up of chemicals. Therefore, chemical identification and standardization must be the primary step in the experiment of modern herbal study. Otherwise, the results are not reliable or accepted.

3. Using the wrong extraction method or solvent, such that the bioactive compounds are not extracted.

Make sure the extraction method will extract the corresponding bioactive compounds. For example, if an extract is for a steroid receptor binding assay, the potential ligands will probably be lipophilic, thus a less polar solvent such as chloroform may be selected. If an extract is for an antivirus experiment, the possible bioactive compounds may be large molecular glycoproteins or polyssacharides; lipophilic solvents or alcohol will not extract them out. The best way is to extract the material with different polar solvents in succession and test them separately in the primary test.

4. Using a dosage for the bioassay or animal test that is too low.

Since the efficacies of bioactive compounds in herbs are relatively weaker than the positive modern drug in most cases, and the concentrations of bioactive compounds are very low in the extract, the negative result of a sample in an assay or animal test may become positive if the concentration of sample is increased. Several dosages at different magnitudes are suggested to prepare for the primary test. Sometimes, the concentration of an herbal extract might be 1000 times higher than that of the positive control. For example, when the estrogenic activity was evaluated for red clover, methanol extract of red clover did not show positive results in the estrogen receptor binding assay until its concentration was increased to 20 μg/mL.

5. Having a test period in an animal study or clinical trial that is not of sufficient length.

Because the effects of bioactive compounds in herbs are relatively moderate in comparison with the positive modern drug, it usually takes a longer time to see the positive result of an herbal extract in animal tests. For example, an estrogenic test for synthetic drug candidates on ovariectomized rats may only need a week, but positive results of a red clover extract were not observed until the third week of the experiment.

6. Using samples that vary in composition, leading to unrepeatable results.

Ideally, the same batch of herbal sample solution should be used for the same assay or test. If not, chemical analysis should be performed for different batches of samples by HPLC to avoid variable results caused by inconsistent quality or quantity of compounds in samples.

1.4 RESEARCH ON TRADITIONAL HERBS SHOULD REFER TO THEORIES AND CLINICAL APPLICATION OF TRADITIONAL MEDICINE

Many traditional herbs are clinically prescribed by practitioners of traditional medicine under the guidance of theories in traditional medicine, such as TCM in China and Ayurveda in India. This aspect has mostly been ignored by scientists in the field of modern research of herbal medicine for product development, particularly in Western countries. Even in Asia, chemists, biologists, and pharmacologists who have been studying herbal medicine with modern knowledge and technology in labs for many years rarely know enough about theories that guide the applications of herbal treatments in the clinic. One of the reasons is that such study is more challenging.

Traditional herbs might treat a disease in a way different from known modern drugs. Take TCM as an example. A disease can be divided into several "*zhengs*" based on TCM differentiation. "*Zheng*" is a Chinese word that is similar in meaning to English symptoms or signs. For example, there is "cold *zheng*," "hot *zheng*," "internal *zheng*," "external *zheng*," "excessive *zheng*," "deficient *zheng*," "yin *zheng*," "yang *zheng*," "damp *zheng*," and "*bi zheng*" (bi means blocked). Different herbs may be used on different patients with same disease but different *zhengs*. Sometimes, no animal models can be found to match these *zhengs* for PD study of herbs that are clinically used for treatment of certain types of *zhengs*. Thus, a new model with a particular *zheng* has to be established first. To do this, scientists have to be knowledgeable in both traditional and modern medicines. Otherwise, the study results are not reliable. Up to now, Chinese scientists have found out the biological and pathological foundation for most of the *zhengs* in TCM and established many animal models for pharmacological study of herbal medicine.[9]

Theories of traditional medicines, such as TCM, cover etiology, pathology, diagnosis, and treatment. Study of these theories can not only help us to explore the mechanisms of herbal treatment, but also help scientists explore possible new etiology and pathology for diseases whose causes are still unknown in modern medicine, thus providing new directions for drug development. For such purposes, a variety of *in vitro* bioassays on different receptors, enzymes, and other targets and *in vivo* animal pharmacological tests should be performed on herbs—not only individual ones, but also herbal formulas.

For example, clinical practice has confirmed that Gui Zhi Fu Ling Wan (Cinnamomi and poria composition), a Chinese herbal formula composed of five Chinese herbs, is very effective in decreasing or eliminating uterine fibroids when their diameter is less than 5 cm. This has been confirmed by comparing ultrasound exam results before and after the treatment in the clinic and by pharmacological study on rats. The uterine fibroids are usually removed by surgery in modern medicine if they cause severe abnormal bleeding or if they are too big. Quite often, the uterus will be removed together with the fibroids in order to prevent the regrowth of fibroids in the uterus at a later date. The Chinese formula can not only stop abnormal bleeding and decrease and eliminate the fibroid, but also prevent the regrowth of the fibroid because it regulates the imbalance of the hormones, the cause of fibroid growth. Female hormones, particularly estrogen and progesterone, are known to be related to stimulation of fibroids. TCM considers fibroid formation to be related to accumulation of stagnated blood (called "*yu zheng*"). Therefore, herbs that invigorate blood circulation are added to the formula. Combining the knowledge about formation of uterine fibroids in modern medicine and TCM, the mechanism of herbal treatment can be explained by chemical, biological, and pharmacological study. To study the treatment mechanism of the formula, not only *in vitro* assays and *in vivo* animal tests related to hormone regulation should be performed; those involved in blood circulation should also be carried out.

Research on traditional herbal medicine should be performed on the basis of clinical application and reference to the corresponding theories in each system. The main systems of traditional medicine from different countries will be briefly introduced in Section 1.5. TCM is mentioned below only as an example.

The application of traditional Chinese herbs is not as simple as Western drugs in that not all doctors prescribe the same medicines for the same disease. Quite often in TCM, one herbal formula consisting of several Chinese herbs (most often 5 to 15) is used for different diseases. On the other hand, one disease can be treated with different formulas by different doctors. For example, if an herb is unavailable, experienced Chinese doctors can easily modify a formula by replacing one or two herbs to give similar treatment results. This makes research scientists perplexed and frustrated because explanations by clinical doctors using terminology of TCM are sometimes difficult to understand. Due to the current meticulous division of research areas and a limited amount of energy, most scientists focus on in-depth study in one field, and have no time to spend on other areas that are not closely related to their research. Even to those familiar with both TCM and modern science, if the knowledge on both sides is not extensive, it is still difficult for them to scientifically explain TCM theories with simple modern medicinal terms.

Many patients turn to TCM treatment after they have tried treatment with Western medicines with no effect. Chinese herbal formulas work better than Western drugs for many diseases, not only chronic ones caused by stress, but also on acute infections such as SARS and the H_1N_1 flu virus. However, research results show that effects of the components isolated from these herbs are mostly less than those of current Western drugs. Thus the question arises: Why or how are the effects of these formulas better?

According to the experimental results, the answer is definitely not the placebo effect. The following might explain the reason.

1. Chinese herbs in a formula can work on different targets, that is, on different receptors and enzymes or other substances in the human body and stimulate the functions of nervous, circulatory, endocrine, immune, digestive, and other systems simultaneously. This is why TCM is a holistic medical system.
2. TCM emphasizes the protection of the digestive function as well as regulation of *qi* (pronounced "chee") and blood (details about the definition and explanation of *qi* and the importance of regulation of *qi* and blood in TCM will be given in Chapter 10). TCM believes that a good digestive system will guarantee an effective supply of essential nutrients from foods to the human body. It also believes that blocked *qi* and blood circulation may cause hundreds of types of diseases. For treatment of chronic diseases with Chinese herbs, there are always herbs that improve blood circulation in the formulas; if the patient has a digestion problem together with other symptoms, herbs that regulate the digestive system are usually given first. These actually emphasize the importance of maintaining cell functions with enough nutrients and excluding metabolites in a timely manner through functional blood circulation.

Scientists are currently trying to find out the relationship of mutant genes as causes of diseases, such as Alzheimer's and Parkinson's. But what are the main causes of the gene mutations? According to TCM, I would propose that the main cause of such diseases or aging is probably poor capillary blood circulation, which can be caused not only by the fats we eat, but also by the accumulation of metabolites from cells or dead cells. My reasoning is based not only on the above TCM theories and my clinical application of Chinese herbs, but also on the confirmation that the disease of age-related macular degeneration (AMD) is pathologically related to the accumulation of aging retina in the photoreceptor outer segment membrane,[18] part of a research program I performed when I worked as a postdoctoral scientist in the group of Professor Koji Nakanishi from Columbia University. No doubt, further experimentation is required.

1.5 BRIEF INTRODUCTION OF DIFFERENT SYSTEMS OF TRADITIONAL MEDICINE

The use of plants for prevention and treatment of diseases is the earliest type of medicine on earth. The practice of traditional medicine developed along with the cultures of ancient China, India, Egypt, and other places. Different species of plants are used as medicines for treatment in different countries because of the different ecological environments. In countries with long histories and cultures, theories of etiology and pathology, methods for diagnosis, and treatment with herbal medicines or other methods under these theories were gradually formed along with the understanding of diseases and accumulated therapeutic experiences, and their own complete medical systems finally established. To fully explore the preventative and therapeutic mechanisms of traditional herbal medicines, it is necessary to have a deep understanding of the theories in their corresponding medical systems. The following are brief introductions of some ancient but still currently popular traditional medical systems in the world. The systems in which herbal medicine is not a key therapeutic tool are not covered here.

As a summary, "holistic" is one of the most common characteristics of these major popular medical systems and their biggest difference from conventional or allopathic medicine. It is not only reflected in the beliefs of importance of interaction and harmonization between the human body and environment and among organs and tissues, but also implemented in treatment with different herbal components.

1.5.1 Traditional Chinese Medicine (TCM)

TCM originated in China thousands of years ago through meticulous observation of nature, the cosmos, and the human body. Today, it not only remains as a form of primary care in health systems throughout most Asian countries, but also as the most popular complementary or alternative form of medicine in most of the Western countries. It has an extremely complex theory system established mainly on the basis of two philosophical views, the integral and dialectical concepts. The major theories include yin and yang, the five elements, Zang–Fu theory, *qi* and blood theory, meridians, collaterals, etiology and pathology, and prevention.[19]

Yin and Yang

Yin and yang reflect all the forms and characteristics existing in the universe. They may represent two separate phenomena with opposite natures, as well as different and opposite aspects within the same phenomenon. While yin is dark, passive, downward, cold, contracting, and weak, yang is bright, active, upward, hot, expanding, and strong.

The basic theory of yin and yang is about their relationship. They are opposing, interdependent, inter-transforming (in a state of constant change), and balanced. The yin and yang theory is applied in TCM for diagnosis and as the principles of treatment. The imbalance and fluctuation of yin and yang are considered the basic causative factors of disease occurrence and development. The goal of clinical treatment is to restore yin–yang balance in the patient. For example, heat syndromes are treated with cold nature herbs, while cold syndromes are treated with hot nature ones.

The Five Elements

In this theory, nature is divided into five elements: wood, fire, earth, metal, and water. Color, taste, emotion, sense, season, organs in human body, and others can all be classified into the five elements. The laws of movement of the five elements are as follows: inter-promoting, interacting, counteracting, and mutual relation. The five elements theory is applied in TCM to explain the physiological and pathological interrelationship among Zang–Fu organs and guide diagnosis and treatment of diseases.

Zang–Fu Theory

In TCM, the heart, lung, spleen, liver, and kidney are known as the five Zang organs, while the gallbladder, stomach, small intestine, large intestine, bladder, and triple energizer are the six Fu organs. The pericardium is a protective membrane of the heart, so it is also considered an organ. The triple energizer is the central body cavity that is connected to Zang organs. There is no biomedical equivalent of the triple energizer in modern medicine. Its function includes transformation, purification, and distribution of air, food, and water. It can be further divided into three parts. The upper part regulates respiration and the circulation of protective *qi*, the middle part governs the *qi* of the various digestive system functions, and the lower part controls the *qi* of the absorption of fluids/nutrients, waste disposal, and sexuality/reproduction. The triple energizer is the central energetic structure and strength of human health and well-being.

The Zang organs are solid and yin in nature. Their physiological functions are to manufacture and store essential substances, including vital essence, *qi*, blood, and body fluid. They are connected with meridians for the transmission of *qi* and blood. The Fu organs are hollow and yang in nature. Their physiological functions are to receive and digest food, and transmit and excrete the wastes. Fu organs are also connected with meridians. Interconnected by the meridian system, the Zang and Fu organs have an internally–externally linked relationship.

Qi, Blood, and Body Fluid

They are considered fundamental substances that maintain the normal vital activities of the human body and physiological functions of the Zang–Fu, tissue, and meridians.

Qi has such a special meaning in TCM for which no English word exists for translation. It denotes both the essential substances of the human body and the functional activities of the Zang–Fu and tissues. Blood is a red liquid circulating in the vessels, similar to blood in modern terminology. Body fluid is a collective term for all the normal fluids of the body, which include saliva, tears, nasal discharge, sweat, and urine, as well as liquids in stomach, intestines, joint, and other cavities.

Meridians and Collaterals

They are pathways in which the *qi* and blood of the human body are circulated. Meridians constitute the main trunks and run longitudinally and interiorly within the body, while collaterals represent branches of the meridians and run transversely and superficially from the meridians.

The functions of the meridians and collaterals include transporting *qi* and blood, regulating yin and yang, resisting pathogens, reflecting symptoms and signs, and transmitting needling sensation to regulate deficiency and excess condition when acupuncture and moxibustion are applied.

Causes of Diseases

TCM believes that the causes of disease include the six exogenous factors (wind, cold, summer heat, damp, dryness, and fire), the seven emotional factors (joy, anger, melancholy, worry, grief, fear, and fright), improper diet, overstrain, lack of physical exercise, traumatic injuries, bites by insects or wild animals, as well as stagnant blood and phlegm fluid.

Diagnostic Methods

TCM diagnosis includes inspection, auscultation and olfaction, inquiry, and palpation.

Treatment Methods

The TCM treatment methods include herbal medicine, acupuncture, dietary therapy, Tui na, and massage. *Qi* gong and Tai ji are also strongly affiliated with TCM.

Information about properties, current researches, and modern pharmacology of Chinese herbal medicines, and the understanding of TCM theories with modern medical terminology will be given in Chapter 10.

1.5.2 Kampo Medicine

Kampo is the Japanese study and adaptation of Chinese medicine. The first Chinese medical works were introduced to Japan around the fourth or fifth century AD. Since then, the Japanese have established their own herbal medical system and diagnosis based on TCM. Kampo utilizes most of the TCM treatment methods, including herbs, acupuncture, and moxibustion.

Kampo is currently integrated into the national health-care system in Japan. Different from modifying formulas applied in TCM clinics, the Japanese Kampo uses standardized, fixed, and precise combinations of herbs. Today, about 75% of Japanese physicians prescribe Kampo formulas. Since 1967, the Japanese Ministry of Health, Labor, and Welfare has approved 148 Kampo formulas for coverage and reimbursement in the national health insurance plan. The formulas are prepared under strict manufacturing conditions with the Ministry's standardization methodology.

1.5.3 Indian Medicines

Indian medicines include three different systems: Ayurveda, Siddha, and Unani. They are different in origin and practice areas, as well as the theory and application.

Ayurveda

The term means "the science of life." It is another one of the oldest systems of medicine in the world. According to the web site of the U.S. National Center for Complementary and Alternative Medicine (NCCAM) in NIH (http://nccam.nih.gov/health/ayurveda/introduction.htm), Ayurveda medicine originated in India several thousand years ago and continues to be practiced in India, where nearly 80% of the population uses it exclusively or in combination with Western medicine. It is also practiced in Bangladesh, Sri Lanka, Nepal, and Pakistan. Two ancient books, *Caraka Samhita* and *Sushruta Samhita,* written in Sanskrit more than 2000 years ago, are considered the main texts on Ayurvedic medicine.

Ayurvedic medicine aims to integrate and balance the body, mind, and spirit; thus, it is also viewed as "holistic." This balance is believed to lead to happiness and health, and to help prevent illness. A chief aim of Ayurvedic practices is to cleanse the body of substances that can cause disease, thus helping to reestablish harmony and balance. Ayurvedic medicine has several key foundations that pertain to health and disease. These concepts have to do with universal interconnectedness, the body's constitution (prakriti), and life forces (doshas).

Interconnectedness This is about the relationships among people, their health, and the universe as the basis for how Ayurvedic practitioners think about problems that affect health. It believes that disease arises when a person is out of harmony with the universe. Disruptions can be physical, emotional, spiritual, or a combination of these.

Constitution (Prakriti) This refers to a person's general health, the likelihood of becoming out of balance, and the ability to resist and recover from disease or other health problems. It is called the prakriti, which means a person's unique combination of physical and psychological characteristics and the way the body functions to maintain health. It is believed to be unchanged over a person's lifetime and influenced by such factors as digestion and how the body deals with waste products.

Life Forces or Energies (Doshas) Different from TCM, the five fundamental elements in Ayurveda that make up the universe and also human physiology are space, air, fire, water, and earth. Ayurveda believes that health is maintained by the balancing of three subtle energies known as doshas. There are three doshas, called Vata, Pitta and Kapha, and each is mainly a combination of two elements. Vata dosha is made up of space and air. Pitta dosha is a combination of fire and water. Kapha dosha is made up of water and earth. Together, the doshas orchestrate all the activities that occur within us. A person's chances of developing certain types of diseases are thought to be related to the way doshas are balanced, the state of the physical body, and mental or lifestyle factors.

Ayurvedic practitioners first determine the patient's primary dosha and the balance among the three doshas by Asking, Observing, and Checking a pulse (each dosha is thought to make a particular kind of pulse). The goals of treatment include eliminating impurities, reducing symptoms, increasing resistance to disease, and reducing worry and increasing harmony in the patient's life.

Ayurvedic treatments rely heavily on plants, including herbs, oils, and common spices. Currently, more than 600 herbal formulas and 250 single-plant drugs are included in the "pharmacy" of Ayurvedic treatments. According to their effects, for example, healing, promoting vitality, or relieving pain, Ayurvedic medicines have been divided into categories.

Siddha

Siddha is mainly practiced in south India. This system of medicine is believed to be developed by the Siddhars, the ancient supernatural spiritual saints of India, who developed methods and medications that are believed to strengthen their physical body and thereby their soul, including intense yogic practices and years of fasting and meditation.

Unani

Unani means Greek. Unani medicine originated around 980 AD in Persia. The basic knowledge as a healing system was collected by Hakim Ibn Sina. This system is based on the theory of the presence of the elements, which are fire, water, earth, and air. These elements are present in different fluids (phlegm, blood, yellow bile, and black bile). The balance of these element leads to health, and imbalance leads to illness. Most medicines and remedies used in Unani are also used in Ayurveda. The base used in Unani medicine is often honey. Real pearls and metal are also used in making Unani medicine, based on the kind of ailment it is aimed to heal. In India, Unani practitioners can practice as qualified doctors.

1.5.4 Tibetan Medicine

Tibetan medicine combines elements of Indian, Chinese, and Greek medical traditions. Dietary modification, medicines composed of herbs and minerals, acupuncture, and moxabustion are applied for the treatment of illness. Tibetan medicine is currently practiced in Tibet, India, Nepal, Bhutan, China, and Mongolia, and is spreading to North America and Europe.

1.5.5 Muti

Muti is a term for traditional medicine in Southern Africa. The word means tree. African traditional medicine makes use of various natural products, many of which are derived from trees. For this reason, medicine generally is known as Muti. In Southern Africa, the word muti is in widespread use in most indigenous African languages, as well as in South African English and Afrikaans, where it is sometimes used as a slang word for medicine in general.

1.5.6 Islamic Medicine (Arabic Medicine)

Islamic medicine (Arabic medicine) refers to medicine developed in the medieval Islamic civilization and written in Arabic. Its development was closely related to the history of Arab tribes.

In the seventh century, the prophet Mohammed united the Arab tribes and declared a new religion, Islam. By the tenth century, the Muslims not only controlled Italy and Spain, but also extended their raids through the Alpine passages into mid-Europe. The Islamic Empire extended from the Atlantic Ocean on the west to the borders of China on the east, and became the most advanced and civilized nation in the world between the ancient civilizations and the Renaissance era in Europe (the seventh to thirteenth centuries).

According to the online book *Islamic Medicine*, a compilation of articles published in the *Journal of Islamic Medical Association* over the last several years edited by Shahid Athar, the Muslims were avid for the wisdom of the world of Galen, Hippocrates, Rufus of Ephesus, Oribasius, Discorides, and Paul of Aegina. By the tenth century, their zeal and enthusiasm for learning resulted in all essential Greek medical writings being translated into Arabic.

Ibrahim B. Syed, the author of the chapter "Islamic Medicine: 1000 Years Ahead of Its Times," believes that Islamic medicine was advanced in the fields of medical education, hospitals, bacteriology, medicine, anesthesia, surgery, pharmacy, ophthalmology, psychotherapy, and psychosomatic diseases; he even thinks that the European medical system is Arabian not only in origin but also in its structure. The Arabs are the intellectual forebears of the Europeans. Meanwhile, Dr. Abouleish, the author for the chapter "Contributions of Islam to Medicine," admits objectively that the Arabs were assimilated by the countries they reached. It is difficult to identify this new breed as Arabs, because although the language was Arabic, not all the scientists were from the Arabian Peninsula. It is also equally difficult to describe it as Islamic, because although the majority of the scientists were Muslims, sponsored by Muslim rulers, and governed by the Islamic law, some scientists were Christians or Jews. The book also mentions that Islamic medicine was integrated with the elements of TCM and Ayurveda by trade with China and India.

In Islamic medicine, the patients were treated through a scheme starting with physiotherapy and diet; if this failed, drugs were used, and last, surgery would be resorted to. The physiotherapy included exercises and water baths. The Arabs had an elaborate system of dieting and were aware of food deficiencies. Proper nutrition was an important item of treatment. Drugs were divided into two groups: simple and compound drugs. They were aware of the interaction between drugs; thus, they used simple drugs first. If these failed, drugs that are made from two or more compounds were used. If these conservative measures failed, surgery was undertaken.

1.5.7 Mexican Folk Medicine

According to Sandoval's 1998 book, *Homegrown Healing: Traditional Home Remedies from Mexico*, Mexican folk medicine is a healing philosophy that significantly influenced the American Indian and the Spanish conquistadors.

Before the arrival of the Spaniards, Meso-America (the area of central Mexico, Belize, Guatemala, and parts of Honduras) had been agricultural for over 3000 years. The Aztecs believe that there is a delicate balance or "harmony" between themselves and nature, and disease was caused by the gods to punish sinners. Tilting one's balance would cause serious illness or death. Similarly, the Spaniards believed that health was "God's will," and could be taken away as rectification.

In the fifteenth century, several thousand medicinal plants were collected in the Huaxtepec garden. Academic priests conducted research with plant derivatives for their pharmacological benefits. Meanwhile, Spain was leading medical advancement in Europe. Its superiority in medicine had been due, in part, to the knowledge acquired while under Arabic rule. When Catholic priests came with the Spanish conquistadors to the New World, they considered ancient codices of the Aztec priests blasphemous, prompting Hernan Cortes to order all works on botany and science to be destroyed. The values, convictions, and traditions of the Aztec people were almost completely eradicated by Spain in a relatively short period of time.

Fortunately, the early missionaries played a paradoxical role in salvaging the remnants of Aztec knowledge. They traveled throughout Nuevo Espana collecting and documenting materia medica, while integrating European healing philosophies. Some of the native remedies have survived the conquest due to quick thinking Indians who renamed the plants used in ancient ceremonial practices, using the names of benevolent saints. Therefore, many remedies survived the conquest and are still used in Mexico.

The Spaniards introduced humoral pathology, Hippocrates' theory of health being dependent on the proper distribution of the body's four humors: blood, phlegm, yellow bile, and black bile. From humoral pathology, the hot and cold theory of disease has survived in Mexico and in the Southwestern United States. In order to restore the body's symmetry, plants with opposing qualities are still taken.

The Mexican Institute for the Study of Medical Plants was established in 1975. Researchers at the institute have been examining the sixteenth century records to determine the validity of indigenous medicines, with great success. The science, which had dismissed traditional remedies, has begun to reevaluate therapeutic values of botanical lore.

1.5.8 Other Indigenous or Folk Medicines

In China, the definitions of Chinese materia medica and folk or indigenous medicines are different. The former refers to those herbs that are recorded in TCM books and prescribed by the Chinese doctors under the guidance of TCM theory after diagnosis;

Folk or indigenous medicine refers to herbs that are used by folks in only certain regions or minority areas without TCM theory.

Similarly, except some major traditional medicine systems from these older countries with longer cultural history such as India and Arabic countries, not every traditional herbal medicine has its own complete theories for supporting the application. Examples include herbs used by Indian American and many other tribes over the world.

1.6 REGULATION OF HERBAL MEDICINES AND THEIR PRODUCTS

With more and more application of herbs all over the world, regulation of herbal medicines and their products is called for due to quality and safety concerns. The most commonly seen quality issue of raw herbal materials is adulteration with counterfeit species or contamination with microorganisms, pesticide residues, and heavy metals. In order to gain higher profit, some merchants may sell moldy materials, or even adopt unethical and unscrupulous means on herbal materials, for example, dyeing Goji berry with sulfur. The manufacturers of herbal products could secretly mix them with Western drugs without giving note on the label and deliberately give false ingredients on the label. Thus, considering the safety of consumers, strict regulation on herbal products is imperative.

Different countries, such as China, the United States, Japan, and Germany, have taken different actions and formulated various regulations on quality control of raw materials of herbs and their products as well as registration of herbal products. World organizations such as WHO and International Conference on Harmonization of Technical Requirements for Registration of Pharmaceuticals for Human Use (ICH) are also involved in performance of research and clinical practice of herbal medicine.

In 2005, WHO published a report of a global survey about National Policy on Traditional Medicine and Regulation of Herbal Medicines. This report indicated that about 50 countries already have their national policy and laws or regulations on traditional medicines or CAM.

ICH is a unique project that brings together the regulatory authorities of Europe, Japan, and the United States, and experts from the pharmaceutical industry in the three regions to discuss scientific and technical aspects of product registration. According to the ICH web site, its purpose is to make recommendations on ways to achieve greater harmonization in the interpretation and application of technical guidelines and requirements for product registration in order to reduce or obviate the need to duplicate the testing carried out during the research and development of new medicines. The objective of such harmonization is a more economical use of human, animal, and material resources, and the elimination of unnecessary delay in the global development and availability of new medicines while maintaining safeguards on quality, safety, and efficacy, and regulatory obligations to protect public health.

In China, good manufacturing practice (GMP) has been applied to herbal medicine. Management of herbal medicine covers raw herbal materials and their processing, herbal products, export and import, registration, and others. Among thousands of herbs, 87 are allowed to be used as healthy foods, defined as foods that are suitable for a specific population and capable of regulating bodily functions but not taken for therapeutic purposes; nearly 600 individual herbs and 600 herbal products are listed in the *Chinese Pharmacopoeia* (2005 edition) with established quality standards. New herbal products must be registered with required chemical, pharmacological, clinical, and safety data depending on its category in the new drug list (there are five categories for Chinese drugs).

Herbal medicine systems used in Japan include the Japanese traditional medicines, Kampo formulas, and combinations of the traditional medicines with vitamins and pharmaceuticals. Except Kampo formulas, regulation of herbal medicines is the same as the approval for both prescription and over the counter (OTC) drugs. Regulation of quality standards of those herbal products was established in Japanese Pharmacopoeia for more than 90% of them.[20]

Regulatory standards of herbal medicine in Europe were introduced by De Smet in his article written in 2005.[21] Germany and France are indisputably in the lead in OTC sales of herbal medicine among European countries, and they have also had noteworthy markets for prescription herbal preparations.

There was no proprietary regulation on herbs in the United States until the 1990s. According to the U.S. Food and Drug Administration (FDA) web site, FDA regulates dietary supplements including herbs under a different set of regulations from those covering "conventional" foods and drug products (prescription and OTC). The U.S. Congress defined the term "dietary supplement" in the Dietary Supplement Health and Education Act (DSHEA) of 1994 as a product taken orally that contains a "dietary ingredient" intended to supplement the diet. The "dietary ingredients" in these products may include vitamins, minerals, herbs or other botanicals, amino acids, and substances such as enzymes, organ tissues, glandulars, and metabolites. It says that the manufacturer is responsible for ensuring that the product is safe and the product label is truthful and not misleading. Manufacturers do not need to register their products with FDA nor get FDA approval, unlike OTC and prescription drugs. The FDA monitors voluntary adverse event reporting, labeling, and claims, and can take action against any unsafe dietary supplement after it reaches the market.

In August 2007, the FDA issued the Dietary Supplement Current Good Manufacturing Practice (CGMP) Final Rule 21 CFR Part 111. In essence, this final rule requires that the proper controls be in place for dietary supplements during manufacturing, packaging, labeling, and holding operations. The move was in response to concerns about substandard dietary supplement manufacturing practices, as well as mislabeling practices. The 21 CFR 111 ruling addresses the quality of the manufacturing processes for dietary supplements and the accurate listing of supplement ingredients. It does not limit consumers' access to dietary supplements, nor does it address the safety of the ingredients in dietary supplements, or their effects on health, when proper manufacturing techniques are used. The rule applies to all domestic and foreign companies that manufacture, package, label, or hold dietary supplements, including those involved with testing, quality control, and dietary supplement distribution in the United States. This regulation has been fully implemented to all involved companies as mentioned since June 2010.

1.7 ACHIEVEMENTS AND CHALLENGES OF RESEARCH ON CHINESE HERBAL MEDICINES

Since TCM is one of the oldest medical systems with complete theories, Chinese herbs have been applied worldwide with the largest population and studied by scientists all over the world. The achievement of research on Chinese herbal medicine is summarized here as a reference to learn the developmental process and current status of herbal study.

Chinese material medica is the official name for Chinese medicines, which include materials from all parts of plants, substances related to animals, and minerals. There are also many folk medicines in China. The difference between Chinese material medica and folk medicines is that the former are usually prescribed by Chinese doctors on the basis of diagnosis along with the guidance of TCM theory, while the latter are used by laypeople based on personal experience or are given by minority practitioners. In this book, Chinese herbal medicine is used to simply refer to Chinese material medica originated from plants.

In 1999, *Zhong Hua Ben Cao*, the most authoritative Chinese book with a complete record of Chinese materia medica, was published.[10] Its editorial board was composed of experts all over the country and organized by the State Administration of Traditional Chinese Medicine (SATCM). This book records a total of 8980 Chinese medicines that are divided into 34 volumes and summarized the contemporary research of Chinese medicine with modern science and technology. The 2005 edition of *Pharmacopoeia People's Republic of China* listed around 600 individual Chinese materia medica used in clinics, most of them with identification methods. Clinically, about 300 of them are most commonly used by Chinese doctors; many others are generally used locally as folk medicines.

Modern study of Chinese herbal medicine started in the 1920s when Chinese scholars found the biological activity of ephedrine from ephedra. The progress of the study was slow before 1949 due to continuous wars in China. Since the 1950s, the Chinese government has given great support for the development of Chinese medicine. Many provinces established TCM colleges with affiliated hospitals. Now, almost all of the provinces in China have their own TCM colleges or universities that include a department of Chinese Pharmacology; many Western medical universities in China also have either a department or a research institute of Chinese Pharmacology. Due to the emphasis on scientific research since the nationwide economic reforms starting three decades ago, the Chinese government has been investing more money on TCM research. Research of Chinese herbs has been performed nationwide, from chemical isolation and identification, analysis for quality control and standardization, to bioassay and pharmacological study, and clinical trial. Meanwhile, scientists in Japan, Korea, Germany, and other countries have also made great contributions to the Chinese herbal study.

1.7.1 Chemical Isolation, Identification, and Standardization of Chinese Herbs

Isolation of chemical compounds from Chinese herbal medicines in China just started in the 1920s, about 100 years after the isolation of salicin from white willow by Western scientists. However, since the 1950s, more and more Chinese chemists have been using all kinds of column chromatography, MPLC or HPLC, SFE, DCC, HSCC, and other methods for isolating compounds from extracts, and different spectral equipments such as IR, UV, NMR, and MS for identification of the isolated compounds. Theses of most graduate students in phytochemistry majors are about chemical studies of Chinese herbs. Because of the contribution from scholars nationwide in the past half century, most commonly used Chinese herbs have been widely studied up to now. Major compounds in many of them have been isolated and identified. Based on these identified major compounds in herbs, HPLC quantitative and qualitative analysis of many herbal materials and their products are available for quality control and standardization.

1.7.2 Biological and Pharmacological Studies of Chinese Herbs

Extracts or major compounds of most commonly used Chinese herbs have been screened for many types of *in vitro* bioassay tests and *in vivo* animal studies. Most of the results, from the levels of enzyme, receptor, gene or cell to tissue and animal, strongly support their preventative and therapeutic functions. So far, pharmacological studies have been performed on most of the commonly used Chinese herbs listed in the herbology textbooks. The modern pharmacological characteristics of all the Chinese herb groups, categorized according to their functions with terminology of TCM, have been documented.[8] Unfortunately, most of the study results were published solely in Chinese journals due to obvious language translation barriers. Nevertheless, some research results are available in English international journals.

1.7.3 Challenge of Studies on Chinese Herbal Formulas

One of the special characteristics of Chinese herbal treatment is that Chinese doctors like to treat diseases with herbal formulas that are usually composed of several herbs. Each herb in one formula plays its own role; herbs within the formula may also interact with each other (see Chapter 10). The aim of the formula is not only to simultaneously stimulate functions of different organs and systems of the body for respective symptoms, but also to seize synergetic effects of the herbs or suppress side effects of each other. For example, licorice is often used in formulas because it contains glycyrrhizin type saponins, which have sweetening and surfactant properties. Thus, it is used to harmonize the taste of other bitter herbs, and to help dissolve components of other herbs when cooked together.

If studies on a single herb are difficult enough, it is even more challenging to do research on an herbal formula. Not only do the components become more difficult for analysis, but the interactions between herbs further complicate the pharmacological study. The study of formulas is currently the biggest challenge to TCM researchers.

Take the Six-Ingredient Pill with Rehmannia as an example. This is a famous Chinese herbal formula that is composed of six herbs. It is often used for diabetes, chronic nephritis, chronic prostatitis, and menopause symptoms in clinics. Pharmacological study showed that this formula could lower the ALT level of mice with liver damage induced by CCl4, thioacetamide, or hydrocortisone. However, when the six herbs were split into three pairs, the results showed that none of the pairs had any deductive effect on ALT levels in the mice with CCl4-induced liver damage. This example demonstrates the synergetic property of herbal formulas—only when the herbs worked together were they able to be effective.

1.7.4 Formulations of Chinese Herbal Preparation

In ancient China, herbal formulas were mostly cooked with water, which is called a decoction in English. For convenience, pills were made by mixing water extract of some herbs with other powdered ones; honey was added as necessary. Few herbs or formulas were soaked in wine or were simply ground for administration. The preparation of such formulas was either time-consuming or easily contaminated.

Now, with recognition of bioactive components from herbs and their pharmacological activity, many Chinese herbal extracts have been developed into many forms of preparation, including capsule; tablet; granule; soft gel; true solution; colloidal-, suspension-, or emulsion-type liquid formulations; suppositories, and ointments for external use. A few have even been made into liquid or powder injections.

1.7.5 Standardization and Quality Control of Chinese Herbal Materials and Products

Standardization of an herb usually refers to the chemical analysis of the characteristic bioactive and main components for identification or comparison of species. At present fingerprints of HPLC chromatograms are the most popular method. Thin layer chromatography (TLC) is another common one with lower cost.

Quality control for an herbal material usually includes not only the quantitative analysis of the main compounds from herbs, but also other analyses related to hygiene or safety examination such as heavy metal, pesticides, and microorganisms. For each individual formulation of the preparation, quality control has different requirements, such as precipitation test of an oral liquid and time for disintegration of a tablet.

Currently, the contamination of heavy metal, pesticides, and microorganisms is still a problem for many raw materials and their products. To a large extent, the efficacy and toxicity of a product depend on quality control. Toxicity or death reports by consumers by herbal products are mostly related to poor quality control during preparation, storage, or transportation. Recently, several deaths caused by Chinese herbal injections have been reported in China. Therefore, more and more strict quality controls are required from the Chinese government.

1.7.6 Clinical Studies of Chinese Herbs and Herbal Formulas

Although many Chinese herbs and herbal formulas have been in practice for hundreds or thousands of years and evidence of their therapeutic effects is ample in many books, there are not enough data to convince modern people, especially scientists in Western countries, to accept them.

In Chinese TCM journals, many more reports about clinical case studies of herbal formulas or products with dozens of patients were found than were those about the standard double-blind clinical trial with several hundred patients, which are usually performed for new drug development. The reason is that there is no requirement of clinic studies for herbal products that are already on the market. Moreover, clinical studies require a greater investment of time, money, and manpower.

1.7.7 Mechanisms of Chinese Herbal Processing Revealed by Chemical Investigation and Pharmacological Studies

Many Chinese herbs need to be processed when used for treatment of certain types of diseases. Methods of processing include boiling, steaming, frying, or heating with some accessories such as wine, vinegar, salt, honey, sand, and mud. The purposes of such processing are to decrease the toxicity and side effects of herbs, increase their therapeutic effects, change their properties or functions, make them easier for preparation and storage, and others. Chemical and pharmacological studies have revealed that the processing can either decompose the structures in herbs by breaking down the bonds of esters or glycosides, or change the solubility or properties of compounds such as alkaloids, acids, and proteins, or decrease the concentrations of components such as essential oil. Processing mechanisms of commonly used herbs are well known now.[22]

REFERENCES

1. Goodman, J. and Walsh, V. (2001) *The Story of Taxol: Nature and Politics in the Pursuit of An Anti-Cancer Drug*. Cambridge, Cambridge University Press.
2. Liu, J.H., et al. (2001) Evaluation of estrogenic activities of plant extracts on the potential treatment of menopausal symptoms. *Journal of Agricultural and Food Chemistry* 49(5):2472–2479.

3. Liu, J.H., et al. (1998) Inhibitory effect of Angelica pubescens f. biserrata on 5-lipooxygenase and cyclooxygenase. *Planta Medica* 64(6):525–529.
4. Liu, J.H., et al. (2004) Isolation of linoleic acid as an estrogenic compound from the fruits of Vitex agnus-castus L. (chaste berry). *Phytochemistry* 11(1):18–23.
5. Powell, S.L., et al. (2008) In vitro serotonergic activity of black cohosh and identification of N_ω-methylserotonin as a potential active constituent. *Journal of Agricultural and Food Chemistry* 56(24):11718–11726.
6. Burdette, J.E., et al. (2003) Black cohosh acts as a mixed competitive ligand and partial agonist of the serotonin receptor. *Journal of Agricultural and Food Chemistry* 51(19):5661–5670.
7. Rhyu, M.R., et al. (2006) Black cohosh (Actaea racemosa, Cimicifuga racemosa) behaves as a mixed competitive ligand and partial agonist at the human mu opiate receptor. *Journal of Agricultural and Food Chemistry* 54(26):9852–9857.
8. Shen, Y.J. and Chen, C.X. (2008) *Traditional Chinese Pharmacology* (5th ed.), Shanghai, Shanghai Science and Technology Publisher.
9. Chen, Q. (2006) *Pharmacological Research Methodology of Chinese Medicine* (2nd ed.), Beijing, People's Health Publishing House.
10. State Administration of Traditional Chinese Medicine (1999) *Zhong Hua Ben Cao*. Shanghai, Shanghai Science and Technology Publisher.
11. Arlt, V.M., et al. (2002) Aristolochic acid as a probable human cancer hazard in herbal remedies: a review. *Mutagenesis* 17(4):265–277.
12. Debelle, F.D., et al. (2008) Aristolochic acid nephropathy: a worldwide problem. *Kidney International* 74(2):158–169.
13. Schmeiser, H.H., et al. (2009) Chemical and molecular basis of the carcinogenicity of Aristolochia plants. *Current Opinion in Drug Discovery & Development* 12(1):141–148.
14. Woelkart, K., et al. (2008) Pharmacokinetics of the main alkamides after administration of three different Echinacea purpurea preparations in humans. *Planta Medica* 74(6):651–656.
15. Bhattaram, V.A., et al. (2002) Pharmacokinetics and bioavailability of herbal medicinal products. *Phytomedicine* Suppl. 3:1–33.
16. Zhang, L., et al. (2005) Advances in clinical pharmacokinetics of herbal medicines. *Journal of US–China Medical Science* 2(6):59–72.
17. Clement, Y.N. (2009) Factors affecting the pharmacokinetics of herbal preparations and their impact on the outcome of clinical trials. *Focus Alternative Complementary Therapies* 14(2):87–91.
18. Liu, J.H., et al. (2000) The biosynthesis of A2E, a fluorophore of aging retina, involves the formation of the precursor, A2-PE, in the photoreceptor outer segment membrane. *Journal of Biological Chemistry* 275(38):29354–29360.
19. Maciocia, G. (1989) *The Foundations of Chinese Medicine: A Comprehensive Text for Acupuncturists and Herbalists*. Hong Kong, Addison Wesley Longman, China.
20. Saito, H. (2000) Regulation of herbal medicines in Japan, Pharmacological Research. 41(5):515–519.
21. De Smet, P.A. (2005) Herbal medicine in Europe—relaxing regulatory standards. *The New England Journal of Medicine* 352(12):1176–1178.
22. Xu, C.J. and Ye, D.J. (1985) *Processing of Chinese Materia Medica*. Shanghai, Shanghai Science and Technology Publisher.

Chapter 2

Collection and Identification of Raw Herbal Materials

Ping Li, Ling Yi, and Hui-Juan Liu

Collection and species identification of the raw plant materials are considered the most important procedures of the entire process of production or study because they directly influence the quality of products or results of study. Location of plant growth, time of collection, and processing of the collected materials are all related to the chemical composition that will determine yield of the herbal product and its preventative and therapeutic functions. Therefore, it is important to ensure a correct selection of appropriate time with adaptable techniques for the collection of herbal materials. The color, odor, taste, shape, texture, and structure of a medicinal plant should all be considered as references for collection because they are reflections of the chemical compositions in the plant.

Good agriculture practice (GAP) has been applied to herbal cultivation all over the world since many big companies realized the importance of the role of raw materials in the quality of products. For instance, more and more GAP-regulated cultivation bases of Chinese herbs have been established in China, aiming at enlarging the cultivation area of good geo-authenticated Chinese herbs and performing investigations related to quality control to lay a good foundation for correct collection. Take the Chinese herb Rhizoma Chuanxiong, a geo-authenticated raw material in Sichuan province, as an example. Results from a successive 3-year analytical study on the contents of alcohol-soluble extracts and three other major constituents in four different collection periods suggested that the proper collection duration is around 10 days after Grain Full in the Chinese lunar calendar (on about May 20), when it possesses the best quality.[1]

Species identification means identifying the origin of raw herbal materials—authenticating the plants sources by their morphologic, microscopic, physical, and chemical properties. Traditional herbal medicines have been used for a long time throughout the world with great varieties in productions and areas. However, due to

Traditional Herbal Medicine Research Methods, Edited by Willow J.H. Liu

many reasons, such as different regional idioms and custom, historic records, similarity in plant morphology, analogs or substitutes of herbal materials, and so on, the isonym and synonym phenomena are prevalent. It happens that one herbal medicine is originated from multiplants origins, for example, the Chinese herb Radix Stemonae from the roots of three species, including *Stemona sessilifolia* (Miq.) Miq, *Stemona japonica* (Bl.) Miq., and *Stemona tuberosa* Lour. In addition, inferior materials or even the counterfeits of some herbs, especially the precious ones, are found in markets, for example, the Chinese herb Sanguis Draconis, Stigma Croci, and Cordyceps. As a result, species identification of herbal materials is of great importance, for it directly determines the efficacy of their clinical practice, the reliability of scientific research results, and the safety to consumers.

Identification of raw herbal materials includes morphological identification, microscopic identification, physical and chemical identification, and DNA molecular marking identification.[2–5] With the fast development of molecular biology, the DNA molecular marking technique has played an important role in species identification and revealed extensive applications in the future. This chapter mainly introduces methods for identification of Chinese herbal materials.

2.1 COLLECTION OF HERBAL MATERIALS

2.1.1 Location of Collection

The location of collection usually refers to the geographic areas of a species of herbs, that is, the cultivation place of the cultivated herbs and the acquisition place of the wild ones. Although most species of herbs are extensively distributed in wildness, their proper places for collection are relatively insufficient. As we know, plant species are closely related to their geographical distributions. The qualities of herbal materials collected from different locations may vary immensely even if they are processed with the same methods. Take *Pogostemon cablin* as an example: the ones collected from Shipai, Guangzhou, China, give out a fine aroma and contains less volatile oil (0.1–0.15% in stem and 0.3–0.4% in leaves) but higher concentrated patchoulenone, while those collected from Hainan Island, China, smell a bit pungent and turbid and contain more volatile oil (0.5–0.7% in stem and 3–6% in leaves) but much less patchoulenone. The famous Chinese *Eight Drugs* in Zhejiang, *Four Great Drugs* in Henan, *Chuan Drug* in Sichuan, *Drugs from the South*, and *Drugs from the North* tell us that qualities of certain herbs from specific areas are better than those from other places. These are reflections of the relationships between locations and qualities of traditional Chinese herbal medicines.[6,7]

2.1.2 Time of Collection

The time of collection refers to the season and exact time for collection of herbs with certain growing years. Different times of collection will give rise to varied quality of raw materials. For example, *Artemisia capillaries* should be reaped when its shoots grow to around 6–10 cm long in early spring. An old Chinese saying goes, "March for drug while April for mugwort, and yet May for firewood only." This means that if it is collected later, as in May, it could only be used as firewood. Another example is the capitulum of the herb *Artemisia cina*, which contains 3% anthelmintic santonin before flowering; as soon as the flowers mature, however, the content of this compound rapidly decreases. The third example is the powder of *Pyrethrum cinerariifolium*, whose therapeutic function of destroying parasites derived from its buds is double of that from its bloom flowers.

The exact time of collection, on the other hand, should be scheduled according to the properties of the objective materials. For instance, the fresh berries of *Ligustrum lucidum* and *Lycium barbarum* are sappy, easily broken by squeezing, and thus should be plucked at dawn or dusk. Flowers of *Lonicera japonica* would have a nicer color with stronger flavor as fine raw material if they were plucked on a sunny morning after the dew dried.

To scientifically determine the appropriate time of collection, it is necessary to first investigate the bioactive components in the herbs. It is known that the contents of active components in plants vary not only in different species and different parts of each plant, but also in different phenological periods, geographical environments, cultivation conditions, and some other factors. In addition, the rules of formation, accumulation, and reaction of chemicals are distinct in different plants. All those factors that may possibly influence the chemical changes in herbs should be taken into account.

Apart from the total yields of an herb, the contents of bioactive components in them that relate directly to the quality of the materials are also important with regard to economic benefits. In order to obtain the highest total count of active components (=single yield × percentage of active components), appropriate time of collection is when the bioactive components reach their maximum amounts in unit area. In fact, the time of collection is different even for the same herb growing in various areas, soil, and cultivation conditions. Moreover, since different parts of one plant are sometimes used as different herbs, the collection of each part should be done carefully to avoid destroying the other medicinally useful parts in order to make the best use of every part of the medicinal plant.

The following are general rules for the time of herbal collection derived from the same type of organs.

1. *Subterranean organs:* include root and rhizome. The cultivated plants are collected generally at 2–5 years old, at the time when the nutrients and active components are mostly aggregated in subterranean organs, that is, between the end of autumn when aerial parts are withered and early spring when plants barely regenerate to green. Collections of wild materials should be done before the aerial parts of plants are withering in autumn or after their regenerations so that they could be recognized in spring. Radix Astragali, Radix Codonopsitis, and Radix Platycodon are good cases in point.
2. *Cortex:* peel in spring or early summer, preferably April and May, when there is abundant sap with maximum amount of active components in the

plant and the xylem is easily peeled off. Prime examples are Cortex Phellodendri, Cortex Eucommiae, and Cortex Magnoliae Officinalis. However, some root barks are better to be collected in autumn.

3. *Leaves:* usually collect in the most prosperous life stage of plants, when their leaves are thick and green with fairly good qualities. Examples are Folium Nelumbinis, Folium Isatidis, and Folium Apocyni Veneti. In certain conditions, leaves should be plucked off or picked up from the defoliated ones after frost descent, such as Folium Eriobotryae.
4. *Flowers:* collect according to the requirements of medication and florescence. In comparison with other organs, flowers have a relatively demanding collecting season—usually during in-the-bud or early bloom phase, neither too early with insufficient odor nor too late when petals have fallen. Flos Lonicerae and Flos Sophorae are such kinds of flowers. Some flowers have special requirements. For instance, Flos Chrysanthemi and chrysanthemum powder should be collected in full bloom; Flos carthami, on the other hand, should be picked when the corolla of the plant are turning from yellow to red.
5. *Fruits and seeds:* usually collect when ripe or near ripening period of life stage. Take Fructus Zanthoxyli as an example: the odor will be light if collected too early or diminished if too late. Some fruits, for example, Fructus Corni, are better picked after having undergone frost when the ripe fruit turns red. Some need to be collected when they are immature, such as Pericarpium Citri Reticulatae Viride. Generally, seeds are collected after full maturation of fruits. As an example of the exceptions, schizocarp is gathered before dehiscence.
6. *Entire herbs:* generally collect when luxuriantly growing or during early stage of florescence in respect of their high yielding and good quality, such as Herba Agastachis, Herba Schizonepetae, Herba Verbenae, and Herba Menthae. Exceptions include Herba Artemisiae Annuae, which should be collected before flower budding, and Herba Artemisiae Scopariae, which should be collected in offspring life stage.[8–10]

2.1.3 Local Process of Materials

Immediately after collection, the fresh raw materials need to be properly processed, for they might deteriorate due to water contained inside. As a result, most raw materials need to be primarily processed in the place where collected. The primary process of raw herbal materials should aim to maintain its efficacy, degrade its toxicity, prevent it from deterioration, as well as enable easier storage and transport. Different methods have been applied to diverse materials. Generally, the materials are first selected, by methods including picking, screening, and cleaning with winnower or by rinsing; then, the nonmedicinal parts are removed, followed by a drying process.

For Chinese herbs, many traditional processing methods are still currently used. If the collected materials are roots or rhizomes, for example, they should be classified, cleaned, and stricken. Outer coating, pith, or other unused parts need to be discarded for some herbs, for example, Radix Dioscoreae and Radix Platycodi. Many are then sliced into pieces or processed with other methods, such as water blanching (e.g., Radix Asparagi and Radix Stemonae), boiling (e.g., Radix Polygoni Muliflori and Rhizoma Gastrodiae), rubbing and kneading (e.g., Radix Dioscoreae, Radix Codonopsitis, and Rhizoma Polygonati officinalis), piling up, smoking, or mixing with lime or oyster powder for drying. The majority of roots and rhizomes are piled up together when half-dried, making the water inside evaporate or condense into drops on the material's surface; this process is called "perspiring." Examples are Radix Scrophulariae, Radix Isatidis, Radix Astragali, Herba Menthae, Cortex Magnoliae Officinalis, and Cortex Eucommiae.

Drying processes can be divided into two types: air drying and heating. Drying in the open air, in the shade, and in the sun are the three ways of air drying. Sun drying is the most common, but is inappropriate for volatile constituents in the materials. Methods of drying by heat include baking, drying in stove, and infrared (IR) drying. As a general rule, 50–60°C is mostly adopted, while 70–90°C is better for succulent fruits. The cross-sections of dried materials should be consistent in color and luster, giving off a clear and melodious sound when knocking against each other. Dried leaves, flowers, and herbs can be easily broken into fragments, while leaves and flowers can be easily kneaded into powder.[11–13]

2.1.4 Other Factors Influencing Collection of Materials

Synonym and Isonym Phenomena

Synonym means one herb used in different areas with different names, and recognized as different drugs. The synonym phenomenon of Chinese herbs has existed for a long time in China. For example, *Siphonostegia chinensis* Benth. in the family of Scrophulariaceae is used as Herba Artemisiae Anomalae in the north of China, while as native Herba Artemisiae Capillaries or Herba Siphonostegiae in the south. Synonym also exists in another situation, that is, when the name and usage of one plant differ between the past and present. For instance, *Aralia cordata* Thunb., in the family Araliaceae, was recognized as native Radix Angelica in ancient China, but as Rhizoma Araliae Cordatae in modern China. All these phenomena influence the authenticity of raw materials; therefore, the identification should involve botanical study.

Isonym refers to different species named with the same name, or recognized as one herb. In China, the isonym phenomenon of Chinese herbs usually appears in the following three situations.

1. One name for different species with similar curative efficacy. They are usually genetically close to each other and have the same types of active

components, and thus could be used as one drug. Sometimes, species from different families indicating similar efficacy are also used as one drug in different regions. For example, *Rheum palmatum* L., *R. tanguticum* Maxim. ex Regel, and *R. officinale* Baill. are all used as Radix et Rhizoma Rhei. The phenomenon in which one Chinese herb has more than one source is quite common.

2. One name for different species with different curative efficacies. They usually come from different origins that have diverse components. For instance, the bulbs of *Iphigenia indica* Kunth from Lijiang city are confused with tubers of *Bolbostemma paniculatum* and *Fritillaria* species from the Yunnan province. Nevertheless, it should be noted that colchicina in the bulbs of *Iphigenia indica* Kunth may cause intoxication after administration. In quite a few cases, more than 20 isonym substitutes or counterfeits can be found with one Chinese name, such as Radix Pulsatillae, Rhizoma Cyrtomii, Radix Angelicae Pubescentis, Herba Taraxaci, Caulis Sinomenii, Caulis Spatholobi, and Semen Vaccarae. Such phenomena severely influence the authenticity of raw materials.
3. Coexistence of synonym and isonym for some herbs. This is an even more complicated phenomenon. Prime examples are Herba Eupatorii and Herba Lycopi, two Chinese herbs that have different functions and indications and are used as an aromatic drug to dissipate hygrosis to improve appetite and digestion and as an emmenagogue drug to promote blood flow and to remove blood stasis in the pelvic area and uterus, respectively. Lycopi was found to come from two genera in ancient time, that is, Lyopus in the family of Labiata and Eupatorium in the family of Compositae. However, the modern medical use of Herba Eupatorii is only sourced from those plants belonging to Eupatorium, even though they are still used as Herba Lycopi in some regions. As a result, the use of these two herbs is still quite confusing.[14,15]

Relationships between Species

In general, for medicinal herbs, the closer the species are in source, the more similar chemical compositions they have and the more similar characters and efficacies they share. There are various species of herbal plants in vast territory, distributed diversely from south to north and from east to west. For many historical and geographical reasons, Chinese doctors have learned to apply closely related species that are available to replace the herb they need when it is unavailable.

For example, Radix Scutellariae originated from *Scutellaria baicalensis* Georgi, the only source according to Chinese pharmacopoeia. However, it is commonly replaced by *S. amoena* C.H. Wright in Yunnan and Sichuan provinces, and by *S. viscidula* Bge. in Shanxi province where it is abundantly grown. The research results show that the chemicals and pharmacological functions are similar between the three species. Therefore, the latter two species are documented as legal local standard substitutes.

However, there are always exceptions. Unlike above, some closely related species contain different chemicals and indicate diverse pharmacological functions. The confused applications of these raw materials have severely debased the validities of herbs, demonstrating the importance of strict classification in clinical practice.

Let us take Radix Bupleuri as an example. Despite the fact that many Bupleurum species are available, *B. chinense* DC. and *B. scorzonerifolium* Willd. are documented in the Chinese pharmacopoeia as Radix Bupleuri and conventionally nominated as "north bupleuri" and "south bupleuri," respectively, because they grow separately in north and south parts of China. However, *B. longiradiatum* Turcz., *B. smithii* Wolff, and *B. smithii* Wolff var. *parvifolium* shan et Li. can never be used as substitutes of this herb because they contain types of chemicals involving some toxic components.

Another example is herbs in the genus Amorphophallus. Konjac mannan (glucomannan) is an active component with a total content up to 60% in the stem tuber of *A. rivieri* Durieu, but it is not found in *A. sinensis* Belval. In addition, the anticancer effects of *A. rivieri* cannot be replaced by *A. sinensis* either, despite being from the same genus.[13,16]

2.2 METHODS FOR SPECIES IDENTIFICATION OF HERBAL MATERIALS

2.2.1 Morphological Identification

Morphological identification, also called descriptive identification, usually includes origin identifications and medical part confirmations.

Origin identification is to verify the geographical information and plant sources with scientific names by the knowledge of plant taxonomy, so as to ensure the correct use of species. The examination of plants' morphous and verification of specimen should refer to literatures.

Medical part confirmation is to verify herbal materials as genuine or false and superior or inferior mainly based on their morphological characteristics, referring to the shape, size, color, surface, texture, cross-section, odor, taste, and so on. This is important and sometimes difficult because most of the herbal materials presented in markets are processed into parts or pieces and no original plants are available for origin identification.

The traditional way of morphological identification for Chinese herbal medicine is to examine with the sense organs, such as observation by naked eyes or anatomical lens, touch by hands, smell by nose, taste by mouth, or by some simple tests with water or fire. The results should be described by the correct terminologies and recorded by drawing pictures or taking photos.

These means can be specified as follows.

1. *Observation* is to observe the features of shape, size, color, surface, and cross-section of raw materials, and to describe with simple terms.

2. *Touch* is to feel the raw material, to estimate its texture, and to determine its fragility. Terms such as hard and soft, tight or loose, light or heavy, pliable, starchy, horny, oily, flossy, woody, and viscosity are usually used.
3. *Smell* is to sense the odor of whole raw materials, or when peeling, kneading, and breaking the materials.
4. *Taste* is to lick the surface of raw materials with the tip of the tongue, or to chew a small portion of the drug in the mouth to discriminate the taste. Toxic drugs should be tasted with caution. Spit out immediately and rinse mouth as soon as possible.

Furthermore, tests by water or fire are also common means for morphological identification. Water tests are used to measure the material's density, solubility, color change, clarity, expansivity, rotation, viscidity, pH, and so on. Fire tests are used to observe fire color, smoke, sound formation, expansion, and melting when burning the materials. For example, immersing Flos Carthami in water will make the water change into golden yellow color without causing the flower to fade. Throwing Lignum Sappan into hot water will make the water change into a bright persicinus color. When Cortex Fraxini is dipped in water, the infusion will show blue fluorescence. Indigo Naturalis will produce violet smog for quite some while when it is burned. When Ligunum Aquilariae Resinatum is lit, it will emit dense smoke with strong aroma, accompanied with effused black oily material.

Apart from descriptions in words, the result of morphological identification should be illustrated by pictures to show all distinct features. Digital photographing technique has become a chief method to show the herbal materials' morphological features, taking advantage of easy access, real imaging, vivid coloration, convenience for processing and storage by computer, and fast transformation through the Internet.

Morphological features and examples of herbal materials from different parts of plants are expressed here.

Materials of Roots

Root materials are mainly derived from angiospermous roots, including whole root or with part of the rhizome. There are usually no nodes, internodes, or buds on roots, but adventitious buds are occasionally observed on dicotyledonous roots. First, it is necessary to distinguish whether the root is derived from dicotyledon or from monocotyledon.

The dicotyledonous root is mostly cylindrical or conic, straight or slightly curved, twisted, branched, often connected with breviate rhizome at the top end; the outer surface is mostly marked with cork, lenticels or branch root scars; the cross-section shows radial structures with distinct cambium ring, no pith, but occasionally allotype structures. The monocotyledonous root is mostly fibrous or swollen as a root tuber with various shapes.

For identification, it is important to describe the shape, size (length and diameter), and characteristics of the main root, the top end and lower part of root or

Figure 2.1 Root of *Panax ginseng.*

rootlet, the fibrous root or radicella, the oncoides of root or rhizome in beaded form, the rhizome of crispation in root part, striation, wrinkle, channel, furrow, and so on. The center part, chrysanthemum marking, wheel marking, moire, and cobwebbing in the cross-section are usually examined. The presence of cinnabar dot (red oil cavity), powder dust (starch granules), and hoar frost (educed crystals) are described for both cross-sections and slices.

Example Radix Ginseng: the dried root of *Panax ginseng.* C. A. Mey. (see Fig. 2.1)

Morphological Features Axial root cylindrical or fusiform, 3–15 cm long, 1–2 cm in diameter; externally pale yellowish white or grayish yellow, exhibiting distinct longitudinal wrinkles and a few transverse and elongated lenticels; the upper part marked with sparse, shallow, and interrupted transverse-striations; the lower part bearing 2–3 branch roots and a few slender rootlets, up to 12 cm in length. Top end carries stunted rhizomes, 1–5.5 cm long, 0.3–1.5 cm in diameter, showing several dented stem scars, arranged alternately, sometimes bearing slender and trans-extended adventitious. Whole sun-dried Ginseng bears numerous fine rootlets marked with indistinct small verruciform protuberances. Axial root relatively hard, fractures even, pale yellowish white; cambium ring brownish yellow, cortex marked with punctiform yellowish brown resin ducts and radial clefts; rootlets texture fragile. Odor slightly fragrant and distinctive; taste slightly bitter and sweet.

Materials of Rhizomes

Materials of rhizomes are derived from subterraneous stems, including root stock, stem tuber, bulb, and solid bulb. The difference of rhizomes from roots is that rhizomes have nodes and internodes on the surface. This phenomenon is more distinguished in monocotyledon. Degenerative scale-like leaves on node are often observed. Adventitious roots or root scars around or below a node can be easily found in monocotyledonous rhizomes, while leaf scar and bud scar are seen occasionally. Pteridophyte rhizome usually has scales or palea on surface, sometimes with dense concinnous leafstalk base around node.

In cross-section, monocotyledonous rhizome can be distinguished from the dicotyledonous by distinct scattered vascular bundles around endothelial layer. Dicotyledonous rhizomes have radial structure and obvious pith in the center. Some pteridophyte rhizomes have wood but no pith in the center, some display integrity xylem ring with pith, and some show interrupted meristeles arranged in a ring. Meristele's pieces and the arrangement in cross-section of leafstalk base can also help to distinguish pteridophyte rhizomes from others.

Example 1 Rhizoma Coptidis: the dried rhizome of *Coptis chinensis* Franch, *C. deltoidea* C.Y. Cheng *et* Hsiao, or *C. teeta* Wall. (Fig. 2.2).

Morphological Features The following are the morphological features.

1. Rhizome of *Coptis chinensis*: Mostly clustered into a bundle; resembling a chicken claw. Single rhizome cylindrical, 2–8 cm in length, 2–8 cm in diameter. Externally yellowish gray, bearing nodular protrudings and remains of rootlets, mostly with scale leaves; top end often with remnants of stems or petioles; internode in the middle part of rhizome slender, surface smooth, 1–4 cm long. Texture hard, fracture uneven; cortex dull brown; xylem golden yellow, radially arranged, with clefts; pith in the center reddish brown, sometimes hollow. Odor faint; taste extremely bitter.

A B

Figure 2.2 Rhizome of *Coptis chinensis* (A) and *C. deltoidea* (B).

Figure 2.3 Tuber of *Pinellia ternata.*

2. Rhizome of *Coptis deltoidea*: Mostly single, somewhat cylindrical, slightly curved, 4–8 cm long, 0.5–1 cm in diameter. Internodes relatively long. Apex with some remains of stems.
3. Rhizome of *Coptis teeta*: Curved hook-like, mostly single, relatively small.

Example 2 Rhizoma Pinelliae: the dried tuber of *Pinellia ternata* (Thunb.) Breit. (Fig. 2.3).

Morphological Features Sub-spherical or flattened sub-spherical, some slightly slanted; 1–1.5 cm in diameter, cultivated relatively big, up to 3 cm in diameter. Externally pale white or pale yellow, top end mostly rounded and flattened, marked with a dented stem scar in the center, surrounded by dense and dotted root scars; base obtuse and rounded, relatively smooth. Texture hard, fracture white and starchy. Odorless; taste pungent with numbing and irritating sensation.

Materials of Caulis

Materials of caulis are derived from the stem or its parts, including the branches of woody plants, stems of woody climbers, and stems or stem piths of herbaceous plants. Caulis are usually in the shape of cylindrical, cubic-columnar, or flattened cylindrical, mostly with distinct nodes and internodes, some swollen nodes with remnants of scars of little branches, leaves, or buds; leaf arrangement can be examined when leaf scar is prominent.

Herbaceous stems commonly form raised ridges and furrows externally after dried and shrink for the existence of vascular bundles and mechanical tissue. Woody

Figure 2.4 Caulis of *Spatholobus suberectus*.

stems are commonly rough on the surface, with cracks in length and breadth presenting in the cork layer occasionally, and easily observed pits.

Dicotyledonous stem can be distinguished from monocotyledonous stem by its radial arrangement of transverse section and distinct pith. Undeveloped wood and porous or hollowed pith are observed in herbaceous stems of dicotyledon, but developed wood and thin cortex are obvious in woody stems.

Example Caulis Spatholobi: the dried vine of *Spatholobus suberectus* Dunn. (Fig. 2.4).

Morphological Features Oblong or irregular oblique slices, 0.3–1 cm thick. Cork grayish brown, sometimes grayish white patches visible and appearing reddish brown when the cork is exfoliated. In the transversely cut surface, xylem reddish brown or brown, showing numerous pores of vessels; phloem with resinous secretion reddish brown to blackish brown, arranged alternately with xylem, forming 3–8 eccentric semicircular rings; pith inclined to one side. Hard texture; odorless; taste astringent.

Materials of Wood

Materials of wood are derived from the inner part of cambium, mainly the heartwood. Wood is sawed into sections, split to strip pieces, or cut to thin slices. The shape, external color, surface marks and plaque, texture, odor, and annual rings or rays viewed from transverse section or longitudinal section are the key points when examining woods.

Example Lignum Aquilariae Resinatum: the resin-containing wood of *Aquilaria sinensis* (Lour.) Gilg and *A. agallocha* Roxb. (Fig. 2.5).

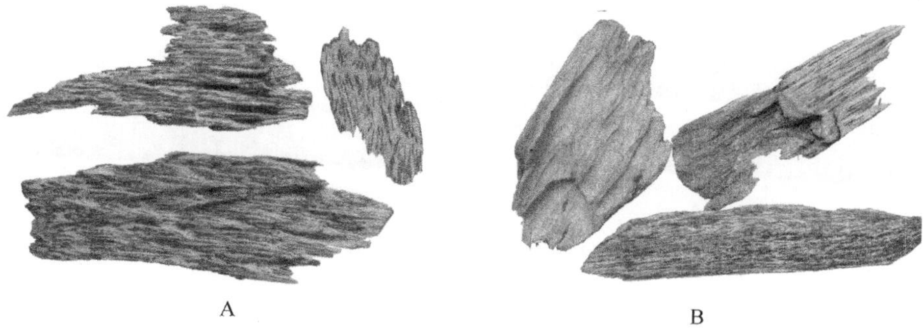

Figure 2.5 Resin containing wood of *Aquilaria sinensis* (A) and *A. agallocha* (B).

Morphological Features These are the morphological features:

1. Wood of *Aquilaria sinensis* (Lour.) Gilg: Irregular lumps and flakes, or helmet-shaped, sometimes in small fragments. Compact and woody, pale brown, with scars of knife cutting on one side; the other side is the solidified surface of the exuded resin, earth yellowish brown, uneven, with fissuring and honeycombed holes. Texture hard, mostly does not sink in water. Odor, with a distinctive fragrance; taste, slightly bitter. Produce dense smoke and a heavy aroma when burning, and ooze a black oily substance.
2. Wood of *A. agallocha* Roxb: Cylindrical or irregular lumps and flakes, both end and outer surface uneven, marked with scars of knife cutting, furrows and holes; pale yellow or grayish black, densely covered with interrupted brownish black longitudinal striation (the xylem ray with resin), sometimes scars of blackish brown resin spotted slightly with luster can be seen. Transverse section with dense and fine brownish black spots. Texture hard and heavy, sink or half-sink in water. Odor with strong fragrance.

Materials of Bark

Materials of bark are derived from truck, branch, or the outer part of root cambium of woody dicotyledon or gymnosperm, containing stem bark, trunk bark, and root bark. Some drugs' outer bark had been peeled off, using only the "inner bark," mainly phloem, as pharmaceutical use. Due to different parts and process methods used, or thickness of barks, materials of bark can be flat or curly sheet, sulciform, tubular or double tubular in shape. Root bark is relatively irregular in shape, curled inhomogenous, boot-like when peeled off near ground.

The outer surfaces of barks are relatively coarse, with clefts in length and breadth and lenticels different in shape and size. Some are peeled off in scaled-shapes; some are cohered with grayish-white lichen speckles, or consolidated with strings or veins; some are fairly smooth if the outer bark is scraped. The inner surface of bark is usually smooth, dark in color, frequently with longitudinal fine striations (fiber bundles) or reticular wrinkles. Whether the bark is easily broken off or not

A B

Figure 2.6 Stem bark of *Magnolia officinalis* (A) and *Magnolia officinalis* var. biloba (B).

depends on the thickness of bark and occupation of fiber layer. Fractures are sometimes even, granular (showing groups of stone cells), fibriform, or splinter-form, which can be torn layer by layer (showing fiber layer). Some others show connection by colloid filaments or powder dust when broken off.

Example Cortex Magnoliae Officinalis: the dried stem bark, root bark, and branch bark of *Magnolia officinalis* Rehd. et Wils. or *M. officinalis* Rehd. et Wils. var. *biloba* Rehd. et Wils. (Fig. 2.6).

Morphological Features The morphological features are as follows.

1. *Stem bar:* Quilled singly or double quilled, 30–35 cm long, 0.2–0.7 cm thick, stem bark near the root with one end spread out like a bell, 13–25 cm long, 0.3–0.8 cm thick. Outer surface grayish brown, rough, sometimes scaly, easily exfoliated, with distinct elliptical lenticels and longitudinal wrinkles, appearing yellowish brown when the coarse bark is peeled; inner surface purple brown or dark purple brown, relatively smooth, with fine and dense longitudinal striations and exhibiting oily trace on scratching. Texture hard, not easily broken, fracture granular, grayish brown in the outer layer and purple brown or brown in the inner layer, oily, sometimes numerous small bright spots visible (crystal of magnolol and honokiol). Aromatic, pungent, and slightly bitter.
2. *Root bark:* Quilled singly or pieced irregularly, some curved like chicken intestines. Texture hard, easily broken, fracture fibrous.
3. *Branch bark:* Quilled singly, 10–20 cm long, 0.1–0.2 cm thick. Texture fragile, easily broken, fracture fibrous.

Materials of Leaves or Leaflets

Materials of leaves or leaflets are mainly derived from simple leaves, or from leaflets of a compound leaf, or twigs with leaves. When examining leaves or leaflets, they should be macerated in water and spread out first to examine whether it is a simple or compound leaf, since they occur in a more or less crumpled condition. If leaf scars are found in the same level on the mixed twigs with no bud scars beside, they should be the phyllopode of compound leaf; if the leaf scars are found in alternative

Figure 2.7 Leaflet of *Cassia angustifolia* (A) and *C. acutifolia* (B).

or opposite position, twig and small stem are confirmed. Furthermore, the shape, size, color, leaf apex, leaf base, leaf margin, vein, upper and lower surface, texture, presence of leafstalk, and the length of leafstalk all should be examined. The surface features of leaves are diverse, such as some with relatively thick cuticle and smooth, some covered with hairs on one or both surface sides, some with glandular scale *sub vitro*, or some with gland spots (oil cavity) when seen through the blade to the light. Whether the leafstalk is straight or twisted is also a key point. The base of leaflet is often dissymmetric.

Example Senna Leaf: the dried leaflet of *Cassia angustifolia* Vahl or *C. acutifolia* Delile. (Fig. 2.7).

Morphological Features The following are the morphological features.

1. Leaflet of *Cassia angustifolia*: Leaflets mostly integrated and even, elongated ovate, ovate-lanceolate or lanceolate, 1.5–5 cm long, 0.4–2 cm wide, margin entire, apex acute or with close needling, base slightly asymmetrical. Upper surface yellowish green, underside pale yellowish green, glabrous or nearly glabrous, veins slightly prominent, petioles about 1 mm long. Texture leathery. Odor is slight and characteristic; taste slightly bitter and mucilaginous.
2. Leaflet of *C. acutifolia*: Leaflets mostly broken, slightly rolled, lanceolate or elongated ovate, 2–3 cm long, 0.6–1 cm wide, apex mucronate or slightly convex, base asymmetrical, both surfaces covered with fine and short hairs. Texture relatively thin and fragile.

Materials of Flowers

Materials of flowers or inflorescences are derived from buds, flowers, part of flowers such as petal, corolla, stigma and pollen, or wholly inflorescence. Most of the flowers

Figure 2.8 Flower of *Matricaria recutita*.

are characteristically with bright color and fragrance. The appearance, shape, size, color, odor, number, arrangement, and presence of hair should be described, if necessary, examining with anatomical lens under moistened condition. For the inflorescences materials, pay attention to the type of inflorescence and the shape of bract and periclinium.

Example 1 Flos Matricariae: the dried flower heads of *Matricaria recutita* Linne (Fig. 2.8).

Morphological Features Flower head is hemispherical, about 6 mm in diameter, composed of a few ray florets and numerous disk florets, carried on a receptacle surrounded by an involucre. Involucre is green, formed of two to three rows of lanceolate, glabrous, and imbricated bracts with blunt apices and scarious whitish edges. Ray florets, which usually have fallen off, have 10 to 20 pistils; corolla is ligulate, white. Disk florets are yellow, perfect; corolla is tubular with five teeth; five stamens are epipetalous and syngenesious. Receptacle is hollow, hemispherical in the young and conical in the old flower head, 3–10 mm in width. Achene is ovoid and has three to five longitudinal ribs.

Example 2 Flos Lonicerae Japonica: the dried flower bud of *Lonicera japonica* Thunb. (Fig. 2.9).

Morphological Features Small clavate bud, slightly curved, 1.3–5.5 cm long; the upper part is thicker, 2–3 mm in diameter. Externally pale yellow or pale yellowish brown, gradually darkening on storage, densely covered with rough

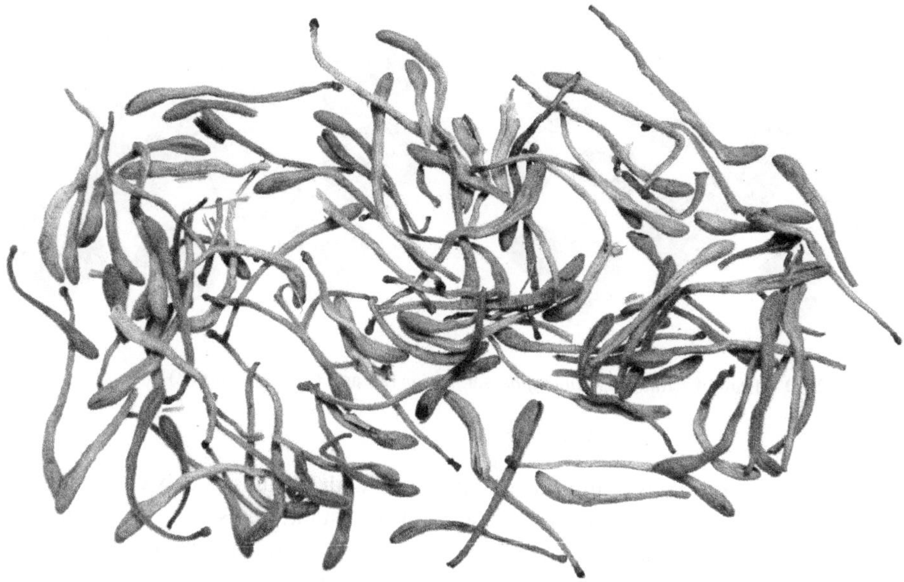

Figure 2.9 Flower bud of *Lonicera japonica.*

hairs and long glandular hairs. Calyx small, calyx tube sub-spherical, about 1 mm long, glabrous, apex five-lobed, calyx teeth ovoid-triangular, covered with hairs. Corolla tubular, two-lipped in apex, about 5 cm long. Stamens five, epipetalous. Pistil one, with a slender style. Odor, fresh and fragrant; taste, sweet and slightly bitter.

Materials of Fruit

Materials of fruit are derived from wholly fruit, pericarp, kernel, pericarp vascular bundles, ear, carpopodium, persistent calyx, and so on. For identification, the type, shape, size, color, top, base, surface, and section features should be observed. The presence of remnant bract, calyx, stamen, stylar foot, or capopodium, whether the outer surface of fruit is lustrous or covered with frosting, or with raised ridges or excavate oil cavities, or covered with hairs should also be examined. For the whole fruits, the number, shape, size, color, and outer surface feature of the seeds are key identifying points.

Example 1 Fructus Crataegi: the dried ripe fruit of *Crataegus pinnatifida* Bge. var. *major* N. E. Br. or *C. pinnatifida* Bge. (Fig. 2.10).

Morphological Features The morphological features are as follows.

1. Fruit of *Crataegus pinnatifida* Bge. var. *major* N. E. Br.: Sub-spherical, 1–2.5 cm in diameter. Externally dark red and lustrous, with small grayish

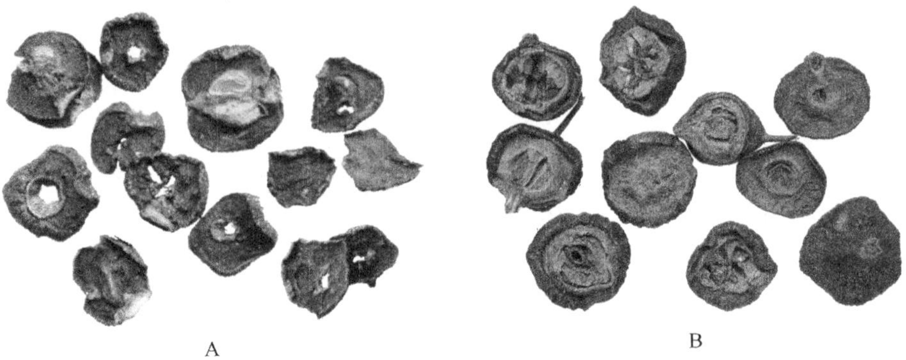

Figure 2.10 Fruit of *Crataegus pinnatifida* var. *major* (A) and *C. pinnatifida* (B).

white spots; apex exhibiting a hollowness with remains of calyx around; base showing a slender fruit stalk or scars of fruit stalk; seeds five, arciform. Most are crosscut slices, orbicular in shape, shriveled and uneven, 1–2 cm in diameter, 2–4 mm in thickness. Sarcocarp dark yellow to pale gray. Cross-section of the middle part yellow, showing five seeds, mostly fallen off. Odor, slightly fresh and fragrant; taste, sour and slightly sweet.

2. Fruit of *C. pinnatifida* Bge.: Fruit 1–1.5 cm in diameter. Externally brownish-red, with small spots; apex exhibiting remains of calyx; base showing a slender fruit stalk. Texture hard. Odor fresh and fragrant; taste slightly sour. Most are crosscut slices, orbicular in shape. Sarcocarp dark yellow to pale brown; middle part showing five pale yellow seeds, mostly fallen off and hollowed.

Example 2 Fructus Agni Casti.: the dried mature fruits of *Vitex agnus-castus* L. (Fig. 2.11).

Morphological Features Spherical to ovoid, 2–4 mm in diameter, very hard fruit. The fruit is reddish brown to black, slightly rough, and covered with glandular hairs. There are four grooves perpendicular to one another, and a slight depression on the apex, more evident on large fruits. The internal appearance of the fruit is yellowish. The internal structure of the fruit includes four compartments, each containing an oblong seed with a high-fat content. The fruit is often covered by tubular, greenish gray, fine tomentous calyx, which is persistent and has five teeth.

Example 3 Fructus Serenoae Repentis: the dried ripe fruit of *Serenoa repens* (Bartram) Small. (Fig. 2.12).

Morphological Features Fruit sub-spherical to ovoid drupes, about 2–3 cm in length and 1.5 cm thick, dark brown to black with a smooth, dull surface and with large, irregular depressions and ridges caused by shrinkage on drying; remains of

Figure 2.11 Fruits of *Vitex agnus-castus*.

Figure 2.12 Fruit of *Serenoa repens*.

style at the summit; base bearing a small depression with the scar of the stalk; epicarp and underlying sarcocarp forming a fragile layer that partially peels off, revealing the hard, pale brown layer of endocarp surrounding the seed. Seed irregularly spherical to ovoid, up to about 1.2 cm in length and 1 cm thick; hard, surface finely pitted and reddish brown with a paler, raised and membranous area over the raphe and micropyle; when cut transversely, it shows a thin testa, a narrow perisperm, and a large area of dense, horny, grayish white endosperm.

Materials of Seeds

Materials of seeds are mainly derived from whole seeds, occasionally including testa, kernel, or aril. When examining, first observe the shape, size, color, and surface characters, including the position and shape of hilum, raphe, chalaza, micropyle, presence of striations, prominence, hairs, calyptra, and the characteristics of transverse and longitudinal sections, then peel testa, take notice of the existence of perisperm. Generally, non-endosperm seed shows developed cotyledon and a layer of transparent membrane as endosperm; albuminous seed has rich oily or corneous endosperm and rich oily or starchy cotyledon.

Example 1 Semen Strychni: the dried ripe seed of *Strychnos nux-vomica* L. (Fig. 2.13).

Morphological Features Button-shaped, usually slightly dented on one side and slightly bulgy on the other, 1.2–3 cm in diameter, 3–6 mm in thickness. Externally

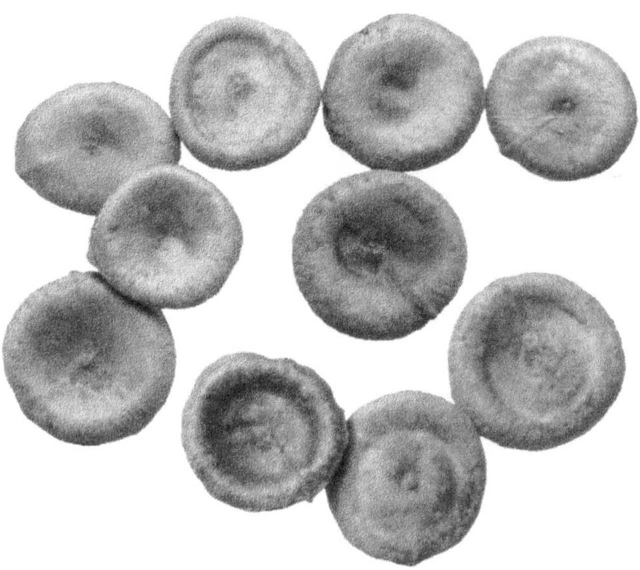

Figure 2.13 Seed of *Strychnos nux-vomica.*

grayish green or grayish yellow, closely covered with silver gray silky hairs radiating from the center. A protuberant, dot-like hilum in the center of the bottom surface, with a protuberant micropyle around, a prominent ridge between hilum and micropyle seen indistinctly. Texture hard. When cutting along the margin, it shows pale yellowish white fleshy and cuticular endosperms; cotyledons two, cordate, 5–6 mm long, with 5–7 palmate veins. Odorless; taste extremely bitter. Extremely toxic; must not be tasted.

Example 2 Semen Hippocastani: the dried seeds of *Aesculus hippocastanum* L. (Fig. 2.14).

Morphological Features Irregularly ovoid or sub-spherical, a few somewhat hemispherical, with an almost planar facet, 2.5–3 cm in diameter; surface smooth, dark, marbled reddish brown, a little uneven and with dull luster; large, nearly circular, rough, grayish brown scar of hilum; in section, testa 1–1.5 mm thick, dark brown, hard and brittle, closely surrounding the embryo of two large, horny, off-white cotyledons and radicle; endosperm absent.

Materials of Herbs

Materials of herbs are mainly derived from the whole parts of the herbaceous plant, the aerial parts, the flowering branches or fruiting branches with leaves, or the herbaceous stems of shrubs. Herbs are mostly dried or obtrite, so they are better examined after moistening. If the material is composed of integral flowers and fruits, it

Figure 2.14 Seeds of *Aesculus hippocastanum.*

Figure 2.15 Aerial part of *Leonurus japonicas.*

can be examined following the plant identification methods; if not, examine according to the morphous of each part.

Example Herba Leonuri: the dried aerial part of *Leonurus japonicus* Houtt. (Fig. 2.15).

Morphological Features Stem quadrangular, frequently branched at the upper part. Externally yellowish green, with longitudinal edges, covered with strigas. Texture fragile, fracture with a white pith in the center. Leaves opposite and shrunken, usually fallen off or remained; verticillaster axillary, corolla mostly fallen off; calyx 8–15, persistent, aggregate to a globe; nutlet brown, prismatic. Odorless; taste slightly bitter.

2.2.2 Microscopic Identification

Different traditional herbal materials have significant different microstructures. Microscopic identification is a method using a microscope to identify the structural features, cells, and ergastic substances of herbal samples with application of the knowledge of plant morphology and anatomy so as to authenticate plant species. The procedure of identification contains three main steps: selection of typical materials, preparation of slides or powder, and observation of features.

The objects of the microscopic observation include slides of transverse section, longitudinal section, surface section, powder and disintegrated tissues. The shape, size, and content of the plant cells under microscope will provide important information for species identification.[17–19]

In addition to traditional microscopic identification, development of modern microscopic technologies has extended the application of microscopic study on herbal materials to microscopic quantification, microscopic histochemistry, electron microscopic scanning combined with technology of image quantitative analysis, systematic cluster analysis, three-dimensional reconstruction through computer, NIR-FT-Raman mapping, fluorescence, and so on.[20–26]

General microscopic features and specific examples of herbal materials from different parts of plants are given here.

Materials of Roots

1. *Tissue characteristics*

By observing the vascular system, the microscope identification is able to determine whether the material is the first or secondary structure of dicotyledonous, or the monocotyledonous root.

The primary structure of dicotyledonous root usually consists of small stele, a few phloem and xylem bundles arranged alternately, primary xylem forming an asterism shape, and mostly without pith. The secondary structure (e.g., Radix Glycyrrhizae) usually consists of cambium layer, phloem outside, xylem inside, rays, and mostly without pith. Moreover, some dicotyledonous roots have abnormal tertiary structures, such as Radix Polygoni Multiflori with several abnormal vascular bundles outside the cambium ring, Radix Acanthopanacis Bidentatae with several concentric vascular bundles, and Radix Phytolaccae with several layers of cambium rings. The monocotyledonous root (e.g., Radix Stemonae) usually consists of numerous phloem and xylem bundles arranged alternately to form a circle, with pith but without cambium layer or rays.

Most dicotyledonous medicinal roots are secondary structures with cork on the surface. Few materials have an epidermis as nourishing and protecting tissue, such as Radix Pulsatillae and Radix Asteris, or a surface layer consisting of metaderm derived from epidermis and part of cortical cell walls through corkification, such as Radix Aconiti and Radix Scrophulariae. Cortex is usually narrow. In general, the primary cortex is absent, and the secondary cortex is accumulated by phelloderm; phloem is well developed. Sometimes the secondary cortex is well developed while phloem fairly narrow, such as Radix Morindae Umbellatae. Xylem consists of vessels, tracheids, fibers, parenchymatous cells and rays, some with well-developed vessels and fibers, and narrow xylem rays, such as Radix Morindae Officinalis and Radix Polygalae, some with sparsely arranged radial vessels and broad xylem rays, such as Radix Paeoniae Alba and Radix Ginseng. Attention should be paid to the shape, diameter, and arrangement pattern of vessels, visible thickened striations in longitudinal section, as well as the shape, size, and distribution of xylem fibers. Most roots have no pith, but a few have an obvious one, such as Radix Gentianae and Radix Aconiti.

Monocotyledonous roots have no cork tissue. The outer walls of epidermal cells sometimes are thickened or present as several arrays of velamina cells through corkification or lignification of cell walls. Walls of some roots, such as Radix Stemonae, show fine, thickened striations. The broad cortex occupies the majority of the root. Distinct casparian dots are usually obvious in the endothelial layer. Pith is fairly large, mostly consisting of parenchymatous cells, with a few walls thickened and lignified, such as Radix Liriopes and Radix Stemonae Parviflorae. Roots usually contain secretory tissues scattered in phloem, such as laticiferous tubes in Radix Platycodi and Radix Codonopsis, resin canals in Radix Ginseng and Radix Notoginseng, oil cavities in Radix Aucklandiae, oil cavities and tubes in Radix Angelicae Sinensis, and oil cells in Herba Asari and Radix Aristolochiae.

Example Microscopic features of Radix Ginseng (root of *Panax ginseng* C.A. Mey.) in transverse section, resin canal, parenchyma cells containing clusters of calcium oxalates (Fig. 2.16).

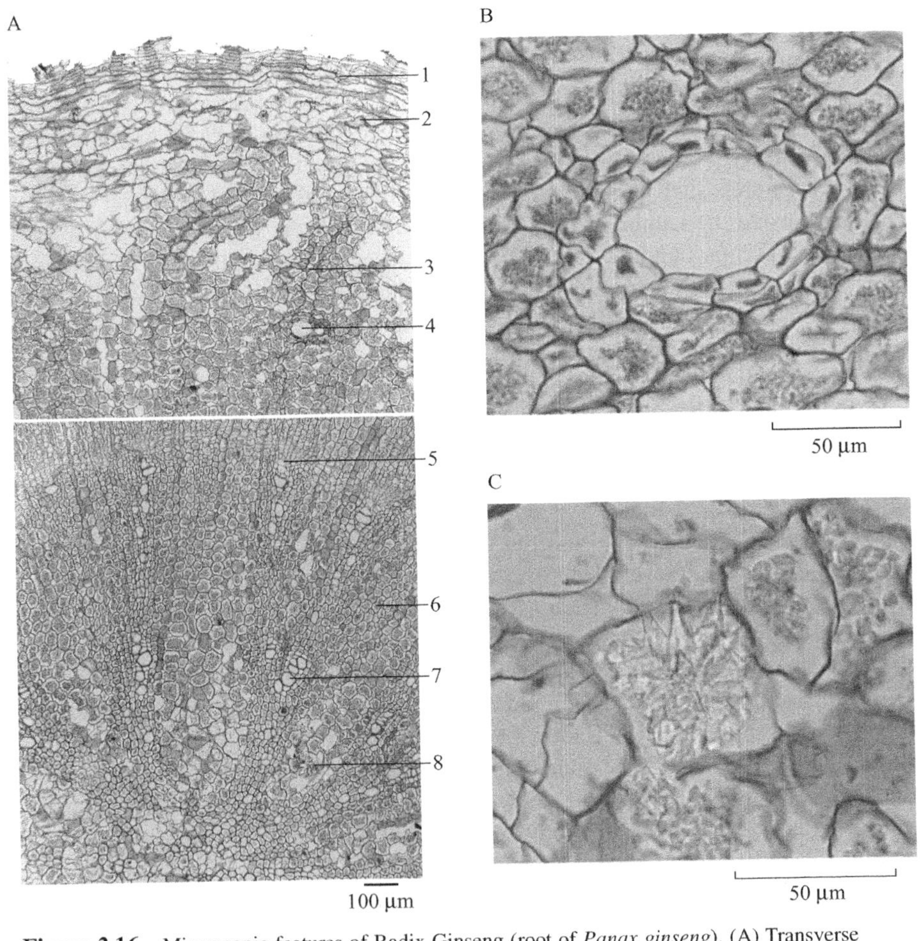

Figure 2.16 Microscopic features of Radix Ginseng (root of *Panax ginseng*). (A) Transverse section (1. cork layer; 2. cortex; 3. phloem; 4. resin canal; 5. cambium; 6. xylem ray; 7. xylem; 8. cluster of calcium oxalate); (B) resin canal; (C) parenchyma cells (containing cluster of calcium oxalate).

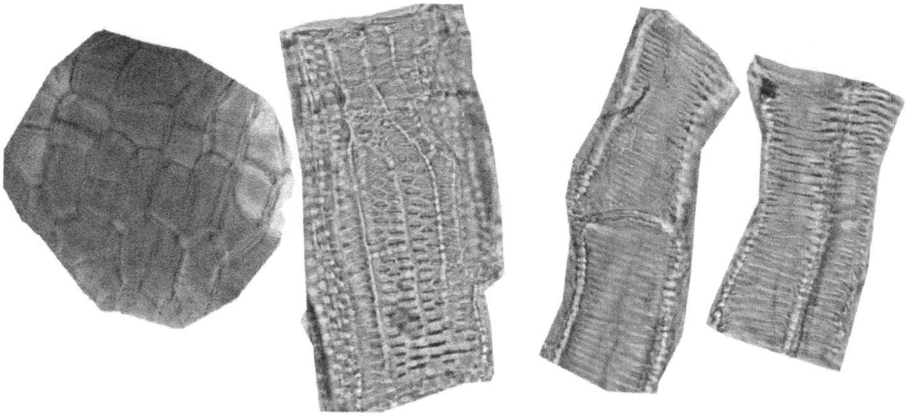

Figure 2.17 Cork cells and vessels of Radix Glycyrrhizae (A); vessels of Radix Aucklandiae (B).

2. *Powder identification*

Powder identification of root materials focuses primarily on the diagnostic characteristics of ergastic substance of cells, sclerenchyma and secretory tissues, as well as the epidermis, hypodermis, velamen or cork tissue, endodermis, vessels, and tracheids. Parenchyma, sieve tube, companion cell, and parenchymatous cells of cambium layer and pericycle are usually not significant diagnostic characteristics. Crystals of calcium oxalate are observed frequently, such as cluster crystals in Radix Polygoni Multiflori and Radix Ginseng, raphides in Radix Asparagi and Radix Ophiopogonis, sandy crystals in Radix Stellariae and Radix Acanthopanacis Bidentatae, and prisms in Radix Glycyrrhizae and Radix Hedysari, which form into crystal fibers. In addition, pay attention to the existence of starch grains, fibers, and stone cells; starch grains are fairly tiny; fibers are frequently observed, but stone cells are infrequent.

Other cells and tissue fragments may be found except mesophyll tissue. Epidermal cells of roots are seldom found. Cork tissue may be frequently seen. Notice the shape, color, and wall thickness of cork cells in surface view (Fig. 2.17).

Vessels are usually thick. Pay attention to the type, diameter, length of vessel elements, perforate in the end wall, pit shape and arrangement, and so on. (Fig. 2.17). Attention should also be paid to stone cells regarding shape, size, shape and extent of sclerosis, shape and size of pits, density of pit canals, and so on. (Fig. 2.18). For fibers, identify the shape, length, caliber, end wall and sclerosis extent and characters, type, canal shape, and arrangement of piths. In addition, note whether or not cells surrounding fiber bundles contain crystals that form crystal fibers (Fig. 2.19).

When observing secretory tissue, attention should be paid to the types of secretory cells, secretory cavity, secretory tube, laticiferous tube, shape of secretory cells, the color of secretory substance, and the arrangement and shape of surrounded cells (Fig. 2.20).

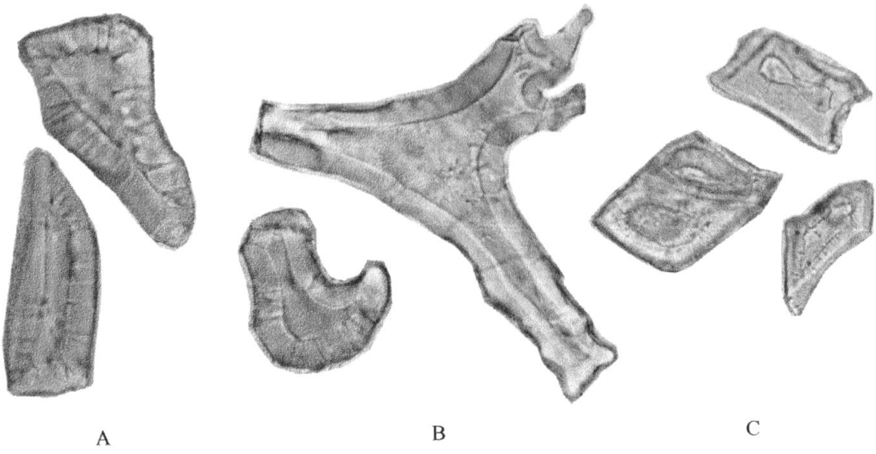

Figure 2.18 Stone cells. (A) Radix Morindae Officinalis; (B) Radix Menispermi; (C) Radix Codonopsitis.

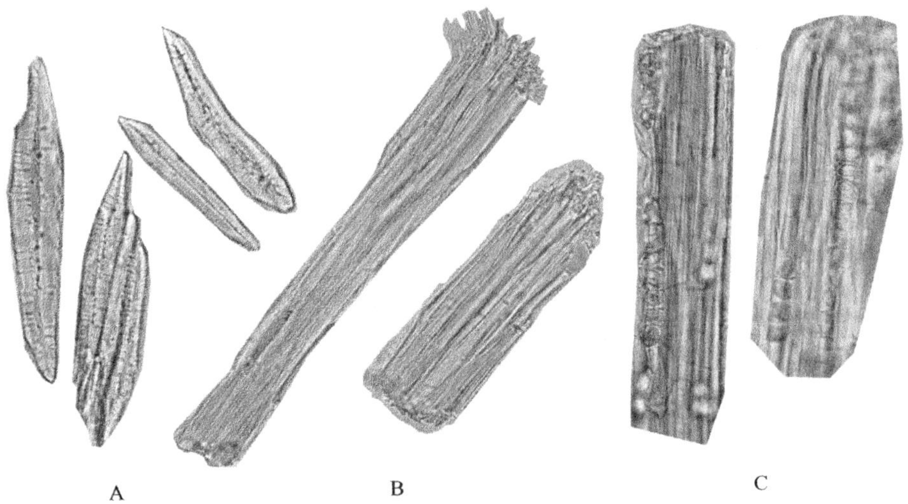

Figure 2.19 Fibers. (A) Radix Scutellariae; (B) Radix Astragali; (C) Radix Glycyrrhizae.

Crystals are mostly calcium oxalates, sometimes inulin, silica bodies, and so on. Pay attention to the type, size, and arrangement of crystals, and the shape of crystalliferous cells (Fig. 2.21).

Starch grains are usually tiny. Pay attention to the amount, shape, type, size, shape and location of hilar spot, striations, and so on (Fig. 2.22).

Figure 2.20 Secretary tissues. (A) Radix Saposhnikoviae; (B) Radix Codonopsitis; (C) Radix Ginseng.

Figure 2.21 Crystals. (A) Prism of calcium oxalate of Radix Glycyrrhizae; (B) raphides of calcium oxalate of Radix Asparagi; (C) cluster of calcium oxalate of Radix Ginseng; (D) sand crystal of calcium oxalate of Radix Achyranthis Bidentatae; (E) inulin of Radix Aucklandiae.

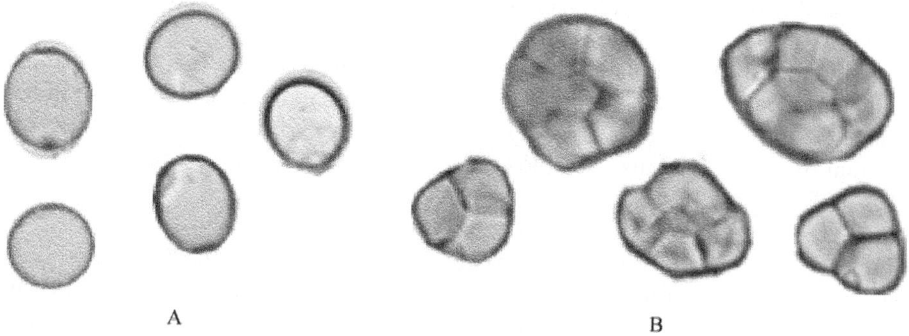

Figure 2.22 Starch grains. (A) Radix Trichosanthis; (B) Radix Notoginseng.

Materials of Rhizomes

1. *Tissue characteristics*

Similar to roots, the identification should first determine whether the rhizome is pteridophyte, dicotyledon, or monocotyledon by identifying the types of steles and vascular bundles.

The outmost of rhizome of pteridophyte is usually composed of epidermal and hypodermal cells with thickened wall, and basic parenchyma well developed. The types of stele are varied. Some are protostele with xylem (tracheid only) in the center, with phloem surrounded, pericycle and endodermis outside, such as Spora Lygodii; some are amphiphloic siphonostele, with xylem forming cylinders, phloem, pericycle, and endodermis existent in exterior and interior parts, such as Rhizoma Cibotii; some are dictyostele with several meristeles arranged interruptedly, forming a ring in transverse section slides, with every meristele forming a protostele, such as Rhizoma Dryopteris Crassirhizomae. Moreover, sometimes, internal unicellular glandular hairs are present in parenchyma cellular spaces that contain secretions, such as Rhizoma Dryopteris Crassirhizomae. The shape and marginal characters of surface scales also have considerable contributions to identification.

Most dicotyledonous rhizomes contain cork tissue or cork stone cells, such as Rhizoma Coptidis, Rhizoma Atracylodis, and Rhizoma Atractylodis Macrocephalae. Root trace bundles can occasionally be seen in cortex. Stele presents infinite collateral vascular bundles arranged in a ring; and pith in the center. A few kinds have tertiary structures with abnormal complex vascular bundles in pith, such as Radix et Rhizoma Rhei.

The outmost layer of monocotyledonous rhizome is mostly epimis. Some cork tissues are formed by outer parts of the cortex, such as Rhizoma Pinelliae and Rhizoma Zingiberis. Some metaderm are formed by suberized cortex cells, such as Rhizoma Veratri Nigri. Leaf trace bundles can be observed in cortex; endodermis is mostly distinct except for some thick rhizomes and tubers; and limited collateral bundles or perixylem bundles scatter in stele, such as Rhizoma Acori calami. Stomata are visible on the epidermis of scale leaf from bulb.

Some rhizomes are present with oil cavities (e.g., Rhizoma Chuanxiong, Rhizoma Atractylodis, and Rhizoma Atractylodis Macrocephalae) or oil cells (e.g., Rhizoma Acori Tatarinowii, Rhizoma Cyperi). For the monocotyledonous rhizomes, raphides of calcium oxalate are often seen in mucilage cells, such as Rhizoma Arisaematis, Rhizoma Pinelliae, Rhizoma Gastrodiae, Rhizoma Bletillae, Rhizoma Polygonati Odorati, Rhizoma Polygonati, and Rhizoma Dioscoreae. Apart from the characteristics mentioned above, sclerenchyma, types of vessels, and crystals of calcium oxalate should also be noted.

2. *Powder identification*

Similar to root materials, large amounts of starch grains often occur in bulbs, stem tubers, and corms. Form, size, hilum, striation, and types of starch grains involving the compound grains, semi-compound grains, and multi-hilum in simple grains are all important identifying characteristics. Stomata are usually visible in the epidermis of scale leaf from bulb. Vessels in annular type are frequently visible in rhizome of monocotyledons while tracheids are visible only in that of pteridophyte.

Example Microscopic features of transverse section of Rhizoma Coptidis (rhizome of *Coptis chinensis* Franch.) (Fig. 2.23).

Materials of Caulis and Stems

The observation points of microscopic identification of caulis tissue structure and the powder of the materials are basically similar to those of roots and rhizomes. The main differences are as follows.

1. *Tissue characteristics*

The outmost layer of dicotyledonous herbal stem is the epidermis. Cortex is the primary formation; its laterals differentiate into collenchyma, sometimes showing endodermis. Pericycle frequently differentiates into fibers, or has scattered stone cells forming a continuous ring. Intradascicular cambium is distinct and secondary phloem presents in bundles or plates; pith is fairly broad with wide medullary rays. The outermost layer of secondary structure is epidermis or cork tissue, and cambium layer presents as a continuous ring.

Xyloid stems of dicotyledons are otherwise composed of the following tissues: cork giving rise externally to secondary cortex, sclerenchyma of pericycle is arranged continuously in a ring or interruptedly, cambium obvious, secondary phloem and secondary xylem showing cylinder structure, relatively narrow rays formed by cells with lignified walls, pith small.

The stems of monocotyledons are covered with epidermis externally, followed by several limited collateral vascular bundles scattered in elementary tissue, without pith.

The structure of gymnospermous stem is similar to xyloid stem of dicotyledonous. However, xylem almost consists of tracheids instead of vessels (except Herba Ephedrae); phloem consists of sieve cells and parenchymatous cells.

2. *Powder characteristics*

The cuticle of epidermal cells is usually thick. Attention should be paid to its color, thickness, surface properties, like smoothness and striations. For example, the surface of cuticle in caulis of *Sinomenium acutum* (Thunb.) Rehd. et Wils. shows indistinct striations, while that of *Sinomenium acutum* (Thunb.) Rehd. et Wils. var. *cinereum* Rehd. et Wils. shows granularity on the cuticle surface. In addition, notice the presence of stoma and trichome on epidermis, and the existence of hypodermal layer or hypodermal fiber under epidermis.

Fibers generally exist. The types include cortical fibers, pericycle fibers (surrounded fibers), phloem fibers, xylem fibers, and perimedullary fibers, such as Caulis Pipers Futokadsurae. Attention should be paid to the characteristics of fibers, such as shape, length, diameter, wall thickness, wall lignification, pit, pit canals, and smoothness or cracks in surface. Tiny sandy or needle crystals may be observed in

Figure 2.23 Microscopic features of transverse section of Rhizoma Coptidis (rhizome of *Coptis chinensis*) (1. cork layer; 2. cortex; 3. group of stone cells; 4. pericycle fiber bundles; 5. phloem; 6. cambium; 7. xylem; 8. pith).

some material, such as Caulis Perillae. Moreover, crystal fibers and silica bodies in the surrounding cells of fiber bundles should also be noted, such as Caulis Dendrobii.

Stone cells can be seen frequently with big and various shapes. Attention should be paid to the shape, size, wall thickness, pit, pit canals, and striations. Some contain prisms such as Ramulus Mori and Caulis Millettiae. Some contain prisms and colored substances such as Herba Taxilli.

Crystals of calcium oxalate are easily seen as clusters and prisms; tiny sandy crystals and raphides may be found. Monocotyledonous raphides of calcium oxalate present in mucilage cells, such as Rhizoma Dioscoreae and Rhizoma Arisaematis.

Starch grains are seldom seen, and are usually tiny; few are large, such as Caulis Cistanchis.

Example Microscopic features of transverse section and powder of Caulis Aristolochiae Manshuriensis (caulis of *Aristolochia manshuriensis* Kom.) (Fig. 2.24).

Materials of Woods

1. *Tissue characteristics*

Materials of wood are usually observed in the following three sections.

a. *Transverse section:* the vertical cross-section of longitudinal axis. Annual rings can be seen in transverse section as several concentric rings. Rays present radially with visible ray width. Vessels, tracheids, xylem fibers, and xylem parenchymous cells between rays show sub-round or polygonal shape with various size and wall thickness.

This section is mainly used to observe the width of xylem rays (cell rows), density, proportion, and arrangement of vessels and xylem parenchymous cell, and shape and diameter of vessels and xylem fibers, etc.

b. *Radial section:* the longitudinal section that crosses the caulinary diameter. Annual rings in this section are seen as vertical horizontal barred. Rays are laterally arranged, vertical with annual rings, and the rays' height is visible. All the longitudinal cells such as vessels, tracheids, and xylem fibers are formed into longitudinal elongated or rhombus shapes, with distinct thickened textures in the secondary wall.

This section is mainly used to observe the height of xylem rays and type of cells (homocellular ray or heterocellular ray). Xylem rays in radial sections form transverse bands, vertical with axial vessels, xylem fibers, and parenchymous cells. Meanwhile, the type and length of vessels, their diameters and thyloses, and the type and size of xylem fibers, wall thickness, pit, and so on can also be examined in this section.

c. *Tangential section:* the longitudinal section that is cut not across the caulinary center but along the arc tangent line. It becomes obvious in this section

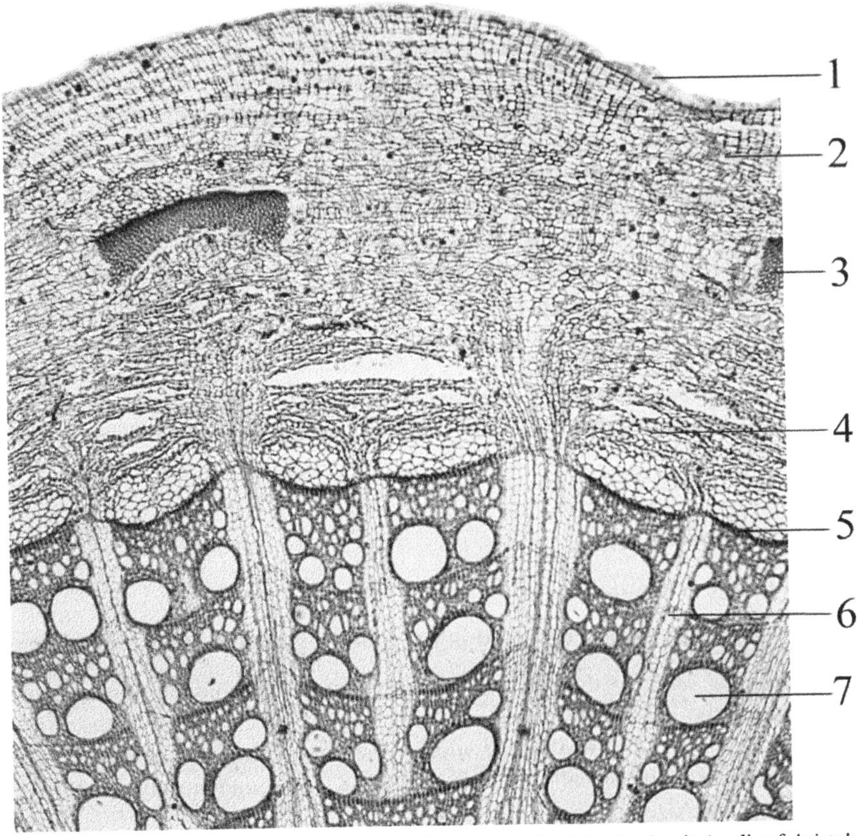

Figure 2.24 Microscopic features of of Caulis Aristolochiae Manshuriensis (caulis of *Aristolochia manshuriensis*). (A) Transverse section (1. cork layer; 2. phelloderm; 3. pericycle fiber bundles; 4. phloem; 5. cambium; 6. secondary ray; 7. xylem; 8. pith ray; 9. pith). (B) Powder (1. vessel; 2. xylary parenchyma cell beside vessel; 3. phloem fiber; 4. xylary fiber; 5. cluster of calcium oxylate; 6. stone cell).

that ray cells form sub-round or rectangular shapes, and are usually integrated as fusiform cell groups, arranged interruptedly. Width and height of rays can therefore be observed. Other appearances of vessels and tracheids are similar to radial section.

This section is mainly used to observe width, height, and type of xylem rays. Xylem rays form rhombuses in radial section; the width refers to cell layers in the widest part, while the height refers to cell layers from the top down. The vessels, xylem fibers, and so on should be noted as well.

Vessels in wood materials are often presented with pits on the border; their type, size, and arrangements should be examined. Xylem fibers are divided into libriform fibers and fiber tracheids. The former has a cell wall without pit or with single oblique pits, while the latter has bordered pits. The pit shape should be noted. Xylem ray cells and parenchymous cells are usually lignified with pits except when phloem exists (e.g., Lignum Aquilariae Resinatrm). In addition, some materials possess oil cells (e.g., Lignum Cinnamomi Camphorae), tubiform sectrtory cells (e.g., Lignum Santali Albi), columnar crystals of calcium oxalate (e.g., Lignum Aquilariae Resinatrm), or prisms forming crystal fibers (e.g., Lignum Sappan and Lignum Santali Albi). Attention should be paid to the width and lignification of crystal cell walls. For gymnosperms, tracheid and xylem ray cell are mainly observed.

2. *Powder characteristics*

The powder identification should focus on the appearances of vessels, libriform fibers, fiber tracheids, and the cell ergastic substances, as well.

Example Microscopic features of three sections of Lignum Aquilariae: resin wood of *Aquilaria sinensis* (Lour.) Gilg. (Fig. 2.25).

Materials of Bark

1. *Tissue characteristics*

The material of bark mainly consists of periderm, pericycle, and phloem.

Cork tissue commonly exists with only few exceptions when endodermis is used. The observation should focus on the number of cork cell layers, color, the thickness of cell walls, and presence of cork stone cells (hard cork). Some cork cells show a thickened inner wall, such as Cortex Eucommiae, some are in the outside wall of innermost layer of cork cells, such as Cortex Cinnamomi, and some are obvious with thickened cork cells, such as old Cortex Acanthopanacis. Rhytidome is visible in some old tissues, such as Cortex Lycii and Cortex Eucommiae.

Cortex is usually narrow. The primary cortex of root bark is absence, and the secondary cortex is formed from phelloderm. Phloem occupies the most part of the cortex, with rays penetrating totally, and can be divided into phloem rays and phloem bundles. The position where rays reach is usually the boundary of phloem and cortex. Attention should be paid to the width of phloem rays (cell layers), whether they are bent, straight, or deviated to one side, and the shape, size, wall thickness, pit, contents of ray cells, and shape of possible expansion of rays. Phloem bundles mainly consist of sieve tubes and parenchymatous cells. The lateral sieve tubes are usually collapsed and form irregular bands. Fibers (e.g., Cortex Phellodendri and Cortex Mori) or stone cells (e.g., Cortex Eucommiae) in some phloem bundles of cortex aggregate tangentially to form several layers and strips (called hard phloem), arranged alternatively with sieve tubes and parenchyma tissue (called soft phloem).

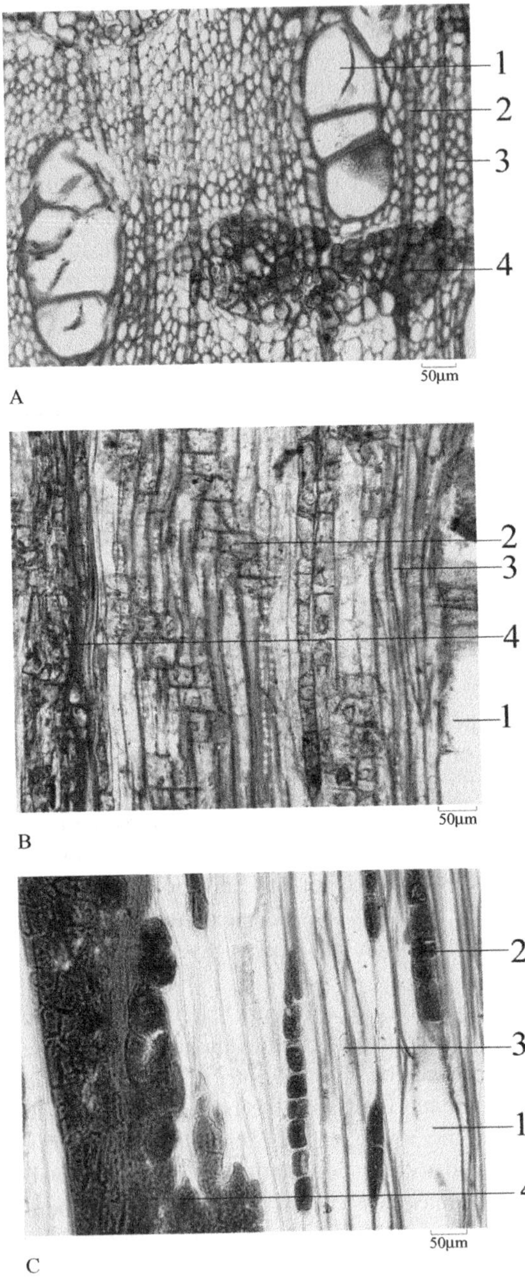

Figure 2.25 Microscopic features of Lignum Aquilariae (resin wood of *Aquilaria sinensis*). (A) Transverse section; (B) radial section; (C) tangential section (1. vessel; 2. ray; 3. xylary fiber; 4. interxylary phloem).

Sclerenchyma often presents in phloem and cortex. Carefully observe the shape, size, and arrangement of fibers and stone cells, and the width, lignified extent, striations, pit canals, and cell lumens of cell walls.

Cortex usually contains secretory tissues, such as resin canals, oil cells, and laticiferous tubes, and crystals of calcium oxalate. Some stone cells contain crystals of calcium oxalate. Some walls of stone cells and fibers parquet with tiny prisms of calcium oxalate, forming intercalary crystal sclerenchymatous cell or intercalary crystal fiber, such as Cortex Kadsurae Radicis. Most cortex tissues contain tiny starch grains.

2. *Powder characteristics*

Xylem tissues such as vessels and tracheids are usually absent. Attention should be paid to the cork cells, fibers, stone cells, secretory tissues, and crystals of calcium oxalate. Some walls of cork cells lignify unevenly, with distinct pits like stone cells. Fibers are scattered singly or in bundles. Observe carefully the color, shape, diameter, length, thickness of wall, and distinctiveness of pits. Some stone cells are bright yellow and branched, or have crystals of calcium oxalate in lumens. The sieve area of the compound sieve plate in the end wall of the sieve element is usually visible; so is the sieve area in the lateral wall of sieve cells in Pinaceae, such as Cortex Pseudolaricis.

Example Microscopic features of transverse section, stone cells, colloid filaments, and powder of Cortex Eucommiae (bark of *Eucommia ulmoides* Oliv.) (Fig. 2.26).

Material of Leaves

The microscopic observation should mainly focus on transverse section, surface section, and powders when doing identification for leaves.

1. *Tissue characteristics*

a *Transverse section:* to observe the structure of epidermis, mesophyll, and veins.

Epidermis Notice the shape, size, side wall, stomata, cuticle width, and presence of cell contents, especially the type and characteristics of trichomes of upper and lower epidermal cells. Some epidermal cells contain cytolites, such as Herba Andrographitis; some have convex papilates on the side wall of upper and lower epidermal cells.

Mesophyll It usually differentiates into palisade and spongy parenchyma, occasionally with no differentiation. Palisade parenchyma consists of one to several layers of column-shaped cells, which are noted with their shape, size, layers, and the proportion and arrangement in mesophyll. Some palisade parenchyma cross veins (e.g., Folium Sennae and Folium Nelumbinis), some present crystal cell layers

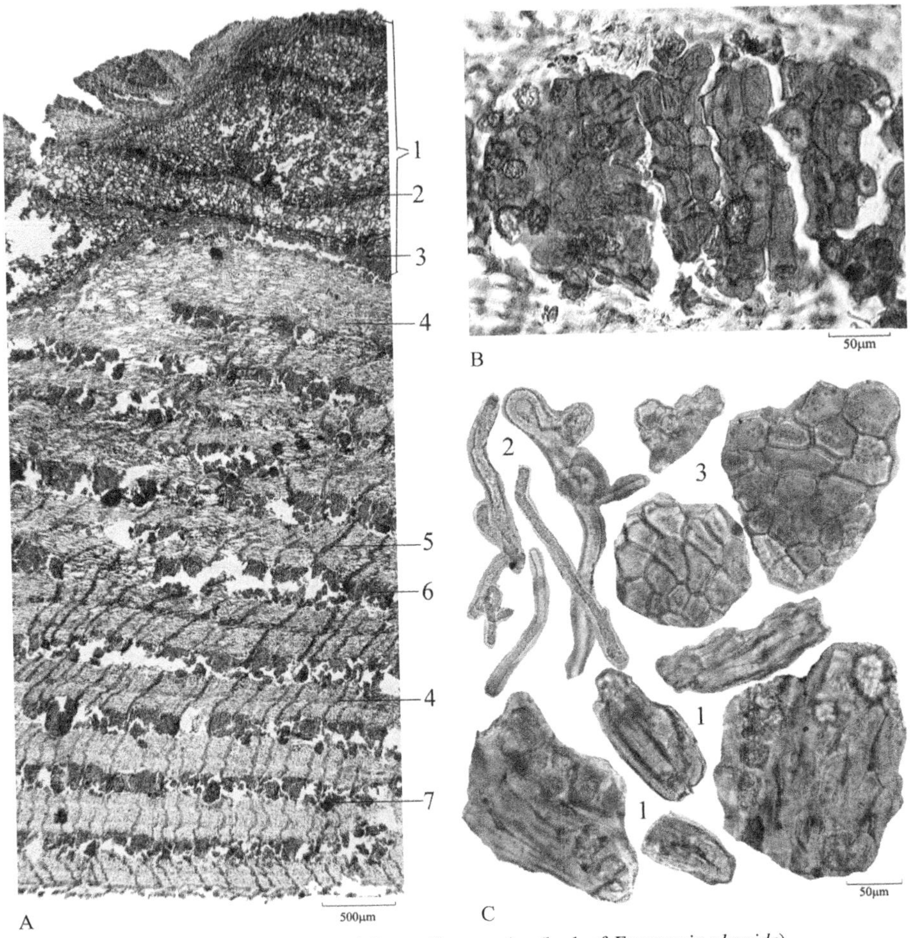

Figure 2.26 Microscopic features of Cortex Eucommiae (bark of *Eucommia ulmoide*).
(A) Transverse section (1. rhytidome; 2. cork; 3. phelloderm; 4. phloem; 5. ray; 6. stone cell; 7. colloid); (B) stone cells and colloid filaments; (C) powder (1. stone cell; 2. colloid filament; 3. cork cell).

below (e.g., Atropa Belladonna and leaves of *Datura stramonium* L.), and some are two sides of palisades (e.g., Folium Eucalypti and Folium Sennae). The layers of spongy parenchyma are indistinct, so attention should be paid to the occupied proportion of mesophyll. Some mesophylls have large oil cavities (e.g., Folium Eucalypti), laticiferous tubes (e.g., Folium Mori), or large branched stone cells (e.g., theae, folium).

Main Vein Observe the shape and types of vascular tissue, and whether fibrous layers exist surrounded b or in the lateral of phloem.

b *Surface section:* mainly to observe epidermal cells, stomata, the total shape of all kinds of trichome, as well as some differential points in mesophyll tissue such as crystals of calcium oxalate and its distribution.

Epidermal Cells and Stomas Attention should be paid to the style of upper and lower epidermal cells, curvature and thickened extent of anticlinal walls with or without pits, striations of cuticle, style of stomas, and the number of accessory cell. The shapes of upper epidermal cells are usually polygonal or irregular, with few or no stoma. Occasionally, convex papilate forms on the outer layer of the wall, such as Folium Nelumbinis. The anticlinal walls of lower epidermal cells undulate, stomatamany; some side walls present as convex papilate, such as leaves of *Epimedium sagittatum* (Sieb. et Zucc.) Maxim.

Trichome Trichome is one of the most important identifying characteristics. Observe carefully the non-glandular hairs, the cell number, layers, color, shape, length, thickness of cell wall, and characteristics on surface. The apical cell of non-glandular hair in leaves from Compositae are transversely elongated, connected to several short cells in the center, or forming a T shape; the lengths of both arms of apical cells are significant in identification. Attention should be paid to the presence of stellate hair, peltate hair, or candelabra hair. If it is glandular hair, observe the shape of head, cell number, size, color of secretions, and length, cell number, or layer number of the petiole. The head of glandular hair in leaves from Labiatae presents as a depressed ball that consists of eight cells, covered with cuticle; the handle cell is single and short, forming into glandular scale. The diameter and color of head and handle are significant for identification. Circumambient epidermal cells in the base are usually different from general epidermal cells in their number, shape, arrangement, and wall thickness.

2. *Powder identification*

Similar to the surface section, the trichomes are usually incomplete, and transverse section of leaf can be seen in powder.

Example Microscopic features of transverse section and surface of Herba Menthae (leaf of *Mentha haplocalyx* Briq.) (Fig. 2.27).

Materials of Flowers

1. *Tissue characteristics*

Based on the different objects, the bract, calyx, corolla, stamen, pistil, or the entire flower is prepared on the slide surface separately for microscope observation. Sometimes a transverse section of calyx tube is also made for identification.

a. *Bracts and calyces:* the structures of bract and calyx are similar to that of leaf. Pay attention to the stomata and trichomes of the epidermis. Mesophyll

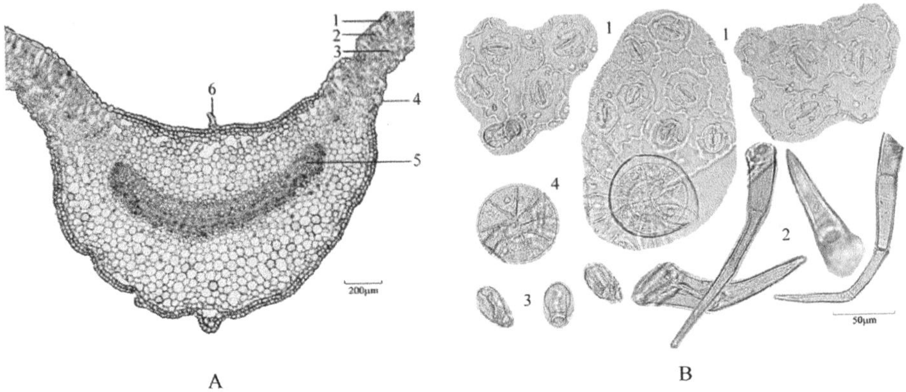

Figure 2.27 Microscopic features of Herba Menthae (leaf of *Mentha haplocalyx*). (A) Transverse section (1. upper epidermis; 2. palisade tissue; 3. spongy tissue; 4. lower epidermis; 5. vascular bundle of main vein; 6. trichome). (B) Surface slide (1. epidermal cells and stomata of leaf; 2. nonglandular hair; 3. small glandular hair; 4. glandular scale).

tissue is non-differentiated, and looks like spongy parenchyma. Some bracts totally consist of fibrous-thickened wall cells.

b. *Corolla:* the structure is similar to that of calyx, but with degenerate, tiny stomata. Upper epidermal cells often form convex papilate or villiform prominence. The positions corresponding to the mesophyll are usually several layers of loose parenchyma cells with tiny vascular tissues, sometimes with only a few spiral vessels. Some corolla contain oil cavities (e.g., Flos Caryophylli), and tubiform secretory cells (e.g., Flos Carthami).

c. *Androecium:* stamen is built by capillament and semet. The structure of capillament is simple, with longitudinal prolong rectangle epidermal cells, possibly presenting trichome, amphicribral vascular bundles, and basic parenchyma containing crystals of calcium oxalate, volatile oil, and others. Semet consists of two lobes, two anther chambers each (called pollen sacs), containing pollen grains. Anther sacs of androecium often show reticulate, strip form, or punctiform thickening in inner walls of cells, and are mostly lignified. Pollen grain is also an important characteristic of flowers, noting the shape, size, germinal aperture or furrow, exhymenine structure, sculpture, and others.

d. *Gynecium:* epidermal cells, especially the cells at the apex of pistillate stigma usually show convex papilate, or differentiate into villiform (e.g., Stigma Croci), and yet some are without prominence (e.g., Flos Daturae). The epidermal layer of the ovary wall usually has trichomes or all kinds of prominences. Some epidermal cells of the ovary contain tiny crystal prisms of calcium oxalate, such as Flos Inulae.

2. *Powder identification*

When observing the powdered flower, pollen grains, inner wall cells of anther sac, non-glandular hairs, and glandular hairs are all major diagnostic features. Crystals of calcium oxalate, secretory tissue, pigment cells, and others should be noted as well.

Example Microscopic features of surface of Flos Lonicerae Japonica (bud of *Lonicera japonica* Thunb.) (Fig. 2.28).

Materials of Fruits

1. *Tissue characteristics*

For identification of fruit herbal materials, characters of pericarp are generally observed. Pericarp of true fruit, which derived and multiplied from ovary wall, can be divided into three parts: exocarp, mesocarp, and endocarp. Endocarp and exocarp resemble the upper and lower epidermis of leaves, respectively, while mesocarp is related to mesophyll.

Exocarp Usually one layer of epidermal cells, some containing crystal of hesperidin (e.g., Pericarpium Zanthoxyli), some with oil cells scattered (e.g., Fructus Schisandrae), some differentiated into non-glandular hairs (e.g., Fructus Mume and

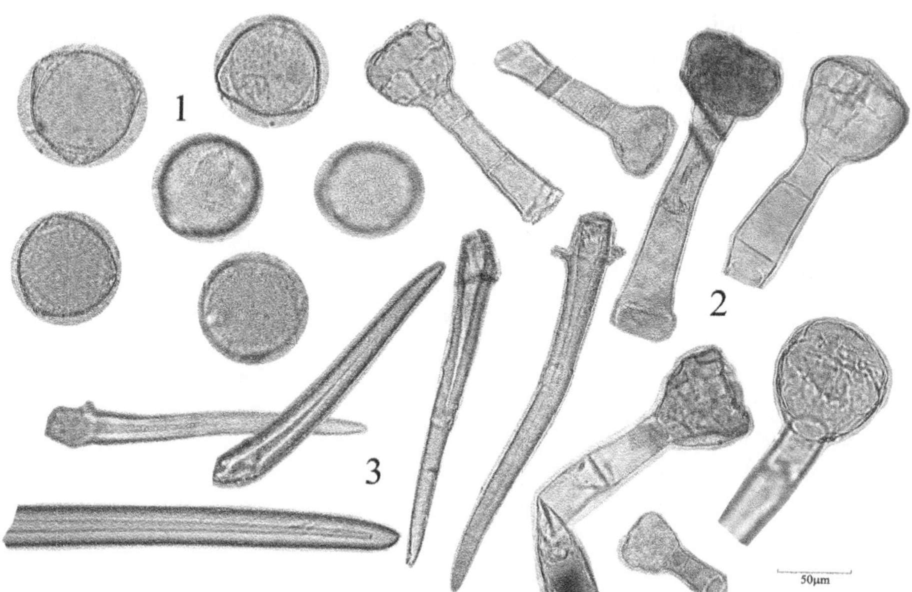

Figure 2.28 Microscopic features of surface slide of Flos Lonicerae Japonica (bud of *Lonicera japonica* (1. pollen grain; 2. glandular hair; 3. nonglandular hair).

Fructus Rubi), glandular hairs (e.g,. Evodiae and Fructus Psoraleae), or glandular scales (e.g., Fructus Viticis), and some showing textures on cuticular layers, involving irregular reticulate (e.g., Fructus Forsythiae), straight (e.g., Fructus Schisandrae), or granular-shaped ones (e.g., Fructus Corni). The thickness and density of the textures are of diagnostic significance. Some exocuticles are also composed of epidermis and hypodermis, which differentiated into stone cells, such as Fructus Piperis.

Mesocarp Composed of several layers of parenchyma cells, scattered with tiny vascular bundles, usually collateral, sometimes bicollateral (e.g., fruits from Solanaceae), or combination of two collateral bundles (e.g., Fructus Foeniculi). Mesocarp usually contains oil cavities (e.g., Pericarpium Zanthoxyli), oil cells (e.g., Fructus Schisandrae), oil tubes (e.g., Fructus Foeniculi), and sclerenchyma distributed. Some mesocarp have sandy crystal cells scattered, or starch grains sometimes (e.g., Fructus Schisandrae). In addition, reticular cells are sometimes visible (e.g., Fructus Foeniculi).

Endocarp Vary greatly. Some are one layer of parenchyma cells, with stone cells scattered; some are crystal cells (e.g., Fructus Leonuri), mosaic cell layers (e.g., fruits from Umbelliferae); some others differentiate into fibrous layers (e.g., Pericarpium Zanthoxyli) or stone cell layers (e.g., Fructus Mume and Fructus Arctii).

2. *Powder characteristics*

Attention should be paid to the shape of exocarp cells, the thickening condition of the anticlinal wall, cuticle texture, and the presence of non-glandular hair and glandular hair and their characteristics; the secretory tissues, sclerenchyma, crystals of calcium oxalate in the mesocarp; the shape of endocarp cells, and the number of mosaic cells and their arrangement for mosaic layers (Fig. 2.29).

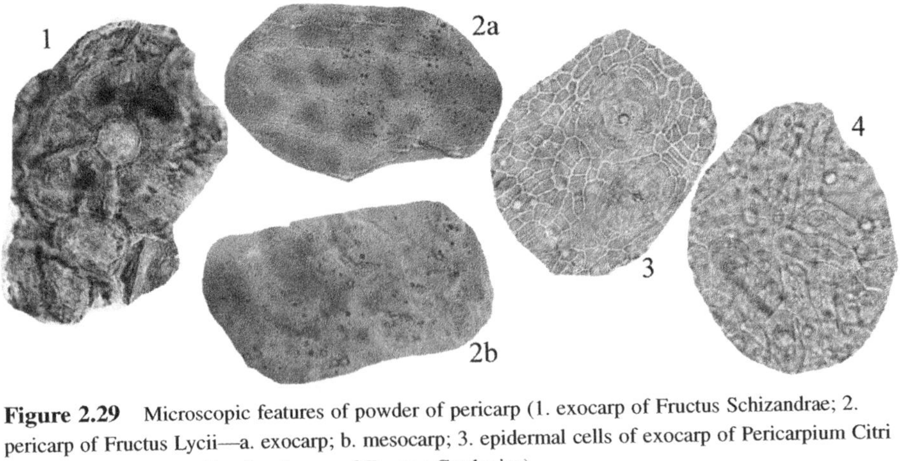

Figure 2.29 Microscopic features of powder of pericarp (1. exocarp of Fructus Schizandrae; 2. pericarp of Fructus Lycii—a. exocarp; b. mesocarp; 3. epidermal cells of exocarp of Pericarpium Citri Reticulatae; 4. stone cells of endocarp of Fructus Gardeniae).

Materials of Seeds

1. *Tissue Characteristics*

Observation of seed herbal materials should first focus on the structure of testa, then prosembryum, endosperm, shapes, and storing substance of cells of seminific leaf as well.

Testa Some are composed of only one layer of cells, while most consist of several kinds of cells. Epidermis and the tissues below are usually observed.

a. *Epidermis of testa:* the shape, size, and arrangement of epidermis cells, the thickening or lignification condition of cell walls, and the presence of containments are all of diagnostic importance. It commonly consists of one layer of parenchyma cells, such as Fructus Bruceae. But some have stomata (e.g., Semen Juglandis); some differentiate to non-glandular hairs (e.g., Semen Pharbitidis) or to pachypleurous non-glandular hairs entirely (e.g., Semen strychni); some display glandular hair of single-cell head or multicells handle (e.g., Semen Impatientis); some are composed of parenchymatous cells and stone cells (e.g., Semen Armeniacae Amarum and Semen Persicae); some consist of stone cells (e.g., Fructus Schisandrea Chinensis, Semen Hyoscyami); some present slime layer formed by mucilage cells (e.g., Semen Sinapis and Semen Plantaginis); some show paliform cells (e.g., seeds in Leguminosae); some contain globular crystals of calcium oxalate in every lumina close to the outside wall of palisade epidermal cells (e.g., Semen Sesami Nigrum).
b. *Tissues below epidermis of testa:* when examining layers of paliform cells (e.g., Semen Pharbitidis and Semen Cuscutae), note layer number, size, thickening condition and presence of "light bands." When observing layers of oil cells (e.g., seeds in Zingiberaceae), note the shape, size, and color of containments. When identifying pigment layer, focus on number of pigments in the layer, and the color of containments. When observing stone cells, fibers, and brace cells, pay careful attention to their distribution, shape, size, and the thickness of walls.

Endosperm Attention should be paid to prosembryum, endosperm, and seminal leaf cells with their shape, wall thickening conditions, and contents such as fatty oil, aleurone grain, or starch grain, for these all have diagnostic significance. For example, endosperm cells of Semen Arecae have a very thick wall with big subround pits; aleurone grains in the endosperm of seeds from Umbelliferae contain tiny clusters of calcium oxalate inside (formed as rosella crystals).

2. *Powder characteristics*

When observing, note the distinct appearances in the surface and lateral view of testa. For example, the testa palisade cell layer is a group of thickened wall cells presenting as tiny polygons in a surface view. In the lateral view, cells are narrow

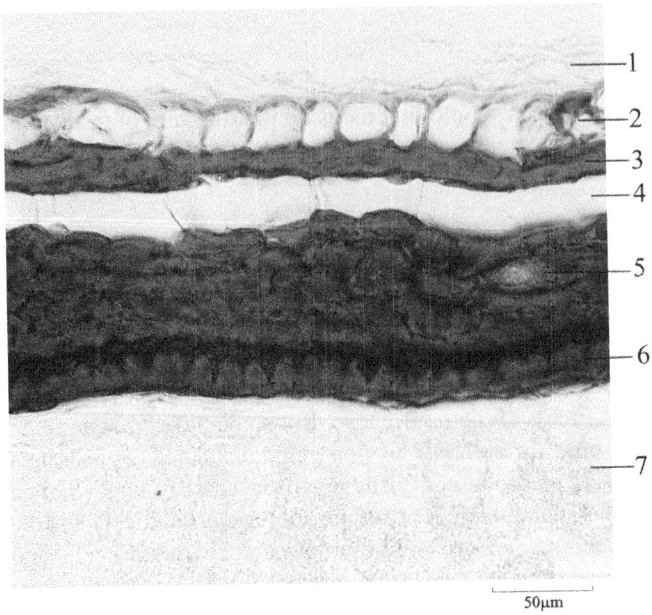

Figure 2.30 Microscopic features of transverse section of Fructus Amomi (seed of *Amomum villosum*) (1. aril; 2. epidermis of testa; 3. hypodermis; 4. oil cell layer; 5. pigment layer; 6. endotesta; 7. prosembryum cells).

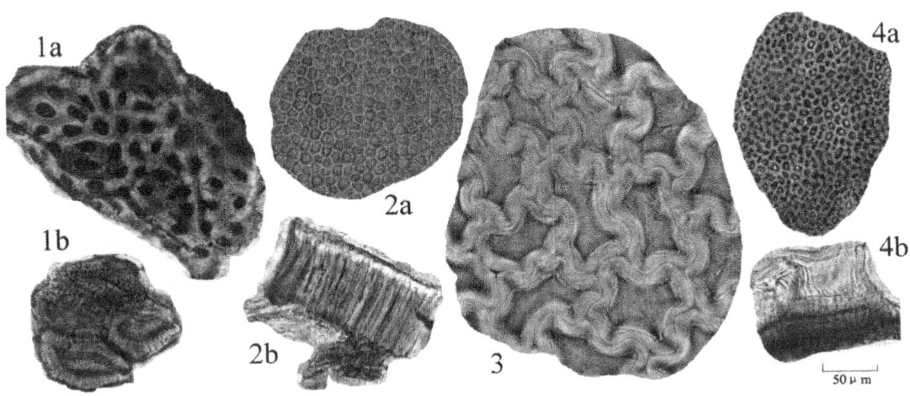

Figure 2.31 Microscopic features of powder of seed (1. stone cells of testa of Fruictus Schizandrae Chinensis—a. outer layer; b. inner layer; 2. grille cells of testa of Semen Zizyphi Spinosae—a. in apical view; b. in lateral view; 3. stone cells of testa of Fruictus Lycii; 4. grille cells of testa of Semen Sinapis—a. in surface view; b. in lateral view).

rectangles, with little thickened walls, showing two light bands, such as Semen Cassiae. Stone cell layers of testa show groups of polygonal cells regularly arranged in a surface view, such as Fructus Schisandrae. Some are irregular in shape, with wavy, anticlinal walls, thickening as crestiform of inner and anticlinal wall in a transverse view, such as Fructus Lycii. Some are sub-round or sub-polygonal, and conchoidal in lateral view, such as Semen Armeniacae Amarum and Semen Persicae. Others such as trichome, secretory tissue, crystals of calcium oxalate, aleurone grains, and starch grains should also be observed.

Example Microscopic features of transverse section of Fructus Amomi (seed of *Amomum villosum* Lour.) (Fig. 2.30); microscopic features of powder of seeds (Fig. 2.31).

Entire Herbs

Entire herbs include all parts of herbaceous plants, thus characteristics for each part described above should be noted when observing entire herbs.

2.2.3 Physical and Chemical Identification

Physical and chemical identifications of herbal materials refer to qualitative and quantitative determination of the active, major, or characteristic components using physical and chemical methods. The physical constants of measurement, chemical reactions, spectroscopy, and chromatography are the four commonly used methods in physical and chemical identification. Detailed information of these methods is available in Chapters 4 and 9. Here is a brief introduction to each method.

The targets of the morphological identification and microscopic identification are the original plants or materials. Microscopic identification is also used for powder of materials, but is not applicable to extracts of the materials. In this case, physical and chemical identification, especially chemical identification, become very powerful.

Physical constants measurements include measuring relative density, optical activity, refractive index, freezing point, and melting point. Different materials show different characteristics that are specific for the identification.

Chemical reactions include simple color reaction and precipitation reaction. By virtue of their diverse structures and functional groups, each type of component will react specifically with certain agents, resulting in the formation of colors or precipitations. For instance, alkaloids produce orange precipitations when reacting with potassium heptaiodobismuthate; anthraquinones show orange, red, and blue in color when reacting with lye. Other chemical reactions include the reaction of flavonoids with hydrochloric acid and magnesium powder, the ferric hydroxamate reaction of coumarin and lactone, Libermann-Burchard reaction of saponins, K-K reaction of cardiotonic glycoside, ferric chloride reaction of phenols, gelatin precipitation reaction of tannin, ninhydrin reaction of amino acids, and hydroxybenzene-sulphuric acid reaction of carbohydrate.

Spectroscopy method is based on the theory that different structures of compounds in herbs absorb light at specific wavelengths. All kinds of spectroscopic methods could be applied to identification. The usual methods are visual ultraviolet

(UV) spectrometry, IR spectrometry, mass spectrometry (MS), nuclear magnetic resonance (NMR), fluorospectrometry, X-ray analysis, and so on. Please refer to Chapters 4 and 9 for more details on these methods.

Chromatographic identification includes thin layer chromatography (TLC), gas chromatography (GC), high-performance liquid chromatography (HPLC), paper chromatography (PC), gel electrophoresis (GE), capillary electrophoresis (CE), and electric chromatography.

For material whose chemical composition has been well known, the identification of a sample can be simply performed by comparing the sample with a standard or a known material. TLC, GC, and HPLC are the three most reliable and commonly used methods for comparison of chemical composition, particularly when the morphological and microscopic identifications of the material are not sure. If the chemical composition of the sample is not clearly known, it usually needs to be extracted first, then isolated by chromatography and identified by spectral characteristics. The principles and detailed introductions on chromatographic methods will be found in Chapters 3, 4, and 9. We will only give two examples of TLC identification in this chapter.

TLC is the most convenient, direct, and economic chromatographic method. Samples could be directly compared and identified by the fluorescence or color reaction of the developed spots. It is often used to confirm morphological and microscopic identifications. In addition, qualitative and quantitative analyses are available through TLC scanning method. This method has been the most frequent and effective way among all physical and chemical identification methods in view of the simple and quick operation.

Example 1 Distinguishing Radix Ginseng from Radix Panacis Quinquefolii by TLC.

As is known, the component ginsenoside R_f is specifically contained in Radix Ginseng, while the Pseudo ginsenoside F_{11} is only in Radix Panacis Quinquefolii. The TLC method can successfully display the difference.

Preparation of test solution: to 1 g of the powder, add 40 mL of chloroform, heat under reflux on a water bath for 1 h, discard the chloroform layer, evaporate the residue to dryness. Moisten the residue with 0.5 mL of water, add 10 mL of *n*-buthanol saturated with water, ultrasonicate for 30 min. To the supernatant liquid add three volumes of ammonia solution, mix well, and allow it to stand. Evaporate the supernatant liquid to dryness; dissolve the residue in 1 mL of methanol as the test solution.

Preparation of reference solution: dissolve ginsenosides R_{b1}, R_{b2}, R_c, R_e, R_d, R_{g1}, R_f, and F_{11} in methanol separately to make a solution with concentration of 2 mg/mL as the reference solution.

Thin layer plate: silica gel G60 prefabricated plate.

Developing agent: bottom layer solution of chloroform-ethyl acetate-methanol-water (15:40:22:10) placed in 10°C.

Coloration and observation: spray 10% solution of sulfuric acid in ethanol on the developed TLC plate, then heat at 105°C for a few minutes. Observe the plate under UV light at 365 nm wavelength. Radix Ginseng and Radix Panacis Quinquefolii show different spots representing different chemical composition.

Example 2 TLC identification of Radix et Rhizoma Rhei.

The plant sources of Radix et Rhizoma Rhei include *Rheum palmatum* L., *R. tanguticum* Maxim. et Balf., and *R. officinale* Baill, according to *Chinese Pharmacopoeia* (2005 edition). The TLC of these three plant extracts all show the same colored spots, demonstrating a basis for the multi-sources of this herbal medicine.

Preparation of test solution: to 0.1 g of the powder, add 20 mL of methanol for 1 h, and filter. Evaporate 5 mL of the filtrate to dryness, then dissolve the residue with 10 mL of water, add 1 mL of hydrochloric acid, heat on a water bath for 30 min, and cool immediately. Extract with ether twice, each of 20 mL. Combine the two extracts of ether, evaporate the solvent, then dissolve the residue with 1 mL of chloroform as the test solution.

Preparation of reference solutions: take 1 g of Radix et Rhizoma Rhei reference material and prepare the solution with the same procedure as above. In addition, dissolve aloe-emodin, rhein, emodin, rheochrysidin, chrysophanol with methanol to make a solution with a concentration of 1 mg/mL as the reference standard solution.

Thin layer plate: self-prepared silica H plate containing sodium carboxymethylcellulose as the coating substance.

Developing agent: upper layer of petroleum ether (30–60°C)-ethyl formate-formic acid (15:5:1) after placed.

Observation: under UV light at 365 nm wavelength, there should be five orange fluorescent spots developed from the test solution that correspond, respectively, in position and color to the spots obtained from the reference solutions. The spots become red under sunlight on exposure to ammonia vapor.

2.2.4 Genetic Identification

Morphological and microscopic identification of herbal medicine are common approaches to determine their validities and qualities by means of their specific characteristics, with the advantages of convenience and less cost. As mentioned before, chemical composition between species within a genus are mostly, but not always, similar. Therefore, these methods are not reliable enough to specifically identify congeneric but multi-source species and some intraspecific variations of geo-authentic medicinal herbs. DNA molecular genetic markers for identification of herbal materials have come out with the rapid development of pharmacognosy and molecular biology technique.

DNA molecular genetic marker of herbal medicine is a method to identify plant origins and determine their scientific species by comparing their genetic diversities. As we know, during reproduction by plants and animals, the parental generation will transmit genetic information, including external appearance, function, and physiological and biochemical characteristics, to the filial generation through the process of cell division. Genetic materials are present in the cell nucleus, in which chromatosomes carry genetic information. The number and morphous of chromatosomes are specific characteristics in animal and plant. Chromosomes consist of DNA, RNA, and protein; DNA is the genetic material for most living beings, except for a few viruses.

DNA is a double helix, long, chain-like molecule including four types of nitrogen bases, that is, guanine, adenine, cytosine, and thymine, in which the specific genetic information of the species are contained. The variations of sequencing of these four nitrogen bases present the different genetic characters between species, also known as genetic diversity. Three kinds of gene maps are present in the DNA genome, including the coding region that is closely related to the survival of species, the coding region that is less related to survival, and the noncoding region. These different regions in DNA genome bore various degrees of selective pressure during the organic evolution process. The coding region received more selection pressure and displayed higher conservation in comparison with the noncoding areas that show more mutations. As a result, different regions in the DNA molecule display various degrees of genetic diversity. Based on this theory, proper DNA molecular genetic markers could be used to identify exact objectives in grouping species, subspecies, populations, and individuals.

There are three main types of DNA molecular marker techniques. The first is based on the techniques of electrophoresis and molecular hybridization, for example, restriction fragment length polymorphism (RFLP). The second is DNA molecular fingerprinting and DNA sequencing technique on grounds of electrophoresis and polymerase chain reaction (PCR), for example, random amplified polymorphic DNA (RAPD) and arbitrarily primed PCR (AP-PCR), simple sequence repeats (SSR) and inter-simple sequence repeats (ISSR). The third is the combination of the above two techniques, for example, amplified fragment length polymorphism (AFLP), RFLP-PCR, RAPD-PCR, and gene chipping.

The DNA molecular genetic marker techniques have shown a unique and extensive prospect in application of herbal medicines. Still in the exploratory level, further standardization and improvement techniques are highly demanded. Some points to note are (1) the validity of objective gene and homology of DNA, (2) the stability of the DNA molecular marker technique, and (3) the cost of molecular biological study.

Nevertheless, the application of molecular biological techniques to pharmaceutical botany research is on a rising stage. Only a small part of genetic techniques as introduced below is employed in the identification of medicinal herbs. It is believed that, with the development of molecular biology, new techniques will be discovered, along with deeper and more comprehensive applications to the study of herbal medicine.[27]

RFLP

During the evolutionary process of organisms, gene mutations and DNA rearrangements occurred for various reasons, consequently causing changes of nucleotide sequence in one or more sites. If the changes, even slight changes, are carried on restriction sites, the length of DNA fragments will be altered upon digestion by restriction enzymes. As a result, the bands of restriction spectrum will produce different patterns; this is called RFLP. Southern hybridization is often used to detect this kind of polymorphism. The procedures usually include transferring DNA from agarose gel to nitrocellulose filter or other filters, hybridizing DNA with labeled DNA probe, and radioautographing on X-ray film in order to reveal DNA polymorphism. Nevertheless, this method is too complex due to the multiple procedures involving Southern blotting, probe labeling, hybridization, and detection. Moreover, this method has to use fresh materials with un-degraded DNA and requires a large amount of DNA sampling since there is no DNA amplification. It is frequently restricted by probe sources. All these factors limit the application of RFLP. Along with the development of PCR, PCR-RFLP has been increasingly applied to authenticity identification of traditional herbal medicines.

RAPD

RAPD is also named AP-PCR. In comparison to normal PCR, RAPD shows its distinctive characteristics in the following aspects.[28]

1. RAPD needs neither designed primers nor the genomic DNA sequence of the objective material. The primer, usually 9–10 bps long, is synthesized randomly or selected arbitrarily.
2. RAPD needs only one primer, which matches a DNA template randomly without specific amplification.
3. RAPD needs low annealing temperatures (usually 36°C), which guarantees the stability of the primer-template pairs and allows for some mis-pairing, so as to enlarge the randomness of pairing in the DNA genome.
4. RAPD is programmed easily. A set of random primers would obtain a great number of DNA markers, which can be analyzed systematically by computer. RAPD has been extensively used in genetic fingerprinting, gene assignment, phyletic evolution, as well as identification of species of animal, microorganism, and plants, including herbal medicine.

AFLP

AFLP is a variation of RAPD, resulting from RFLP and PCR. The AFLP technique was first developed by the Dutch scientist Marc Zabeau and his colleagues in the early 1990s.[29,30] The polymorphism of AFLP lies in the variations of restriction endonucleases' recognition sites, or in the insertion and depletion of DNA sequences in restriction fragments, using essentially the same theory as RFLP. Based on the selective PCR amplification of restriction fragments from a total digest of genomic DNA, the AFLP technique has the advantages of randomness and specificity in combination. It selects different DNA sequences and numbers from the conversion of the primer types or combinations. Various restriction endonucleases are available to achieve the selective specificity by this technique.

However, an AFLP experiment requires superior instruments and expensive reagents. It needs radioactive isotopes for detection, which is harmful for the environment and human beings to some extent.

SSR and ISSR

SSR, also called microsatellites, are stretches of 1–6 nucleotide units repeated in tandem and widely dispersed in eukaryotic genomes, such as (CA)n, (GAG)n, (GACA)n, and so on.The typical microsatellite consists of a 1-, 2-, 3-, or 4-bp units repeated 10–20 times. These monomer units display tandem organization representing hypervariable loci where the variations arise due to the differences in the number of repeating units forming the repeat array. Both sides of the SSR have a conservative segment of DNA sequence, by which oligonucleotide primers of complementary sequence can be designed to amplify the SSR by PCR. Since the SSR polymorphism is made simply based on the differences of the repeat times of sequence, this phenomenon usually shows codominance. Such length-polymorphisms of SSR can be easily detected on high-resolution gels, for example, agarose gel or polyacrylamide GE, by running PCR amplified fragments obtained using a unique pair of primers flanking the repeat.

As a novel marker system, ISSR has recently been developed to access variation in the numerous microsatellite regions dispersed throughout the various genomes. The first study employing ISSR markers was published in 1994.[31,32] The basic principle underlying the ISSR technique is the presence of SSR in genomic DNA. On the basis of wide existence of SSR in plant material, ISSR markers utilize the common SSR in plant genome to design PCR primers without cloning and sequencing in advance.

DNA Sequencing Identification

DNA sequencing identification of herbal medicine is based on the development of a gene-sequence database of authentic species, their counterfeits, and their confused species or variations by DNA sequencing. The identification of herbal medicine can be achieved by comparison of the gene sequences of test samples with those in the database. This method can provide accurate and reliable results because of its good reproducibility. The sequences mentioned in this method are usually the versatile marker sequences. To identify herbal medicine by using DNA sequencing, the appropriate target genes are generally chosen to design universal primers by its conservative sequence and then to amplify the region between the conservative sequences recognized by the primers. Furthermore, the DNA of organism groups of different classification levels can be amplified without previously knowing the sequence of target genes, which make the herbal identification by DNA sequencing possible.

However, it is troublesome in real sample analysis by comparison of complete sequences. For this reason, a more convenient identification method by PCR-RFLP and site-specific diagnostic PCR has been developed on the grounds of DNA sequencing.[33]

DNA Chip Technology

DNA chip, or DNA microarray, on which interactions are based on Watson-Crick base pairing, is a large-scale gene analysis technology that allows thousands of gene probes to be integrated on a solid surface known as an array. After hybridizing with the genes needed to be detected, thousands of genes can be identified in parallel simultaneously.

DNA chip technology provides the ability to detect and analyze large-scale nucleotide sequences with a small volume sample quickly and cost-effectively. The advantage of high throughput of simultaneous bio-information collection makes it superior to other present analytical techniques. The prerequisite of herbal identification is to acquire the specific gene sequence of different medicine samples, i.e., genotyping. After finding out the specific gene or DNA sequence of certain samples, researchers can make the DNA chip by immobilizing these specific sequences as probes on a glass slide. Once a sample has specific gene fragments complementary to the probes, the DNA chip can recognize it immediately. Therefore, DNA chip is applicable to identification of many real samples so long as there are enough specific gene sequences from different samples immobilized on a single chip.[34]

Application

With these great technological advances in the life sciences, molecular biology and gene engineering is ever advancing. It has become practical to classify and identify herbal materials based on their genetic diversity on the molecular DNA level. The genetic identification technology has been applied to the evaluation of the authenticity of herbal materials in the following aspects:

1. *Identification of raw materials in closely related species*

The DNA molecular genetic marker technique has been extensively employed in the research of raw materials in closely related species. For example, the analysis of gene sequences in the internally transcribed spacer (ITS) region and 5.8S-rRNA of twelve species in the genus of *Panax* indicated that *Panax quinquefolius* Linn. growing in the northeast of America showed a closer relationship with *Panax gensing* C.A. Mey., *P. japonicus* C. A. Mey. and *P. notoginseng* (Burk.F. H. Chen) from East Asia.[35] However, the ITS sequence indicated that *Panax gensing* C.A.Mey., *P. quinquefolius* Linn, and *P. notogenseng* (Burk.) F. H. Chen were not in a monophyly. Gene identification of three species in Panax (including *P. ginseng, P. quinquefolius*, and *P. notogengeng*) used for herbal medicines and the four fake species (including *Platycodon grandiflorum* A. DC., *Mirabilis jalapa* L., *Talinum paniculatum* (Jacq.) Garetn., and *Phytolacca acinosa* Roxb) was also successfully performed by means of RAPD marker method.[36] Another reported successful identification was carried out between six *Panax species,* i.e., *P. ginseng* C.A.Mey, *P. quinquefolius*

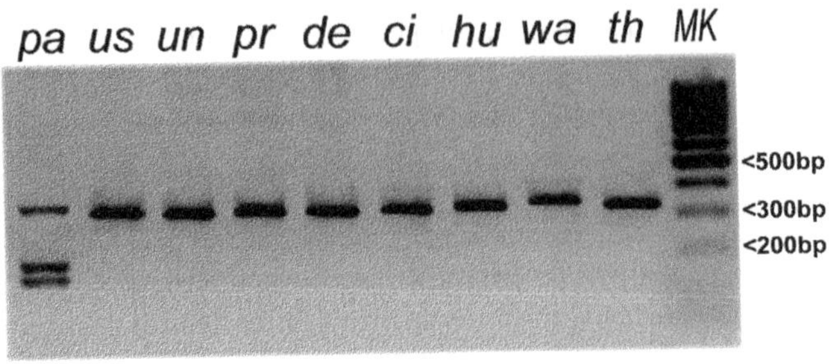

Figure 2.32 Enzymolysis electropherogram of amplified PCR products in rDNA1 ITS regions; PCR-RFLP assay for Eco81I digestion of the PCR products was amplified with primers ITS-P1, ITS-P3 (pa—*F. pallidiflora*; us—*F. ussuriensis*; un—*F. unibracteata*; pr—*F. prezawalskii*; de—*F. delavayi*; ci—*F. cirrhosa*; hu—*F. hupehensis*; wa—*F. walujewii*; th—*F. thunbergii*; MK—DNA markers).

Linn., *P. notoginseng* (Burk.F. H. Chen), *P. japonicus* C. A. Mey, *P. trifolius* L, *and P. major* (Burkill) K. C. Ting ex C., and two common fake species, i.e., *Mirabilis jalapa* L. and *Phytolacca acinosa* Roxb, using amplification of DNA sequence in ITS1-5,8S-ITS2 region.[37] The RFLP marker technique is also available to distinguish or identify those species mentioned above.

2. *Identification of valuable raw materials in easily confused species*

As is known, valuable and rare raw herbal materials are extremely precious for a nation, but they are usually accompanied with many counterfeits. For some precious Chinese medicines such as Cordyceps, Bulbus Fritillariae Cirrhosae, Stigma Croci, and Caulis Dendrobii which are limited in source, the use of the DNA molecular genetic marker technique only needs small a amount of sample and could provide valid identification without the high expense of the precious specimen.

PCR-RFLP analysis of the nuclear ribosomal ITS1 region was applied to differentiate four species of Bulbus Fritillariae Cirrhosae from the other five species of Bulbus Fritillariae through sequencing ITS1 regions of the nuclear ribosomal DNA (rDNA) of these nine species documented in *Chinese Pharmacopoeia*. After comparing two molecular authentication methods based on the sequences of rDNA ITS regions, a pair of species-specific framers to authenticate *Fritillaria pallidiflora* Schrenk was designed. A distinctive site was recognized by the restriction endonuclease Eco81I and consequently used to differentiate the *F. pallidiflora* Schrenk from the other species of Fritillaria. As displayed electropherogram (Fig. 2.32), the differences were quite obvious and the method was well acceptable.[38,39]

Caulis Dendrobii is another example. Among various species, *Dendrobium candidum* Wall.ex Lindl. was proven to show immunologic enhancement function. To achieve a species-specific authentication, eleven specific sites in the rDNA ITS region of five Dendrobium species were determined, seven at ITS1, one at 5.8S, and three at ITS2. The *Dendrobium officinale* was successfully distinguished from the other four common fake herbal materials. In addition, based on rDNA ITS sequences of *D. officinale* and the other thirty-seven species of Dendrobium, a pair of allele-specific diagnostic primers, TP-JB01S (5′-GTGCTGCTGAGATAAAATCCACTG-3′) and TP-JB01X (5′-CCACCATGCACATCCGAGCCTTAGT-3′), were designed to authenticate *D. officinale* from the other species. The results showed that when the annealing temperature was set at 66°C, only the template DNA of *D. officinale* could be amplified, whereas the diagnostic PCRs of all the other Dendrobium species were negative.[40,41] Moreover, a pair of allele-specific diagnostic primers for authenticating *D. devonianum* Paxt. has also been designed as JB-Chiban-01S (5′-CGCCAA GTGAATAACGAAACAATAT-3′) and JB-Chiban-01X(5′-TGCGTTCAAAGACT CGATGGTTCA-3′), based on rDNA ITS sequences database of the Dendrobium species of "Feng dous" group (shaped like a dipper) and "Huang caos" group (with relatively soft and golden caulis).[42]

3. *Identification of geo-authenticated herbal medicines*

Geo-authenticity of herbal materials has always been a distinct and comprehensive criterion to evaluate plant material quality during the long practice of traditional Chinese medicine. Because of the similarities in morphouses, descriptions, and chemicals, it is difficult to distinguish the geo-authenticated herbal materials from the non-geo-authenticated ones. The modern genetic marker technique provides tools that help to achieve the identification of geo-authenticity on the level of geo-groups and molecules and make it possible to clarify the biological characteristics of herbal materials.

For instance, different geo-groups of Cordyceps were investigated by the RAPD technique. The results demonstrated that specimens from identical locations in the same region showed little genetic variations, while ones from the different locations in the same region varied in some extent; the greatest differences were exhibited by the ones collected from different regions. This conclusion supported the geo-authenticity research of Cordyceps on the molecular level.[43]

Flos Lonicerae Japonicae, a commonly used anti-inflammatory herbal medicine, is sourced from *Lonicera japonica* Thunb. as documented in *Chinese Pharmacopoeia*. It was found that a mutation site in the ITS region from geo-authentic *L. japonica* can be recognized by the restriction endonuclease EcoN I after sequencing. The cleavage rate of PCR products by EcoN I was determined to be more than 70% in all geo-authentic *L. japonica* and less than 20% in non-geo-authentic *L. japonica* and other species from the genus Lonicera. The rate correlated remarkably with the geographical origin of *L. japonica*.[44]

REFERENCES

1. Jia, M.R. (2007) *GAP Researches on Radix Chuanxiong and Radix Angelicae Dahuricae*. Chengdu, Sichuan Publishing House of Science and Technology.
2. Li, P. (2005) *Pharmacognosy*. Beijing, People's Medical Publishing House.
3. Li, P. (2006) *Modern Pharmacognosy*. Beijing, Science Press.
4. The United States Pharmacopeial Convention (2006–2007)*The United States Pharmacopeia* (30th ed.), Rockville, MD, The United States Pharmacopeial Convention Publisher.

5. The Japanese Pharmacopeia Committee (2001) *The Japanese Pharmacopeia* (14th English version), Tokyo, Society of Japanese Pharmacy.
6. Zhang, H.Y. and Yuan, C.Q. (1995) *Resources of Chinese Materia Medica*. Beijing, Science Press.
7. China Pharmaceutical Committee (1993) *Encyclopedia of Traditional Chinese Medicines*. Beijing, Chinese Medical and Technological Press.
8. Chen, R.L. (1985) *Pharmacognosy*. Taibei, Ho-Chi Book Publishing Corporation.
9. Lou, C.Z. and Li, S.H. (1997) *Morphological and Microscopic Identification of Medicinal Herbs*. Beijing, Beijing Medical University and Xiehe Medical University.
10. Wang, G.Y. and Tian, J.W. (1996) Collection and processing methods of Flos Lonicerae Japonica and comparisons of their materia medica products. *Chinese Herbal Medicine* 27(4):233–235.
11. Li, J.H., et al. (2001) Continuous exploitation and sustainable development of raw materials resources. *Journal of Shanghai Traditional Chinese Medicine* 35(7):44–46.
12. Li, P. (2005) *Pharmacognosy*. Beijing, People's Medical Publishing House.
13. Hong Kong University of Science and Technology (2000) *Research and Development Reviews of Traditional Chinese Medicines*. Beijing, Science Press.
14. Xie, Z.W. (1990) *Discussion of Species of Chinese Materia Medica*. Shanghai, Shanghai Scientific and Technological Publishers.
15. China Pharmaceutical University (1993) *Encyclopedia of Traditional Chinese Medicines*. Beijing, Chinese Medical and Technological Press.
16. Dong, S.L. (1994) *Plants Resources*. Harbin, Publishing House of Northeast Forestry University.
17. Xu, L.S., et al. (1996) *Microscipic Identification of Traditional Chinese Medicines*. Nanjing, China Pharmaceutical University Publisher.
18. Li, J.L., et al. (1997) Opinions on microscopic identification of Chinese patent medicines. *Journal of Chinese Medicinal Materials* 20(1):46–48.
19. Department of Health of the Government of the Hong Kong Special Administrative Region of the People's Republic of China (2005) *Standards of Chinese Materia Medica in Hong Kong*. Hong Kong, Department of Health of the Government of the Hong Kong Special Administrative Region of the People's Republic of China.
20. Liu, G.Z. (2000) Research of microscopic quantitation methods on traditional Chinese medicines. *Heilongjiang Journal of Traditional Chinese Medicine* 4:58, 60.
21. Gu, J. and Liu, P. (2002) Progresses of microscopic identification techniques of traditional Chinese medicines. *Pharmaceutical Journal of Chinese People's Liberation Army* 18(4):223–225.
22. Xiao, X.H., et al. (1997) A model atlas of the crude drug Radix Ophiopogonis in dimensional computer reconstruction from their serial transections. *Acta Pharmaceutica Sinica* 32(6):461–466.
23. Pei, Y.Y. (2003) Histochemical method and its applications to pharmacognosy. *Jiangsu Pharmaceutical and Clinical Research* 11(5):41–43.
24. Techen, N., et al. (2004) Authentication of medicinal plants using molecular biology techniques to compliment conventional methods. *Biomedical Chromatography* 11(11):133–140.
25. Baranska, M. (2004) Identification of secondary metabolites in medicinal and spice plants by NIR-FT-raman microspectroscopic mapping. *Analyst* 129(10):108–115.
26. Liang, Z.T., et al. (2006) Distinguishing the medicinal herb Oldenlandia diffusa from similar species of the same genus using fluorescence microscopy. *Journal of Chromatographic Science* 39(2):277–282.
27. Shaw, P.C., et al. (2002) *Authentification of Chinese Medicinal Materials by DNA Technology*. Singapore, World Scientific Publishing Co.
28. Shinde, V.M., et al. (2007) RAPD analysis for determination of components in herbal medicine. *Evidence-based Complementary and Alternative Medicine* 4(Suppl. 1):21–23.
29. Zabeau, M. and Vos, P. (1993) *Selective Restriction Fragment Amplification: A General Method for DNA Fingerprinting*. European Patent Application 94202629.7 (Publication No.0534858A1). Paris, European Patent Office.
30. Vos, P., et al. (1995) AFLP: a new technique for DNA fingerprinting. *Nucleic Acids Research* 23(21):4407–4414.
31. Zietkiewicz, E., et al. (1994) Genome fingerprinting by simple sequence repeat (SSR)-anchored polymerase chain reaction amplification. *Genomics* 20:176–183.

32. Gupta, M., et al. (1994) Amplification of DNA markers from evolutionarily diverse genomes using single primers of simple-sequence repeats. *Theoretical and Applied Genetics* 89(7–8):998–1006.
33. Pui, Y.Y., et al. (2007) DNA methods for identification of Chinese medicinal materials. *Chinese Medicine* 2:9 online. doi: 0.1186/1749-8546-2-9.
34. Preeti, C., et al. (2006) DNA microarrays in herbal drug research Evid based complement. *Alternative Medicine* 3(4):447–457.
35. Wan, J. and Zimmer, E.A. (1996) Phylogeny and biogeography of Panax L. (the Ginseng genus, Araliaceae): inferences from ITS sequences of nuclear ribosomal DNA. *Molecular Phylogenetics and Evolution* 6(2):167–177.
36. Shaw, P.C. and But, P.P. (1995) Anthentication of Panax species and their adulterants by random-primed polymerase chain reaction. *Planta Medica* 61(5):466–469.
37. Ngan, F., et al. (1999) Molecular authentication of Panax species. *Phytochemistry* 50(5):787–791.
38. Wang, C.Z., et al. (2007) Simultaneous identification of Bulbus Fritillariae cirrhosae using PCR-RFLP analysis. *Phytomedicine* 14(9):628–632.
39. Wang, C.Z., et al. (2005) Identification of Fritillaria pallidiflora using diagnostic PCR and PCR-RFLP based on nuclear ribosomal DNA internal transcribed spacer sequences. *Planta Medica* 71(4):384–386.
40. Ding, X.Y., et al. (2002) Authentication of stems of Dendrobium officinale by rDNA ITS region sequences. *Planta Medica* 68(2):191–192.
41. Ding, X.Y. (2003) Allele-specific primers for diagnostic PCR authentication of Dendrobium officinale. *Planta Medica* 69(6):587–588.
42. Ding, X.Y., et al. (2002) Specific PCR identification of Dendrobium devonianum. *Acta Pharmaceutica Sinica* 37(11):897–901.
43. Cheng, K.T., et al. (1998) Differentiation of genuines and counterfeits of Cordyceps species using random amplified polymorphic DNA. *Planta Medica* 64(5):451–453.
44. Wang, C.Z., et al. (2007) Discrimination of Lonicera japonica Thunb. from different geographical origins using restriction fragment length polymorphism analysis. *Biological and Pharmeutical Bulletin* 30(4):779–782.

Chapter 3

Extraction and Isolation of Compounds from Herbal Medicines

Hong-Wei Liu

Plants, particularly medicinal herbs, constitute the foundation of traditional pharmacopeias, and have produced many currently important pharmaceutical drugs, for example, taxol from *Taxus brevifolia*, vinblastin and vincristine from *Catharanthus roseus*, and huperzine A from *Huperzia serrata*. There are at least 250,000 species of higher plants on earth, but only about 5–10% of them have been investigated so far. With the development of modern molecular biotechnology, there is an increasing demand for extraction and isolation of compounds from herbs for the purpose of screening bioactive chemical molecules for new drug development, exploring therapeutic and preventative mechanism of herbs, as well as establishing quality control and standardization of herbs and herbal products.

The chemicals in plants can be divided into primary metabolites and secondary metabolites based on the range of molecular weight, distributions in species, and biological roles to plants. The ubiquitous macromolecules (mol wt > 2000 amu) of primary metabolism in plants, for example, polysaccharides, proteins, lipids, and nucleic acids, are called primary metabolites. They provide nutrients, and thus are essential for growth and survival. In contrast to primary metabolites, small organic compounds (mol wt < 2000 amu) are produced often by a particular species, usually possessing important biological activity. They are not necessary for growth and survival and are called secondary metabolites.

A number of secondary metabolites in plants serve as chemical messengers and defensive chemicals, and play significant biological and ecological roles. There is growing interest in the study of plant secondary metabolites as they represent a tremendous library of potentially useful leading compounds for new drug development. The secondary metabolites comprise of a range of chemically diverse com-

Traditional Herbal Medicine Research Methods, Edited by Willow J.H. Liu

pounds. Based on the chemical structure features, they are classified as alkaloids, flavonoids, coumarins, lignans, quinones, terpenoids, and so on.[1–5]

Isolation of chemical compounds from herbs is an important step for a systematic study of herbal medicine. It provides compounds not only for structural identification or elucidation and standards for quantitative and qualitative analysis for quality control of herbal extracts or products, but also for *in vitro* bioassay screen and *in vivo* pharmacological and toxicological study and clinical trials. Before isolation of an herbal medicine, the material should first be identified using methods introduced in Chapter 2, to ensure it is the right species and was collected and stored properly. Extraction and isolation should be carried out with a well-planned scheme based on knowledge of the sample. Different extraction and isolation methods are suitable for different types of compounds. Thus, it is important to do a complete literature search and know the characteristics of compounds in the sample before making the plan. This chapter will mainly introduce the characteristics of major types of secondary metabolites in plants and commonly used methods for extraction and isolation.

3.1 COMPOUNDS IN PLANTS AND THEIR STRUCTURES AND PROPERTIES

3.1.1 Alkaloids

Definition and Distribution in Plants

Alkaloids are a large class of nitrogen-containing secondary metabolites of plants, microbes, or mammals. Alkaloids are famous for their manifold pharmacological activities. Since the discovery of morphine from the opium poppy, *Papaver somniferum*, in 1806, more than 10,000 alkaloids have been purified and identified from natural resources. Many modern drugs are produced from naturally occurring alkaloids or their synthetic analogs.

Alkaloids are mainly distributed in higher plants such as the *Taxus* genus, the *Pinus* genus, the *Picea* genus, the *Ketelearia* genus, the *Ephedra* genus, the *Cephalotaxus* genus, the Ranunculaceae family, the Berberidaceae family, the Menispermaceae family, the Magnoliaceae family, and the Liliaceae family, but rarely found in lower plants.

Biological Activities of Alkaloids

Alkaloids are a group of substances possessing various remarkable biological activities, such as antibacterial (e.g., berberine), anti-malaria (e.g., quinine), analgesia (e.g., morphine), anesthesia (e.g., cocaine), anticancer (e.g., vincrinstine), cardiant (e.g., dl-demethylcoclaurine), antihypertention (e.g., resepine), cholinomimeric action (e.g., galatamine), relieving cough (e.g., codeine), spasmolysis (e.g., atropine), vasodilatation (e.g., vincamine), anti-arhythmia (e.g., quinidine), and anti-asthma (e.g., ephedrine). For example, the morphine alkaloids are powerful pain relievers and narcotics, and vincristine, isolated from *Vinca rosea* (now *Catharanthus roseus*), is one of the most potent anti-leukemic drugs in use today.

Properties of Alkaloids

Alkaloids in plants exist as free states, as salts, or as N-oxides. The nitrogen compounds may exist as a primary amine, as a secondary amine, as a tertiary amine, or as a quaternary amine. They are generally white or yellowish solids with a few exceptions (i.e., nicotine is a brown liquid) and have bitter taste. They can produce a precipitate when reacted with heavy metal iodides. Most alkaloids are able to form cream-colored precipitate with Mayer's reagent (potassiomercuric iodide solution). The reaction of Dragendorff's reagent (solution of potassium bismuth iodide) with alkaloids gives an orange-colored precipitate, which is sensitive and can be used for the detection of alkaloids on thin layer chromatography (TLC).

Structures and Classification

Alkaloids are commonly subclassified according to their biosynthetic origin. Most alkaloids can be covered in the following subgroups.

1. Alkaloids derived from the nonprotein amino acid L-ornithine including pyrrolidine/tropane and pyrrolizidine alkaloids (see Fig. 3.1A).
2. Alkaloids derived from the amino acid L-lysine including piperidine, quinolizidine, and indolizidine alkaloids (see Fig. 3.1B).

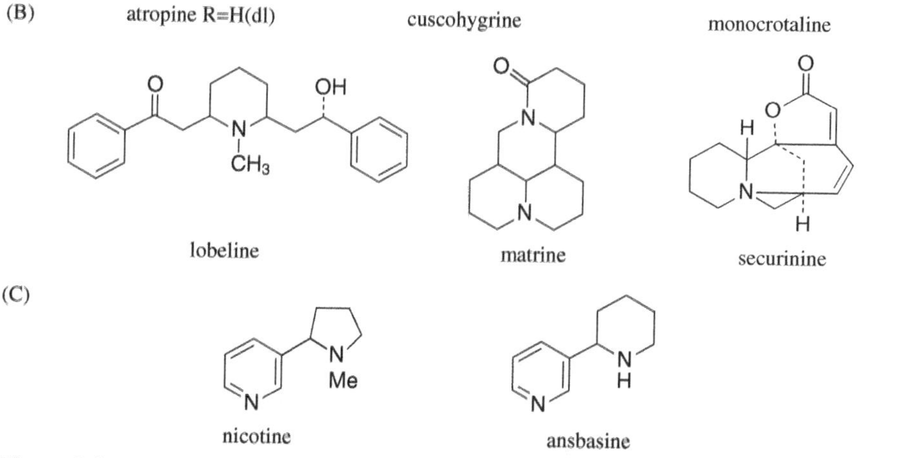

Figure 3.1 Alkaloids derived from L-ornithine (A), L-lysine (B), and nicotinic acid (C).

3. Alkaloids derived from nicotinic acid including pyridine alkaloids (see Fig. 3.1C).
4. Alkaloids derived from the amino acid L-phenylalanine and L-tyrosine including phenyethylamines, tetrahydroisoquinoline, benzyltetrahydroisoquinoline, phenethylisoquinoline, and terpenoid tetrahydroisoquinoline alkaloids (see Fig. 3.2A).

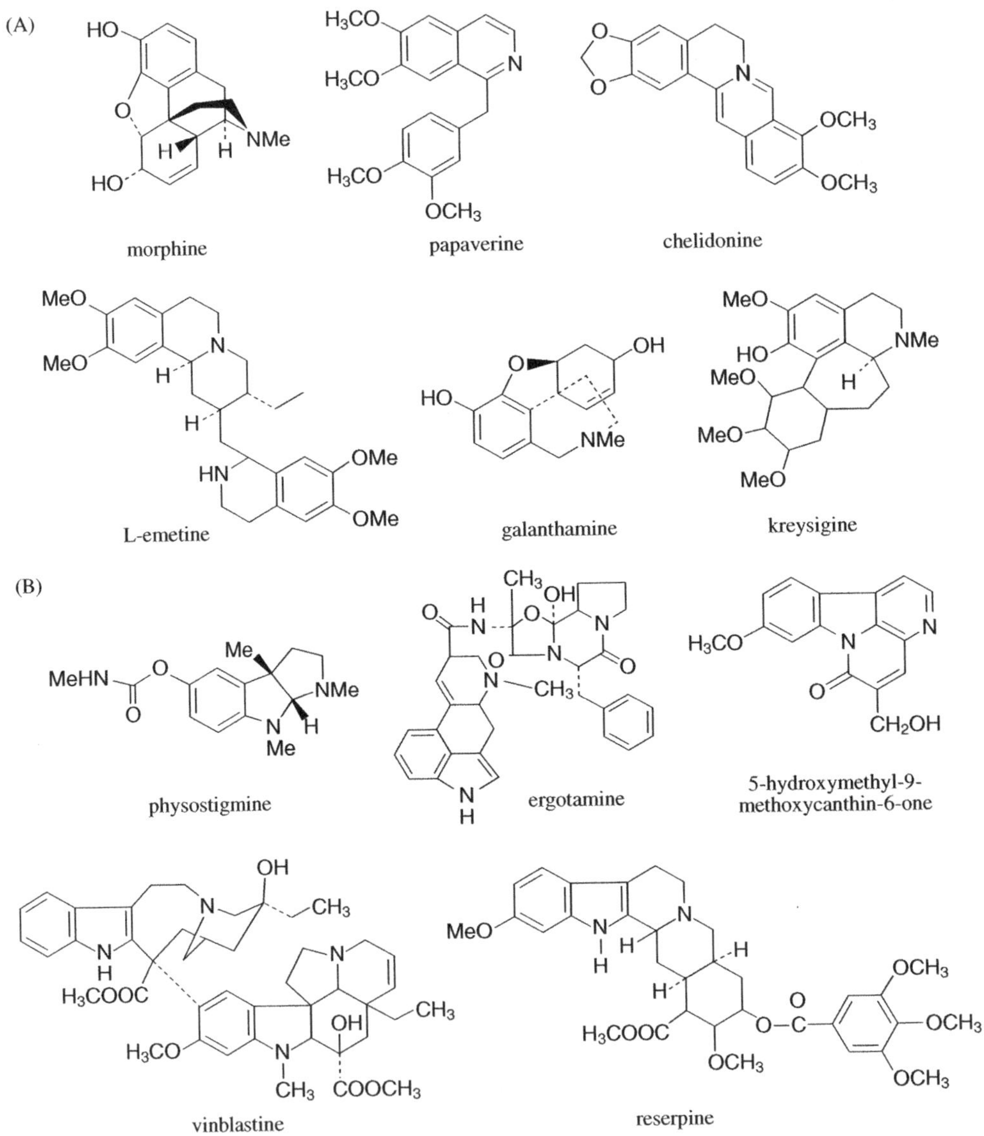

Figure 3.2 Alkaloids derived from L-phenylalanine, L-tyrosine (A), and L-tryptophan (B).

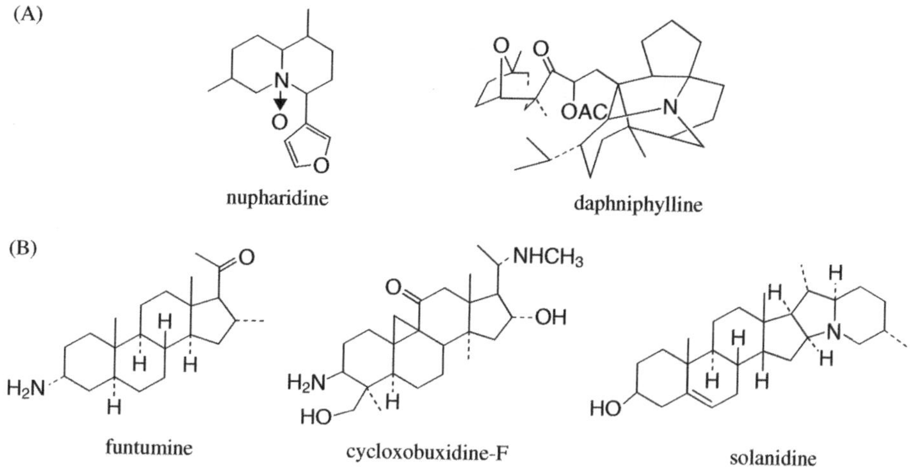

Figure 3.3 Alkaloids derived from terpenoids (A) and steroids (B).

5. Alkaloids derived from the amino acid L-tryptophan including indole, terpenoid indole, quinoline, pyrroloindole, and ergot alkaloids (see Fig. 3.2B).
6. Alkaloids derived from the origin of terpenoids (see Fig. 3.3A).
7. Alkaloids derived from the origin of steroids (see Fig. 3.3B).
8. Alkaloids derived from anthranilic acid (see Fig. 3.4A).
9. Alkaloids derived from histidine (see Fig. 3.4B).
10. Alkaloids derived from purine derivatives (see Fig. 3.4C).

3.1.2 Flavonoids

Definition and Distribution in Plants

Flavonoids are polyphenolic compounds that are ubiquitous in nature. It is one of the most important groups of secondary plant metabolites. The chemical structure of flavonoids is based on a C_{15} skeleton with a chromane ring bearing a second aromatic ring B in position 2, 3, or 4 (see Fig. 3.5).

More than 6000 flavonoid compounds have been purified and identified, many of which occur in fruits, vegetables, and beverages. Flavonoids constitute one of the most characteristic classes of compounds in higher plants, and are responsible for many of the plant colors that dazzle us with their brilliant shades of yellow, orange, or red.

Biological Activities of Flavonoids

Flavonoids have attracted considerable interest recently because of their potential beneficial effects on human health. Flavonoids are most commonly known for their

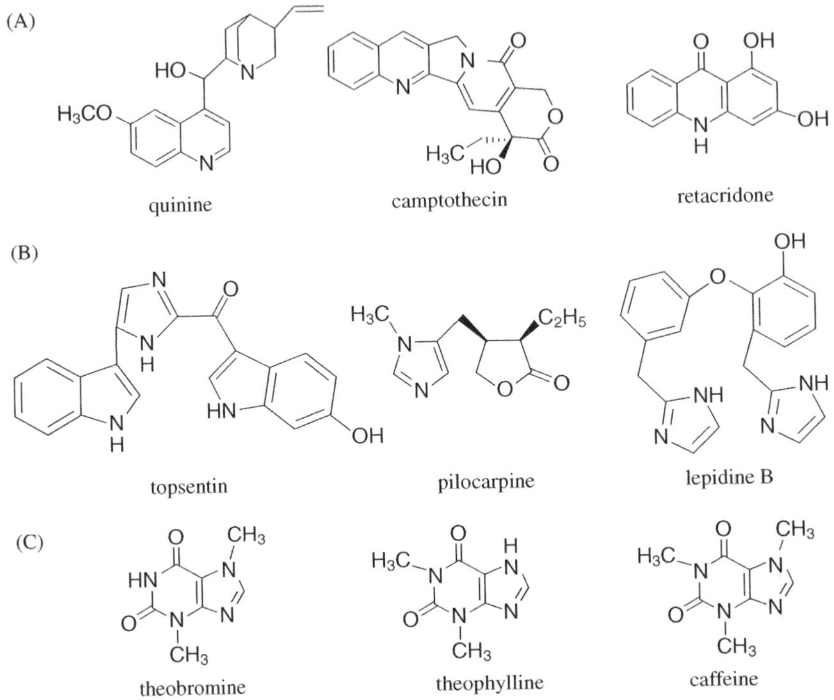

Figure 3.4 Alkaloids derived from anthranilic acid (A), histidine (B), and purine (C).

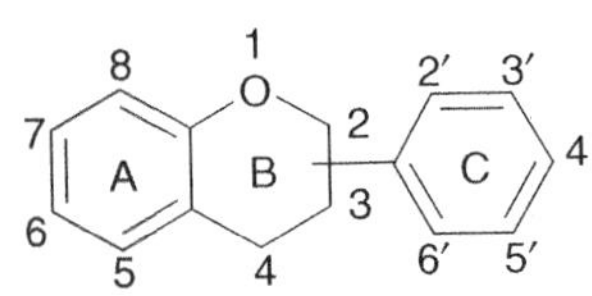

Figure 3.5 Structural skeleton of flavonoids.

antioxidant activity. The capacity of flavonoids acting as antioxidants depends upon their molecular structures. The positions of hydroxyl groups and other features in the chemical structure of flavonoids are important for their antioxidant and free radical scavenging activities. Quercetin, the most abundant dietary flavonol, is a potent antioxidant because of its proper structural features for free radical scavenging activity. It has been discovered that flavonoids also provide other important biological activities such as antibacterial, antiviral, antiallergic, antiplatelet, anti-inflammatory, and antitumor activities. The intake of certain subclasses of flavonoids is demonstrated to be associated with lower occurrence of coronary heart disease. The antiviral function of flavonoids has been demonstrated with the HIV virus, and also with HSV-1, a herpes simplex virus.

Properties of Flavonoids

Most flavonoids are crystalline solids, and only a few are amorphous powders. The colors of flavonoids are dependent on the conjugated system and the number and position of substituents of auxochromes. For example, the hydroxyl and methoxy at 7- or 4′- position will deepen the color of compounds due to their acceleration to the electron rearrangement. In general, flavone, flavanol, and their glycosides are grayish yellow or yellow solids, while flavanone and flavanonol have no color due to lack of conjugated system. But with the treatment of aqueous ammonia, they feature characteristic color or fluorescence and change color. The color of isoflavone is pale yellow due to short conjugated structure. The color of chalcone ranges from yellow to orange. There is a strong relationship between the color of anthocyanins and pH value. The color is red, purple, and blue at $pH < 7$, $pH=8.5$, and $pH > 8.5$, respectively.

The solubilities of flavonoids in solvents depend on their existing forms. Aglycones of flavonoids are less soluble in water, but easily soluble in methanol, ethanol, trichloromethane, and other organic solvents. Flavonoid glycosides are easily dissolved in hot water, methanol, and other polar solvents, but they are solid in benzol, trichloromethane, and other organic solvents. The more sugars connected to the aglycone, the more soluble the glycoside is in the water. Most flavonoids containing phenolic hydroxyl groups are soluble in alkaline aqueous solution (such as sodium carbonate solution) and alkaline organic solvents (such as picolinamide and dimethylformamide).

Flavonoids are usually weakly acidic due to the presence of phenolic hydroxyl groups, and can dissolve in basic solutions. Because of the presence of phenolic hydroxyl groups and γ- pyrone, flavonoid compounds have the capacity to produce various colors when reacting with some reagents, as listed in Table 3.1.

Structures and Classification

Based on chemical structures, flavonoids are categorized into flavones, flavonols, flavanones, isoflavones, chalcones, catechins, anthocyanidins, xanthones, and aurones.

Table 3.1 Color Reaction of Flavonoids

Regents	Final color				
	Flavone	Flanonol	Flavanone	Isoflavone	Chalcone
HCl-Mg	Yellow red	Red mauve	Red, blue	—	—
HCl-Zn	Red	Mauve	Mauve	—	—
$NaBH_4$	—	—	Blue-mauve	—	—
$AlCl_3$	Yellow	Olivine	Blue	Yellow	Yellow
$MgAcO_3$	Yellow	Yellow	Blue	Yellow	Yellow
NaOH aq.	Yellow	Deep yellow	Yellow orange	Yellow	Orange red
H_2SO_4	Yellow orange	Yellow orange	Orange purple	Yellow	Orange

1. Flavones: refer to flavonoids that share structural features, which include the B ring substituted at C_2, a double bond at C_2-C_3, and carbonyl group at C_4 (see Fig. 3.6A).
2. Flavonols: refer to flavonoids that share structural features with flavones including the B ring substituted at C_2, a double bond at C_2-C_3, and a carbonyl group at C_4, but with an addition of a hydroxyl group substituted at C_3 (see Fig. 3.6B).

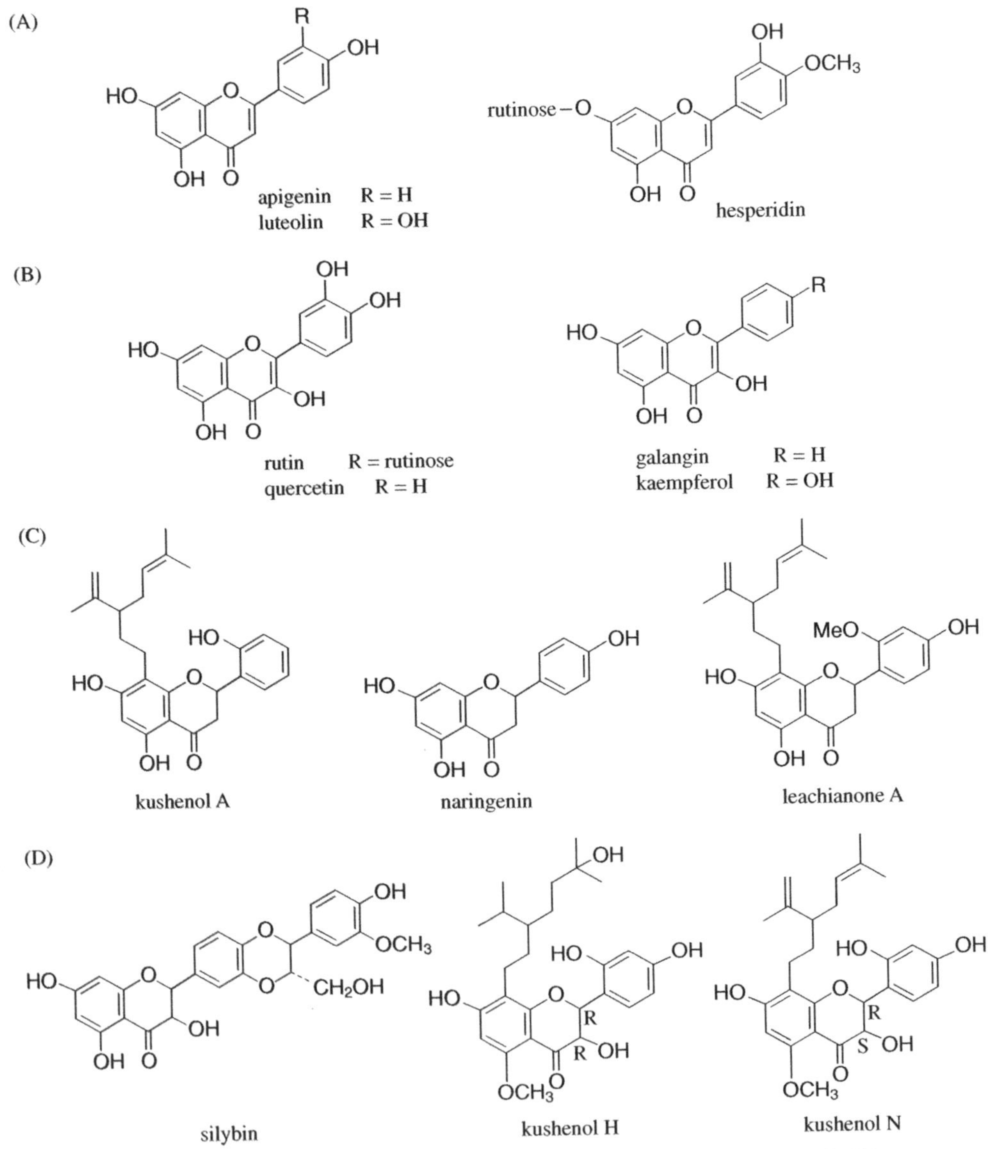

Figure 3.6 Compounds of flavones (A), flavonols (B), flavanones (C), and flavanonols (D).

3. Flavanones: refer to flavonoids that share structural features with flavones including the B ring substituted at C_2, and a carbonyl group at C_4, but without a double bond at C_2-C_3 (see Fig. 3.6C).
4. Flavanonols: refer to flavonoids that share structural features with flavanones including the B ring substituted at C_2 and a carbonyl group at C_4, and without a double bond at C_2-C_3, but with an addition of a hydroxyl group substituted at C_3 (see Fig. 3.6D).
5. Isoflavones: refer to flavonoids that share structural features of flavones including a double bond at C_2-C_3 and a carbonyl group at C_4, but the B ring is attached at C_3 (see Fig. 3.7A).

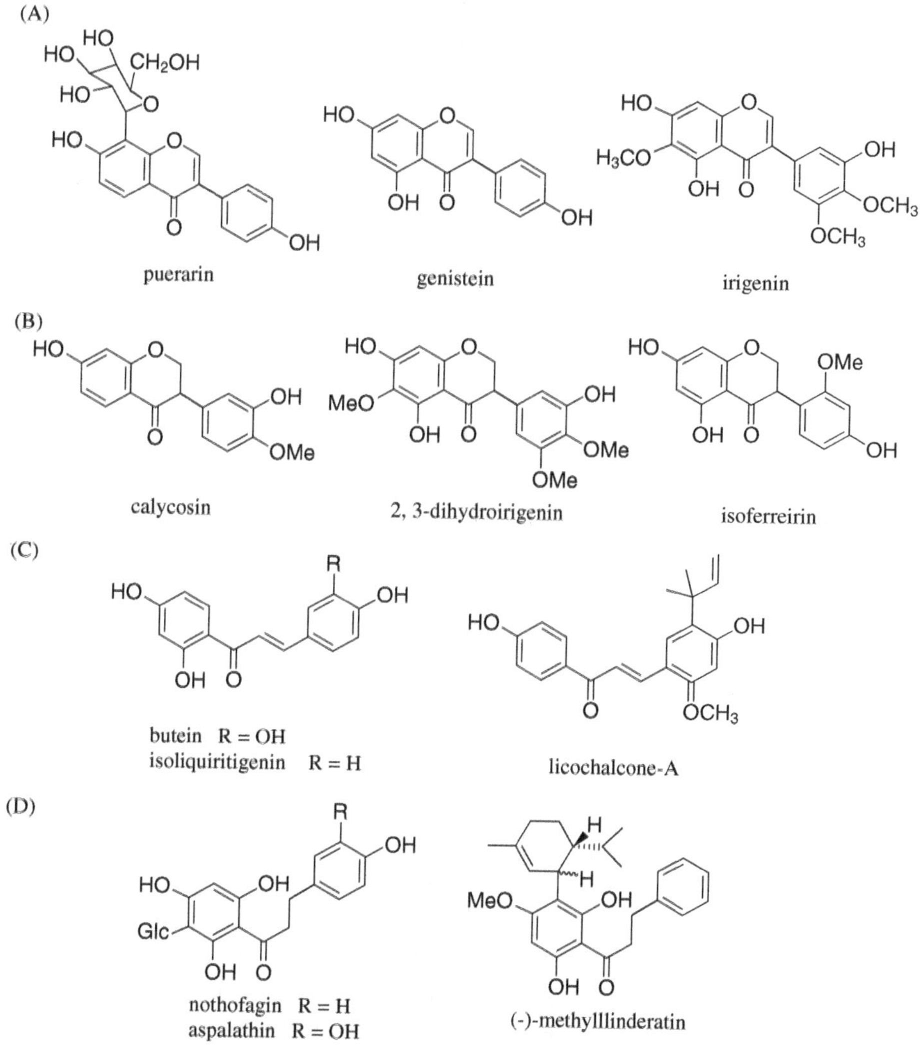

Figure 3.7 Compounds of isoflavones (A), isoflavanones (B), chalcones (C), and dihydrochalcones (D).

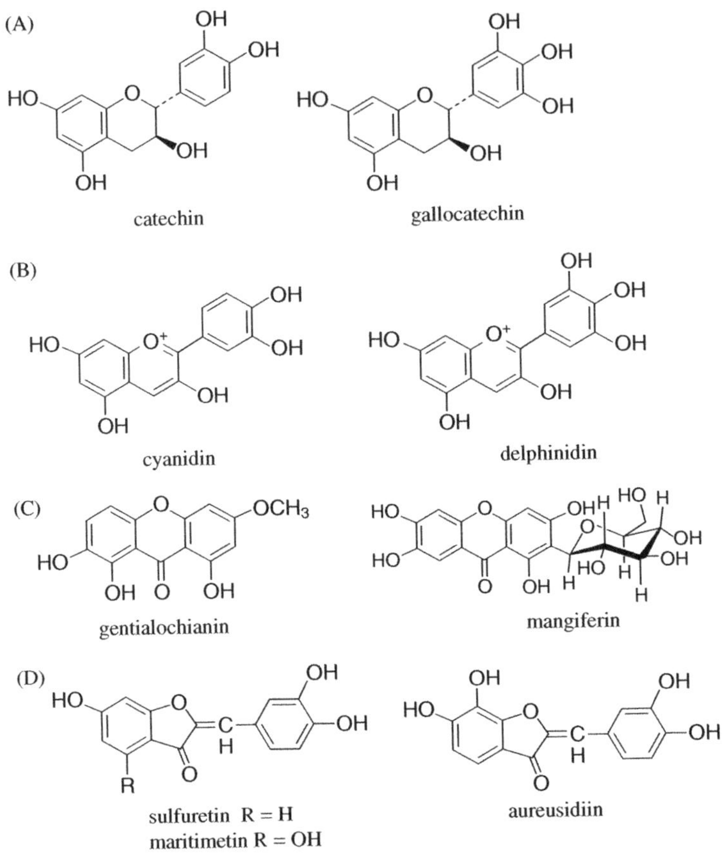

Figure 3.8 Compounds of catechins (A), anthocyanidins (B), xanthones (C), and aurones (D).

6. Isoflavanones: refer to flavonoids that share structural features of isoflavones including the B ring substituted at C_3 and a carbonyl group at C_4, but without a double bond at C_2-C_3 (see Fig. 3.7B).
7. Chalcones: refer to flavonoids that share structural features of flavones including a double bond at C_2-C_3 and a carbonyl group at C_4, but the C ring is opening at position 1 (see Fig. 3.7C).
8. Dihydrochalcones: refer to flavonoids that share structural features of chalcones including the opening of the C ring at position 1 and a carbonyl group at C_4, but without a double bond at C_2-C_3 (see Fig. 3.7D).
9. Catechins: refer to flavonoids that share structural features of flavones including the B ring substituted at C_2, hydroxyl groups substituted at C_3 or/and C_4, but without a carbonyl group at C_4 (see Fig. 3.8A).

10. Anthocyanidins: refer to flavonoids that have 2-phenylbenzopyrylium salts structures (see Fig. 3.8B).
11. Xanthones: refer to flavonoids that have benzochromnone skeleton xanthones (see Fig. 3.8C).
12. Aurones: refer to flavonoids that have the 2-benzylidene coumaranone skeleton; its C ring is a five-member ring (see Fig. 3.8D).

3.1.3 Coumarins

Definition and Distribution in Plants

Coumarins are a group of 1-benzopyran derivatives that mainly exist in higher plants. Coumarins serve as growth inhibitors (anti-auxins) as well as defense compounds in plants and are found in almost every plant family, but are more concentrated in the families of Leguminoseae (bean family), Rutaceae (citrus family), and Umbelliferae (a.k.a. Apiaceae, parsley-fennel family).

Biological Activities of Coumarins

Coumarins have been found to have multi-biological activities such as anti-HIV, antitumor, antihypertension, anti-arrhythmia, anti-inflammation, anti-osteoporosis, pain relief, and prevention of asthma and antisepsis. Coumarin derivatives are used widely as anticoagulants for the treatment of excessive or undesirable blood clotting due to their competitive binding to vitamin K reductase and vitamin K epoxide reductase, which are necessary for blood clotting. 7-hydroxy coumarins are used to absorb ultraviolet (UV) rays in sunscreen cosmetics and for synthesis of anticancer drugs.

Properties of Coumarins

Coumarins are mostly white or faint yellow crystallines with fragrant scent and noticeable melting points. A few have the property of sublimation. Most of them are water-insoluble. However, the 4-hydroxy substitution confers weakly acidic properties to these compounds that make it water-soluble under slightly alkaline conditions. Coumarins readily dissolve in less polar solvents such as petroleum ether, ether, hexane, and chloroform. But coumarin glycosides can dissolve in polar solvents such as methanol, ethanol, and hot water. Most coumarins show strong blue or green fluorescence under UV light. Coumarins undergo base hydrolysis easily due to the existence of lactone group. They can react with hydroxyamine hydrochloride and ferric ion to form red hydroxamates, and are positive to ferric chloride reagent. Some coumarins possessing a reactive hydrogen atom at the para position of phenolic hydroxy can react with Gibbs and Emerson reagents to form color substance.

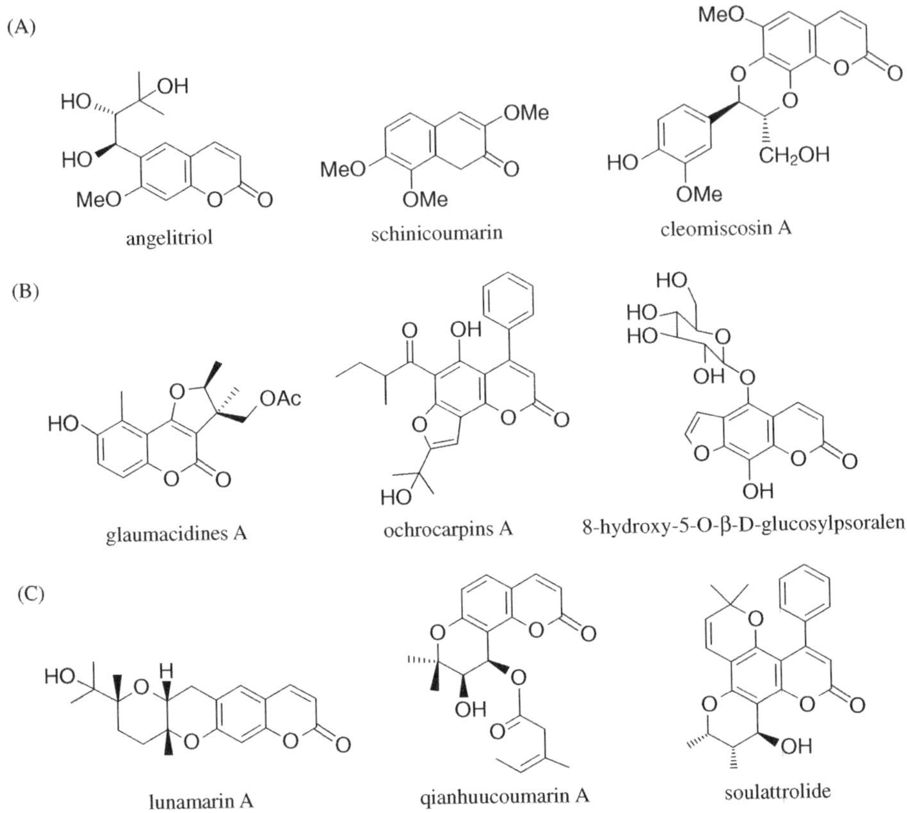

Figure 3.9 Compounds of simple coumarins (A), furanocoumarins (B), and pyranocoumarins (C).

Structures and Classification

Based on the structural characteristics, coumarins are subdivided into five groups.

1. Simple coumarins: (see Fig. 3.9A).
2. Furanocoumarins: refer to coumarins that have an additional furan ring system attached to the phenyl ring system of coumarins (see Fig. 3.9B).
3. Pyranocoumarins: refer to coumarins that have an additional pyran ring system attached to the phenyl ring system of coumarins (see Fig. 3.9C).
4. Dimeric and trimeric coumarins: respectively refer to coumarins that exist as a dimmer and trimmer (see Fig. 3.10A and Fig. 3.10B).
5. Isocoumarin: (see Fig. 3.10C).

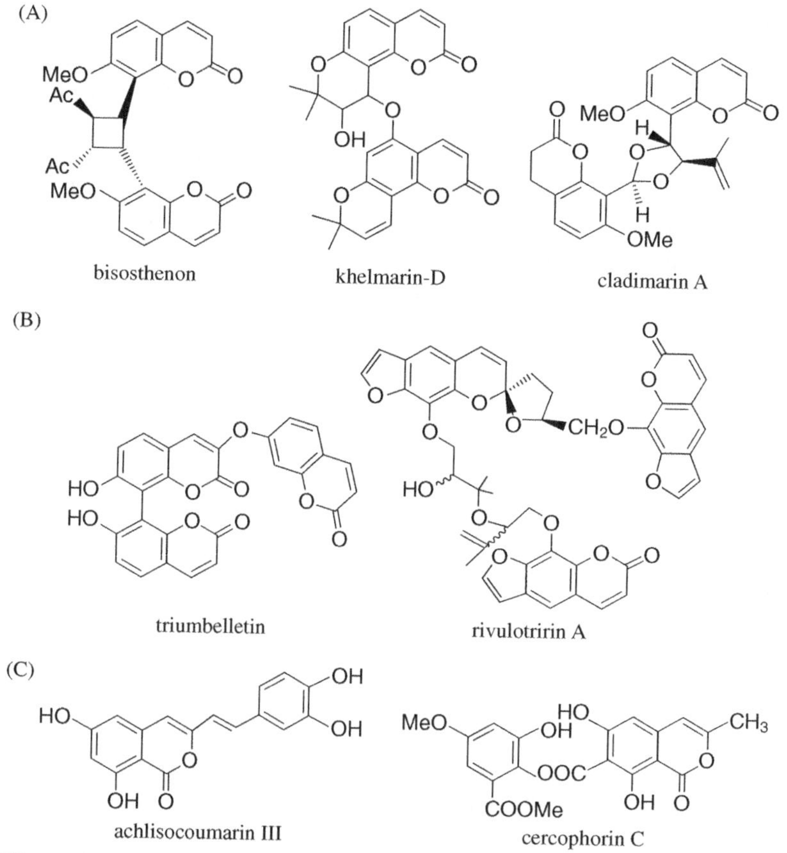

Figure 3.10 Compounds of dimeric (A), trimeric coumarins (B), and isocoumarin (C).

3.1.4 Lignans

Definition and Distribution in Plants

Lignans are widely occurring compounds in plants and are closely related to lignin, which forms the woody component of trees and other plants. The lignans are characterized by their dimeric composition of two or more C_6-C_3 units from cinnamic acid. For nomenclature purposes the C_6-C_3 unit is seen as propylbenzene, numbered from 1 to 6 in the ring, starting from the carbon that connects to the propyl group, and from 7 to 9 on the propyl group, starting from the carbon that connects to the benzene ring. With the second C_6-C_3 unit, the numbers are primed.

Biological Activities of Lignans

Lignans are found in flax seeds, pumpkin seeds, rye, soybeans, broccoli, some berries, and many traditional Chinese herbs such as *Magnolia officinalis*, *Schizandra chinensis*, and *Podophyllum peltantum*.

Lignans are one of the major classes of estrogen-like chemicals called phytoestrogens because they are capable of binding to estrogen receptors and interfering with the cancer-promoting effects of estrogen on breast tissue, by which they may inhibit the growth of breast, prostate, and colon cancer, and improve bone density. For this reason, lignans are currently being widely studied for their potential anti-cancer properties. A number of preliminary human and animal studies have confirmed their effects for cancer prevention. Podophyllotoxin, the well-known lignan, is a mitotic inhibitor that was first isolated from *Podophyllum peltatum* with strong anti-neoplastic activity. Etoposide is a podophyllotoxin derivative now used to treat lung cancer, testicular cancer, and acute lymphocytic leukemia. Lignans are also known as good antioxidants. These potent antioxidants scavenge free radicals that can damage tissue and are thought to play a role in the treatment of many diseases, such as liver inflammation.

Properties of Lignans

Most lignans are colorless solids; few have a property of sublimation. They are readily dissolved in lipophilic solvents such as hexane, chloroform, ether, ethanol, and ethyl acetate. Their water solubility increases after combining with glucose to form glycoside. Most lignans have optical activity due to the several asymmetric carbon atoms in their molecular structures.

Structures and Classification

1. Lignans: refer to those compounds in which the two C_6-C_3 units are linked by a bond between positions 8 and 8′ (see Fig. 3.11A).
2. Neolignans: refer to the lignans with two C_6-C_3 units linked by a chemical bond apart from C-8 to C-8′ bond (see Fig. 3.11B).
3. Norlignans: refer to those molecules that are composed of phenylpropanoid units, but with the loss of one or more carbon atom in the unit (see Fig. 3.11C).
4. Oligomeric lignans: refer to lignans that are composed of three, four, or more phenylpropanoid units (see Fig. 3.12).

3.1.5 Quinones

Definition and Distribution in Plants

Quinones refer to a class of compounds containing one of the benzoquinone isomers in their structures. Quinones occur as pigments in bacteria, fungi, and certain higher

(A)

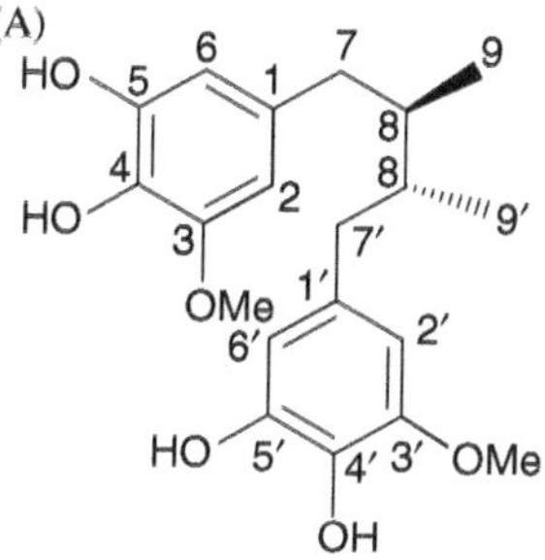

sauriol A

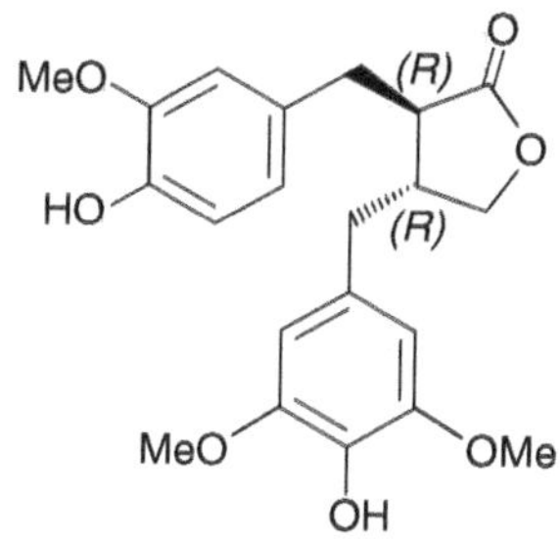

(-) - 4′- demethyltraxillagenin

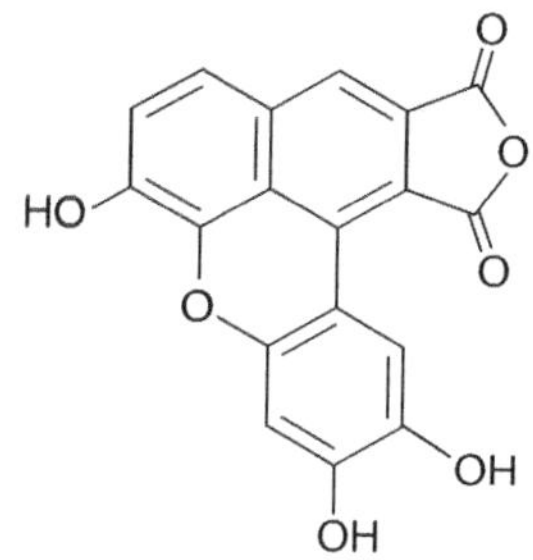

rufescidride

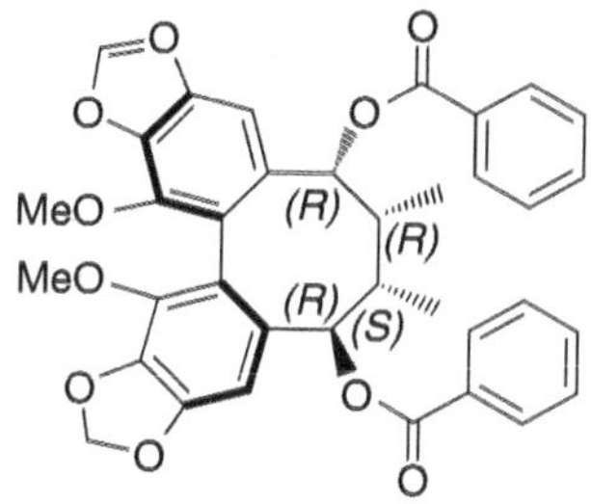

angustifolin A

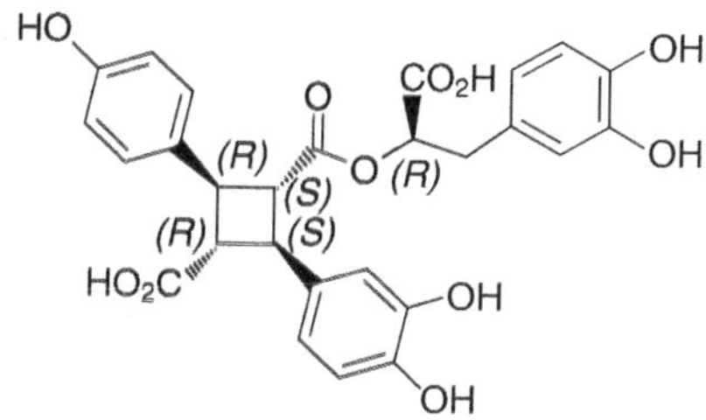

anthocerotonic acid

(B)

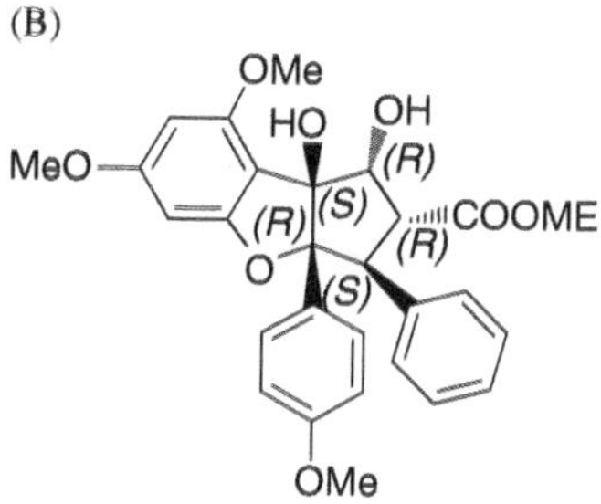

methylrocaglate

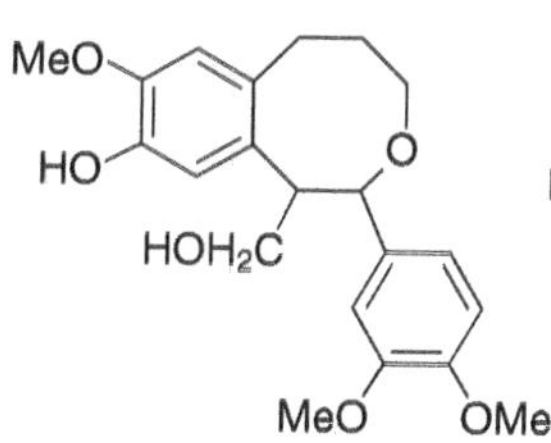

biondinin A

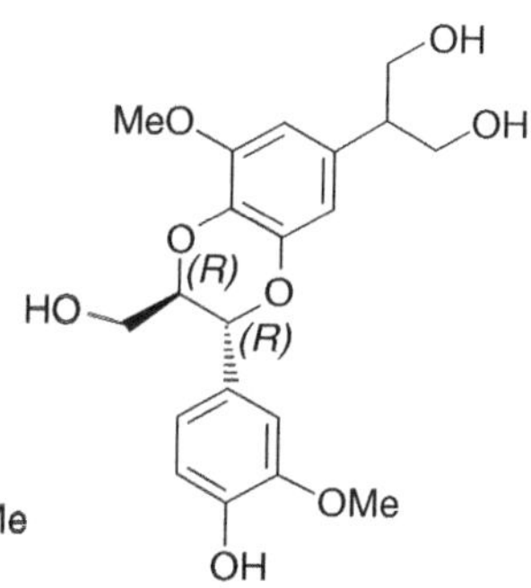

taxriyunin B

(C)

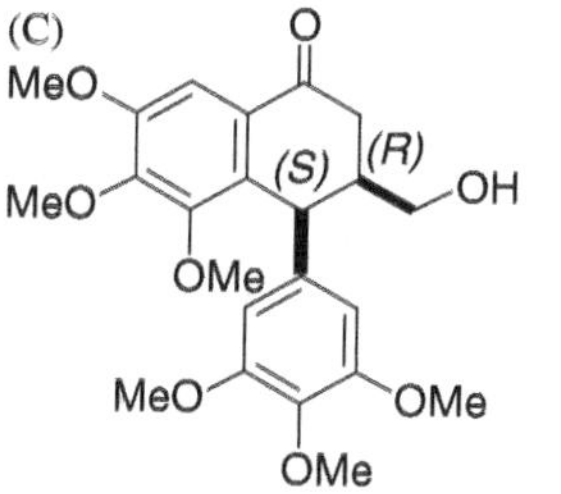

aglacin H

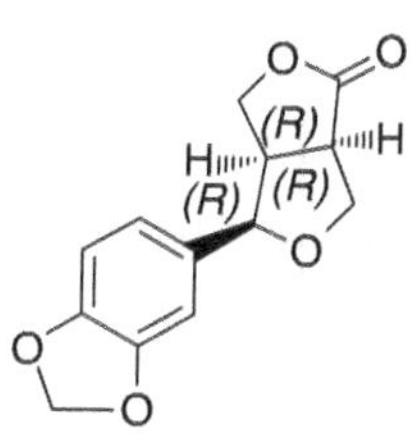

acuminatolide

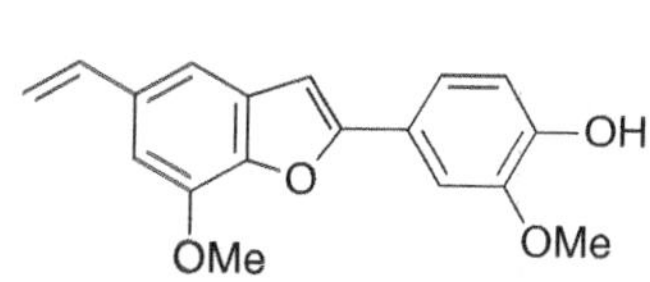

dehydroxymethylailanthoidol

Figure 3.11 Compounds of lignans (A), neolignans (B), and norlignans (C).

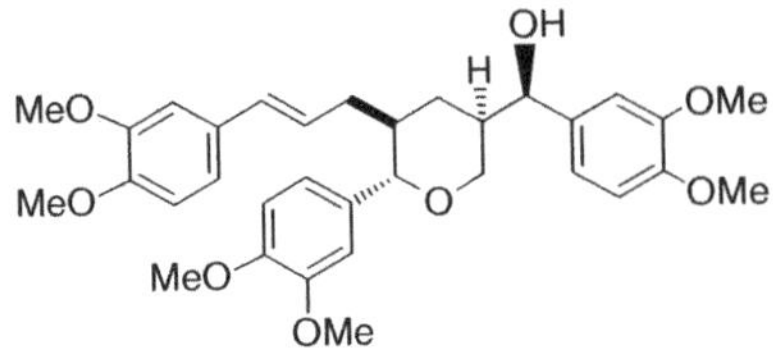

morinol A

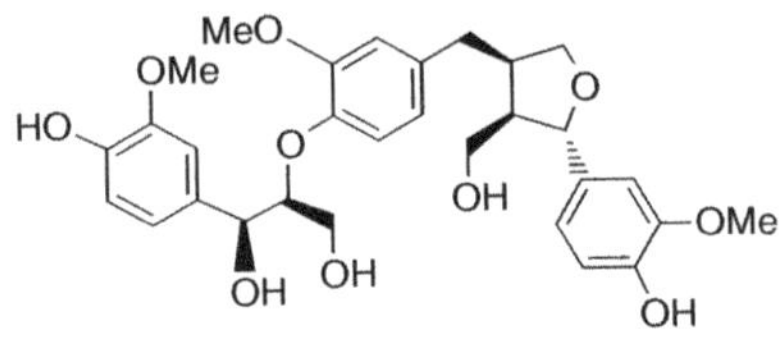

ehletianol C

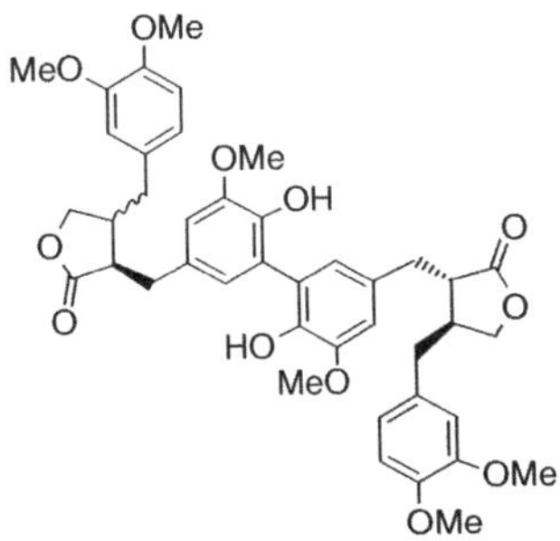

diarctigenin

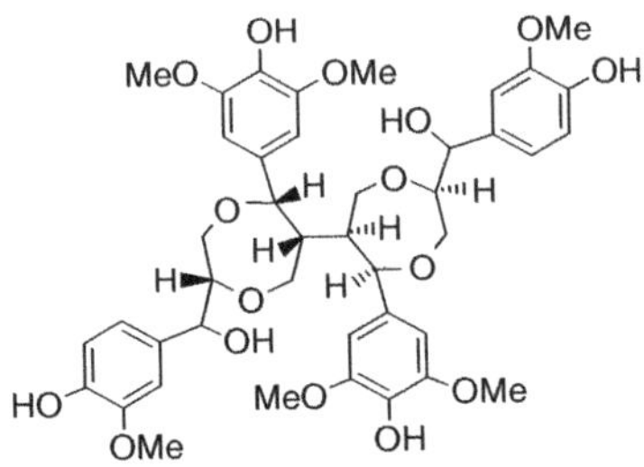

phyllostadimer A

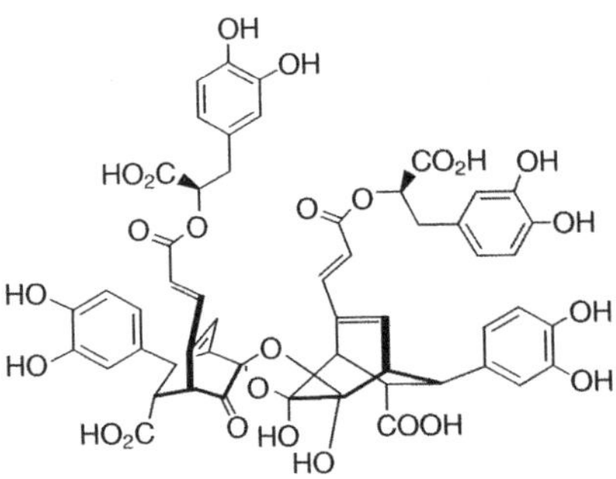

yunnaneic acid A

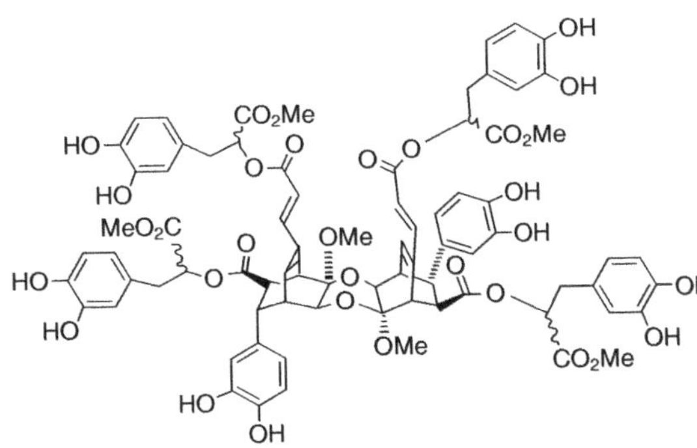

helicterin A

Figure 3.12 Compounds of oligomeric lignans.

plants. A number of quinone derivatives have been isolated from plants and animals, such as juglone in unripe walnuts, spinulosin from the mold *Penicillium spinuhsum*, arnebinone and arnebifuranone from *Arnebia euchroma*, tanshinone derivatives from *Salvia miltiorrhiz*, and sennoside A–D from *Rheum palmatum*.

Biological Activities of Quinones

Quinones possess various biological activities, including purgative actions (e.g., sennosdes), antimicrobacterial (e.g., rhein and saprorthoquinone), antitumor (e.g., emodin and juglone), inhibition of PGE_2 biosynthesis (e.g., arnebinone and arnebifuranone), and anti-cardiovascular disease (e.g., tanshinone II_A). Coenzyme Q10 is benazoquinone derivative that is used for treatment of cardiovascular diseases, hypertension, and cancer in clinics. Vitamin K compounds like K_1 and K_2 belong to naphthoquinones. They can promote blood coagulation, and thus are used for the treatment of natal bleeding.

Properties of Quinones

Quinones containing phenolic hydroxyl groups usually present beautiful colors such as yellow, orange, and red. Quinones in free-state form easily dissolve in organic solvents such as ether and benzene, and its glycosides can dissolve in hot water, methanol, and ethanol. Quinones are usually acidic due to the existence of phenolic hydroxyl groups in structures, thus can dissolve in basic aqueous solution and undergo Borntrager's reaction to form red or purple products. Quinine derivatives undergo Feigl reaction to produce purple products. Benzequinone and naphthaquinone derivatives can turn the colorless ethanol solution of leucomethylene blue into blue.

Anthraquinones are an important group of bioactive compounds in some herbal medicines. They are soluble in toluene, trichloromethane, ether, acetone, glacial acetic acid, and alkaline solution; slightly soluble in ethanol; and insoluble in water. Most anthraquinones are yellow or orange crystal solids, and some anthraquinones are sublimatory. The acidic property of anthraquinone derivatives are caused by numbers and position of carboxyl groups and phenolic hydroxyl groups on them.

Structures and Classification

1. Benzoquinones: refer to quinones containing a cyclohexadienedione in structure. So far, only para-benzoquinone and ortho-benzoquinone are found in nature, the meta isomer has not been reported (see Fig. 3.13A).
2. Naphthoquinone: refer to quinones derived from naphthalene. They are further divided into three groups: 1,4-naphthoquinone, 1,2-naphthoquinone, and 2,6-naphthoquinone. The important natural naphthoquinones are vitamins K_1 and K_2 that are found in blood and are responsible for proper blood clotting reaction (see Fig. 3.13B).
3. Phenanthraquinones: refer to the quinones derived from phenanthalene. They can be divided into two groups based on the substitution position of dione group: ortha and para-phenanthraquinones (see Fig. 3.13C).
4. Anthraquinones: refer to the quinones derived from anthracenes with the dione group substituted at the middle aromatic ring (see Fig. 3.13D).

3.1.6 Terpenoids

Definition and Distribution in Plants

Terpenoids are compounds made up of five-carbon units, often called isoprene units, that assemble in a regular pattern, usually head-to-tail in terpenes up to 25 carbons.

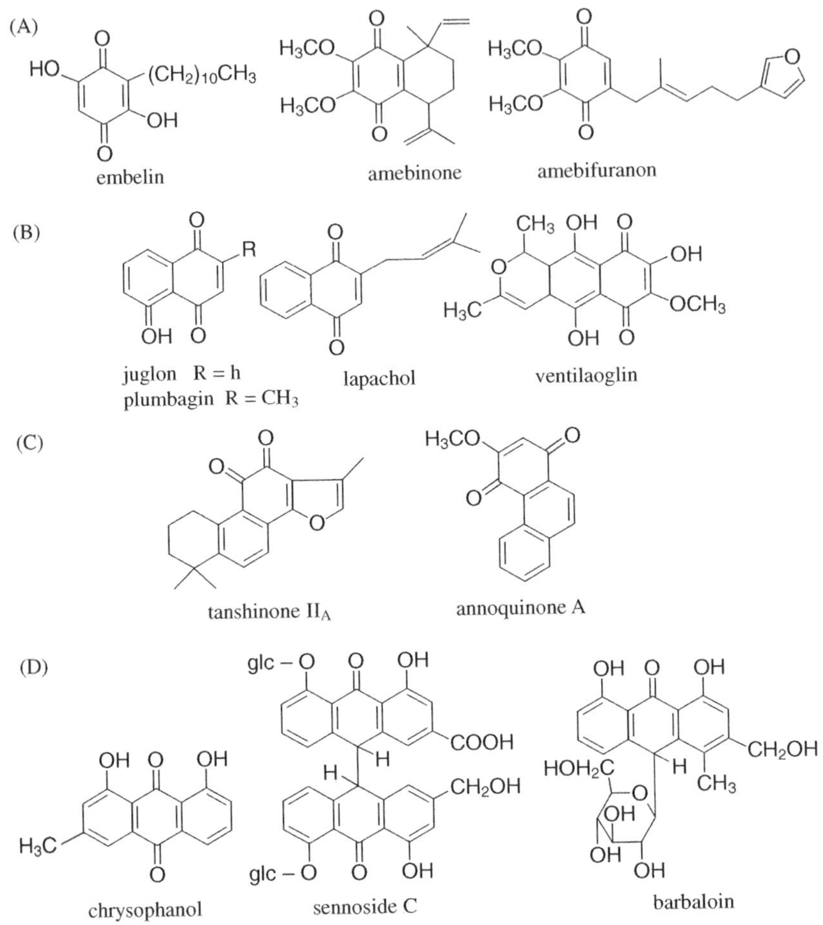

Figure 3.13 Compounds of benzoquinones (A), naphthoquinonoes (B), phenanthraquinones (C), and anthraquinones (D).

Terpenoids with 30 carbons or more are usually formed by the connection of two smaller terpene precursors, so they do not follow the head-to-tail rule. Terpenoids are widely distributed in nature, mainly in plants. The total number of terpenoids discovered has been over 22,000, and many of them have become important clinical drugs or been used as leading compounds for new drug developments.

Classification, Properties, and Biological Activities of Terpenoids

Terpenoids are further classified into hemiterpenoids (C_5), monoterpenoids (C_{10}), sesquiterpenoids (C_{15}), diterpenoids (C_{20}), sesterterpenoids (C_{25}), triterpenoids (C_{30}), tetraterpenoids (C_{40}, carotenoids), and polyterpenoids (C_{5n}) based on the number of carbon atoms in the same manner as are isoprene.

Most terpenoids are optically active, and have bigger refractive indexes (RI) because they contain asymmetric carbon atoms. Most terpenoids are hydrophobic, readily dissolve in liposoluble solvents, but difficult to dissolve in water. The solubility of terpenoids can be improved by increasing the number of oxygen-containing functional groups. When terpenes are linked with sugars to become glycosides, such as saponins of triterpoids in ginseng and licorice, they are easily dissolved in methanol, ethanol, and hot water.

Monoterpenoids and sesquiterpenoinds usually exist in plants as an oily liquid with a characteristic odor, and are the main components in volatile or essential oils that can be prepared by distillation. The melting point of monoterpenoids is lower than that of sesquiterpenoids. The melting point and boiling point of monoterpenoids and sesquiterpenoids increase with their molecular weights, numbers of double bonds and functional groups increasing, but their volatility decreases. They are easily soluble in ether and chloroform, fairly soluble in alcohol, and slightly soluble in water.

Diterpenoids are mostly crystal solid. Many of them are bitter. With the increase of the number of isoprene units in the structures, that is, increase of the molecule, diterpenoids become less volatile in comparison with the mono- and sesquenterpenoids. Most of them are insoluble in water, but soluble in most organic solvents. Diterpenoids and sesterterpenoids are commonly obtained as a crystalline solid.

Triterpenoids and sterols are naturally occurring as free states or glycosides in plants. Their solubility in solvents depends on their existence form. In general, their glycosides can dissolve in hot water, methanol, and ethanol. Triterpene and sterol hardly dissolve in water, but readily dissolve in organic solvents. The glycosides of both triterpenes and sterols are usually called saponins due to their ability to cause foaming in water and to hemolyze blood cells.

Tetraterpenoids are usually called tetraterpenoid pigments due to their conjugated polyene feature. They are lipid-soluble compounds, and readily dissolve in nonpolar or less polar organic compounds, for example, ether, chloroform, petroleum, and benzene.

Terpenoids are a group of bioactive natural compounds. Because the bioactive activities are too wide, they are separately introduced under each subgroup below.

1. *Monoterpenoids*

Monoterpenes are a class of terpenes that consist of two isoprene units and have the molecular formula $C_{10}H_{16}$. Monoterpenes may be linear (acyclic) or contain rings. Biochemical modifications such as oxidation or rearrangement reaction produce the related monoterpenoids. Examples from this class include camphor, menthol, thujone, thymol, nerol, linalool, limonene, geraniol, and perillyl alcohol, found in citrus peels, mint leaves, lavender, and thyme. Among them, menthol is a useful topical pain reliever and anti-puretic; thujone is a toxic agent found in wormwood (*Artemisia absinthium*), from which the liqueur and absinthe are made; borneol is derived from pine oil and used as a disinfectant and deodorant; camphor is used as a counterirritant, anesthetic, expectorant, and antipruritic, among many other uses (see Fig. 3.14A).

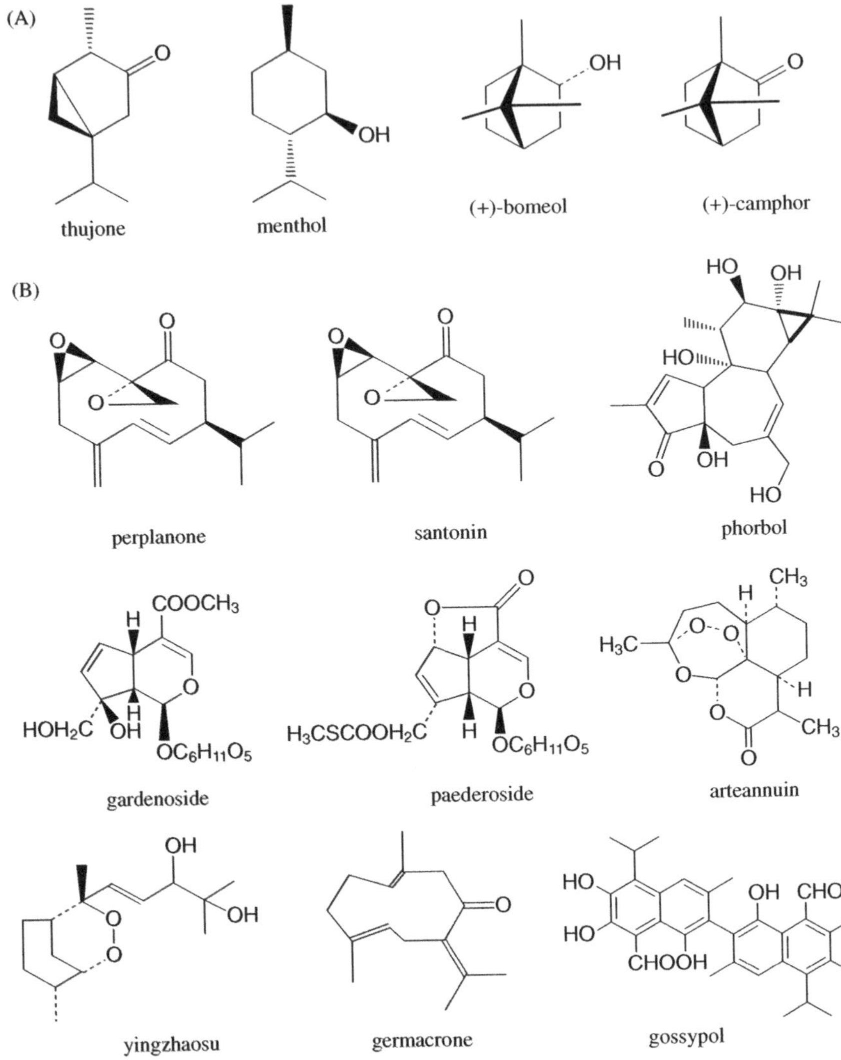

Figure 3.14 Compounds of monoterpenes (A) and sesquiterpenoids (B).

2. *Sesquiterpenoids*

Sesquiterpenoids refer to those terpenoids having a C_{15} skeleton derived from isoprene units. They are found mainly in higher plants. Examples include nerolidol, farnesol, and ylangene in neroli, mints, sandalwood, ginger, and German chamomile. These phytochemicals commonly have antiallergen and anti-inflammatory properties. In addition, periplanone B is a female sex attractant of a species of cockroach, santonin is a photosensitizer found in wormwood, and gossypol is a dimeric sesquiterpene isolated from the seeds of cotton plants that has been clinically used in China as a male contraceptive. Artemisinin is the most valuable medicinal sesquiterpenoids that was first isolated from traditional Chinese herb *Artemisia annua*, with strong anti-malaria activity. Its derivatives, artemether and artesunate, have cured many malaria patients all over the world (see Fig. 3.14B).

3. *Diterpenoids*

Diterpenoids refer to these terpenoids having a C_{20} skeleton derived from isoprene units. They can be divided into two groups: acyclic and cyclic diterpenoids.

Vitamin A, the retinoids, is a group of monocylcoditerpenes rich in fish oil. Phytol is a naturally linear diterpene alcohol used in the preparation of vitamins E and K_1. It is also part of the composition of chlorophyll as a side chain.

The common but important diterpenoids in terms of bioactivities are dicyclo- and tricyclo-diterpenoids. One of the most well-known medicinally valuable diterpene is paclitaxel (Taxol®). Paclitaxel was first isolated from the bark of the Pacific yew, *Taxus brevifolia* in the early 1960s. By the late 1980s, its value as an anticancer drug had been determined by various clinical studies. Paclitaxel and its analog docetaxel (Taxotere®) have been approved by U.S. FDA to treat various cancers, including non-small-cell lung cancer, small-lung cancer, ovarian cancer, breast cancer, and head and neck cancers. Another example of the important medicinally diterpenoids are ginkgolides discovered from *Ginkgo biloba*, which show strong bioactivity against the aggregation of platelet (see Fig. 3.15A).

4. *Sesterterpenoids*

Sesterterpenoids refer to these terpenoids having a C_{25} skeleton derived from isoprene units (see Fig. 3.15B). They are narrowly distributed, only found in several species of plant, marine organisms, and insects. Up to now, the number of isolated natural sesterterpenoids is less than 40.

5. *Triterpenoids and steroids*

Triterpenes are made of 30 carbons, derived essentially from the coupling of two sesquiterpene precursors. The biological activities of triterpenoids have attracted much interest. Triterpenoids are known as cancer chemopreventive, antiulcer, and antidiabetic agents, inhibitors of angiogenesis and eukaryotic DNA polymerases, and so on. They are the biological active components in several famous Chinese herbal medicines, for example, ginseng, licorice, and bupleurum. Triterpenes of the quassinoid class, such as bruceantin, have been shown to have significant antineoplastic activity in animal studies and investigated for the treatment of cancers. Arbruside E, a relatively nontoxic triterpene, was isolated from the extreme toxic jequerity (*Arbrus precatorius*) as a potential sugar substitute because it is 30–100 times sweeter than sucrose.[6] Two triterpenoid saponins with insulin-like activity, named assamicin I and II, were isolated from the roots of *Aesculus assamica* Griff. They inhibited release of free fatty acids from epinephrine-treated rat adipocytes and enhanced glucose uptake into 3T3-L1 adipocytes[7] (see Fig. 3.16).

Steroids are triterpene derivatives that are based on the cyclopentane perhydrophenanthrene ring system. They are well known for their biological role as

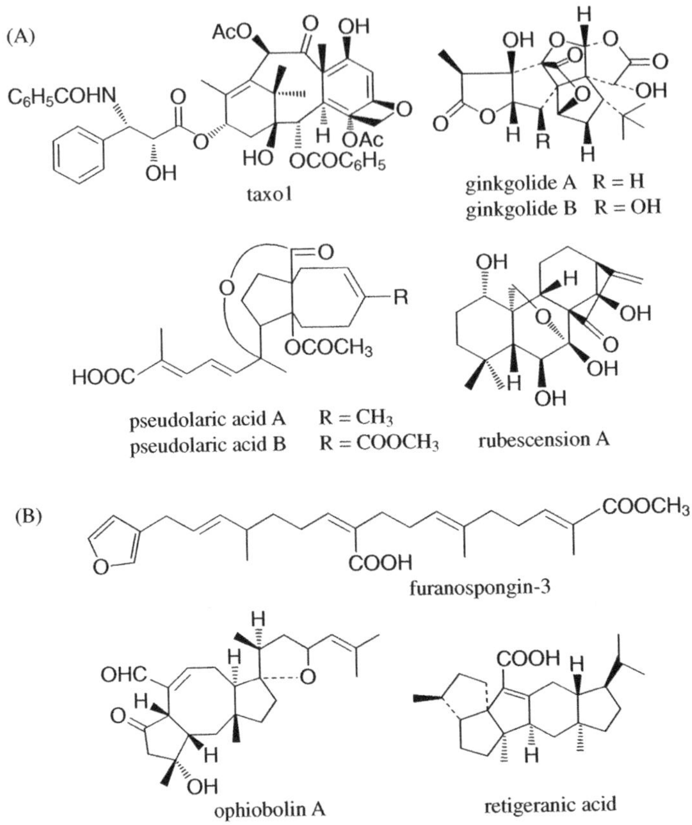

Figure 3.15 Compounds of diterpenoids (A) and sesterterpenoids (B).

hormones, testosterone, and progesterone. Sitosterol, stigmaterol, and campesterol are three ubiquitous "phytosteroids" in higher plants. Recent chemical investigation and pharmacological studies reveal various biological roles such as anti-inflammation, anticancer, anti-cardiovascular, hypoglycemic, and antifungal activities for steroidal compounds in plants. A serial of furostanol glycosides isolated from the traditional Chinese herb *Dioscorea colletti Hook. var. hypoglauca* were confirmed to possess strong cytotoxicity *in vitro* in a National Cancer Institute's anticancer drug screening program[8] (see Fig. 3.17A).

6. *Tetraterpenoids*

Tetraterpenoids are made of 40 carbons, derived essentially from the coupling of two diterpene precursors. The widely distributed lipid-soluble pigments and essential dietary required carotenoids belong to this group of chemicals. The important C_{20} isoprenoid vitamin A is mainly from the hydration and splitting of β-carotene. Lycopene is another bioactive carotenoid found in tomatoes and other red fruits and has been considered a potential agent for prevention of some types of cancers (see Fig. 3.17B).

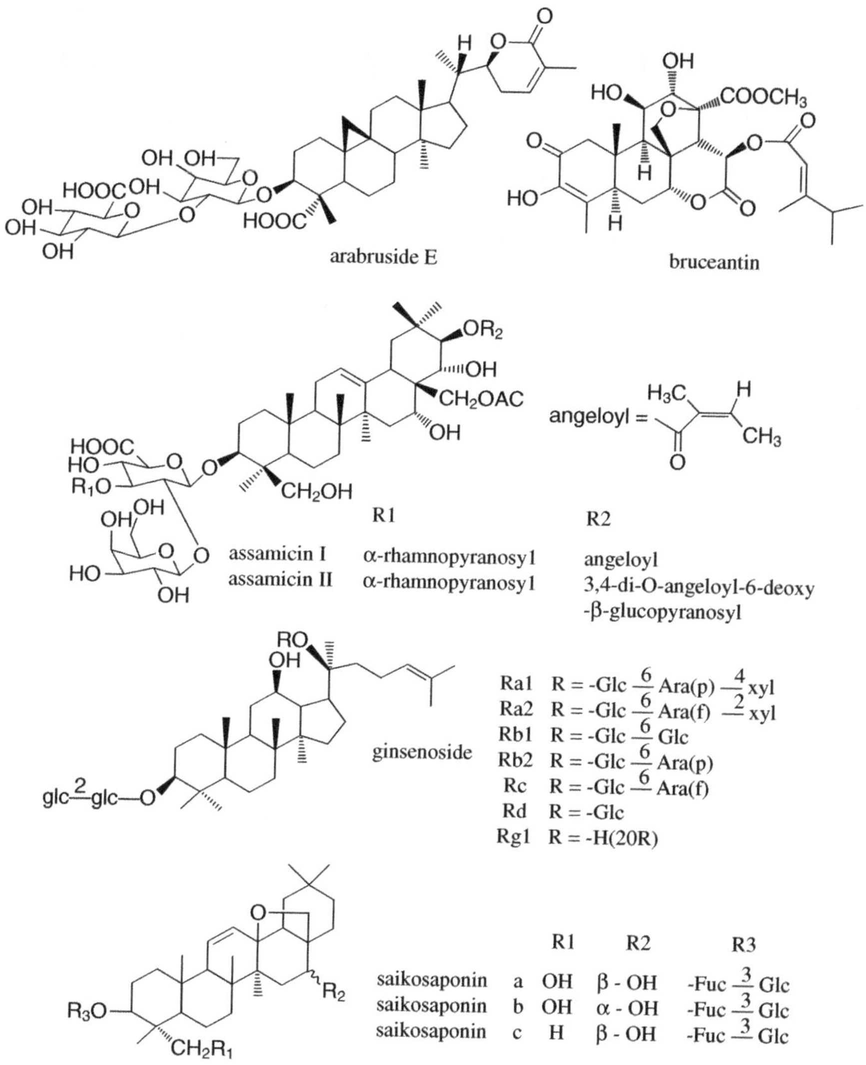

Figure 3.16 Compounds of triterpenes.

3.1.7 Cardiac Glycosides

Cardiac glycosides (also called cardenolides) are named from the impact of this group of compounds on the heart. The structures of cardiac glycosides are composed of two parts: cardiac aglycone and sugar moieties. Cardiac aglycone is biosynthesized from sterol in nature. Cardiac glycosides are categorized into two groups (I and II) based on the chemical structures of their cardiac aglycones. For cardiac glycosides in group I, the side chain in the sterol skeleton is a five-member $\Delta^{\alpha\beta}$-γ-

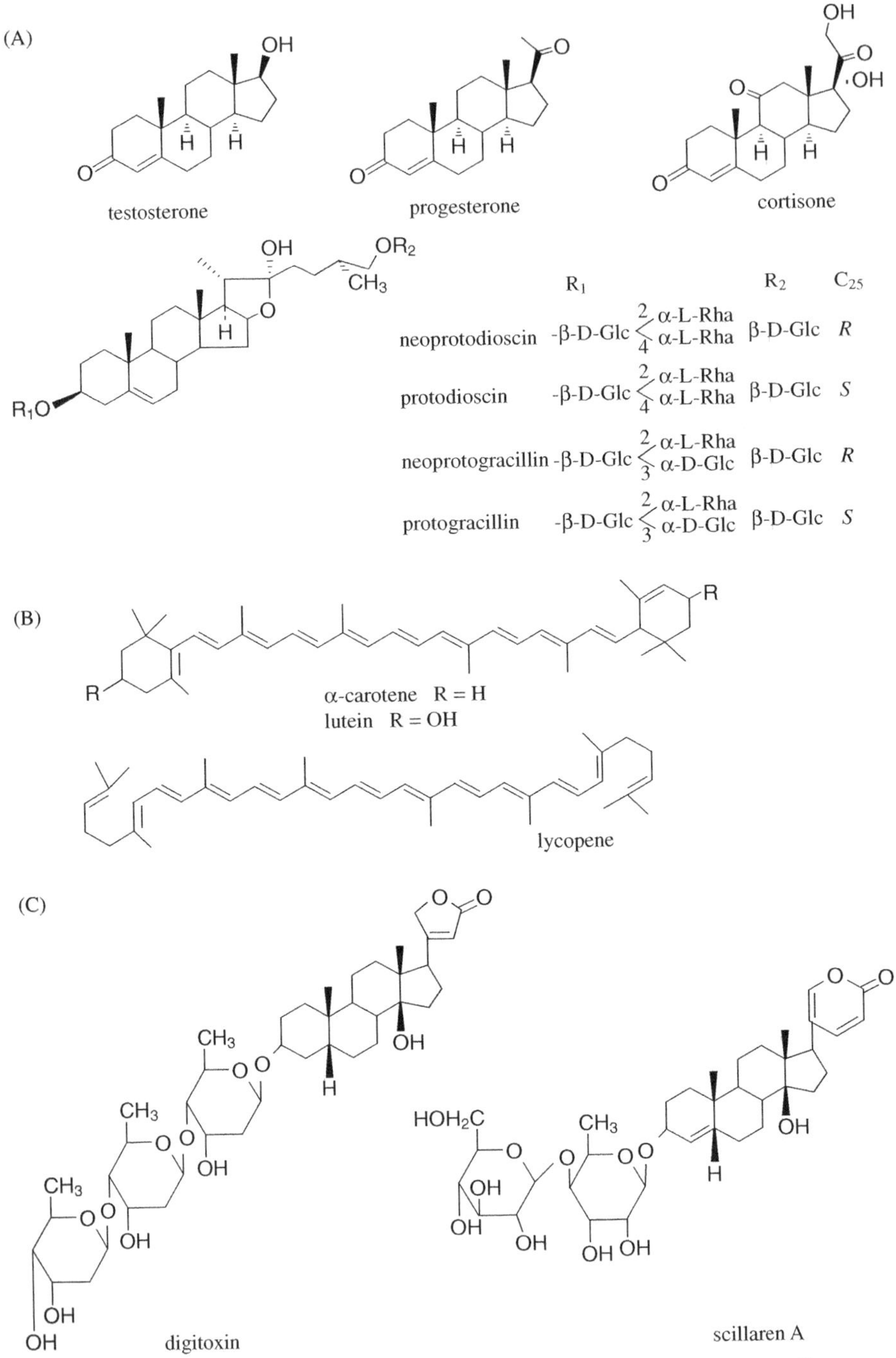

Figure 3.17 Compounds of steroids (A), tetraterpenoids (B), and cardiac glycosides (C).

lactone that is replaced with six-member $\Delta^{\alpha\beta,\gamma\delta}$-δ-lactone in group II. Most cardiac glycosides are toxic and have many pharmacological activities. One example is oleandrin, the toxic component from the leaves of the oleandrin (*Nerium oleander*), which has an unusual structural feature with special sugar substitutes that are uncommon in the plant kingdom. Another example is digitoxin, one of the components of the heart drug digitalis (see Fig. 3.17C). Cardiac glycosides are used for the treatment of cardiac failure in clinic.

Cardiac glycosides are neutral compounds. They are usually obtained as colorless crystalline or amorphous powder with a bitter taste. Like other glycosides, they can dissolve in hot water, methanol, and ethanol, but do not dissolve in nonpolar organic solvents. The cleavage of lactone ring in structures easily occurs in treatment with alkaline solution (sodium hydroxide solution).

3.2 METHODS FOR EXTRACTION OF HERBAL MEDICINES

In general, plant material should be dried and milled prior to extraction. The temperature is better maintained below 30°C to avoid destruction of thermolabile compounds when drying. The surroundings for drying should be far away from sunlight exposure and should be in a room with good air circulation. If fresh material is needed for extraction, it is necessary to extract it as soon as possible after the material is collected or to snap-freeze the material for long-distance transportation.

To enhance the efficiency of extraction, the plant materials should be cut into small pieces or ground into smaller particles, depending on individual experiment requirements. A number of techniques and instruments have been developed and employed to extract plant materials. Some new techniques developed recently for extraction include supercritical fluid extraction (SFE), pressurized liquid extraction (PLE), and microwave-assisted extraction (MAE). In comparison with Soxhlet extraction, SFE, PLE, and MAE are much more efficient and can significantly shorten the extraction time. Among these extraction techniques, solvent extraction, steam distillation, and SFE are among the most often encountered extraction methods, and are described in detail below.

3.2.1 Methods of Solvent Extraction

Organic and/or aqueous solvents are widely used in the extraction of plant materials due to their effectiveness and low costs. It relies on the principle of "solid–liquid" extraction. In the solid–liquid extraction, solvents diffuse into plant cells first, and the metabolites are then dissolved in solvents and transferred out of the cells.

The whole extraction process is dynamic, and can be facilitated by grinding and heating the plant material (for thermostable compounds only). Methods of solvent extraction can be divided into two classes on the basis of solvent flow state: continuous (percolation, soxhlet extraction) or discontinuous methods (maceration, sonication-assisted solvent extraction, and extraction under refluxing).

When selecting a solvent or solvent system to extract plant materials, the factors that need to be considered include solubility of the target compounds, safety (low toxicity, low flammability, low risk of explosion), cost, ease of solvent recovery, and concentration. The grade and purity of solvent should be taken into consideration. The physicochemical properties of some common solvents used in natural products extraction can be tailored in many chemical handbooks.

The selection of extraction method is intended for interesting compounds. Following the principle of "like dissolves like," it is not difficult to find the solvent to maximize the yields of interesting compounds. For most lipophilic components in plants, such as alkanes, fatty acids, sterols, some terpenoids, and alkaloids, the nonpolar solvents (pet-ether, *n*-hexane, chloroform, and so on) should be selected. Medium-polarity solvents (ethyl acetate, dichloromethane, and so on) are chosen to extract compounds with intermediate polarity, such as flavonoids and quinones. For more polar compounds, such as glycosides, polyphenols including tannins, and anthocyanins, more polar solvents (water, methanol, ethanol, acetone, and so on) should be used.

For organic acids, bases, and amphoteric compounds, the extraction yields differ for various pH values of the extraction aqueous solution.

When whole chemical components in plants are to be extracted, a polar organic solvent (methanol or ethanol) or an aqueous alcoholic mixture is employed. The aqueous alcoholic mixture (60% or 70% ethanol–water, v/v) is used frequently due to its safety and ability of extracting large amounts of polar, medium-, and low-polarity constituents. However, it is unable to extract polysaccharides and proteins.

In practice, bioactivity-guided isolation in bioassay screening, a series of extractions of a plant material with different solvents, usually in an order of increasing polarity, is commonly applied, not only to avoid missing any bioactive component, but also to exclude the interference of components in bioassay. Some solvents used for a series of extraction are ether, chloroform, acetone, methanol, and water.

No matter what extraction methods are employed, the extracted liquid should be filtered first to remove any particulate insoluble residues, and then concentrated under reduced pressure on rotary evaporators (temperature is recommended below 40°C) or freeze-dried for thermolabile components in lab. The recovered solvents from rotary evaporators can be recycled for further extraction of the same material. Thus, the ideal solvent for extraction should have good solubility for extraction efficacy, and a low boiling point for easy solvent recovery.

In comparison with the discontinuous methods, the continuous extraction methods are performed with a larger volume of solvent or by repeated extraction, thus they usually provide higher yield because of the limited solubility or saturation of compounds in a certain volume of solvent. A complete extraction is very important for a quantitative analysis. A reliable analysis closely depends on the selection of the extraction solvent and method. If a discontinuous method is applied for quantification of an herbal sample, the extraction should be repeated at least three times with a sufficient volume of solvent in each extraction.

In practice, it is better to search literature about chemical research on the related genera and families before starting extraction of a plant whose composition is not well known yet. This will help predict the types of secondary metabolites that might be present in the plant and select the solvents for extraction.

Maceration

Maceration is a simple-approach extraction method, in which the plant material is placed in a closed container full of extracting solvent. The solvent should be added to cover the top of the sample. The material should be macerated with solvent in a covered container at room temperature for a certain period of time, from a few hours to several weeks depending on the properties of the material and solvent as well as the purpose of the experiment. The maceration method is easy but time-consuming and less efficient. It is suitable for thermolabile compounds.

Sonication-Assisted Solvent Extraction

Sonication-assisted solvent extraction is a modified maceration, in which ultrasound is utilized to improve the extraction efficiency. The plant material is placed in a closed container like in maceration. The container is then placed in an ultrasonic bath. In such a condition, ultrasound transfers the mechanical power onto the plant cells, leading to the breakdown of cell walls and increase of solubilization of metabolites in the solvent. The frequency, length, and temperature of sonication are the main factors affecting the extraction yields. It is an easy and efficient method commonly used in the lab.

Percolation

The percolator is a cylindrical or conical container made from glass or metal with a tap at the bottom. The ground plant material is first added into the container, leaving enough room on top to allow expansion later after solvent is added. The solvent is added into the percolator till its surface is above the top of the material. The percolator is sealed with a cover and the materials are allowed to soak for 24 h. If the solvent is completely absorbed by the materials, add more solvent until it is above the top of the material. After 24 h, let the liquid slowly flow out of the bottom of the percolator with a certain flow rate by adjusting the switch. Meanwhile, fresh solvent is continuously added on top of plant material. The percolation will continue until the recovered solid residue of the extraction solution turns out to be little. Sometimes, the fine-powdered materials, especially some roots with excessive starch and sugar, could be expended excessively to stop the percolation. In addition to the solvent used for the extraction, the percolation rate and the temperature of solvent will influence the extraction yields. In comparison with the above two methods, percolation usually provides higher yield.

Extraction under Refluxing

Extraction under refluxing is suitable for heat-stable compounds. This method can be used for both small- (several grams) and large-scale (tens of kilograms) extraction for herbal study. In the lab, the material is immersed in an appropriate solvent in a round-bottomed flask, which is vertically connected to a condenser. The solvent is heated by water bath or electric heater to boiling. The evaporated solvent goes up into the condenser, and then is condensed into liquid form and drops down to the flask. Generally, the powders or small pieces of plant material are first macerated in the solvent for 24 h before extraction, and then extracted under refluxing for several (3–5) hours until the solvent is saturated with solutes. The volume of solvent should at least be sufficient to cover the top of the sample. The materials are usually exhaustively extracted after three extractions. Since the heat can help the solvent penetrate into plant cells and the solubility of compounds in a solvent will increase when the temperature is increased, the extraction efficiency of this method is much higher than for maceration. This is one of the most popular extraction methods employed for herbal extraction in the lab.

Soxhlet Extraction

Soxhlet extraction is a convenient method for the extraction of herbal materials of small to medium volumes. The commercially available soxhlet instrument is composed of an extraction chamber with reflux condenser and a collecting flask (Fig. 3.18). The chamber is placed between the collection flask and refluxing condenser. The plant powder or pieces are kept in a cellulose thimble in the extraction chamber. A suitable solvent is added to the flask and heated under refluxing. The solvent will first be evaporated up into the condenser, then liquefied into the chamber. When the condensed solvent in the chamber reaches a certain height, it is siphoned into the flask, and the next extraction is initiated. Usually, 50–60 times of recycling are necessary for an extraction. Because of the repeated extraction, this method is usually more efficient than refluxing and produces a higher yield of extract with less volume of solvent.

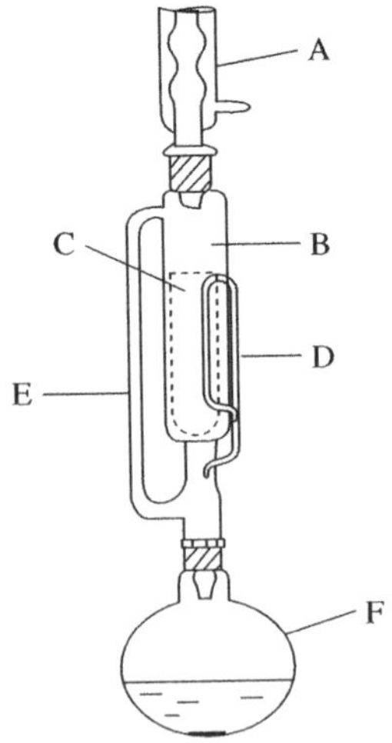

Figure 3.18 The soxhlet instrument. A, Reflux condenser; B, extraction chamber; C, filter-paper cylinder (sample-containing); D, siphon tube; E, vapor tube; F, collecting flask (solvent-containing).

For systematic chemical study or bioassay-guided isolation, different solvents can be consecutively used in an order of the polarity of solvents, such as *n*-hexane, chloroform, acetone, and methanol, to get several groups of extracts, each consisted with different types of compounds.

3.2.2 Steam Distillation Extraction

Steam distillation sees limited use, and is usually applied to the plant material enriched with volatile oils. In steam distillation extraction, a vapor mixture of essential oil and water is produced by heating the plant material immersed in water. The vapor mixture is condensed, and the distillate is separated into two or three immiscible layers. If the density of the volatile oil is less than that of water, it will be in the upper layer; if it is greater than that of water, it will be in the lower layer. If the volatile oil is composed of multiple compounds with various densities, then it will appear both above and below the water layer. Steam distillation is an effective method in preparation of fragrance and flavoring water-insoluble natural products (Fig. 3.19).

3.2.3 Supercritical Fluid Extraction (SFE)

A supercritical fluid is a fluid at a temperature and pressure above its thermodynamic critical point. Carbon dioxide and water are the most commonly used supercritical

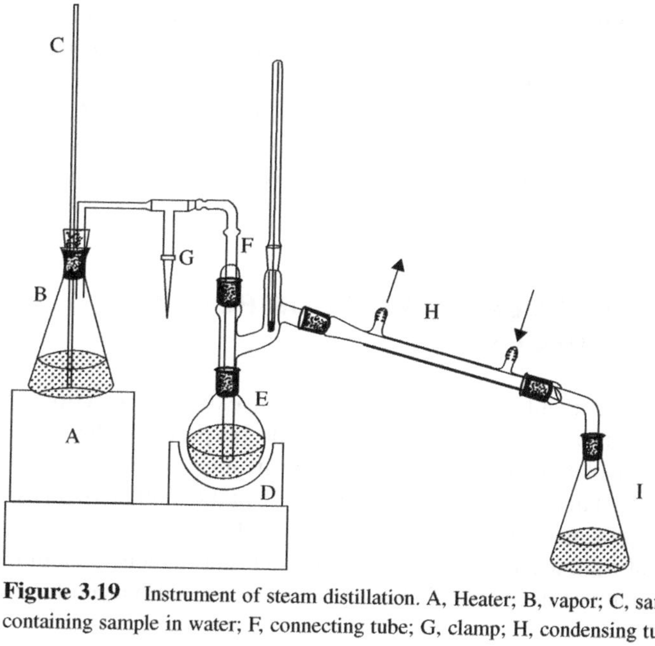

Figure 3.19 Instrument of steam distillation. A, Heater; B, vapor; C, safe tube; D, heater; E, flask containing sample in water; F, connecting tube; G, clamp; H, condensing tube; I, collecting bottle.

fluids. The supercritical fluids have solvating properties that are parallel to organic solvents and enable the dissolution of the metabolites in plants but with advantages of higher diffusivities, lower viscosity, and lower surface tension. A supercritical fluid can readily change in density upon minor changes in temperature or pressure. Details about the principles and practice of SFE are available from the book written by McHugh and Krukonis.[9]

The SFE technology, especially the one using supercritical CO_2 (SCF), has been widely applied in extraction and isolation of natural products in the past two decades because it has the unique ability to diffuse through solids like a gas, and dissolve materials like a liquid. In the early development stage, SCF is mainly employed in the preparation of lipophilic compounds and volatile oils from plants. Their solvating property can be easily adjusted via altering the pressure and temperature, or by the addition of modifiers to the SCF. The commonly used modifier is methanol, which enhances the polarity of SFC and makes it suitable for compound extraction from medium or high polarity.

In contrast with the conventional solvent extraction, SFE is an environmentally friendly and controllable method, and has been successfully applied in the extraction of many kinds of compounds, such as essential oils and flavonoids. Examples include capsaicinoids and polyphenols from grape marc, hyperforin from *Hypericum perforatum*,[10] parthenolide from feverfew plant,[11] paclitaxel from Pacific yew tree,[12] and resveratrol from *Vitis vinifera*.[13] However, SFE requires more sophisticated and specialized equipment, which to some extent hinders its spread and application.

During the development of SFE as an extraction method for natural products, the solubility of the target compounds in supercritical CO_2 or other SCF should first be determined. The solubility test is conducted to determine the influence of temperature and pressure on the solubility of the target compounds. The temperature for any given pressure is to be strictly controlled so as to control the density; its change can greatly influence the solvating power of SCF. The flow rate of SCF can affect the partition coefficient of the analytes between CO_2 and plants, and should be optimized in experiment.[14] In order to extract compounds with moderate polarity, suitable modifiers must be chosen carefully.

3.3 METHODS FOR ISOLATION OF COMPOUNDS FROM HERBAL EXTRACTS

There are many methods available for isolation of herbal extracts. Selection of methods depends on the properties of the compounds and purposes of the study. Compounds in herbs are isolated for many reasons: for chemical identification or structure elucidation, for obtaining standards for qualitative and quantitative analysis, for bioassay screening or mechanism study, for pharmacological and toxicological study, and for clinical trial. Methods for isolation of natural products, including herbs, are available in many sources. The following methods are introduced based on the isolation principles.

3.3.1 Isolation Based on Difference of Solubility

Crystallization

Crystallization and recrystallization methods can be applied to purify a highly concentrated major compound in an extract or isolated mixture. In the crystallization process, the extract or mixture containing the target compound is first dissolved in a solvent or solvent mixture (neither excessively soluble nor insoluble) to make a saturated solution; heating is often necessary. The saturated solution is then kept at room temperature or in a refrigerator, with a cover, but usually not completely sealed, to allow the compound of interest to gradually crystallize while the impurities and the other compounds remain in the solvent. An environment with well-circulated air and low temperature usually helps the formation of the crystallite. The compound crystal can be separated from solution by careful filtration. In traditional structure identification before the wide application of high-performance liquid chromatography-UV-mass spectroscopy (HPLC-UV-MS) or HPLC-nuclear magnetic resonance (HPLC-NMR), crystallization is usually performed to provide purified compounds for melting point measurement and structure identification by infrared (IR), MS, and NMR. Crystallization is not only a method useful for natural product purification but also a necessary step for structure determination by single-crystal X-ray diffraction.

Precipitation

This method is applied based on the solubility of isolated compounds in different solvents. For example, when adding 95% ethanol into a concentrated aqueous herbal extraction solution (usually the volume ratio of the two liquids is 3:1, or the final ethanol concentration is between 70% and 80%), macromolecular substances, such as polysaccharides and proteins, can be precipitated and separated from small molecule pigments or other soluble compounds by filtration. For some herbal materials enriched with saponins, the total saponins can be obtained by mixing the alcoholic extract solution with a certain amount of acetone or ether.

3.3.2 Isolation Based on Difference of Partition Coefficient

Partition Coefficient (K)

In the solvent partition method, compounds are partitioned between two immiscible solvents according to their different partition coefficients. In general, one of the solvents is water, sometimes with a low percent of methanol or ethanol in it; another one is a water-insoluble organic solvent, such as chloroform, hexane, ethyl acetate, or butanol. This method is easy to manage and suitable for isolation of natural products from crude extract.

The partition coefficient (K) of a compound is the ratio of its concentration in two immiscible solvents ($K = C_U/C_L$, U: solvent in upper layer; L: solvent in lower layer). The separation efficiency depends on the difference of partition coefficients, which is called the separation factor ($\beta = K_A/K_B$). The partition times can be initially deduced from the β value. When it is equal or bigger than 100, only one time of partition is enough. When it lies between 10 and 100, 10 to 12 times may be needed. If the acid, alkali, or amphoteric compounds are the targeted compounds, the pH modifications of partition solvent can facilitate the separation process as shown below. Taking the acid compounds as an example, the disassociation equilibrium and disassociation constant (K) can be obtained as follows:

$$HA + H_2O \rightleftharpoons A^- + H_3^+O$$

$$Ka = \frac{[A^-][H_3^+O]}{[HA]}$$

$$pka = pH - \log[A^-]/[HA]$$

$$pH = pKa + \log[A^-]/[HA]$$

Obviously, when the pH value of water is two times bigger than the pKa, acid compounds remain to be completely disassociated in water. Conversely, when the pH value of water is two times smaller than pKa, it presents as a free state. A reverse phenomenon was observed for alkali compounds.

Solvent Partition Method

The crude extracts of herbal material are generally highly complex mixtures of various secondary metabolites with different chemical and physical characteristics. As mentioned before, for systematic study on chemical composition or bioassay-guided isolation, the first step of separation is usually intended to divide the compounds into several groups based on their similarity in chemical and physical properties. This can be carried out in several ways. One is by partition with a series of solvents. Herbal extracts are usually suspended in aqueous solution and partitioned with organic solvents in an order of increasing polarity, such as in *n*-hexane, chloroform, ethyl acetate, and *n*-butanol. As a result, compounds with different polarity can be partitioned into different fractions according to the principle of "like dissolves like." A typical liquid–liquid partition protocol is described in Figure 3.20.

As indicated in Figure 3.3, the herbal extract is dissolved in an appropriate volume of methanol aqueous solution, which is first defatted by extraction with the same volume of *n*-hexane. The methanol aqueous solution is concentrated to remove methanol, then suspended in water. The concentration of suspension or solution is usually kept at about 0.1 g of extract per milliliter (or 1 g/10 mL).

The suspension or solution is partitioned with the same volume of $CHCl_3$, EtOAC, and *n*-BuOH successively to four fractions. Because of the slight solubility of water in butanol, the *n*-BuOH is better saturated with water prior to partition to prevent dramatically decreasing the volume of the water layer. During each partition

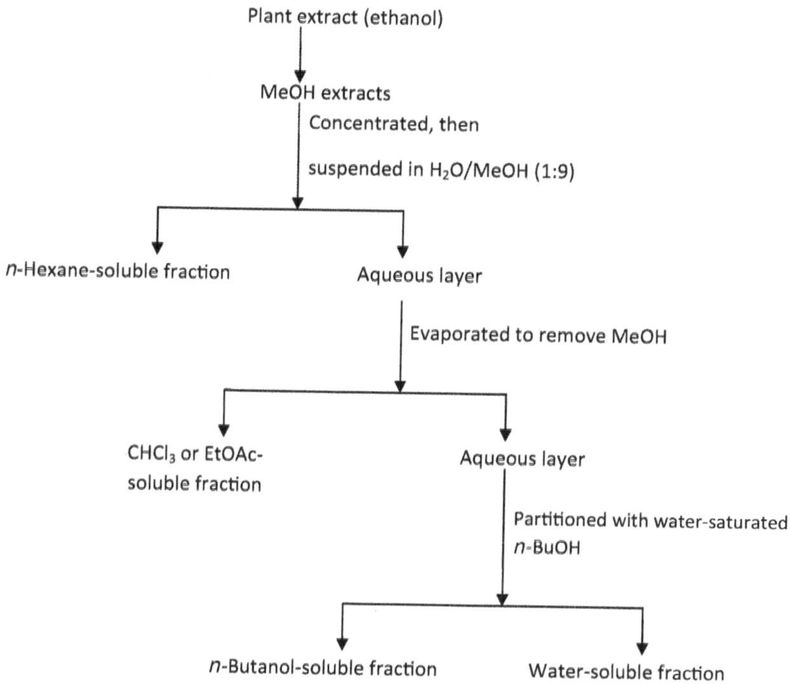

Figure 3.20 A typical partitioning scheme using immiscible solvents.

after defatting, the aqueous layer should be evaporated first to remove the remaining organic solvent. This procedure ensures repeatability of results and decreases the overlapping of compounds in the successive fractions. Occasionally, miscible solvents are also used for partition on addition of water.[15] In combination with our experience in isolation of bioactive compounds in marine organisms, the general protocol with miscibility is outlined in Figure 3.21.

The solvent partition method has been successfully applied in the area of natural product separation, and has a long history. A number of biologically active secondary metabolites have been isolated from plants by this method, such as alkaloids from *Chincohna succirabra* and anthraquinones from *Rheum palmatum* L.

A phenomenon commonly seen in the partition method is the formation of an emulsion layer between the two solvent layers. This is related to the solvent used and the components of the extract. Heating with a hot tower or tapping on the outside of the glassware may help to eliminate the emulsion layer.

Droplet Countercurrent (DCC) and High-Speed Countercurrent (HSCC)

Application of the countercurrent separation method began in the 1950s. The early instrument involves a sequential separating cell in which the stationery phase with

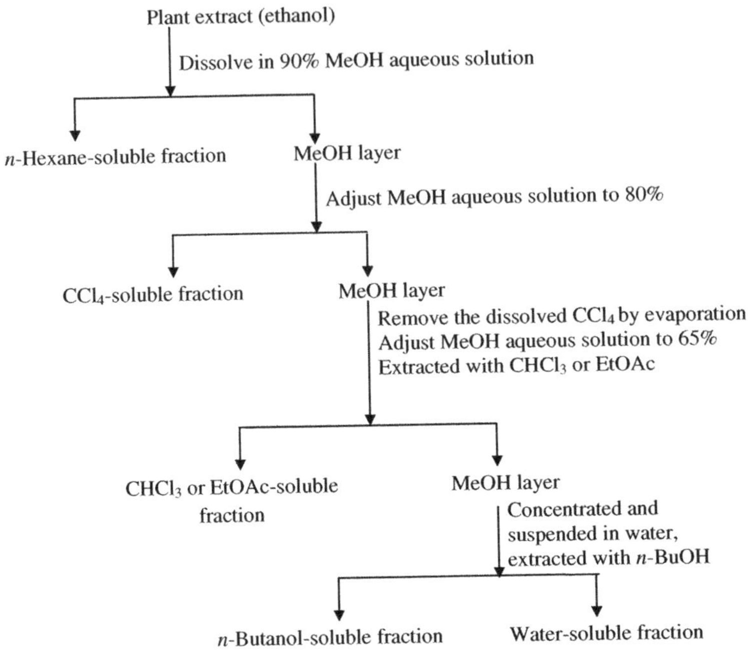

Figure 3.21 A typical partitioning protocol using miscible solvents.

relatively lower density could be equilibrated with the mobile phase with relatively high density in successive cells, and therefore carrying the solute along on the basis of its partition coefficient. DCC and HSCC technologies are currently available methods developed from countercurrent method.[16]

The DCC instrument consists of a number of vertical tubes, in which the droplets of one phase rise or fall through the other phase. The DCC method has higher separation efficiency and higher recovery of load sample, but is time-consuming. Modern HSCC chromatography has become commercially available since 1980, sharing all the advantages and overcoming all the drawbacks of early countercurrent instruments. There are two types of HSCC instruments: the centrifugal partition chromatography (CPC) and the coil planet centrifugal partition chromatography. Since HSCC is a liquid–liquid partition separation method, the solvents system must be carefully selected and optimized to get the highest separating efficacy for isolation of natural products from herbal extracts. Partition coefficient (K) is the most important parameter in solvent system selection, which should be in the range of 0.5–2.5 to get an efficient separation and a suitable run time. Two phases should be separated from each other in 5 s.

When isolating acidic, alkali, and amphoteric compounds, small amounts of acid or base should be added into the solvent system to keep the target compounds at the same state (free or ionizable) during separation. Taking the chemical and physical property of the target compounds into account, selection of a proper solvent

Table 3.2 Commonly Used Solvent Systems for the Application of HSCC

Solvent system (v/v)
n-Butanol-water (1:1)
n-Butanol-acetic acid-water (4:1:5)
n-Butanol-methanol-water (4:1:5)
n-Butanol-ethanol-water (4:1:5)
Chloroform-methanol-water (5:6:4)
Chloroform-methanol-water (4:3:3)
Chloroform-methanol-acetic acid-water (5:6:1:4)
Chloroform-methanol-acetic acid-water (2:1:1:2)
Ethyl acetate-*n*-butanol-acetonitrile-water (5:1:7:10)
Ethyl acetate-ethanol-water (3:1:2)
Ethyl acetate-methanol-water (0.1% HAC) (3:1:2)
Ethyl acetate-*n*-butanol-acetonitrile-water (5:2:5:10)
Ethyl acetate-*n*-butanol-acetonitrile-0.1% HCL (5:2:5:10)
n-Hexane-ethyl acetate-methanol-water (1.5:5:1.5:5)
n-Hexane-ethyl acetate-methanol-water (70:30:15:6)
n-Hexane-ethyl acetate-methanol-water (12:3:3:2)
n-Hexane-ethyl acetate-methanol-water (1:1:1:1)
n-Hexane-ethyl acetate-methanol-water (1:6:1.5:7.5)
n-Hexane-ethyl acetate-methanol-acetic acid-water (1:6:1.5:1:4)
n-Hexane-ethyl acetate-methanol-acetic acid-water (1.6:1.5:2:3)
n-Hexane-chloroform-methanol-water (1:1:1:1)
n-Hexane-DCM-methanol-water (5:1:1:1)

system is the most important step for HSCC. The commonly used three- or four-component solvent systems are presented in Table 3.2.

The analytes separated by HSCC are usually detected by a UV detector, an evaporative light scattering detector (ELSD), or TLC analysis to guide the purification of interested compounds and the collection of fractions. This method has been recently applied to isolation and purification of bioactive compounds from plants, microbial origins, and marine organisms, providing a number of useful references.[17,18]

CPC was first used by Sanki Eng in 1982. A CPC column is composed of a series of channels linked in cascade by ducts and aligned in cartridges or disks in a circle around a rotor; setting the rotor in motion submits this assembly to a constant centrifugal field that keeps the stationary phase in place. As a liquid–liquid separation instrument, CPC has advantages for separation, isolation, and purification of lipid samples such as fatty acids and phospholipids in comparison with liquid–solid separation methods, such as regular column chromatography and HPLC. Advantages of CPC include its mild operation condition that avoids decomposition of unstable compounds, low-solvent consumption, and its suitability for large-scale continuous separations. The solvent can be recovered and recycled.

CPC utilizes any biphasic liquid–liquid system as mobile and stationary phases. The correct selection of solvent system is the most important step for a successful CPC separation. The physical and chemical properties of sample components, such as their polarity, solubility, charge state, and partition coefficient in different solvents, are considered important factors. Herein, two basic solvent systems of chloroform/methanol/water (polar) and *n*-hexane/ethyl acetate/methanol (less polar) are mentioned as examples. In practice, the solvent systems used in other countercurrent chromatographic separations can be referred to.

Solvent phase diagrams have been used for the selection of solvent system.[19] The CPC method has been successfully applied in separation of eicosapentaenoic acid (EPA) and docosahexaenoic acid (DHA) from micro-algal oil and purification of phosphatidylcholine to homogeneity from a crude phospholipid extract of squid.[20,21]

3.3.3 Dialysis for Macromolecular Compounds

Dialysis is an important method for isolation of macromolecular compounds, for example, protein, glycoprotein, and polysaccharide. In practice, protein is usually first extracted with a buffer solution and centrifuged to obtain supernatant fluid. The protein is then precipitated by adding ammonium sulfate in the supernatant fluid and obtained by filtration. For dialysis, the protein is made into a highly concentrated solution first and put into a dialysis tubing that allows solvent and small molecular compounds to pass through while protein is retained. The dialysis tubing is placed into a container containing enough volume of buffer or distilled water, and stirred for hours to let the solutions in two sides of the bag reach equilibrium. Change the buffer in the container and repeat the dialysis several times, until all of the small molecules in the tubing are completely dialyzed out.[22] The purified proteins are then freeze-dried for study.

Polysaccharides have been proven as active components in many traditional Chinese herbs, for example, isatis root, ganoderma, and ginseng. To isolate polysaccharide, the materials are usually first extracted with hot water, and then centrifuged to get supernatant fluid. The supernatant fluid is precipitated by adding volumes of isopropanol or ethanol (usually 1:3 in volume). After centrifugation, the collected precipitate is dissolved in an appropriate volume of water and extensively dialyzed against water to get rid of small molecular compounds. The nondialyzable solution, after filtration, is freeze-dried to give a crude polysaccharide.[23]

Further purification of proteins and polysaccharides is usually performed by ion-exchange chromatography and gel-permeation chromatography.

3.3.4 Isolation with Column Chromatography

In general, crude extracts are initially separated into several fractions by the isolation methods described in Section 3, such as liquid–liquid extraction. The chemical isolation of herbal medicines can be divided into two classes according to the study purpose: systematic chemical-guided isolation and bioactivity-guided isolation. No matter which case, the goal is to get pure compounds and clarify their structures. In most situations, isolation and purification of compounds from a complex herbal extract depend on successful application of several isolation methods. TLC and column chromatography are usually necessary. Herein, the commonly used TLC and column chromatography methods are discussed in detail.

TLC

TLC is an easy, economic, and quick method that can be used for qualitative and quantitative analysis as well as purification of natural products. Practically, it is useful in identifying a known compound and detecting the presence of various types of secondary metabolites in herbs, for example, phenols, steroids, alkaloids, flavonoids, and coumarins. As an auxiliary tool, TLC is used to help optimize conditions of column chromatography, including selecting the mobile and solid phases, identifying compounds isolated from column, as well as confirming purity of the isolates.

Quantitative analysis by TLC was popular several decades ago, but it has been replaced by the more convenient and accurate HPLC analysis. To some scientists, preparative TLC (PTLC) is still the favorite method for small quantitative isolation.

When applying TLC for analysis or isolation, the solution of herbal extract in a suitable solvent (methanol or ethanol) is spotted as a single round spot or thin line on a TLC plate, near the edge of one side of the plate. The TLC plate is then obliquely placed into a well-closed tank containing sufficient volume of solvent at the bottom, letting the spotted edge of the plate merge with the solvent, but maintaining the sample spot or line above the surface of the solvent. The developing solvent will slowly migrate up the plate due to adsorption of the sorbent (solid phase) on the plate (see Fig. 3.22A).

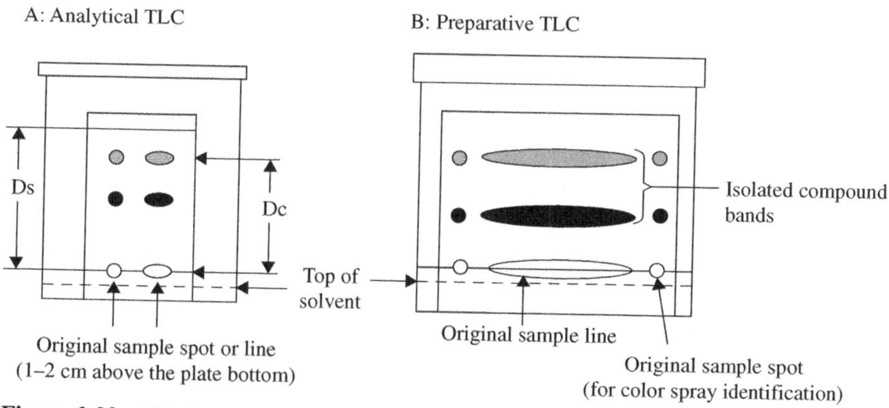

Figure 3.22 Thin layer chromatography. A, Analytical TLC; B, preparative TLC. Ds, Distance from original sample spot to developed solvent front; Dc, distance from origin sample spot to the developed compound spot.

Table 3.3 Commonly used TLC-Developing Solvent Systems

Solvent system	Sorbent	Application
Hexane:ethyl acetate (EtOAc)	Silica gel	Separation of low and medium polar compounds
Petroleum ether:chloroform	Silica gel	Separation of low polar compounds, for example, cinnamic acid derivatives, coumarins
Chloroform:acetone	Slica gel	Separation of medium polar compounds
Benzene:acetone	Silica gel	Separation of aromatic compounds
Chloroform:methanol	Silica gel	Separation of polar compounds, for example, glycosides and polyphenolic compounds
Butanol:acetic acid:water	Silica gel	Separation of glycosides and sugars
Methanol:acetonitrile:water	C_{18}/C_8	A universal system for reversed phased TLC. For acidic or basic compounds, the addition of small amounts of acid or base may improve chromatography
Methanol:ethanol:water	Polyamide	A universal system for polyamide
Methanol:ethanol:water	Cellulose	Separation of highly polar compounds, for example, sugars and glycosides

The commonly sorbents used as solid phase for TLC plates include silica, alumina, octadecasilica (ODS), cellulose, dextran gels, polyamide, or other ion exchange polymeric resin. Among them, alumina is often used for alkaloids isolation, cellulose for sugars, polyamide for flavonoids, while ODS and dextran gels are widely used for many types of compounds; silica gel is the most commonly used, although it has some drawbacks in analyzing the acidic or basic compounds and is unable to isolate highly polar compounds (sugar, glycosides, and quaternary amine aklaolids).

The interactions between acidic groups (e.g., –COOH, –OH) and silanols will result in a "tail" for acidic compounds on TLC of silica gel, which can be conquered by the addition of a small amount of volatile acid, for example, 1% trifluoroacetic acid (TFA) or acetic acid to the developing solvent. For alkali compounds, adding of a small amount of base, for example, 1% diethylamine or triethylamine, would improve the TLC behavior. Due to the strong adsorption of highly polar compounds to silica gel, their development needs a more polar solvent system, usually including a small amount of water or acid. The common solvent systems are listed in Table 3.3.

An important qualitative parameter in TLC of a particular sorbent and solvent system is the R_f value. It is defined as:

$$Rf = \frac{\text{Distance of compound from origin spot to the developed spot}}{\text{Distance of solvent from origin spot to the developed front}}$$

To identify a compound on TLC, the standard reference compound is usually developed on the same plate. The two developed spots must have the same R_f value on the same TLC plate that developed with at least two different solvent systems (see Fig. 3.5A). The isolated compound that showed a single spot in three developing solvent systems with different composition (R_f) is presumed to be pure. The solvent system that gives the best separation of compounds on TLC plate is generally suitable as a mobile phase for column chromatography using the same stationary phase.

For both analytical TLC and PTLC, effective detection is required. Most of compounds are colorless under regular sunlight. Thus, they must be detected under other conditions. Detection could be nonstructure-destructive (under UV light) or structure-destructive (by color reaction with a reagent). Most UV lamps have two wavelengths of UV light, 254 and 366 nm. So, compounds that absorb UV light at 254 or 366 nm will be visible on TLC plates under the UV lamp. Other compounds will require spray detection. The mechanism of spray detection is the color reaction between the compound of interest on the TLC and the spray reagent. Some color reactions on TLC plates may need heating. The spray reagents can be divided into two types, the universal one and the specific one. For known compounds, the specific spraying reagents for the compounds of interest should be used. In the case of crude extracts with unknown compounds, a universal spraying reagent is recommended. The most popular spray reagents are listed in Table 3.4.

TLC is very helpful for the determination of the stationary and mobile phases in column chromatography separation. The TLC behavior of a sample on plates

Table 3.4 Commonly Used TLC-Spraying Regents for Detection of Natural Compounds

Spray regent	Recipe	Treatment	Detection
Vanillin/sulfuric acid	Dissolve vanillin (4 g) in conc. H_2SO_4 (100 mL)	Heat at 100°C until colored spots appears	A universal spray Many terpenoids, sterols, and saponins give red and blue colors.
Ferric (III) chloride	Dissolve ferric chloride (1.5 g) in ethyl alcohol (10 mL)		Phenolic compounds, for example, coumarins, flavonoids, tannins. A blue or red coloration will be observed.
Dragendorff's reagent	5 mL A* + 5 mL B* + 20 mL acetic acid + 70 mL water.	Generally, color reaction occurs rapidly, but heat is required occasionally.	Alkaloids give a dark orange to red coloration
Ninhydrin	Dissolve ninhydrin (0.3 g) in butanol (100 mL), then add 0.3 mL acetic acid.	Heat at 100°C until coloration	Amino acids, amines, and peptides

* Solution A: 1.7 g basic bismuth nitrate in 100 mL of water/acetic acid (4:1); Solution B: 40 g potassium iodide in 100 mL of water.

coated with a certain stationary phase can reflect the separation results on column chromatography with the same stationary phase. The selection of the stationary phase is determined by comparing the separation effects on TLC plates. Once the stationary phase is chosen, the mobile phase should be optimized by developing the TLC plate with various solvent systems.

The solvent system that retains the compounds of interest at R_f value of 0.2–0.3 is suitable for purification on column chromatography. The reason not to select the solvent system retaining R_f around 0.5 is because gravity in column chromatography will accelerate the descent of the compounds in the column, thus washing out without being well separated from other compounds. In a step gradient elution of column chromatography, the solvent system that develops the first obvious spot visualized by UV or spraying reagents at R_f=0.2~0.3 is usually regarded as the initial mobile phase; the solvent system that develops the last obvious spot at R_f=0.2–0.3 is used at the final mobile phase. In the separation with silica gel or bonded silica gel, methanol is often adopted as the last mobile phase to elute out all of the compounds in the column.

PTLC was a popular isolation method before preparative HPLC and HSCC became available. PTLC technique is still applied by some researchers because of its low cost and its speed. In comparison, PTLC plates (0.5–4 mm) are much thicker than the analytical TLC plates (0.1–0.2 mm), and so are able to load more sample, up to 1 g. PTLC is generally applied for purification of compounds in a relatively simple fraction that have been separated by several chromatographic columns. The developing solvents are usually referred from analytical TLC condition. If possible, the target compounds on PTLC plates are detected by UV. When a spray reagent needs to be used, it is usually applied on the edge of the developed line while most of the line is covered by a glass. The separation process of PTLC is outlined in Figure 3.22B.

Analytical TLC and PTLC plates are commercially available. They can also be prepared in the lab. Details about how to prepare the TLC and PTLC plates at lab and how to use the PTLC in the isolation of natural products are available in the book of *Natural Product Isolation*.[24]

Adsorption Column Chromatography

The mechanism of adsorption chromatography lies on the physical and chemical interaction between the stationary phase and the solute molecules. The interaction may originate from hydrogen bonding, van der Waals forces, or dipole–dipole forces. The separation mainly depends on the difference between the adsorption strength of the material in stationary phase to solute molecules in mobile phase. The solvent of the mobile phase competes with the solute molecules for adsorption sites. The eluting power of the mobile phase is proportional to its competence for adsorption sites on the stationary phase. The solute molecules that possess strong interaction with the stationary phase will retain on the column. The solute molecules with weaker adsorption are easily eluted from the stationary phase by the mobile phase solvent.

Table 3.5 Commonly Used Silica-Based Stationary Phases of Chromatography

Adsorbent	Application
Silica gel ⩾Si–OH	Normal phase material; suitable for isolation of most nonpolar and less polar small molecule compounds, used with pH < 7.
Alkyl-bonded phase ⩾Si–C_nH_{2n+1}	Reversed phase material, including C_1, C_2, C_4, C_5, C_6, C_8, C_{10}, C_{18}; suitable for isolation of most compounds from the less polar to the polar.
Amino ⩾Si–$(CH_2)_3$–NH_2	Reversed phase or normal phase material; suitable for isolation of polar compounds such as sugars.
Phenyl ⩾Si–$(CH_2)_3$–Ph	Reversed phase material; suitable for isolation of aromatic compounds.
Cyanopropyl ⩾Si–$(CH_2)_3$–CN	Reversed phase or normal phase material; suitable for polar compounds.

Several techniques are often used in the application of column chromatography, which include vacuum liquid chromatography (VLC), flash chromatography (FC) (air or nitrogen pressure-driven), low-pressure (75–300 psi) liquid chromatography (LPLC), medium-pressure (75–300 psi) liquid chromatography (MPLC), and high-pressure (>300 psi) liquid chromatography (HPLC).

In terms of stationary phase, normal phase silica gel is the most commonly used adsorbent. Silica gel is regarded as a typical polar sorbent. Its surface is composed of free silanol groups that can form strong hydrogen bonds or strong dipole–dipole interaction with various compounds. The surface of silica gel is weakly acidic, so it is not preferable for separation of strongly alkali compounds because of strong acid–base interaction. Polar compounds can form strong hydrogen bonds with silica gel, thus needing a high polar solvent system for elution. Nonpolar solvents can be employed to elute nonpolar compounds. In the application of silica gel column chromatography, a binary solvent system consisting of a nonpolar and a polar solvent is often recommended as the mobile phase. The change of solvent polarity can be achieved by a step gradient change of polar solvent in mobile phase. When the hydroxyl group attached to the silica is replaced by the alkyl derivative group, the material can be used as normal or reversed phase adsorbent, depending on the property of the substituted alkyl group (see Table 3.5).

Alumina is a polymer of aluminum oxide (Al_2O_3). There are three types of alumina available for liquid column chromatography: the acidic alumina, which is suitable for the separation of acid compounds; the basic alumina, which is useful for the separation of basic compounds; and the neutral alumina, which is used for nonpolar compounds. However, when applying alumina for separation, one should keep in mind that alumina can catalyze the degradation reactions of many compounds.

Packing Up a Suitable Column for Highly Effective Isolation A proper column packed with enough sorbent and eluted by a mobile phase with suitable polarity are necessary to obtain an ideal separation of herbal extract. The amount of sorbent used for separation is usually 20–50 times of the weight of crude extract to be isolated. In the earlier time of column chromatography application, the ratio of sorbent weight to that of the isolated sample was usually 100:500; it turned out later that such high-weight ratio of sorbent to sample resulted in relatively longer time of isolation and lower recovery of low-concentrated bioactive compounds. In recent years, with the rapid development of new separation techniques, researchers prefer to decrease the amount of silica gel or alumina. In most cases, silica gel or alumina is first employed to cut the crude extract into 10–20 fractions that commonly further undergo partition chromatography or size exclusion chromatography (SEC) or HPLC separation. The bed height for separation is usually controlled at 20–30 cm for a proper glass column. The bottom of the column should have either a sintered glass frit or a plug of glass wool/cotton to support the stationary phase.

The column is usually packed in two ways: slurry packing and dry packing. In slurry packing, the sorbent of stationary phase is first suspended in the initial mobile phase, and then poured into the column that already contains a small amount of initial solvent. After the sorbent is all packed into the column, allow the eluent to constantly flow through the column bed for several hours at a certain rate until the stationary phase becomes solid-compacted, when the top line of the column bed no longer goes down. Be sure to always keep the eluent level above the top of the column bed. To ensure better isolation, try to avoid air bubbles within the column bed. Bubbles can easily form when the column bed becomes dry due to the running out of eluent. If this happens, try removing the bubbles by stirring with a stick. In dry packing, the dry stationary phase, usually silica gel or bonded silica gel, is directly poured into the column with essential tapping. The mobile phase is then allowed to flow through the stationary phase for equilibration. This method has been widely used in VLC, as well as low- and medium-pressure column chromatography.

Load and Elute Sample There are two ways of loading crude extract on the stationary phase. If the sample can dissolve in a minimum of the initial mobile phase, dissolve the sample using as little volume of solvent as possible. Transfer the sample solution evenly onto the top of the column bed directly to allow the sample to get into the bed. If the sample is unable to dissolve in the initial mobile phase, dissolve it in an appropriate solvent, and then mix the solution with two to three times of silica gel. Let the solvent in the mixture of silica gel and sample solution evaporate, then evenly scatter the mixture on the top of the bed in the column, where a small amount of the initial mobile phase remains. The flow of the eluent should be stopped when the mixture is loaded. Add some silica gel above the sample mixture to amortize the impulsive force of the poured solvent.

Care must be taken to make sure the column bed is not disturbed and is not allowed to turn dry during the whole separation process.

A step gradient of mobile phase flows through the column bed, resulting in the elution of compounds with different polarity. Usually, one bed volume of the mobile phase is collected as an isolation fraction. The mobile phase should be optimized

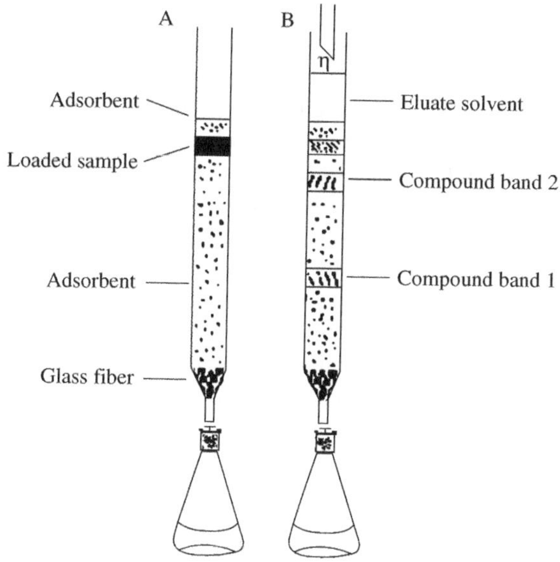

Figure 3.23 Column chromatography. (A) Before isolation; (B) after a period of isolation showing isolated compound bands.

based on TLC plate analysis as mentioned in the TLC section prior to separation. A typical step gradient elution might be outlined as follows: started with 100% chloroform, followed by 98%, 96%, 90%, 85%, 80%, 70%, 60%, and 50% chloroform in methanol. A typical column chromatography experiment is presented in Figure 3.23.

For isolation of particularly compounds, the solvent system needs to be specified.

Partition Column Chromatography

The partition chromatography separation is based on the solubility differences of the solutes between the stationary phase and the mobile phase. Silica gel can be modified by various chemical groups to form stationary phases with different physical and chemical properties, and therefore meet the demands for the separation of various natural products. The most popular bonded group is an organic function group (R), such as alkyl, alcoholic or phenolic hydroxyl, amine, phenyl, carbonyl, and nitrile.

According to the functionality of bonded groups, partition chromatography readily falls into two classes: the reversed phase class and the normal phase class.

Reversed phase partition chromatography is the most commonly used separation method, which has lipid stationary phases and aqueous mobile phases. The lipid stationary phases are usually chemically bonded to an inert support material such as porous silica gel to give chemically bonded stationary phases. The reversed phase stationary is formed by treatment of silica gel with chlorodimethyl alkylsilanes of different chain length, including C_2, C_4, C_6, C_8, C_{10}, C_{18}, and C_{20}. The mobile phase used for reversed partition chromatography consists of alcoholic or acetonitrile aqueous solution.

For open column chromatography and low- or medium-pressure column chromatography, the C_8- and C_{18}-bonded silica gel is the most popular stationary phase. In comparison with C_{18}-bonded silica gel, C_8-bonded silica gel is used to isolate less polar compounds. In separation of compounds from herbal extract, a low- or medium-pressure reversed partition column chromatography separation is commonly operated after initial silica gel open column chromatography.

The bonded normal phases include functional cyanopropyl, nitro, aminopropyl, and diol groups (see Table 3.5). The mobile phase used for normal partition chromatography is a mixture of organic solvents. In comparison with reversed phase chromatography, normal phase chromatography is applied to isolate lipid or nonpolar compounds.

Size Exclusion Column Chromatography

SEC is also known as gel-permeation or gel-filtration chromatography. In SEC separation, the natural products are isolated according to their hydrodynamic volume. The materials used as stationary phases are non-adsorbing porous beads with various pore sizes. Before separation, the stationary phase needs to be thoroughly immersed first in solvents to form a matrix. The secondary metabolite compounds in herbal extracts are with different dimensions in solution.

Macromolecular compounds are too big to get inside of the matrix, and thus will elute out of the column first. Intermediate size compounds can only enter part of the matrix, so will be eluted second. Small molecular compounds are eluted out last due to their ability to freely get into the matrix (Fig. 3.24). The commonly used stationary phases for SEC are dextran gel, sephadex G gels, and sephadex LH-20 gel.

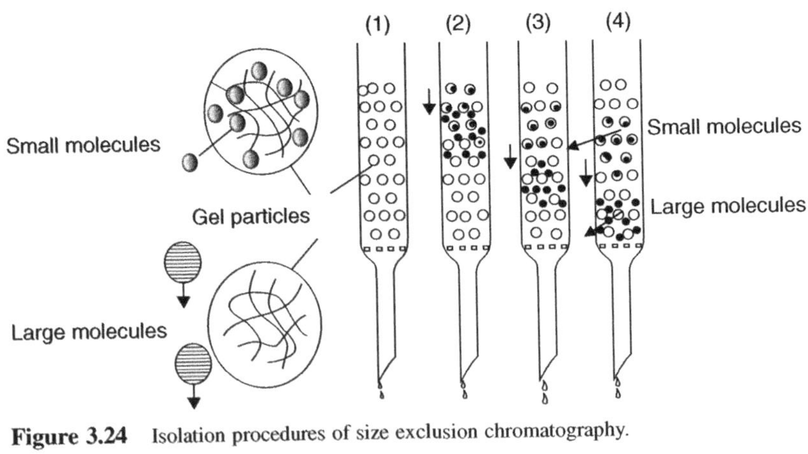

Figure 3.24 Isolation procedures of size exclusion chromatography.

Sephadex is formed by cross-linking water-soluble dextran with epichlorohydrin. Commercially available sephadex G-type gels usually include G-10, G-15, G-25, G-50, G-75, G-100, G-150, and G-200. They are able to separate compounds with mass ranging from 10 to 100,000 amu. Each sephadex G gel has its own fractionation range of molecular weight. The number attached represents 10 times the volume of solvent embedded in every gram of dry beads. For example, Sephadex G-50 and Sephadex G-200 indicate that each gram of dry beads can absorb up to 5 or 20 mL water, respectively. For purification of secondary metabolites from herbal extracts, G-10 and G-15 are the two most popular stationary phases. Sephadex G gels are hydrophilic. They swell in water and require aqueous solution as mobile phase.

Most herbal extracts are mainly composed of lipophilic, especially nonpolar or intermediate polar compounds, so Sephadex G gels are not suitable for their separation. To conquer the drawback of Sephadex G gels, an alternative gel Sephadex LH-20 was made by adding hydroxypropyl groups onto the Sephadex G-25. As a result, Sephadex LH-20 possesses both hydrophilic and lipophilic properties. The lipophilicity characteristic allows the gel to be used in organic solvents to isolate or purify organic-soluble natural products.

When one-component mobile phase is employed, the separation mainly works by gel filtration. When two-component mobile phase is used (commonly a mixture solution of nonpolar and polar solvents), the separation is based on both partition and gel filtration. The partition process occurs because gel tends to take up more polar solvent, and results in the composition difference between the stationary phase and the mobile phase. In addition to mechanisms of filtration and partition, Sephadex gel separation is also involved in adsorption, the formation of hydrogen bonds, which occurs in the separation of phenolic, heteroaromatic, and cyclic compounds, such as tannins.

Sephadex LH-20 is particularly ideal for the removal of chlorophyll from the crude herbal extracts and purification of flavonoids, tannins, and other phenols. The mobile phase for LH-20 separation is usually composed of one solvent with good solubility and another with poor solubility for sample.

Sephadex gels are chemically inert and stable in all solvents except for strong acid and oxidizing agents. In storage, dextran gel should be kept away from microbial contamination.

More information about application of Sephadex LH-20 is available from the book *Preparative Gel Chromatography on Sephadex LH-20*, written by Hans Henke.[25]

Pack Up a Suitable Column for Highly Effective Isolation For SEC column chromatography, a thin (diameter 10–20 mm) and long (length 40–100 mm) glass column is preferred. The size of the column depends on the required bed volume. The gel (Sephadex LH-20) should be expanded with proper solvent thoroughly before use. As described in the silica gel column preparation, suspend the gel with the initial eluent and pour into the column. Let the mobile phase flow through the column bed at the rate of 2 mL/min for several hours, while tapping the column occasionally to remove air bubbles for better isolating efficacy.

Load and Elute Sample The sample dissolves better in a mobile phase with a volume less than 5% of the column bed volume. The solvent in the column is first adjusted to a level just above the column bed. Add the sample solution carefully onto the top of stationary phase, and then allow the solvent to flow into the stationary phase. After that, add more mobile phase into the column. A small piece of cotton or glass fiber can be inserted on the top of the gel to amortize the impulsive force when the mobile phase is added. Adjust the flow rate at 2–4 mL/min. Collect each 5 mL or more as a fraction, depending on the size of the column.

Ion-Exchange Column Chromatography

The ion-exchanging chromatography has been widely applied in the isolation of charged or ionized natural products in the past decades, and still remains an important isolation technique. Its separation mechanism is based on the reversible binding of charged molecules to an oppositely charged insoluble matrix. During separation, the ion-exchanging resin matrix is used as stationary phase, the acidic or basic water solution and high-salt concentration water solution are employed as mobile phase. The ion-exchanging resin is classified as cationic or anionic resins. The separation process can be described as the following balance:

For strong acid cationic resin,

$$\text{Resin matrix-SO}_3\text{H} + \text{M}^+ \rightleftharpoons \text{Resin matrix-SO}_3^-\text{M}^+ + \text{H}^+.$$

For strong base anionic resin,

$$\text{Resin matrix-N}^+(\text{CH}_3)\text{Cl}^- + \text{M}^- \rightleftharpoons \text{Resin matrix-N}^+(\text{CH}_3)\text{M}^- + \text{Cl}^-.$$

Ion-exchange resins are produced by covalently bonding the negative- or positive-charged groups to the insoluble matrix material. The supporting matrix is usually constructed by polystyrene polymer and carbohydrate polymers. Polystyrene resins have good resin capacity due to the high substitution degree of the charged functional group. They are thermally and chemically stable, able to be used over wide pH range (1–14). Carbohydrate polymer-based ion-exchanging resins are hydrophilic, and suited to the isolation and purification of biological macromolecules.

The alkaloid compounds in natural products can become positively charged by treatment with acid. They may be separated by strong or weak cationic exchanging resins. The typical strong cationic resin comes from sulfonic acid group, which is negatively charged over the full pH range. The resins bonding the carboxylic acid are defined as weak cationic resins, being dissociated at above pH 6.

The carboxylic acid derivatives in natural products can be purified by the application of either strong or weak anionic exchanging resins. These compounds will be ionized at pH over 6. The induction of quaternary ammonium groups into the resin forms the strong anionic resin. Weak anionic resins contain protonated primary, secondary, or tertiary amines.

The selection of resin largely depends on experience and the availability of the resin at hand.

Pretreatment of the Resin for Separation The commercially available resins need pretreatment before use. The resin polymer should be washed thoroughly to get rid of impurities or "fines" produced from the manufacture and storage with solvent; the procedure is usually available from the manufacture. The next step is to let the volume of resin expand with the initial solvent system that will be used for separation. This procedure is to let the mobile phase solvent penetrate into the resin to ensure the exposure of functional groups inside the resin matrix to the mobile phase. The commercially available resin is provided in a counter-ion form, which needs a cycle conversion. For cationic exchangers, the conversion is usually from hydrogen to sodium; for anionic exchangers, it is usually from chloride to acetate group. The details of the pretreatment are usually available from the providers.

Select and Pack Up a Proper Column for Separation According to our experience, the short column with a wider diameter is suited for operation. The height of column bed is preferred at 10–30 cm. The bed volume should be determined as three to five times of the required exchange capacity. The thoroughly swollen resin is slurry packed into a column and allowed to form a uniform bed.

Load and Elute the Sample When applying the ion-exchanging technique for separation, pH stability of the targeted compounds must be taken into consideration in the first place. Samples containing acidic or basic compounds are usually dissolved in water in ion forms that can be obtained by adjusting the pH value of the aqueous solution. The sample solution should be clarified by centrifugation or filtration. There are two ways of loading samples. The resin and sample solution can be premixed together for adequate time or applied onto the top of already packed column. The flow rate is usually kept at 5–10 mL/min. The compounds will be eluted by adjustment of ion strength or changes in pH. In some cases, the water-miscible co-solvent such as methanol or acetonitrile needs to be added in the mobile phase to break the strong hydrophobic bonding of neutral compounds on resins. The separation process will be enhanced by the application of gradient elution.

The ion-exchanging chromatography technique provides an alternative method for isolation of strong and weak acids and bases, as well as amphoteric compounds. It plays very important roles in discovering the bioactive products from natural resources, such as plants, soil microorganisms, and marine organisms.[26,27]

3.3.5 Isolation with High Performance Liquid Chromatography (HPLC)

HPLC is a popular method for chemical analysis and separation that is applied in many fields, such as the food and pharmaceutical industries, agriculture, herbs, cosmetics, and the environment. Its applications include separation, purification, identification, and quantification of various compounds. Chemical separations can be accomplished using HPLC by utilizing the fact that different compounds have different migration rates given a particular column and mobile phase.

Preparative HPLC refers to the process of isolation and purification of small quantitative compounds using HPLC. Preparative HPLC has been applied to isolation of most classes of compounds in herbal medicine, especially minor bioactive compounds. In recent years, HPLC has been available in most research labs and is playing more and more important roles in purification of biologically active compounds. When compounds with close similar structures in a mixture need to be isolated, HPLC is usually the only and best method due to its rapid separation, high resolution, online detection, and versatile available columns.

Crude herbal extracts usually consist of hundreds of various compounds; thus, isolation of target compounds is time- and energy-consuming. Considering the efficiency and economy, although HPLC method is proven to be a fast and efficient means for separation of natural products, it is usually not recommended for isolation of crude extracts or complex mixtures directly. The crude herbal extracts should be initially separated using the solvent partition, open column chromatography, low- and medium-column chromatography, or other techniques introduced in early sections. Preparative HPLC methods are particularly suitable for separation and purification of compounds in fractions or compound mixtures obtained from these separation procedures.

This section mainly introduces the practical application of preparative HPLC technique on separation of natural products, particularly stationary phases of HPLC columns, mobile phases, and detection techniques, as well as method development.

Stationary Phases

The stationary phase in HPLC refers to the solid support contained within the column over which the mobile phase continuously flows. According to the chemical and physical properties of compounds of interest, there are mainly four types of HPLC chromatography: normal phase columns,[28,29] reversed phase columns,[30,31] gel-permeation columns,[32,33] and ion-exchanging columns.[34,35] The typical stationary phases available for HPLC are illustrated in Table 3.6.

1. *Normal phase HPLC chromatography:* uses normal phase columns to isolate compounds on the basis of hydrophilicity and lipophilicity of the targets by using a polar stationary phase (silica or diol) and a less polar mobile phase (nonaqueous). Thus hydrophobic compounds are eluted out more quickly than hydrophilic compounds. Normal phase HPLC is most suitable for the separation of lipophilic compounds.
2. *Reversed phase HPLC:* uses reversed phase columns to isolate compounds also on the basis of hydrophilicity and lipophilicity. The stationary phase consists of silica covalently bonded with *n*-alkyl chains. The mobile phase used in reversed phase HPLC generally comprises a mixture of water and miscible organic solvents. Thus hydrophilic compounds are eluted out more quickly than hydrophobic compounds. Reversed phase HPLC provides good separation for isolation of medium and high polar compounds.

Table 3.6 Typical Stationary Phases for HPLC

Stationary phase	Mobile phase	Separation
C_{18}, C_{10}, C_8 (reversed phase)	Methanol/water Acetonitrile/water	Medium and high polar compounds
CN (normal & reversed phase)	Heptane/ethanol Acetonitrile/water	Polar compounds
NH_2 (normal phase)	Acetonitrile/water	Carbohydrates
Silica (normal phase)	Heptane/ethanol Heptane/chloroform Heptane/ethyl acetate	Less polar compounds
Benzene sulfonic acid (strong cation exchanging)	Phosphate buffer	Alkaloids
Quanternary ammonium (strong anion exchanging)	Phosphate buffer	Carboxylic derivatives
Macroporous polystyrene/divinyl benzene copolymers (reversed phases)	Water buffers	Proteins and oligosaccharides

3. *Ion-exchange chromatography:* uses ion-exchange columns to isolate compounds based on the selective exchange of ions in sample with counter-ions in the stationary phase. Ion exchange columns contain charge-bearing functional groups attached to a polymer matrix. Compounds in samples are first retained in the column by replacing the counter-ions of the stationary phase with its own ions and then eluted from the column by another mobile phase (e.g., with different pH) that is able to replace the ions of compounds from the stationary phase. Ion-Exchange is useful for acidic or basic compounds preparative separation.
4. *Gel-permeation HPLC:* uses gel-permeation columns to isolate compounds on the basis of the molecular size of compounds being separated. The stationary phase consists of porous beads. The larger compounds will be excluded from the interior of the bead and thus will be eluted out of the column first. The smaller compounds will be allowed to enter the beads and thus will be eluted out later according to their ability to exit from the same sized pores they were internalized through. The column can be either silica or non-silica-based. Gel-permeation HPLC is the effective method for isolation of tannins, peptides, and polysaccharides from herbal materials.

Mobile Phase

The mobile phase in HPLC refers to the solvent that runs through the column (the stationary phase) to wash compounds out of the column. The mobile phase acts as a transporter for a sample. After a sample dissolved in a proper solvent is injected into the HPLC column through the injector, the sample solution is pumped through

Table 3.7 Properties of Solvents Used in Normal and Reversed Phase HPLC

Solvents	Polarity	Viscosity	b.p. (°C)	UV cutoff (nm)
Methanol	6.6	0.6	65	205
Ethanol	4.3	1.2	79	205
Acetonitrile	6.2	0.37	82	190
Acetone	5.4	0.32	57	330
Chloroform	4.4	0.57	61	200
Hexane	0.06	0.33	69	200
Ethyl acetate	4.3	0.45	77	260
i-Propanol	4.3	2.37	82	205
Tetrahydrofuran	4.2	0.55	66	215
Water	10.2	1	100	—

the column, carried by the mobile phase, and the compounds migrate according to their different non-covalent interactions with the stationary phase. The interactions of compounds between the mobile phase and stationary phase determine the speed of compound migration and separation of the sample. For example, the compounds having stronger interactions with the mobile phase than those with the stationary phase will be eluted from the column faster and thus have a shorter retention time, while the reverse is also true.

The mobile phase is usually composed of two solvents. To reversed phase chromatography, water and an organic solvent such as methanol or acetonitrile are usually combined. The elution ability can be adjusted by changing the ratio of the two solvents in the system to manipulate the interactions between the isolated compounds and the stationary phase. Isocratic and gradient mobile phases are commonly applied in HPLC performance. In isocratic elution, compounds are eluted using constant mobile phase composition. In gradient elution, compounds are eluted by increasing the ratio of the organic solvent. The strength of the mobile phase is increased in a stepwise or linear manner, and subsequently results in elution of retained components. As an example, to optimize the isolation condition of a compounds mixture using acetonitrile:water system on a reversed phase column, the ratio of the gradient elution can be initially set up as 10:90 (v/v) for 5 min, then changed from 10:90 to 95:5 within 30 min. The gradient can be adjusted according to the chromatogram to ensure the best separation within a shorter time.

The solvents used in HPLC system must be of high purity, and compatible with the stationary phase and the detector. The solvents should be degassed before use. The properties of solvents commonly used in normal and reversed phase HPLC are listed in Table 3.7.

A small amount of strong acid or base is usually applied as a modifier solvent for the mobile phase to maintain the free state of the isolated compounds or convert them into ionic states in reversed phase HPLC. The acid and base should be easily removed after separation. For example, when the compound is a carboxylic acid derivative, the addition of 0.1% of TFA will suppress the disassociation of acid, and achieve good resolution. Lipophilic alkaloids, however, cannot be separated on reversed phase HPLC column due to their strong interaction with the stationary phase. But addition of a strong acid in the mobile phase will convert them to ionic state and enable them to be eluted on reversed phase HPLC. Peak tailing caused by free silanol groups can lead to a poor chromatogram; addition of triethylamine (0.05–0.1%) can usually overcome the tailing phenomenon.

Different columns have different pH application limitations, so check the instructions of column application or call the manufacture when you are planning to add strong acidic or basic reagent to the mobile phase, especially when you plan to use the reversed phase column for ion-exchange based separation.

Detectors

The detector for HPLC is the component that emits a response to the eluting sample compound and subsequently signals a peak on the chromatogram. It is positioned immediately posterior to the column in order to detect the compounds right after being eluted out from the column. For some detectors, such as MS, the sensitivity can usually be adjusted by tuning the detection parameters using the coarse and fine-tuning controls.

There are many types of detectors that can be connected with HPLC: UV, diode array detector (DAD), RI, fluorescent, radiochemical, electrochemical, near-IR, MS, NMR, and ELSD. They will be introduced in Chapter 4 or 9. For preparative HPLC, the nondestructive detectors such as UV, DAD, and RI are preferred in practice.

UV detectors measure the light-absorbing ability of the isolated compound. The primary UV detectors usually have only one fixed wavelength for measurement, commonly at 254 nm. Since different compounds absorb light at variable wavelengths, some compounds may not be detected at one fixed wavelength. There are two types of multi-wavelength detectors available, the dispersion detector and the DAD. Today, very few dispersion instruments are sold in the market but many are still used in the field. DAD is the more popular one currently, because it measures a spectrum of wavelengths simultaneously. UV detectors are the most popular detectors, with a high sensitivity to approximately 10^{-8} or 10^{-9} gm/mL. Compounds detected by UV must contain a chromophore, which limits the application of UV detectors.

The RI detector measures the ability of sample molecules to bend or refract light. All compounds in solution have the ability to refract light. This property for each molecule or compound is called its RI. Thus, the RI detector is referred to as a universal detector, compared with UV detectors. For the RI detector, light proceeds through a bi-modular flow cell to a photodetector. One channel of the flow cell directs the mobile phase passing through the column, while the other directs only the mobile phase. Detection occurs when the light is bent when compounds are eluting from the column, and this is read as a disparity between the two channels. However, RI detectors have limited sensitivity and are not suitable for chromatography running with a gradient elution.

Method Developments

Method development for preparative HPLC will be carried out more easily if literature references for the same or similar compounds are available. For unknown compounds, the preliminary physical and chemical properties of isolates are usually known by chemical investigation before the HPLC separation. Knowing the nature of the targeted compounds helps to select a suitable HPLC column and an appropriate detector. An exploratory separation on high-resolution TLC or analytical HPLC column is necessary to decide the mobile phase composition. As the reversed phase HPLC using C_8 or C_{18} is the mainstream for the isolation of secondary metabolites from plant extracts, here is a brief introduction to their application.

For analysis by reversed phase HPLC, dissolve the sample with the solution of mobile phase if possible; if not, use an organic solvent with a volume as small as possible. A preparatory column usually has a large column diameter and loading capacity that is designed to facilitate large volume injections into the HPLC system. To ensure reproducibility, an isocratic mobile phase is usually preferred in preparative HPLC separation.

Before isolation using a preparative HPLC column, the mixture should be first analyzed by high-resolution TLC, then an analytical HPLC column, to assess the complexity of components, identify the compounds of interest, and help to select the separation solvent system.

The solvent system (methanol/water or acetonitrile/water) that develops the last visualized spot at R_f 0.2 on a high-resolution TLC can be initially tried as the mobile phase for the analytical HPLC column with the same stationary phase, then further optimized by stepwise changing (5–10%) of the organic solvent to achieve the best separation results. A gradient solvent system can also be used for the analytical HPLC column as an initial step to optimize the separation isocratic condition. The typical gradient manner is to increase the organic solvent ratio from 5% to 100% within 30 min. The isocratic solvent system can be optimized by adjusting the organic solvent proportion. When the isocratic condition obtained from the preliminary analysis by analytical column is adapted on the preparative HPLC column, the condition may still need a slight adjustment. It should be kept in mind that the stationary phase used in the preparative HPLC column must be the same brand with the same materials inside the analytical column.

If a two-component solvent system fails to give good separation results, the adjustment of pH value or the use of a three-component solvent system may be considered. Once a solvent system has been determined, the amount of sample loaded onto the column can be optimized by increasing the injection volume until it reaches the column loading and separation limits.

3.4 AN EXAMPLE OF EXTRACTION AND COMPOUND ISOLATION FROM HERBAL MEDICINE

Isolation of bioactive compounds from the traditional Chinese herb "Shi Hu"[36] is given here as an example.

"Shi Hu," derived from the dried or fresh stems of several Dendrobium species (Orchidaceae), is widely used for treatment of various diseases, such as chronic atrophic gastritis, diabetes, skin aging, and cardiovascular disease. *Dendrobium nobile* Lindl. is one of the most popular Dendrobium species recorded in the *Chinese Pharmacopoeia* (2005 edition) as one of the original materials of "Shi Hu." The ethyl acetate-soluble fraction of a 60% ethanol extract of the stems of *D. nobile* was found to exhibit significant antioxidant activity in the 1,1-diphenyl-2- picrylhydrazyl (DPPH) free radical scavenging assay (IC_{50}=32 μg/mL). By bioactivity-guided separation, 53 compounds were obtained from the 60% ethanol extract of the stems of this plant. Their structures were elucidated on the basis of physical and chemical properties and spectroscopic analysis.

The 53 compounds from *D. nobile* are nobilin D (**1**), nobilin E (**2**), nobilone (**3**), crepidatin (**4**), chrysotobibenzyl (**5**), dendrobin A (**6**), chrysotoxine (**7**), moscatilin (**8**), gigantol (**9**), dendroflorin (**10**), dendronobilin A (**11**), dendronobilin B (**12**), dendronobilin C(**13**) dendronobilin D (**14**), dendronobilin E (**15**), dendronobilin F (**16**), dendronobilin G (**17**), dendronobilin H (**18**), dendronobilin I (**19**), 6α,10,12- trihydroxypicrotoxane (**20**), dendrobiumane A (**21**), bullatantirol (**22**), dendrodensiflorol (**23**), dendrobane A (**24**), nobilin A (**25**), nobilin B (**26**), nobilin C (**27**), fimbriatone (**28**), confusarin (**29**), flavanthrinin (**30**), 2,5-dihydroxy-4,9-dimethoxyphenanthrene (**31**), 3,7-dihydroxy-2,4-dimethoxyphenanthrene (**32**), 10,12-dihydroxypicrotoxane (**33**), 10β,13,14-trihydroxyalloaromadendrane (**34**), syringaresinol (**35**), pinoresinol (**36**), medioresinol (**37**), lirioresinol-A (**38**), 2-methylanthraquinone (**39**), nobilate (**40**), syringic acid (**41**), 2-hydroxyphenylpropanol (**42**), vanillin (**43**), apocynin (**44**), coniferyl aldehyde (**45**), syringaldehyde (**46**), syringylethanone (**47**), *p*-hydroxybenzaldehyde (**48**), 3-hydroxy-4-methoxyphenylethanol (**49**), α-hydroxysyringylethanone (**50**), dihydroxyconiferyl alcohol (**51**), *p*-hydroxybenzoic acid (**52**), and *p*-hydroxyphenylpropionic acid (**53**), respectively.

Among these compounds, **1–3** and **11–19** were determined as new (see Fig. 3.25). The following are brief extraction and isolation procedures as well as antioxidation of the isolated compounds.

3.4.1 Extraction and Isolation of Compounds

The powdered, air-dried stems of *D. nobile* (5 kg) were refluxed with 60% ethanol three times. After evaporation of solvent *in vacuo*, the extract (210 g) was suspended in water, and partitioned with ethyl acetate and *n*-butanol successively. The ethyl acetate-soluble fraction (63 g) that exhibited significant antioxidant activity in DPPH assay, was first subjected to column chromatography on silica gel (200–300 mesh, 700 g) and eluted with choloroform/methanol (100:0→0:100) to afford 12 fractions. Fractions 5, 6, 7 were found to be active in DPPH assay (IC_{50}=33.6, 29.7, 31.8 μg/mL).

Fraction 5 (11 g) was further chromatographed on silica gel MPLC by gradient elution with cyclohexane/ethyl acetate (95:5→0:100) to give 13 subfractions.

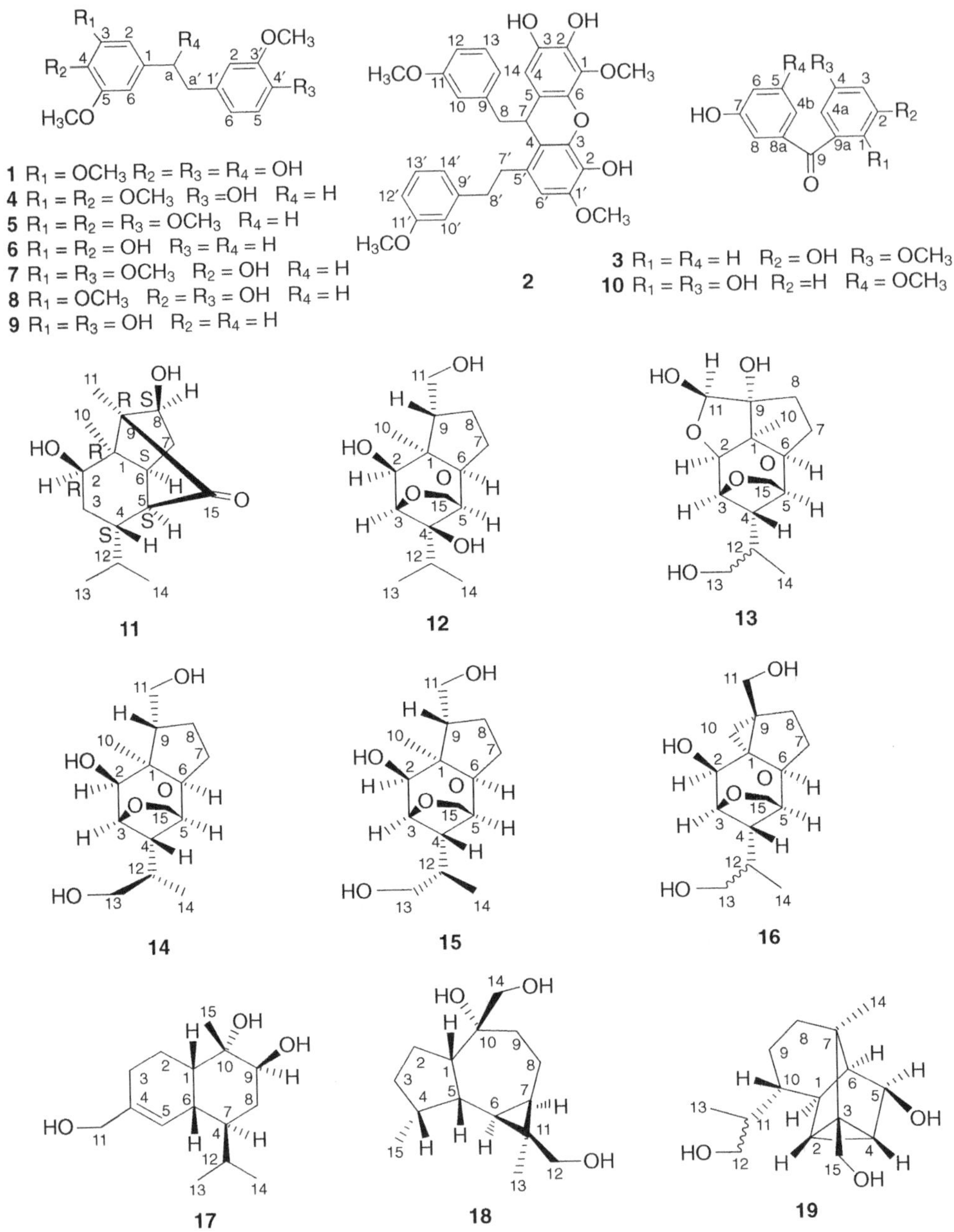

Figure 3.25 Compounds isolated from the stem of *Dendrobium nobile*.

Subfraction 7 (1.5 g) was separated by a Sephadex LH-20 column with choloroform/methanol (1:1) as eluent and then further isolated by an ODS column eluted with methanol/water (4:6→8:2). The eluent of 60% methanol was purified by preparative HPLC with 55% methanol and detected at UV 254 nm to yield compounds **4** (181.7 mg) and **5** (208.1 mg). Subfraction 8 (276 mg) was separated by a Sephadex LH-20 column (choloroform/methanol, 1:1) and then further isolated by an ODS column with methanol/water (4:6→8:2). Compound **6** (2.4 mg) was obtained from the eluent of 60% methanol by purification with preparative HPLC using 55% methanol and detected at UV 254 nm. Subfraction 10 (1.4 g) was separated by a Sephadex LH-20 column (choloroform/methanol, 1:1) and then further isolated by an ODS column with methanol/water (4:6→7:3). The eluent of 50% methanol was purified by preparative HPLC using 50% methanol and detected at UV 254 nm to yield compound **8** (455.4 mg). Compound **7** (259.4 mg) was recrystallized from the eluent of 60% methanol using methanol as a solvent.

Fraction 6 (6.9 g) was chromatographed on silica gel MPLC by gradient elution with cyclohexane/ethyl acetate (85:15→0:100) to give 10 subfractions. Subfraction 5 (727 mg) was separated by a Sephadex LH-20 column with choloroform/methanol (1:1) as eluent and ODS column eluted with methanol/water (4:6→8:2), then finally purified by preparative HPLC using 50% methanol and detected by UV at 254 nm to yield compound **9** (10.0 mg). Subfraction 6 (927 mg) was separated by a Sephadex LH-20 column with choloroform/methanol (1:1) and then further isolated by an ODS column with methanol/water (3:7→8:2). Compound **10** (28.2 mg) was recrystallized from the eluent of 50% MeOH using MeOH. The eluent of 60% MeOH was purified by PTLC developed by cyclohexane/ethyl acetate (6:4) to yield compound **3** (48.5 mg). Subfraction 9 (559 mg) was separated by over a Sephadex LH-20 column (choloroform/methanol, 1:1) and then applied to ODS column (methanol/water, 2:8→8:2). Compound **1** (2.6 mg) was obtained from the eluent of 30% methanol by purification with preparative HPLC using 30% methanol and detected at UV 254 nm.

Fraction 7 (6.9 g) was separated by a Sephadex LH-20 column (choloroform/methanol, 1:1) and then chromatographed on silica gel MPLC by gradient elution with cyclohexane/ethyl acetate (8:2→0:1) to give 9 subfractions. Subfraction 5 (1.0 g) was further isolated by an ODS column eluted with methanol/water (3:7→8:2). The eluent of 80% methanol was further purified by preparative HPLC using 65% methanol and detected at UV 254 nm to yield compound **2** (15.2 mg). Subfraction 4 (1.3 g) was further isolated by an ODS column and eluted with methanol/water (4:6→7:3). The fraction eluted with 40% MeOH was further purified by preparative HPLC using 35% methanol and detected by a RI detector to yield compound **11** (25.7 mg).

Fraction 9 (3.8 g) was separated by a Sephadex LH-20 column with choloroform/methanol (1:1) as eluent and then chromatographed on silica gel MPLC eluting with cyclohexane/acetone (8:2→0:1) to give 7 subfractions. Subfraction 5 (1.3 g) was further isolated by an ODS column using gradient elution with methanol/water (3:7→6:4). Compound **17** (11.4 mg) was obtained from the fraction obtained with

50% methanol by purification with preparative HPLC with 45% methanol and detected by a RI detector.

Fraction 10 (4.7 g) was separated by Sephadex LH-20 (choloroform/methanol, 1:1) and then chromatographed on silica gel MPLC by gradient elution with cyclohexane/acetone (8:2→0:1) to give 8 subfractions. Subfraction 3 (380 mg) was further isolated by an ODS column eluting with methanol/water (3:7→7:3). The fractions obtained with 40% and 50% methanol were purified by preparative HPLC using 48% and 55% methanol and detected by a RI detector to yield compounds **12** (4.3 mg) and **18** (31.4 mg). Subfraction 5 (263 mg) was further isolated by an ODS column with methanol/water (3:7→7:3) as eluent, then purified by preparative HPLC with 30% methanol and detected by RI detector to yield compounds **13** (5.8 mg) and **14** (11.2 mg). Subfraction 6 (948 mg) was further isolated by an ODS column eluting with methanol/water (3:7→7:3). Compounds **15**, **16**, and **19** (15.1, 148.6, and 76.0 mg) were obtained from the eluent of 30% methanol by purification with preparative HPLC using 30%, 30% and 37% methanol respectively and detected by a RI detector.

3.4.2 Antioxidation Bioactivity of Isolated Compounds

In the DPPH assay, compounds **1**, **2**, **4**, **7**, **8**, and **10** exhibited significant scavenging activity that is either higher than or equivalent to vitamin C. In the oxygen radical absorbance capacity (ORAC) assay, compounds **1**, **3**, **4**, and **7–10** displayed potent peroxyl radical scavenging activity that are higher than vitamin C, and compound **6** showed weak activity. In addition, compounds **1–3**, **5**, and **8–10** showed inhibitory effects on NO production in murine macrophages (RAW 264.7) activated by lipopolysaccharides (LPS) and interferon (IFN)—γ.

REFERENCES

1. Mann, J., et al. (1994) *Natural Products: Their Chemistry and Biological Significance*. Harlow, UK, Longman Scientific & Technical.
2. Rahman, A.U. (1995) *Studies in Natural Products Chemistry: Structure and Chemistry* (Vol. 15). Amsterdam, Elsevier Science Publishers.
3. Rahman, A.U. (1995) *Studies in Natural Products Chemistry: Structure and Chemistry* (Vol. 17). Amsterdam, Elsevier Science Publishers.
4. Rahman, A.U. (1997) *Studies in Natural Products Chemistry: Structure and Chemistry* (Vol. 19). Amsterdam, Elsevier Science Publishers.
5. Rahman, A.U. (1998) *Studies in Natural Products Chemistry: Structure and Chemistry* (Vol. 20). Amsterdam, Elsevier Science Publishers.
6. Fullas, F., et al. (1990) Sweet-tasting triterpene glycoside constituents of Abrus fruticulosus1. *Planta Medica* 56:332–333.
7. Sakurai, T., et al. (2002) Assamicin I and II, novel triterpenoid saponins with insulin-like activity from Aesculus assamica Griff. *Bioorganic & Medicinal Chemistry Letters* 12(5):807–810.
8. Namikoshi, M., et al. (2004) Antifungal and antimitotic substances discovered by the bioassay using conidia of pyricularia oryzae. *Drug Design Reviews—Online* 1(3):257–271.
9. McHugh, M.A. and Krukonis, V.J. (1994) *Supercritical Fluid Extraction: Principles and Practice* (2nd ed.), Boston, MA, Butterworth-Hienemann.

10. ROMPP, H., et al. (2004) Enrichment of hyperforin from St. John's Wort (Hypericum perforatum) by pilot-scale supercritical carbon dioxide extraction. *European Journal of Pharmaceutical Sciences* 21(4):443–451.
11. SMITH, R.M. and BURFORD, M.D. (1992) Supercritical fluid extraction and gas chromatographic determination of the sesquiterpene lactone parthenolide in the medicinal herb feverfew (Tanacetum parthenium). *Journal of Chromatography* 627(1–2):255–261.
12. JENNINGS, D.W., et al. (1992) Supercritical extraction of taxol from the bark of Taxus brevifolia. *The Journal of Supercritical Fluids* 5:1–6.
13. PASCUAL-MARTI, M.C., et al. (2001) Supercritical fluid extraction of resveratrol from grape skin of Vitis vinifera and determination by HPLC. *Talanta* 54(4):735–740.
14. VIJA, H., et al. (1997) Lipase-catalysed esterification in supercritical carbon dioxide and in hexane. *Bioorganic & Medicinal Chemistry Letters* 7(3):259–262.
15. TAKEDA, Y. and FATOPE, M.O. (1988) New phenolic glucosides from Lawsonia inermis. *Journal of Natural Products* 51(4):725–729.
16. SARKER, S.D., et al. (2005) *Separation by High-Speed Countercurret Chromatography. Natural Products Isolation* (2nd ed.), Totowa, NJ, Humana Press Inc.
17. LU, H.T., et al. (2004) Preparative high-speed counter-current chromatography for purification of shikonin from the Chinese medicinal plant Lithospermum erythrorhizon. *Journal of Chromatography A* 1023(1):159–163.
18. LIU, R.M., et al. (2005) Preparative isolation and purification of three flavonoids from the Chinese medical plant Epimedium koreamum Nakai by high-speed counter-current chromatography. *Journal of Chromatography A* 1064(1):53–57.
19. ARLT, W. and MACEDO, E.A. (1980) *Liquid–Liquid Equilibrium Data Collection, Ternary and Quaternary Systems*. Frankfurt, Dechema.
20. BOUDIMANT, G., et al. (1996) Purification of phosphatidylcholine with high content DHA from squid by centrifugal partition chromatography. *Journal of Liquid Chromatography & Related Technologies* 19:1793–1804.
21. BOUSQUET, O., et al. (1994) Characterization and purification of fatty acids from microalgae by GC-MS and countercurrent chromatography. *Chromatographia* 39(1–2):40–44.
22. AIKEN, G.R. and LEENHEER, J.A. (1993) Isolation and chemical characterization of dissolved and colloidal organic matter. *Journal of Chemical Ecology* 8(3):135–151.
23. URBAS, B., et al. (1964) Isolation and some structural features of a polysaccharide from birch sap. *Canadian Journal of Chemistry* 42:2093–2100.
24. SATYAJIT, D.S., et al. (2005) *Natural Products Isolation* (2nd ed.), Totowa, NJ, Humana Press.
25. HENKE, H. (1995) *Preparative Gel Chromatography on Sephadex LH-20*. Heidelberg, HÜthig GmbH.
26. BERGSTROM, J.D., et al. (1995) Discovery, biosynthesis, and mechanism of action of the zaragozic acids: potent inhibitors of squalene synthase. *Annual Review of Microbiology* 49:607–639.
27. KINNEL, R.B., et al. (1993) Palau'amine: a cytotoxic and immunosuppressive hexacyclic bisguanidine antibiotic from the sponge Stylotella agminata. *Journal of American Chemical Society* 115:3376–3377.
28. PANFILI, G., et al. (1994) High-performance liquid chromatographic method for the simultaneous determination of tocopherols, carotenes, and retinol and its geometric isomers in italian cheeses. *Analyst* 119(6):1161–1165.
29. MANZI, P., et al. (1996) Normal and reversed-phase HPLC for more complete evaluation of tocopherols, retinols, and sterols in dairy products. *Chromatographia* 43(1–2):89–93.
30. OLSON, C.V., et al. (1994) Preparative isolation of recombinant human insulin-like growth factor 1 by reversed-phase high-performance liquid chromatography. *Journal of Chromatography A* 675(1–2):101–112.
31. MANT, C.T. and HODGES, R.S. (1991) *High Performance Liquid Chromatography of Peptides and Proteins: Separation, Analysis and Conformation*, 11–22. Boca Raton, FL, CRC Press.
32. PORATH, J. and FLODIN, P. (1959) Gel filtration: a method for desalting and group separation. *Nature* 183(4676):1657–1659.
33. EISENSTEIN, M. (2006) A look back, adventures in the matrix. *Nature Methods* 3(5):410.

34. YANG, Y.B., et al. (1996) Characterization of a novel stationary phase derived from a hydrophilic polystyrene-based resin for protein cation-exchange high-performance liquid chromatography. *Journal of Chromatography A* 723(1):1–10.
35. YANG, Y. and HARRISON, K. (1995) Influence of column types and chromatographic conditions on the IEX chromatography of antibodies. The 15th ISPPP, Boston, pp. 18–20.
36. ZHANG, X. (2006) Chemical investigation and biological activity evaluation on *Dendrobium nobile* Lindl. Dissertation. Shenyang Pharmaceutical University..

Chapter 4

Identification and Structure Elucidation of Compounds from Herbal Medicines

Xin-miao Liang, Yu Jin, Jia-tao Feng, and Yan-xiong Ke

Chemical compounds are the foundational substances of the biological and pharmacological activities in herbal medicine. Without knowing the chemical composition in herbs, it is impossible to elucidate their preventative and therapeutic functions. The numerous secondary metabolites in medicinal plants have been found to possess a wide range of biological and pharmacological activities, which have drawn the interest of scientists from all over the world in the past decades. More recently, biological activities of macromolecules in herbal medicines, such as glycoprotein and polysaccharides, were also found to carry biological or pharmacological activities.

Chinese phytochemists like to call traditional Chinese medicine (TCM) a "black box" because most of the chemical composition and biological activities in Chinese herbs still remain unclear. Other traditional herbal medicines hold a similar status. Thus, it is necessary to carry out systematic research on their chemical compounds in order to uncover this "black box." The isolation and identification of chemical compounds in herbal medicines should be performed first to provide substances for further biological or pharmacological studies. Identification of compounds will also help to standardize and control the quality of herbal products.

Identification of chemical compounds in herbal medicine requires scientists to know the structure and properties of compounds in herbs. Modern technologies and advanced equipment accelerate not only the speed of isolation, but also the identification of compounds. Popular application of all kinds of spectroscopic methods, such as ultraviolet (UV), infrared (IR), nuclear magnetic resonance (NMR), circular

Traditional Herbal Medicine Research Methods, Edited by Willow J.H. Liu

dichroism (CD)/optical rotatory dispersion (ORD), and mass spectrometry (MS) has greatly helped scientists to elucidate the structures, including stereo-configuration, of compounds in herbs.

Traditionally, compounds in herbs have to be first isolated and purified using different methods with several steps, mostly with columns, then identified with different spectra. The application of hyphenated techniques of high-performance liquid chromatography (HPLC) with MS and NMR makes the identification of compounds possible without traditional column isolation, thus saving investigators both money and time. This chapter will mainly introduce these methods.

4.1 STRUCTURAL CHARACTERISTICS AND CHEMICAL IDENTIFICATION OF COMPOUNDS IN HERBAL MEDICINES

Based on the structural skeletons, chemical compounds in herbs can be divided into several types, such as flavonoids, saponins, alkaloids, terpenoids, anthraquinones, and phenylpropanoids. Every type of compound can be further divided into subtypes based on differences in the skeleton, while diversified substituent groups and substituent sites lead to structural variety. This section briefly introduces the structural characteristics and their corresponding chemical identification methods.

4.1.1 Flavonoids

Structural Characteristics

Flavonoids exist as both aglycone forms and glycoside forms. Flavonoids include several classes of compounds sharing a common C_6–C_3–C_6 flavone skeleton. The major classes are flavone, isoflavone, flavanone, isoflavanone, flavonol, flavanonols, chalcones, dihydrochalcone, anthocyanins, xanthone, aurones, and homoisoflavone (Fig. 4.1) (see detailed introduction in Chapter 3). They are differentiated on the degree of unsaturation and oxidation of the three-carbon segment. Within different subclasses, further division is based on the number and nature of substituent groups attached to the rings. The most common substituent groups are hydroxyl, methoxyl, and isoamylene.

Flavonoids occur in nature mainly in the form of glycosides, which can be divided into two types: flavonoid O-glycosides and flavonoid C-glycosides. Occasionally, flavonoid C-, O-glycosides can also be found. In flavonoid O-glycosides, one or more hydroxyl groups of the aglycone are bound to a sugar, forming a glycosidic O–C bond. Flavonoid C-glycosides result from direct linkage of the sugar to the flavonoid's basic nucleus with a glycosidic C–C bond. C–glycosylation site is limited to the C-6 or C-8 position of the aglycone (C-3′ or C-5′ position for chalcone). The structure identification of parts of sugars includes the nature of the sugar and the interglycosidic linkages. The common monosaccharides are glucose, rhamnose, and glucuronic acid. Disaccharides are also common in association with fla-

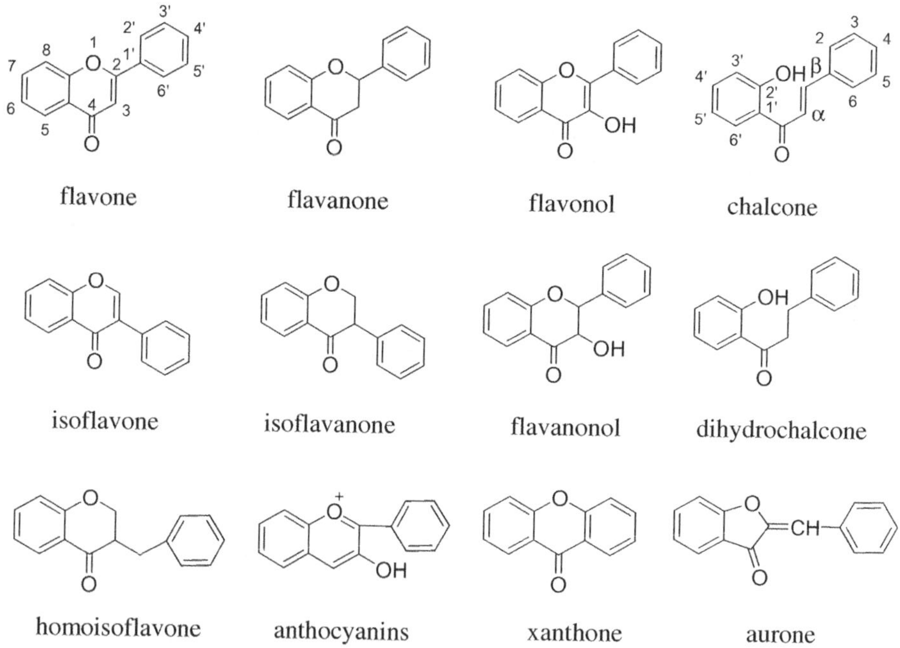

Figure 4.1 Structural skeletons of flavonoids.

vonoids, such as rutinose [rhamnosyl-(1→6)-glucose] and neohesperidose [rhamnosyl-(1→2)-glucose)], and occasionally tri- and even tetra-saccharides can be found. It is well known that the flavonoid glycosides have many isomers with the same molecular weight but different aglycones and sugars conjugating at multiple linkage positions. It is quite often that the difference is only in the positions of the substitutes in the aglycones, for example, different positions of hydroxyl or methoxyl groups.

Chemical Identification

Hydrochloric Acid–Magnesium Reaction This reaction is the most common method to identify flavonoids. A sample is dissolved in methanol or ethanol and a little magnesium or zinc powder is added to the solution. After adding several drops of concentrated hydrochloric acid, a color change of the solution will appear within minutes. Most solutions of flavone, flavanone, flavonol, and flavanonols will turn reddish purple then purple, and a few of them will become purple then blue, but chalcones, aurones, and most of isoflavone are negative to this reaction.

$NaBH_4$ (or KBH_4) Reaction This reaction has high selectivity for flavanone. The sample is dissolved with ethanol, then an equal volume of 2% $NaBH_4$ in methanol solution is added. After a few minutes, several drops of concentrated hydrocholoric acid or concentrated sulfuric acid are added, and a red to purple color will be observed in the solution.

Complex Reaction with Metal Salt Reagent Flavonoid can react with aluminum salt, lead salt, and zirconium salt, and produce a colored complex when flavonoids have two ortho-phenolic hydroxyl groups, or a 3-, or 5-hydroxyl and 4-keto. The commonly used aluminum salt is 1% $AlCl_3$ or $AlNO_3$, and yellow color will be observed due to the formation of complex compound. Flavonoid can react with 1% $PbOAC_2$ or lead subacetate to produce yellow to red precipitate. When there is a free 3–OH or 5–OH in flavonoid, it can react with 2% zirconyl dichloride in methanol solution to produce yellow complex.

4.1.2 Saponins

Structural Characteristics

Saponins are complex molecules in which aglycones link with sugar chain units. Based on the structure of the aglycones, saponins can be divided into two classes: triterpenoidic saponins and steroidal saponins (Fig. 4.2). Triterpenoidic saponins include tetracyclic triterpene (lanostane, dammarane, cucurbitane, meliacane, and protostane) and pentacyclic triterpene (oleanane, ursane, lupane, hopane, and isohopane). The aglycones of steroidal saponins are spirostane derivatives, which can be divided into spirostanols, isospirostanols, furostanols, and pseudo-spirostanols, according to the configuration of C_{25} and the cyclization status of F-ring. Both triterpene and steroidal aglycones have a number of different substituents (e.g., –OH, $-OCH_3$, and –COOH). The sugars can be linked to the aglycone either as one, two, or three side chains with ether or ester glycosidic linkage. The most common sugar residues are hexoses (glucose, galactose), 6-deoxyhexoses (furanose, quinovose, rhamnose), pentoses (arabinose, xylose), and uronic acids (glucuronic acid, galocturonic acid). Thus, great natural diversities of saponin structures are the result of the number of substituents, different sugar chain compositions, and attachment sites. Even within one plant species, different parts of an herb may contain saponins with different structures.

Chemical Identification

Liebermann–Burchard Reaction This is the most commonly used method for chemical identification of saponins. Small amount of samples are dissolved in acetic anhydride, and then concentrated sulfuric acid is slowly added along the cuvette wall. The reaction is positive as the color of interface turns reddish purple.

Kahlenberg Reaction Small amount of samples are dissolved in trichloromethane or alcohol solution, and the sample solution is spotted on the filter paper. After

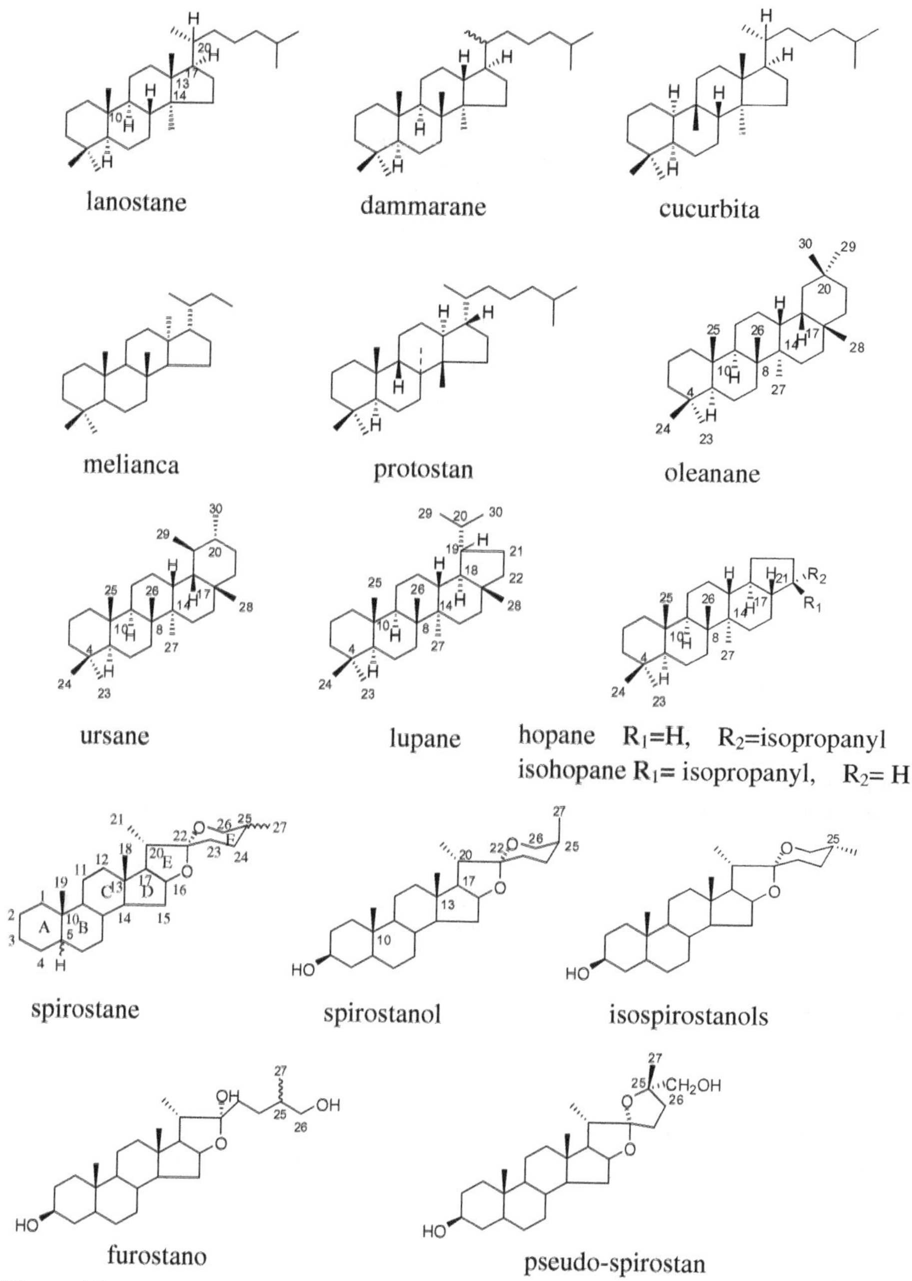

Figure 4.2 Structural skeletons of saponins.

- Derived from amino acids
  - ornithine
    1. Pyrrolidines
    2. Pyrrolizidines
    3. Tropanes
  - lysine
    4. Piperidines
    5. Indolizidines
    6. Quinolizidines
  - anthranilic acid
    7. Quinoline
    8. Acridone
  - phenylalanine\ tyrosine
    9. Phenylalkylamines
    10. Tetrahydroisoquinolines
    11. Benzyltetrahydroisoquinolines
    12. Phenethyltetrahydroisoquinolines
    13. Benzylphenethylamines
    14. Emetines
  - tryptophan
    15. Simple indoles
    16. Simple β-carbolines
    17. Semiterpenoid indoles
    18. Monoterpenoid indoles
- Derived from prenyl
  - terpenes
    19. Monoterpenoid alkaloids
    20. Sesqueterpenoid alkaloids
    21. Diterpenoid alkaloids
    22. Triterpenoid alkaloids
  - steroid
    23. Pregnane alkaloids
    24. Cyclopregnane alkaloids
    25. Cholestane alkaloids

Figure 4.3 Classification of alkaloids.

spraying 20% antimony pentachloride in trichloromethane on the spot, and heating at 60–70°C, the color spot will change into blue, grayish blue, and grayish purple.

Rosen-Heimer Reaction Sample solution is spotted on the filter paper, and 25% trichloroacetic acid in ethanol is sprayed on. When temperature is increased to 100°C, the color of spot becomes red and gradually turns purple.

Salkowski Reaction Sample is dissolved in trichloromethane solution, and the color of the chloroform layer becomes red or blue when concentrated sulfuric acid is added in.

Tschugaeff Reaction Sample is dissolved in glacial acetic acid, and acetyl chloride and zinc chloride are added. Upon slight heating, the color of solution turns light red or reddish purple.

4.1.3 Alkaloids

Structural Characteristics

Alkaloids are a large group of nitrogen-containing phytochemicals with diverse structures and a wide range of biological activities. The alkaloids can be classified

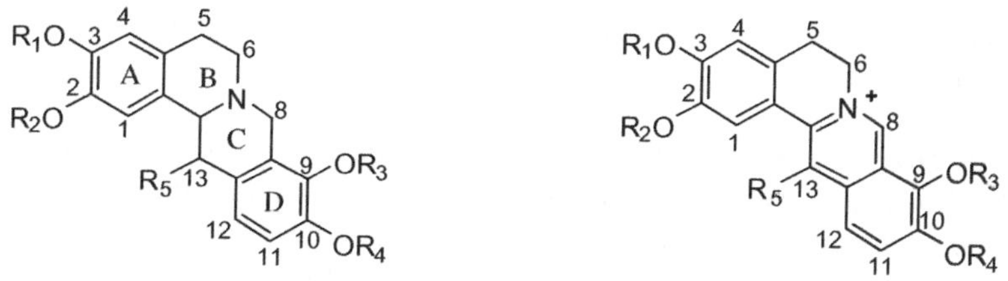

Figure 4.4 Structures of protoberberines and berberines.

based on their biological source or structures (Fig. 4.3) (see detailed introduction in Chapter 3). In general, diverse alkaloids naturally exist in many forms, for example, free states, salt, or amide forms. But there is no one common basic structural skeleton like flavonoids. In this chapter, we only introduce the most familiar protoberberines (tertiary alkaloids) and berberines (quaternary alkaloids) (Fig. 4.4) as examples. The difference between these two alkaloids is that the C-ring in protoberberines is saturated, for example, tetrahydropalmatine, but the C-ring in berberines is unsaturated, for example, berberine, palmatine, and jatroherrine. Their common substituent groups are methylenedioxy, hydroxyl, methoxyl, and methyl, and the substitution sites are usually at positions 2, 3, 9, 10, 5, or 13. The substituent site of methylenedioxy group is at positions 2 and 3 or positions 9 and 10 because one methylenedioxy group needs two neighbor substitute sites. The substituent groups appearing in positions 5 and 13 are always hydroxyl and methyl.

Chemical Identification

Alkaloids can react with many reagents to show different colors or to form precipitates with colors. Some reagents are specific to certain types of alkaloids, and some reagents change into different color when reacting with different types of alkaloids, which is helpful for the identification of alkaloids. Commonly used color reactions are introduced in the following.

Mandelin Reagent One percent ammonium vanadate in concentrated sulfuric acid. When the Mandelin reagent reacts with atropine, the color turns red; when it reacts with quinine, the color turns orange; when it reacts with morphine, it turns blue-violet; when it reacts with codeine, it turns blue; when it reacts with strychnine, it changes from blue-violet to red.

Fröhde Reagent One percent sodium molybdate or 5% ammonium molybdate in concentrated sulfuric acid. When Fröhde reagent reacts with aconitine, the color becomes yellow brown; when it reacts with berberine, it becomes greenish brown; when it reacts with colchicines, it becomes yellow; when it reacts with morphine, it turns purple at the beginning and then to brown; when it reacts with codeine, it changes into dark green, then to light yellow. This reagent does not react with atropine and strychnine.

Marquis Reagent A mixture of 0.2 mL 30% formaldehyde solution and 10 mL concentrated sulfuric acid. When Marquis reagent reacts with morphine, the color becomes orange, then purple; when it reacts with codeine, it turns pink, then yellow brown. Marquis reagent does not react with cocaine and caffeine.

Concentrated Sulfuric Acid When concentrated sulfuric acid reacts with colchicines, the color of solution turns yellow; when it reacts with codeine, it becomes light blue; when it reacts with berberine, it turns into green. This reagent does not react with atropine, strychnine, cocaine, and morphine.

Concentrated Nitric Acid When concentrated nitric acid reacts with morphine, the color of the solution changes from blue-red to yellow; when it reacts with codeine, it turns yellow; when it reacts with strychnine, it becomes yellow; when it reacts with aconitine, it turns into reddish brown. This reagent does not react with atropine, cocaine, and caffeine.

Concentrated Hydrochloric Acid When concentrated hydrochloric acid reacts with veratridine, the color turns into red; when it reacts with berberine with addition of aqueous ammonia, it turns into red.

Precipitation Reaction Precipitation reaction is often used for identification of alkaloids. The common precipitation reagents include Dragendorff's solution, Wagner's reagent, Mayer's reagent, and Bertrad's reagent, among which Dragendorff's solution is most often used. Alkaloids can react with Dragendorff's reagent to form orange or brown yellow colored precipitate, which is sensitive and can be used for the detection of alkaloids on thin layer chromatography (TLC).

Preparation of Dragendorff's reagent: solution I: 0.85 g of bismuth subnitrate is dissolved in 10 mL glacial acetic acid and 40 mL water; solution II: 8 g of bismuth potassium iodide is dissolved in 20 mL water. Equal volumes of solutions I and II are mixed, and 1 mL mixture, 2 mL acetic acid, and 10 mL water are mixed together to prepare Dragendorff's solution.

4.1.4 Terpenoids

Structural Characteristics

Terpenoids are isoprene oligomers $(C_5H_8)n$ and their derivatives. Terpenoids can be classified into monoterpenoid, sesquiterpenoid, diterpenoid, triterpenoid, tetraterpenoid, and polyterpenoid, based on the number of isoprene units, and then further into some subclasses by the number and nature of carbocyclic ring (Fig. 4.5) (see detailed introduction in Chapter 3). In general, terpenoids are characterized as complex

Monoterpenoid:

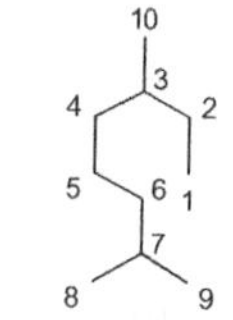

chain

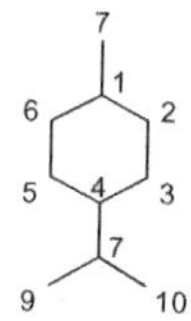

menthane

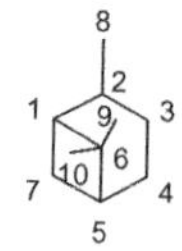

pinane

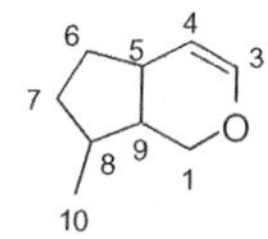

iridoid

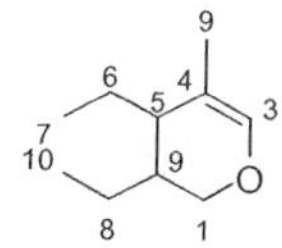

seco-iridoid

Sesquiterpenoid:

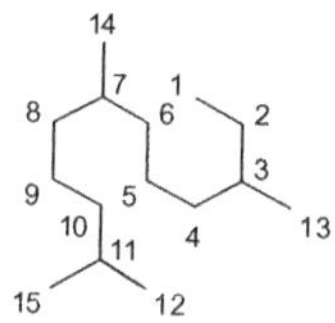

chain

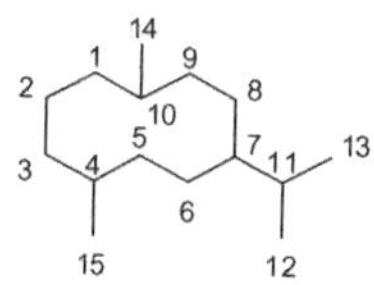

germacrane

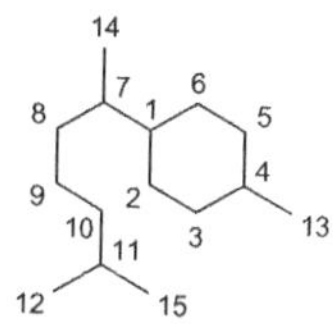

bisabolane

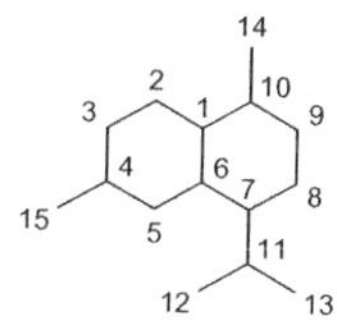

cadinane

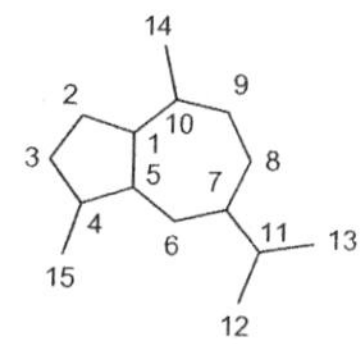

cuaiane

Diterpenoid:

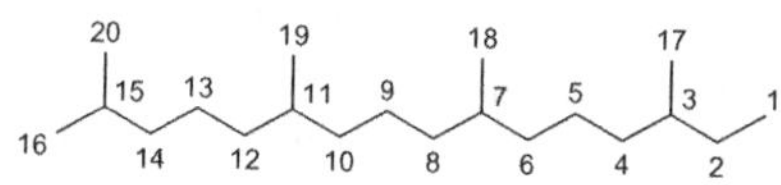

chain

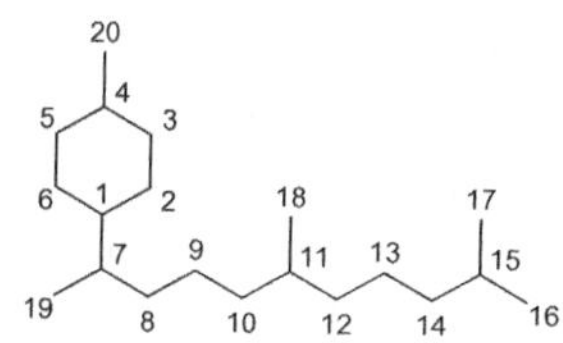

bisabolane

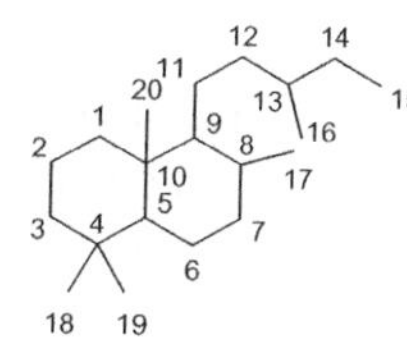

labdane

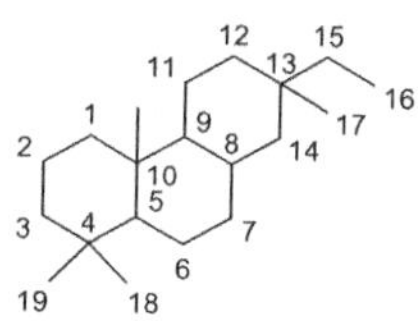

pimarane

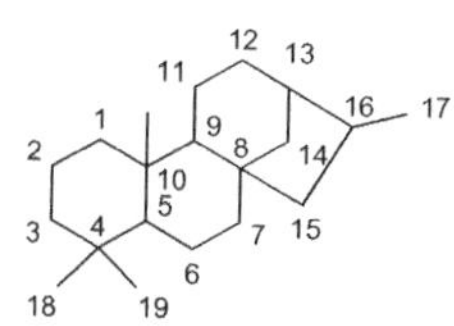

kaurane

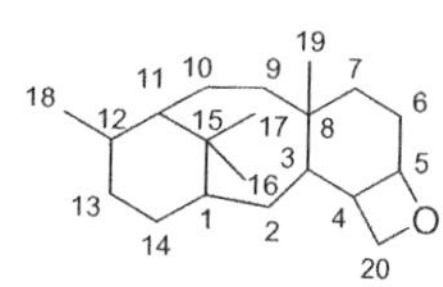

taxane

Figure 4.5 Structural skeletons of terpenoids.

skeletons, diverse structures, and numerous compounds. For example, sesquiterpenoids have more than 200 skeletons and several thousands of compounds.

Chemical Identification

Color reaction introduced in saponins can also be applied to terpenoids. In addition, Tilden reagent is commonly used to identify and separate unsaturated terpenoids. Here is the experimental procedure: the unsaturated terpenoid sample is dissolved in isoamyl nitrite solution, and then concentrated sulfuric acid is added under ice cold conditions with sufficient stirring and shaking. Blue-yellow crystallization of nitroso chloride derivatives is precipitated after small amounts of glacial acetic acid or ethanol are added.

4.1.5 Anthraquinones

Structural Characteristics

Quinones mainly include benzoquinones, naphthoquinones, phenanthraquinones, and anthraquinones, among which anthraquinones are the most important biological components. Anthraquinones (Fig. 4.6) are usually substituted with hydroxyl (–OH), hydroxymethyl (–CH_2OH), methoxyl (–OCH_3), and carboxyl (–COOH). Almost all anthraquinones have hydroxyl groups in their structures. According to the substituent sites of hydroxyl groups, anthraquinones can be classified into emodin type and alizarin type. The hydroxyl groups in the emodin type of anthraquinones are distributed on two benzene rings, while those in the alizarin type of anthraquinones are only on one benzene ring. Besides the derivatives of anthraquinones, the term anthraquinones also includes anthranol, anthrone, and oxidation anthranol (Fig. 4.6). Anthrone can form dimer by linkage at C_{10}–C_{10}.

Chemical Identification

Color Reaction with Alkaline Reagent Common alkaline reagents used for identification of anthraquinones include ammonia, 10% potassium hydroxide in methanol solution, 3% sodium hydroxide, sodium carbonate solution, 50% piperi-

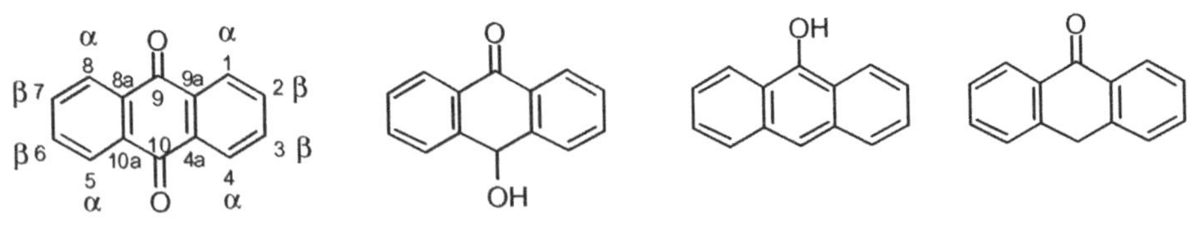

anthraquinones oxidation anthranol anthranol anthrone

Figure 4.6 Structural skeletons of anthraquinones.

dine in toluene solution, and saturated lithium carbonate solution. Alkaline solution reacts with hydroxyl anthraquinones to turn the color of solution from red to purple by forming a conjugated system of phenolic hydroxyl group and carbonyl group (Bornträger's reaction). Anthranol, anthrone, and dianthrone can be oxidized into anthraquinones first and then reacted with the alkaline reagent.

Color Reaction with Acidic Reagent Concentrated sulfuric acid reacts with anthraquinones to generate a red to reddish purple color.

Magnesium Acetate Reagent When 0.25–0.5% magnesium acetate in methanol is dropped into a sample solution of anthraquinones in methanol, the color change can be immediately observed. A positive color reaction will indicate that there is at least one α-hydroxyl in anthraquinones. Anthraquinones with hydroxyl groups at different position can react with the reagent to form different colors. For example, when the reagent reacts with α-hydroxyl anthraquinone having the second hydroxyl group at the para position of the α-hydroxyl, purple color is generated; when it reacts with α-hydroxyl anthraquinone having a hydroxyl group at the ortho position of the α-hydroxyl, a blue color is generated; when it reacts with other α-hydroxyl anthraquinone, a color change from orange to red is observed.

4.1.6 Phenylpropanoids

Structural Characteristics

Phenylpropanoids are a kind of component with the basic C_6–C_3 unit, constructed by a benzene ring and three linear carbon atoms. Coumarin and lignans are two main classes of phenylpropanoids (Fig. 4.7). Coumarins have a skeleton of benzopyranone. The common substituent groups are hydroxyl, alkoxy, prenyl, and phenyl. Coumarin can be divided into four subclasses: simple coumarins, furocoumarins, pyranocoumarins, and other coumarins. Lignans are dimers of phenylpropanoids, which include cinnamic acid, cinnamyl alcohol, propenyl benzene, and allyl benzene.

Chemical Identification

Blue or green luminescence can be observed under UV lamp when phenylpropanoids react with some reagents and after the treatment of aqueous ammonia. These reagents include (1) methanol solution of ferric chloride (1–2%); (2) Pauly reagent (diazotized sulfanilic acid); (3) Gepfner reagent (1:1 of 1% sodium nitrite and 10% acetic acid); and (4) Millon reagent. These reactions can be used for identification of phenylpropanoids on TLC. When Gepfner reagent is used, the reagent is sprayed and dried in the air first, then the plate needs to be treated with 0.5 mol/L caustic alkali in methanol. Lignans can react with different reagents (such as 5% phosphomolybdic acid in ethanol and 30% sulfuric acid) and show different colors.

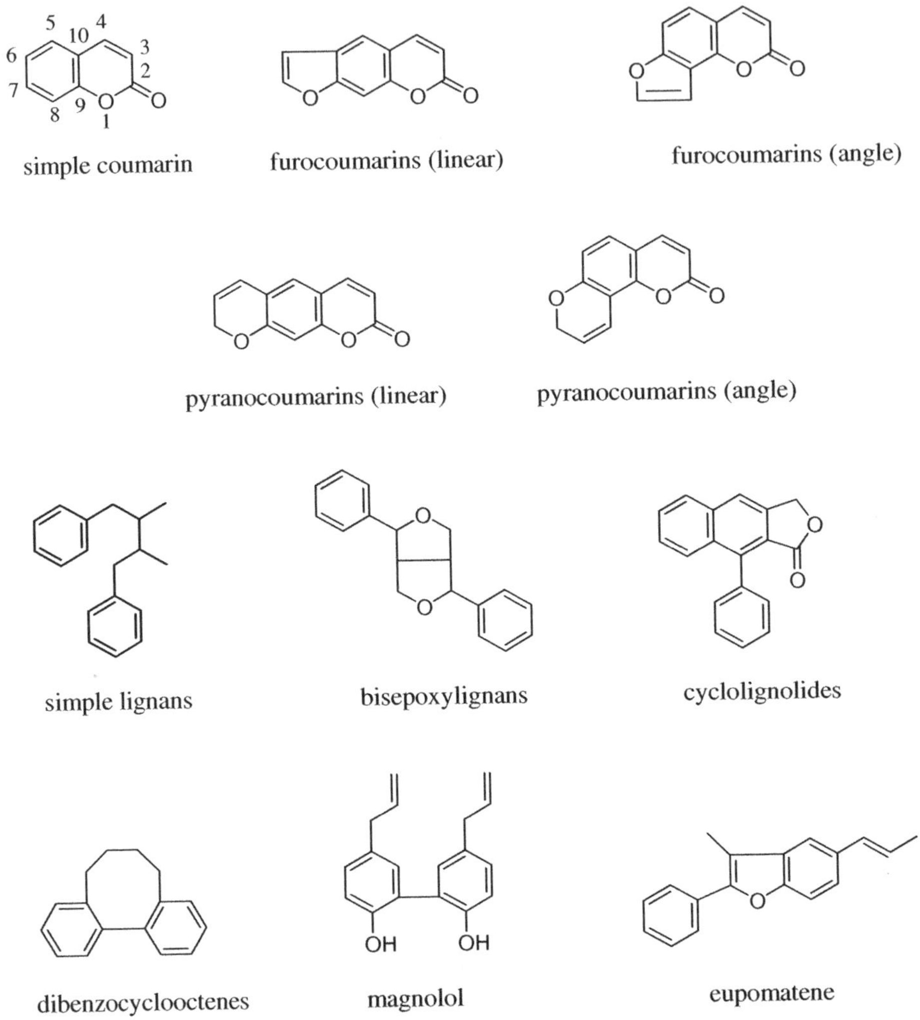

Figure 4.7 Structural skeletons of coumarins and lignans.

4.2 BRIEF INTRODUCTION OF UV, IR, NMR, MS, AND OTHER SPECTRA

4.2.1 UV Spectrum

UV is a kind of molecular absorption spectrum that is produced by the transition of bonding electron between different electronic energy levels. The wavelength of UV commonly ranges from 200 to 400 nm. For natural organic molecules, the main types of electrons involved in transitions are the σ electron, π electron, and n electron. Several important types of transitions are $\sigma \rightarrow \sigma^*$, $n \rightarrow \sigma^*$, $\pi \rightarrow \pi^*$, and $n \rightarrow \pi^*$. Their transition energies are shown in Figure 4.8.

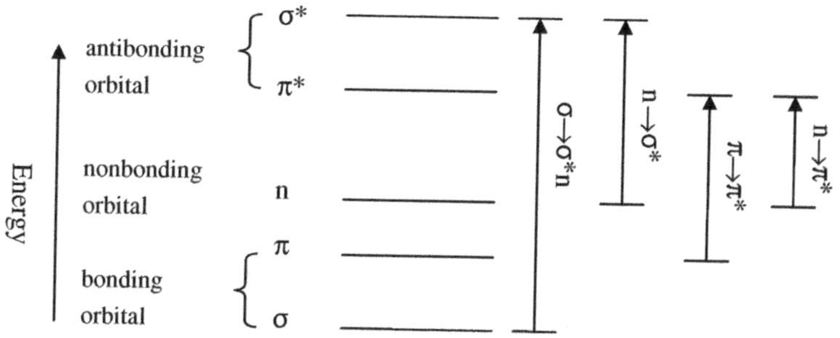

Figure 4.8 Electronic energy level and electronic transition.

It is known that the electron transition is complex when a molecule is excited by UV radiation. Each electronic level has many vibrational and rotational states. The UV spectra are the overlapping of these vibrational and rotational transitions on electronic transition. The spectra always appear as a continuous absorption band, rather than a discrete absorption line.

σ→σ* Transition Refers to a transition in which an electron in a bonding σ orbital is excited to an antibonding state of σ*. The energy gap between σ and σ* is large. The absorption of σ→σ* transition ranges between 150 and 160 nm and is thus not seen in typical UV spectra.

n→σ* Transition Refers to an electron transition from a nonbonding n orbital to an antibonding state of σ*. This transition happens to molecules containing atoms with nonbonding electrons, such as N, O, and S. It needs less energy than the σ→σ* transition. The absorption is within the range of 150–250 nm. The absorption peak of n→σ* transitions is highly related to the type of atoms that provide n electrons.

π→π*, n→π* Transition The two electronic transitions are important for UV spectra because the absorption peaks for these transitions fall into the range of typical UV spectra. The π→π* transition needs an unsaturated group in the molecule to provide the π electrons. The conjugation of multiple π bonds always results in red shift of absorption peaks. The π→π* transitions normally give molar absorptivities between 1000 and 10,000 L $mol^{-1} cm^{-1}$. When the unsaturated bond contains atoms with a lone pair of electrons (e.g., C=O, $-NO_2$), the nonbonding n electron can be excited to π* orbital. The transition needs even less energy than π→π* transition; thus the absorption peaks fall into the range of 270–300 nm. Similar to π→π* transition, conjugation with other double bonds will result in a red shift of the absorption peak. Molar absorbtivities from n→π* transitions are relatively low, within the range of 10–100 L $mol^{-1} cm^{-1}$.

The change in light intensity caused by light absorption and/or light scattering is described by the Lambert–Beer law. According to this law, when a monochromatic light with intensity passes through an absorbing medium, absorption takes place and the light intensity will decrease when leaving the sample. The transmittance (T) is defined as the ratio between the transmitted light (I) and the incident light (I_0):

$$T = \frac{I}{I_0}$$

The absorption A is defined as

$$A = \log\frac{1}{T} = \log\frac{I_0}{I}$$

The absorption A of a dissolved substance is a linear function of its concentration (c). The length of the light path (d) and the extinction coefficient (ε) determine the slope of the linear.

$$A = \varepsilon c d$$

UV spectrum is widely used to judge if a chemical compound has conjugated double bonds, α, β-unsaturated carbonyl (aldehydes, ketones, acids, and their esters), or aromatic groups in its structure. Thus, the function of UV spectra is mainly to deduce the skeleton type of the chemical compound. The UV spectra of some compounds, such as coumarins and flavonoids, will change upon addition of diagnostic reagents. The nature, number, and arrangement of substituent groups of a compound can result in differences in UV spectra, so that further structural identification can be achieved. The UV characteristics of the common chromophores and their maximum absorption wavelength are summarized in Table 4.1.

- *Spectra having no UV absorption:* Such spectra indicate that the compound might be aliphatic hydrocarbons, alicyclic hydrocarbons, or their derivatives (chloride, alcohol, ether, acid, etc.). They might also be non-conjugated alkenes.
- *Spectra having strong absorption within 220–250 nm:* Such spectra indicate the existence of two conjugated unsaturated bonds (conjugated diene or α, β-unsaturated aldehyde, ketone) in the structure.

Table 4.1 λ_{max} and ε_{max} Data of UV Adsorption of Simple Unconjugated Chromophores

Chromophores	Example	Solvent	λ_{max}/nm	ε_{max}	Transition type
Alkenyl	$C_6H_{13}CH{=}CH_2$	*n*-heptane	177	13,000	$\pi\rightarrow\pi^*$
Alkynyl	$C_5H_{11}C{\equiv}CCH_3$	*n*-heptane	178	10,000	$\pi\rightarrow\pi^*$
Carbonyl	CH_3COCH_3	Isooctane	279	13	$n\rightarrow\pi^*$
	CH_3CHO	Isooctane	290	17	$n\rightarrow\pi^*$
Carboxyl	CH_3COOH	Ethanol	204	41	$n\rightarrow\pi^*$
Amide	CH_3CONH_2	Water	214	60	$n\rightarrow\pi^*$
	$CH_3N{=}NCH_3$	Ethanol	339	5	$n\rightarrow\pi^*$
Nitro	CH_3NO_2	Isooctane	280	22	$n\rightarrow\pi^*$
Nitroso	C_4H_9NO	Ether	300	100	$n\rightarrow\pi^*$

- *Spectra having moderate absorption within 250–290 nm with fine structure at different degrees:* Such spectra indicate the existence of benzene or aromatic heterocyclic.
- *Spectra having moderate and weak absorption within 250–350 nm:* Such spectra indicate the existence of carbonyl or conjugated carbonyl group.
- *Spectra having strong absorption above 300 nm:* Such spectra indicate the existence of a long conjugated system. Condensed aromatics and condensed aromatics heterocyclic or their derivatives show strong absorption and obvious fine structure.

Currently, UV is the most popular detector coupled to HPLC for qualitative and quantitative analysis. Besides a UV lamp, a tungsten lamp can be equipped to record the visible spectrum ranging between 400 and 800 nm. UV-Vis detector has undergone development from fixed wavelength to variable wavelength. At present, the diode array detector (DAD) is a popular detector equipped to the HPLC. DAD can record any chromatograms within the range of 190–800 nm, which gives great convenience to optimize the detection wavelength. The UV-Vis spectrum at any point of a peak in HPLC can be recorded to judge the kind of compound. The purity of an HPLC peak can also be judged by comparing the UV-Vis spectra at multiple points of this peak.

4.2.2 IR Spectrum

IR is a technique based on the vibrations of the atoms of a molecule. The wavelength of IR covers wide range from 0.76 to 1000 μm. This wide IR region is divided into three regions, respectively named near-IR (0.76–2.5 μm), mid-IR (2.5–25 μm), and far-IR (25–1000 μm), among which the mid-IR is widely used for structure identification. The peak position can be represented with wavelength λ (μm) or wavenumber σ (cm^{-1}). Wavenumber is the more commonly used form and calculated by the following equation.

$$\sigma\left(\mathrm{cm}^{-1}\right)=\frac{10^4}{\lambda(\mu\mathrm{m})}$$

Two necessary conditions must be satisfied to produce an IR spectrum. One is that the energy of electromagnetic radiation should be equal to the energy difference of two vibrational states. Another is that the dipole moment must change during the molecular vibration.

Molecular vibration can be classified into two main types: stretching vibration and bending vibration. Stretching vibration can be further classified into symmetric stretching vibration (υ_s) and asymmetric stretching vibration (υ_{as}). Bending vibration can be further divided into in-plane bending vibration (δ) and out-of-plane bending vibration (γ), which includes deformation, rocking, wagging, and twisting forms. The basic vibration produces a basic frequency peak. Besides basic frequency absorption bands, the existence of some other bands, such as overtone and combination bands, Fermi resonance, coupling, and vibration–rotation bands makes it complicated to interpret the IR spectra.

A sample in any state can be studied with IR. Liquids, solutions, pastes, powders, films, fibers, gases, and surfaces can all be examined with different kinds of sampling techniques. Because IR is closely related to the molecule structure, it is a powerful and common tool for determination and confirmation of organic structures. Many organic functional groups such as methyl, methylene, carbonyl, cyano, hydroxyl, and amino have a characteristic absorption in IR, which is very helpful for structure elucidation of unknown samples. A characteristic frequency of organic functional groups may be shifted under different chemical environments. When an impact comes from intramolecular and intermolecular interaction, the wavenumber of an organic functional group can be altered by inductive effect, conjugated effect, as well as space effect, including steric hindrance, tension of ring, and hydrogen bond. All these factors should be considered in structure elucidation.

Attention should be paid to the position, intensity, and shape of a peak when identifying structure by IR. The peak position (wavenumber) is the most important feature of IR. The advantage of IR in structure identification for compounds in herbs lies in its ability to determine certain functional groups. With the popular use of MS and NMR, the application of IR in structure identification is mainly used for confirmation of the identification of carbonyl, alkynyl, and cyano, as well as hydroxyl and amino, groups. In addition, IR is often used to identify compounds by comparing them with the IR spectrum of known compounds due to its fingerprint feature. If two compounds have identical absorptions, especially in a sensitive fingerprint area (1500–500 cm^{-1}), it is certain that the structures of these two compounds are the exactly same.

According to absorption wavenumber, the IR spectra can be divided into six characteristic regions. The common absorptions in these six regions are briefly introduced, and more detailed introductions to the absorption of some functional groups are listed in Table 4.2.

4000–2500 cm^{-1} Stretching vibration of X–H (X, including C, N, O, and S)

The wavenumber of 3000 cm^{-1} is taken as a separatrix for differentiating C–H stretching vibration of unsaturated carbon and saturated carbon. The wavenumber of C–H stretching of unsaturated carbon (double bond and benzene ring) is above 3000 cm^{-1}, while that of C–H stretching vibration of saturated carbon (exception for three member rings) is below 3000 cm^{-1}.

The absorption of hydroxyl (O–H) is within 3650–3200 cm^{-1}. In this area, the absorption of free hydroxyl appears at a high wavenumber (3640–3610 cm^{-1}) as a sharp peak. When hydroxyl is in association, the IR absorption shows a broad and blunt peak near 3300 cm^{-1}. With trace water in the sample, hydroxyl absorption appears in the vicinity of 3300 cm^{-1}.

The behavior of the amino group is similar to that of hydroxyl. Free amino (N–H) shows absorption between 3500 and 3300 cm^{-1}, and decreases about 100 cm^{-1} after association.

2500–2000 cm^{-1} Stretching vibration zone of triple bond (–C≡C–, –C≡) and cumulative double bond (>C=C=C<, –N=C=O, –N=C=S)

Table 4.2 Wavenumbers of IR Absorption of Common Functional Groups (1)

Functional groups	Wavenumber (cm^{-1})	Intensity
Alkane		
Methyl ($-CH_3$)	2962 ($\upsilon_{as\ CH}$)	s
	2872 ($\upsilon_{s\ CH}$)	s
	1460 ($\delta_{as\ CH}$)	m
	1380 ($\delta_{s\ CH}$)	s
Methylene ($-CH_2-$)		
	2926 ($\upsilon_{as\ CH}$)	s
	2853 ($\upsilon_{s\ CH}$)	s
	1470 (δ_{CH})	m
Methenyl ($-CH-$)		
	2890 ($\upsilon_{s\ CH}$)	w
	~1340(δ_{CH})	w
Alkene		
	3100–3000 ($\upsilon_{=CH}$)	m
	1680–1600 ($\upsilon_{C=C}$)	m-w
Alkyne		
	3300 ($\upsilon_{\equiv CH}$)	m-s
	2100 ($\upsilon_{C\equiv C}$)	w
Aromatic		
	3100–3000 (υ)	v
	1650–1450 ($\upsilon_{C=C}$)	m-w
	900–650 ($\gamma_{=CH,\ out\ of\ plane}$)	
With five adjacent hydrogen atom	~750,~700	v,s
With four adjacent hydrogen atom	~750	v,s
With three adjacent hydrogen atom	~780	v,m
With two adjacent hydrogen atom	~830	v,m
With one hydrogen atom	~880	v,w
Alcohol, phenol, and ether		
Hydroxyl	3600 ($\upsilon_{OH\ (free)}$)	m
Hydroxyl	3300 ($\upsilon_{OH\ (association)}$)	s
Primary alcohol	1050 (υ_{C-O})	s
Secondary alcohol	1100 (υ_{C-O})	s
Tertiary alcohol	1150 (υ_{C-O})	s
Phenol	1300–1200 (υ_{C-O})	s
R-O-R′	1210–1050 ($\upsilon_{as\ C-O-C}$)	s
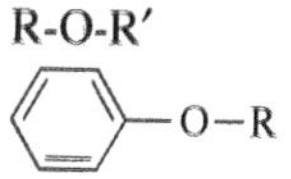	1300–1200 ($\upsilon_{as\ C-O-C}$)	s
	1055–1000 ($\upsilon_{s\ C-O-C}$)	m

Table 4.2 Wavenumbers of IR Absorption of Common Functional Groups (2)

Functional groups	Wavenumber (cm^{-1})	Intensity
Carbonyl group ($\upsilon_{C=O}$)		
Aliphatic ketone	1730–1700 ($\upsilon_{C=O}$)	s
Aliphatic aldehyde	1740–1720 ($\upsilon_{C=O}$)	s
	2850,2740 (C-H Fermi resonance)	m
Carbolylic acid	1720–1680 ($\upsilon_{C=O}$)	s
	3200–2500 (υ_{OH})	
Ester	1750–1730 ($\upsilon_{C=O}$)	s
	1300–1150 ($\upsilon_{as\ C\text{-}O\text{-}C}$)	s
	1140–1030 ($\upsilon_{s\ C\text{-}O\text{-}C}$)	w
Anhydride	1825–1815,1755–1745 ($\upsilon_{C=O}$)	s
Amide	1690–1650 ($\upsilon_{C=O}$)	s
	3500–3050 (υ_{NH}, double peak)	
	1649–1570 (δ_{NH})	
Acyl halide	1819–1790	
Amine and ammonium salt		
Primary amide	3500–3300 (υ_{NH})	w
	1640–1560 (δ_{NH})	m-s
	900–650 (γ_{NH})	m
Primary ammonium salt	3000 (υ_{NH})	m
	1600–1570 ($\delta_{as\ NH}$)	s
	1500 ($\delta_{s\ NH}$)	m
Secondary amide	3350–3310 (υ_{NH})	w
	1580–1490 (δ_{NH})	w
Secondary ammonium salt	2700–2250 (υ_{NH})	s
	1600–1570 (δ_{NH})	m
Tertiary ammonium salt	2700–2250 (υ_{NH})	s
Halogen compounds		
C-F	1150–1050 (υ_{as})	s
	1100–1000 (υ_{s})	s
C-Cl	750–700 (υ)	s
C-Br	670–400 (υ)	m
C-I	550–400 (υ)	m
Sulfide		
S-H	2600–2550 (υ)	w
C=S	1200–1050 (υ)	s
S=O	1070–650 (υ)	s

This is the stretching vibration region of triple bond and cumulative double bond. For example, the absorption of alkynyl group is at 2100 cm^{-1} and that of cyano group is at 2200 cm^{-1}. In this region, the absorption of carbon dioxide in air can be observed at 2365 and 2335 cm^{-1} sometimes, if background deduction is not adequate.

2000–1500 cm^{-1} Stretching vibration of double bond

The absorptions of most carbonyl groups fall into 1990–1650 cm^{-1}. Except for carboxylates, most of carbonyls (C=O) show the strongest or sub-strong absorption with a sharp peak. The absorption of C=C is between 1680 and 1600 cm^{-1} with moderate or low intensity.

The vibrations of the benzene skeleton are at 1600, 1580, 1500, and 1450 cm^{-1}. It is not easy to distinguish the absorption at 1450 cm^{-1} since it is close to that of $-CH_2$ and $-CH_3$. The existence of benzene can be confirmed by either absorption around 1500 or 1600 cm^{-1}. Aromatic heterocyclic has a similar absorption as benzene; for example, there are three absorptions of 1600, 1500, and 1400 cm^{-1} for furan. The absorption of C=N and N=O are also within these regions.

1500–1300 cm^{-1} Bending vibration of C–H and symmetrical stretching vibration of benzene

Absorption in this region provides information on the bending vibration of C–H. The methyl ($-CH_3$) group has absorptions at 1460 and 1380 cm^{-1}, and methylene ($-CH_2-$) has one absorption at 1470 cm^{-1}. Absorption of the symmetrical stretching vibration of benzene is at 1500 and 1450 cm^{-1}, and that of aromatic heterocycle and nitryl are also within this region.

1300–910 cm^{-1} Stretching vibration of single bond and skeleton vibration

This region is also called fingerprint region because absorptions of all the stretching vibration of single bond, skeleton vibration, and some bending vibrations of groups with hydrogen are observed in this region. It is particularly useful for compound identification by comparing the absorption patterns, that is, both the wavelength (or wavenumber) and absorbance (or percent transmittance) of each peak with that of known compound.

910–650 cm^{-1} This is an important region for determining the positions of substituted groups at benzene within the range of 900–650 cm^{-1}.

The development of the Fourier transform (FT) IR spectrometer greatly improves sensitivity and analytical speed. In recent years, the successful combination of IR with separation tools has extended its application. Chromatography-Infrared Spectroscopy extends the application of IR in a complex sample. Gas chromatography (GC), liquid chromatography (LC), and supercritical fluid chromatography (SFC) have all been used to hyphenate with IR for analysis. When a complex sample is separated by chromatography, compounds eluted from the chromatography are directed into IR to acquire structure-related information. In addition, the combination of IR spectrometers and thermal analysis instrumentation provides information about the temperature-dependent physical properties of materials. Combining an IR spectrometer with a microscope facility makes it possible to study small amounts of samples for multiple purposes. More and more new techniques are emerging in the development of IR.

4.2.3 NMR Spectrum

NMR is a very powerful tool for structure identification of natural compounds. The theory of NMR involves nuclear spin, magnet moment, angular momentum, energy of nuclei in magnetic fields, resonance condition, nuclear shielding, and relaxation. There are many reference books introducing the theory, principle, and application of NMR. Because this section introduces NMR only as part of the structural identification for natural compounds, it will not give detailed explanation on NMR principles, due to space constraints. The information provided here is mainly to provide a general guide for scientists who do not work in this area.

Briefly, NMR is a physical phenomenon based on the quantum mechanical magnetic properties of an atom's nucleus. ^{1}H and ^{13}C are the most common nuclei measured in NMR because only nuclei that contain odd numbers of protons and/or neutrons have an intrinsic magnetic moment and angular momentum that can satisfy the resonance condition, producing a signal in the receiver channel of the spectrometer.[1]

In the past half century, a great deal of data has been accumulated in the structure identification of all kinds of chemical compounds in herbal medicines using ^{1}H-NMR and ^{13}C-NMR. The development of two dimensional (2-D) NMR makes it easy to attribute the signals of ^{1}H-NMR and ^{13}C-NMR and provide more precise structural information. In this section, basic concepts and the development of NMR will be briefly introduced. The hyphenate LC-NMR technique for identification of compounds in herbal mixtures will be introduced in Section 4.5.

The Basic Concepts of NMR

There are three important parameters in NMR spectrum: chemical shift (δ), coupling constants (J), and peak area. Chemical shift (δ) refers to the signal position in a spectrum that can provide information about the type of proton or carbon and the chemical environment. Coupling constants (J) show the relationship between nuclei and reflect the stereochemistry of the structure. In the ^{1}H-NMR, the absorption of energy of the nuclei is proportional to the number of protons coming into resonance at the frequency of the signal, meaning the peak area is proportional to the number of protons being detected. This helps to infer the number of hydrogen attached to each functional group and the number of hydrogen attached to functional groups with the same environment. In combination with the information of chemical shifts from the spectrum, the types and numbers of each type of functional groups (–CH–, $-CH_2-$, $-CH_3$, $-OCH_3$, –OH, –COOH, –CHO, etc.) may be judged.

Chemical shift (δ) is defined as follows:

$$\delta = \frac{\nu_s - \nu_{TMS}}{\text{operation frequency}} \times 10^6$$

(ν_s—frequency of sample; ν_{TMS}—frequency of tetramethylsilane [TMS]).

TMS is a commonly used internal standard for NMR measurement. TMS is chosen as an internal standard because it is inert, volatile, nontoxic, cheap, and has only one signal. From the equation, it can be seen that the chemical shift δ is a field-independent number. The advantage of using TMS as an internal standard is that the fluctuation from machine to machine and from day to day can be minimized. The main range of chemical shift δ is from 0 to 10 ppm for ^{1}H-NMR and from 0 to 200 ppm for ^{13}C-NMR. The signal of TMS occurs at the extreme right on zero. The left of the spectrum is high-field and low-frequency shielded. On the contrary, the right of the spectrum is low-field and high-frequency shielded. The chemical shift is affected by factors so that the change of chemical shift reflects the chemical environment of the measured nucleus.

Overlapped signals are often seen in ^{1}H-NMR due to the narrow range of the chemical shift and the splitting peaks arising from the coupling effect. Without coupling with hydrogen or each other, the carbon signals present as sharp and isolated lines in ^{13}C-NMR. As the range of chemical shift of ^{13}C-NMR is 10 times larger than that of ^{1}H-NMR, the minor change in structure is easily observed from ^{13}C-NMR. For the asymmetrical structure with molecular weight at 300–400 Da, each carbon signal is differentiated by its chemical shift of ^{13}C-NMR. There are a variety of methods and techniques of carbon decoupling, such as proton broadband decoupling, off-resonance decoupling, selective proton decoupling, gated decoupling, and distortionless enhancement by polarization transfer (DEPT). These methods can provide rich information to distinguish the type of carbon atom (primary, secondary, tertiary, and quaternary). The attribution of signal of ^{13}C-NMR facilitates the determination and validation of structural skeleton.

The interference of spin–spin coupling between two or two groups of protons with different magnetic moments in a certain distance will lead to splitting, and the distance of peak in the splitting is called the coupling constant (J). The coupling constants are closely related to the number of chemical bonds between two coupling nuclei, because spin–spin coupling is transferred through the bonding electrons. Vicinal coupling is a three-bond coupling (^3J, H–C–C–H), and geminal coupling is a two-bond coupling (^2J, H–C–H). Coupling constants decrease rapidly with the increase in distance between two nuclei. Commonly, ^4J is less than 0.5 Hz. The coupling relates to the geometric array of the nuclei. In some special arrangements, the ^4J and even ^5J can be observed. Because it is hard to find the long-range coupling of protons far beyond four bonds, it should be closely related to the stereochemistry if ^5J is not zero.

Two-Dimensional NMR

Two-dimensional NMR originates from the concept of FT. Two-dimensional NMR produces the structural information by detecting the interaction in space between nuclei in the applied magnetic field. Through the detection of spatial dipole interaction between nuclei, the linkage and spatial configuration of nuclei will be elucidated. Two-dimensional NMR is an effective tool for identification and validation of structure, especially for understanding the stereochemistry of the molecule.

Two-dimensional NMR spectra can be divided into J-resolved spectra (δ-*J* spectra), chemical shift correlation spectra (δ-δ spectra), and multiple quantum spectra. *J*-resolved spectra can be used to distinguish the spin–spin coupling of nuclear, which includes heteronulear *J* and homonuclear *J* spectra. Chemical shift correlation spectra show the relationship of resonance signals, including homonuclear coupling, heteronuclear coupling, nuclear Overhauser effect (NOE), and chemical exchange. Multiple quantum transitions often occur when the quantum transition (△m) is an integer more than 1. The commonly used 2D-NMR techniques are introduced in this section.

Homonuclear Chemical Shift Correlation Spectroscopy (^{1}H–^{1}H COSY) ^{1}H–^{1}H COSY is a simple 2D-NMR. It is used to determine the chemical shift, coupling relationship, and connecting sequence of protons. In the COSY spectrum, two essentially identical chemical shift axes are plotted orthogonally. These two chemical shift axes are frequently labeled f_1 and f_2; the f_2 axis is usually better resolved. The peaks on the diagonal are called diagonal peaks or autocorrelated peaks, and the peaks out of the diagonal are called cross-peaks or correlated peaks. Moreover, the cross-peaks are distributed on both sides of the diagonal symmetrically. The coupling relationship of two groups of peaks can be determined from any cross-peaks on the COSY spectrum. ^{1}H–^{1}H COSY mainly reflects the coupling relationship of vicinal coupling (^3J), but sometimes long-range coupling can also be observed.

^{1}H–^{1}H Nuclear Overhauser Effect Spectroscopy (NOESY) Similar to ^{1}H–^{1}H COSY, the spectrum of NOESY also includes diagonal peaks and cross-peaks. The normal spectrum appears on the diagonal. The cross-peaks indicate those protons that are close in space rather than trough–bond interactions, so a NOESY spectrum provides important information about the geometry of the molecules. NOESY has two important functions in the structure identification. First, the atoms and fragments in a molecule can be linked according to the information provided by NOESY, which is called connectivity through space. NOESY can be applied to resolve some problems that ^{1}H–^{1}H COSY is unable to address. For example, when the dihedral angle of adjacent protons is close to 90°, the coupling constant is nearly at zero, so it is difficult for ^{1}H–^{1}H COSY to get the coupling information at this condition. Second, when the coupling is blocked by a quaternary carbon atom, ^{1}H–^{1}H COSY is unable to work in this condition. NOESY can provide the important information about stereochemistry and solution conformation by analyzing the connectivity through space of atoms in a molecule, so NOESY is an important tool for investigating molecular configuration, conformation, and movement.

Total Correlation Spectroscopy (TOCSY) TOCSY is a homonuclear experiment that produces a COSY-like plot. In a TOCSY spectrum, cross-peaks are found for all protons of a spin system. This allows the easy recognition of complete families of coupled spins. The observed signals are in positive absorption mode, so the TOCSY spectrum has high resolution. Another advantage of this technique lies in the interpretation of spectra where there are overlapping resonances. Correlations from a single resolved proton may be used to trace the coupling network in order that all the protons in a spin system can be identified.

^{13}C–^{1}H COSY ^{13}C–^{1}H COSY is used to correlate the proton spectrum and carbon spectrum and completely assign them in order to elucidate the structure. On a 1-D plot of ^{13}C–^{1}H COSY, the chemical shift axis f_2 is the ^{13}C spectrum, and the chemical shift axis f_1 is the ^{1}H spectrum. The ^{13}C–^{1}H COSY shows the correlation of the carbon atom to the directly connected proton; thus, the carbon signal connected to this proton can be found from a known proton signal, and vice versa. Quaternary carbon has no correlated ^{1}H peaks. Methenyl and methyl have one correlated ^{1}H peak. The peak of methylene is decided by whether its directly connected protons are magnetically equivalent. ^{13}C–^{1}H COSY is useful in identifying the signals. However, it has been substituted by heteronuclear single quantum coherence (HSQC) and heteronuclear multiple quantum coherence (HMQC) due to its low sensitivity.

HMQC and HSQC HMQC and HSQC are 2-D inverse H, C correlation techniques that allow for the determination of carbon (or other heteroatom) to hydrogen connectivity. Differing from ^{13}C–^{1}H COSY that detects ^{13}C signal, HMQC and HSQC detect the proton signal, so they have higher sensitivity. The differences between HMQC and HSQC are hidden in the pulse sequence. HSQC has more pulses and delays than HMQC, and the period between the excitation pulse and the acquisition is longer in the case of HSQC. The elucidation of HMQC and HSQC is the same as ^{13}C–^{1}H COSY, and quaternary carbon has no correlated peak.

^{1}H Detected Heteronuclear Multiple Bond Correlation (HMBC) HMBC is an experiment that identifies proton nuclei with carbon nuclei that are separated by more than one bond. HMBC allows the determination of $^{n}J_{CH}$ ($n \geq 2$) correlations, especially $^{2}J_{CH}$ and $^{3}J_{CH}$ correlations, and provides carbon long-range coupling information unambiguously. Often the one-bond correlations are explicitly filtered out of the spectrum. Quaternary carbon and even heteroatom can be bypassed to acquire useful information about the carbon chain skeleton of a molecule. This technique is of great importance since it is now possible to connect units together.

NMR can measure each active nuclear to deduce the molecular skeleton. Unambiguous structure identification by NMR includes not only identification of the planar structure, but also elucidation of stereochemistry information (such as molecular configuration and conformation) according to the relationship of nuclei in molecules. Structures of some simple compounds may be easily elucidated by analysis of ^{1}H- or/and ^{13}C-NMR spectra.

The accurate arrangement of each group in structure is based on the attributions of signals of all protons and carbons in ^{1}H-NMR and ^{13}C-NMR. Stereochemistry information such as configuration and conformation of a molecule may be deduced from ^{1}H-NMR based on the coupling constants (J). The chemical shift of protons (δ_H) at saturated carbon (–C–H) is between 0 and 2 ppm; acetylenic carbon (≡C–H) is between 2 and 3 ppm; ene-carbon (C=C–H) is at 4.5–6.5 ppm; benzene ring (Ar–H) is between 6 and 8 ppm; aldehyde (–CO–H) is between 9 and 10 ppm;

Table 4.3 Chemical Shifts of Common Functional Groups in ^{1}H-NMR

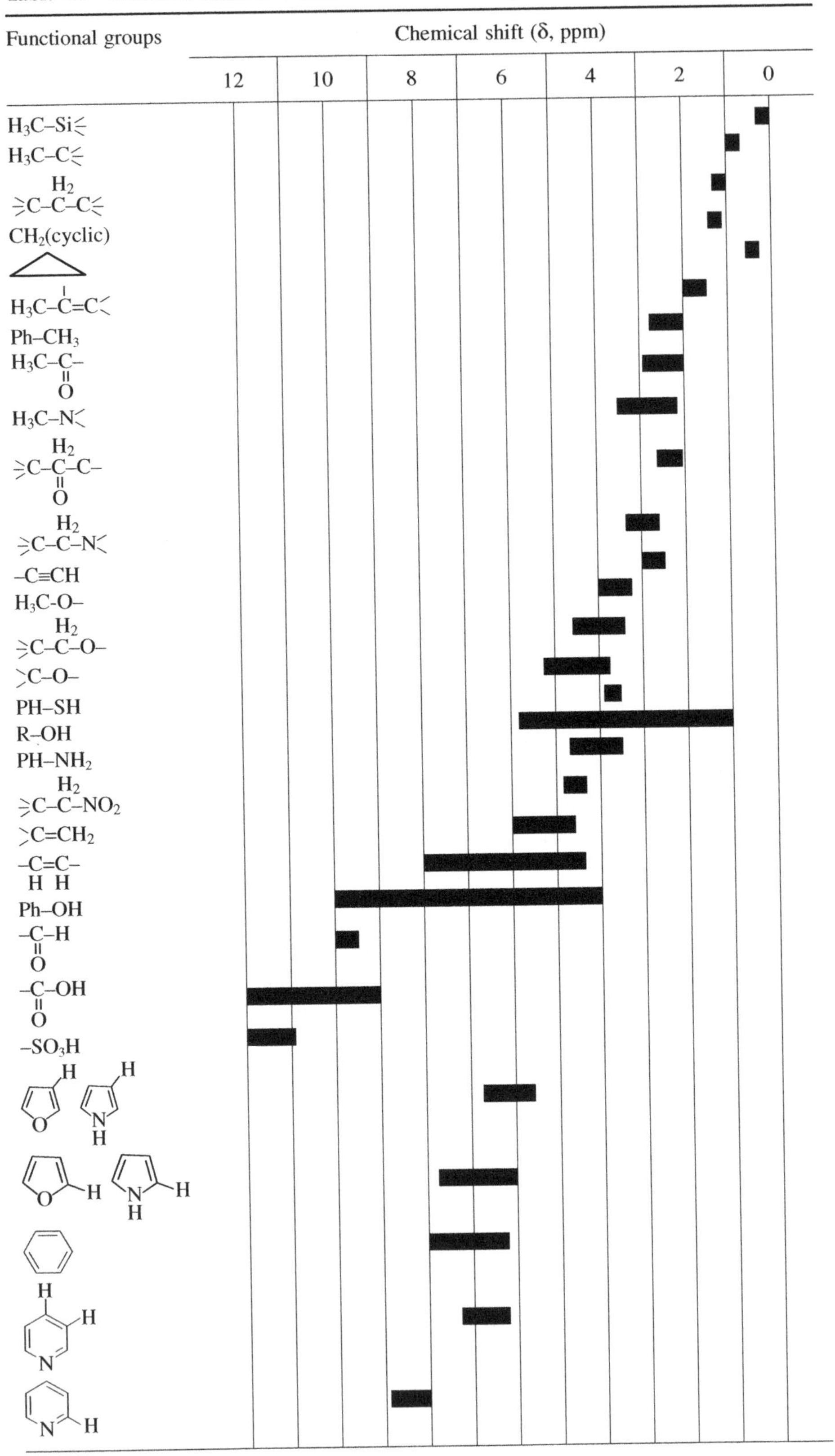

Table 4.4 Effect Factors to the Chemical Shift of Proton Signal in ^{1}H-NMR

Effect factor	Electron density	Shift direction	Chemical shift
Inductive effect	↓	Low field	↑
Anisotropic effect	↑(Shielding effect)	High field	↓
	↓(Deshielding effect)	Low field	↑
van der waals effect	↓	Low field	↑
Hydrogen bond effect	↓	Low field	↑
Solvent effect	Solvent effect is mainly induced by anisotropic effect of solvent and hydrogen bond effect between solvent and solute.		
Exchange reaction	Rapid exchange between multiple forms can counteract the difference in chemical shift.		

carboxyl group (–COOH) is between 10 and 13 ppm; and enolic hydrogen [R_2C=C(OH)R] is at δ_H 11–16 ppm. More detailed information about chemical shifts of all kinds of functional groups is in Table 4.3. Factors affecting chemical shifts are summarized in Table 4.4.

Functional groups also possess characteristic chemical shifts in ^{13}C-NMR. For example, the carbon signal of keto carbonyl is at δ_C 190–220 ppm; aldehyde carbonyl is at δ_C 185–205 ppm; quinone carbonyl is at δ_C 180–190 ppm; and carbonyl in carboxylic acid and its derivatives is at δ_C 160–180 ppm. The carbon signal of methyl is important information. The carbon signal of methoxyl in ether is at δ_C 55–62 ppm; methoxyl in ester is near δ_C 52 ppm; N–CH_3 is at δ_C 30–45 ppm; S–CH_3 is at δ_C 25 ppm; and methoxyl connecting to the double bond (C=C or C=X, X=N, O, S) is about δ_C 20 ppm. In addition, some functional groups containing cyano can be easily differentiated. More detailed information on chemical shifts in ^{13}C-NMR of all kinds of functional groups is listed in Table 4.5. The chemical shift of carbon is also affected by some factors, such as hybrid forms of carbon, anisotropic effect, and solvent effect.

As a variety of NMR methods are available for structure analysis, a specific, reliable, and informative method should be chosen based on the amount of samples, the complexity of structure, property of compounds, and the resolution of instruments. As to the enormous potential of NMR for structural identification of organic compounds, it has been widely used as a very powerful tool to elucidate the structures of unknown compounds in herbal medicines.

Structure identification by NMR usually needs sampling with high purity. However, isolation of a compound from a complex extract is difficult and tedious. The development of NMR-based metabonomics utilizing the multivariate data analysis made it possible to elucidate an NMR spectrum of a sample containing a mixture of compounds without an isolation procedure. Application of this method will increase the efficiency in structure identification of chemical compounds in extracts of herbal medicine. Detailed information about this method is available in literature.[2]

Table 4.5 Chemical Shifts of Common Functional Groups in ^{13}C-NMR

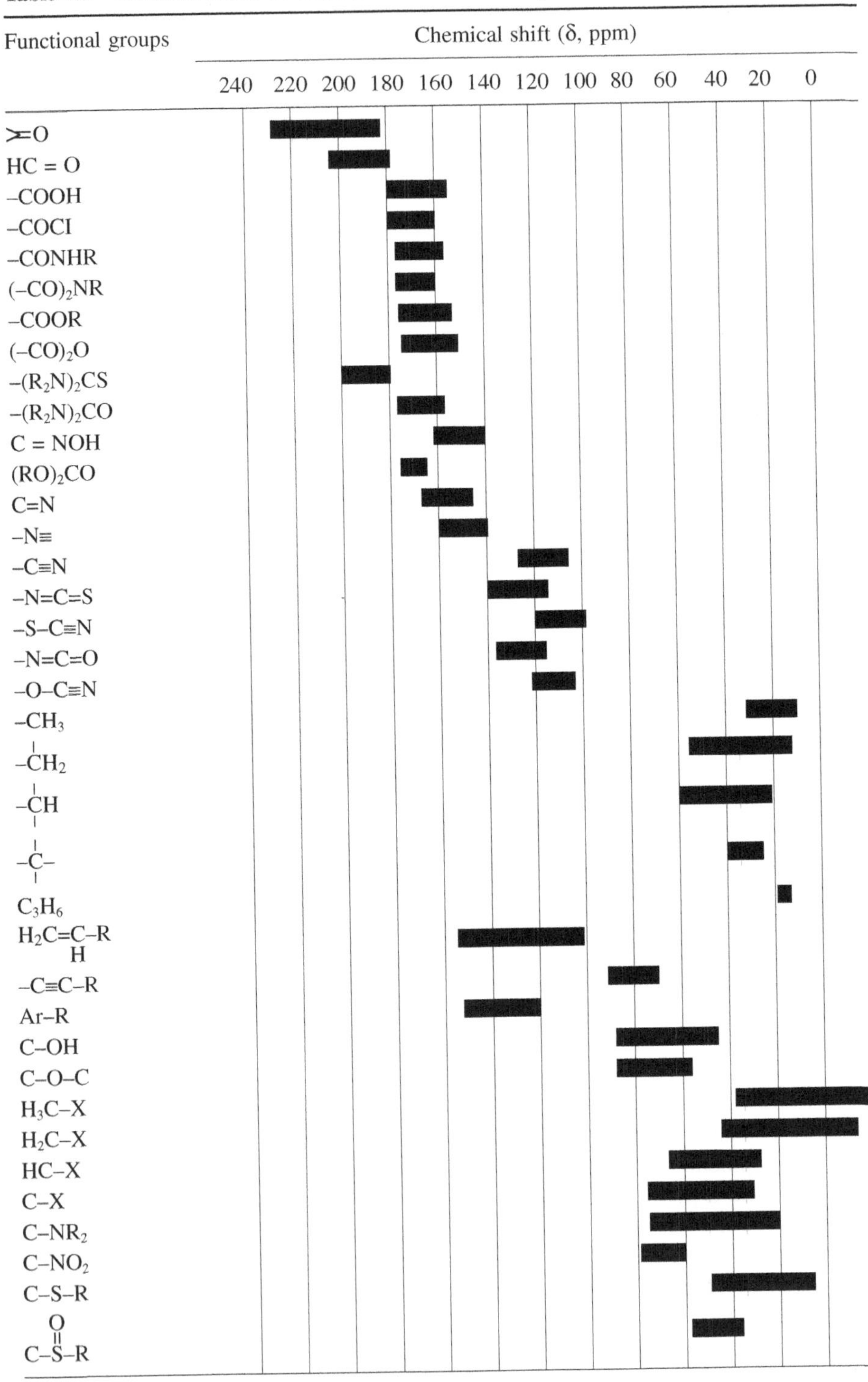

4.2.4 MS Spectrum

MS is an analytical technique that measures the molecular masses of individual compounds and atoms precisely, by converting them into charged ions. The evolution of MS is mainly characterized by the development of an ion source and a mass analyzer. With the development of ionization technology and mass analysis technology, MS has expanded its application scope gradually. Since MS had been used for structural elucidation of complex organic compounds in combination with NMR and IR in the 1960s, it has proven to be one of the most powerful tools to study the structure of organic compounds. The inventions of electrospray ionization (ESI) and matrix-assisted laser desorption (MALDI) enable MS to analyze biomolecules, and biomass spectrometry was rapidly developed in the late 1980s.

The advantages of MS in qualitative analysis also attract the attention of phytochemists. MS can be widely used for structural elucidation because of its unique ability to accurately measure molecular mass and to provide structural related information. At the same time, it also has the advantage of ultrahigh-detection sensitivity, rapid analytical speed, and a wide application range for all types of sample. The hyphenated technique combined with the highly effective separation equipment prompted the development and application of MS. MS is considered a powerful tool for characterization of complex samples.

The Development of Ion Source

Ionization of the sample is the first step of MS analysis, and the ionization technique is one of the focuses in the development of MS. Many ionization techniques have been invented and widely applied. Those most familiar to us are electron ionization (EI), chemical ionization (CI), MALDI, thermospray ionization (TSP), and atmospheric pressure ionization (API), including ESI and atmospheric pressure chemical ionization (APCI). These ion sources played important roles during the development of MS. In this section, the application ranges and traits of different kinds of MS are briefly introduced in terms of the requirement of application. Detailed introduction about theory and principle of MS can be found in many MS specialized books.[3]

EI EI is famous as one of the oldest ionization modes with wide application in GC-MS. EI operates in a high vacuum state (10^{-5}–10^{-6} torr), and the mass range of EI is less than 1000 Da. The sample analyzed by EI must be thermally stable and volatile because the analytes must be available as gas phase molecules before ionization. This restriction limits the application of EI to some degree. The ions produced in EI are molecular ion or radical cation, represented by $M^{+\bullet}$. The biggest advantage of EI is that the standard spectrum is obtained by bombarding a molecule with a beam of 70-eV electron so as to enable construction of the spectral library of organic chemicals. Several spectral libraries are now commercially available, with which rapid search and match can be realized in the structure identification.

CI CI is accomplished through gas phase acid–base reactions between the sample molecules and the reagent gas ions. It is easy to measure the molecular mass information due to the relatively small energy used in CI, so it is particularly suitable for compounds that fail to yield molecular ion signals in EI. The sample analyzed with CI must be volatile, but thermally unstable compounds are also suitable. The mass range of CI is the same as for EI, and the upper mass limit of the compound that can be accessed with CI is 1000 Da. The ions generated by CI can be represented by $[M+H]^+$, $[M+CH_5]^+$, or $[M+C_2H_5]^+$, which is an adduct of ions with methane as the CI reagent gas. The disadvantage of the CI spectrum is the lack of structural information, which is very adverse to the structure elucidation.

TSP TSP is a convenient ionization technique for liquid phase samples. As a gentle mode of ionization, TSP has the advantage of measuring molecular weight by intact molecular ion and its adduct ion in the spectrum. However, gentle ionization also results in the absence of structural information. It is an effective interface for LC/MS applications, and conventional columns with flow rate up to 2 mL/min can be coupled to this interface. The hyphenation with HPLC improves its application.

MALDI MALDI, a landmark development for MS, makes it possible to analyze proteins and other high molecular mass biocompounds with an improved sensitivity of several orders of magnitude. The mass range of MALDI is unlimited in theory, so that biomolecules can be analyzed with MALDI without limitation of mass range. The ions generated by MALDI are mainly singly protonated molecules. The disadvantage of MALDI is its inability to analyze low molecular mass compounds. But some new innovation methods such as suppression of the matrix signal and the use of high-mass matrices make it applicable to the analysis of low molecular mass compounds.

API API includes *ESI* and *APCI*. Both these two ion sources are important interfaces for LC/MS. APCI is applicable to relatively less polar and thermally stable compounds. ESI is not only suitable for low molecular mass compounds, but also applicable to high molecular mass compounds by forming a series of multiple charged ions, represented by $[M + nH]^{n+}$ or $[M - nH]^{n-}$, which show a bell-shaped spectrum. The wide application range of ESI makes it the most popular interface. ESI and APCI can be performed at both positive and negative ion modes and produce $[M+H]^+$ and $[M-H]^-$ ions, respectively. In the positive ion mode, some metal adduct ions, $[M+Na]^+$ and $[M+K]^+$, are usually observed. The metal ions are artificially added to form some adduct ion so that adduct ion can produce some special fragmentation information for the structural elucidation. $[2M+H]^+$ and $[M+H_2O+H]^+$ in positive ion mode and $[2M-H]^-$ and $[M+HCOO]^-$ in negative ion mode are also usually observed. These adduct ions are helpful for the determination of molecular weight.

In a word, ESI and APCI are convenient and sensitive ion sources. Their combinations to HPLC enable their wide application in many fields, including structure elucidation and quantitative analysis of chemical compounds in herbal medicine.[4] The wide application range of ESI has made it the most popular interface. The development of ion sources has empowered more ability to resolve some difficult tasks and satisfy more and more analytical requirements. For example, the invention of atmospheric pressure photoionization (APPI) has expanded the applicability of API techniques toward less polar compounds.

The Development of Mass Analyzers

The mass analyzer is an important part of a mass spectrometer. The magnetic sector mass spectrometer is the oldest type. High resolution has always been one of the goals for the development of mass analyzers. Currently, the most commonly used mass analyzers are the quadrupole (Q) and the ion trap. The Q MS has the advantage of high sensitivity, high selectivity, and multiple scan modes, so it is particularly suitable for quantification of low concentrated compounds in samples with heavy matrix. The ion trap MS has the advantage of obtaining multistage MS; thus, it is ideal for structure identification. The application of new techniques has greatly improved the performance of these two kinds of mass spectrometers.

The Q ion trap (QIT), also called a 3-D ion trap, is a 3-D analog of a Q mass filter.[5] The Q linear ion trap (LIT) is a linear or 2-D QIT. Resonance ejection and excitation are utilized to enhance mass analysis and allow isolation and activation of ions for MS^n capability. LIT improves the trapping and ion-ejection efficiency and increases the ion-storage capacity, scan speed, and detection sensitivity.[6] Recently, LITs have been combined with Q, time-of-flight (TOF), and FT ion cyclotron resonance (FT-ICR) MS and applied to analysis of small molecules and macromolecules.[7]

TOF is a well-known mass instrument for fulfilling the accurate measurement of molecular weight. The combination of TOF with MALDI is an important technique for the analysis of biomolecules. The successful combination of ESI to TOF makes it not only important in the analysis of small molecular compounds, such as the characterization of herbal medicines,[8] but also important in the analysis of protein and other biomolecules, by forming a series of multiple-charged ions.

FT-ICR-MS[9–11] and orbitrap mass analyzers[12–14] have been recently developed for acquiring high-resolution MS. The mass accuracy of FT-ICR-MS for large molecules is 1 ppm, and much better for ions whose mass–charge ratio is less than 300. In an orbitrap, the trapped ions move radially and oscillate along a central spindle electrode. The frequency of these harmonic oscillations is independent of the ion velocity and is inversely proportional to the square root of the mass-to-charge ratio (m/z). The mass resolution of orbitrap can be up to 150,000, and the mass accuracy is 2–5 ppm. The orbitrap mass analyzer also has the advantage of large-space charge capacity and wide dynamic range.

Tandem MS (MS/MS) is an effective tool to acquire fragment ions information. The tandem MS can be realized both in-space such as a triple-Q and in-time mode such as an ion trap. With the development of mass spectrometer, the tandem mass spectrometer also includes TOF-based instruments and hybrid instruments, such as Q/TOF, Q/FT-ICR,[15,16] LIT-FT-ICR instrument,[17,18] and LIT-orbitrap mass spectrometer.[19]

The main advantage of ion trap instruments is the possibility to perform multiple steps of analysis (MS^n), which provides important structural related information to MS experts for structure elucidation. The number of steps, n, is not often indicated, but if the value is specified, for example, MS^3, it indicates three stages of separation. Collision-induced dissociation (CID) is an important technique in tandem mass. In this technique, collision gases such as helium or argon are introduced into the mass instrument and collided with the target atom to result in ion activation and dissociation. The collision energy is the most important parameter during the tandem mass experiment and only the appropriate collision energy will produce abundant and stable fragmentation ions. The relative intensities of fragment ions will change with the alteration of collision energy.

There are several scan modes to acquire signals. Selection of scan mode depends on the purpose of analysis. The main scan methods are briefly introduced in the following.

Product-Ion Scan This scan method only monitors the product ions (also called daughter ions) that are formed exclusively from a mass-selected precursor ion.

Precursor-Ion Scan This scan method only monitors all precursor ions that might fragment to a common, diagnostic product ion.

Neutral-Loss Scan All precursors that undergo the loss of a specified common neutral are monitored.

Selected-Reaction Monitoring (SRM) This scan method monitors one or more chosen precursor–product pairs of the analyte. Monitoring more than one reaction is termed *multiple-reaction monitoring* (MRM).

MS has become an important and indispensable tool for structure identification of chemical compounds in herbal medicines. In comparison with NMR, IR, and UV, MS has the advantage in high-sensitivity and low-sample consumption. In addition to measurement of molecular weight and elementary composition, MS can supply abundant structure-related information. Elucidation of the MS information based on the fragmentation rule can deduce the possible structure. In a word, the development of MS makes it possible to characterize the complex sample with fast speed, high sensitivity, and abundant information. All these advantages prompt the online structure elucidation of chemical compounds in a complex sample with hyphenated technique. More novel methods have also been developed utilizing the traits of each mass spectrometer.

4.2.5 ORD and CD Spectra

ORD and CD are special absorption spectroscopies in the UV and VIS region of the spectrum. Both are used to measure the conformation of natural compounds for study of their 3-D structure. The basic principle of the two methods is the interaction of polarized light with optical active substances.

Linear polarized light can be viewed as a superposition of opposite circular polarized light of equal amplitude and phase. When a linearly polarized light wave passes through an optical active substance, the left-hand circular and right-hand circular components of the linear polarized light travel in the substance with different speeds, resulting in a change of polarization. This change depends on wavelength. This phenomenon is called ORD. If the substance has different absorption coefficients for the two circularly polarized components, the differences can be described as CD.

Since chiral compounds are widespread in nature, ORD and CD spectroscopies, particularly the latter, are very useful tools for determination of the configuration of amino acid, nucleotides, steroids, carotenoids, and other compounds with chiral carbon(s). It is also an important tool for the determination of the secondary structure of macromolecules, such as peptides, proteins, and DNA.

Before further description of the ORD and CD spectra, we first need to discuss some primary physical properties of light polarization.

Light and Polarization

In classic electromagnetic theory, light is a kind of electromagnetic wave. If the electric field vector oscillates in only one direction, then the light is described as linearly polarized. Ordinary lights, such as sunlight, are made up of waves fluctuating in all possible directions. Linearly polarized light can be obtained by passing ordinary light through a polarization filter (or polarizer). The filter allows only light with a specific angle of vibration to pass through.

The following diagrams (Fig. 4.9) show the two sinusoidally varying electric waves propagating in the *Y* direction. If the two waves are 0 or 180° out of the phase, their vector sum leads to one wave, linearly polarized at 45 or 135, respectively.

For linearly polarized light, the electric field is always orientated in one direction. For circularly polarized light, the orientation of the electric field rotates around the direction of travel, in which the electric field rotates 360° around the axis defined by the travel direction of the light. A circularly polarized light wave can be generated by combining two linearly polarized waves that are 90 or –90° out of phase, resulting in an electric field vector from the sum of the two waves rotating around the direction of travel. The following diagram (Fig. 4.10) shows the sum of the electric

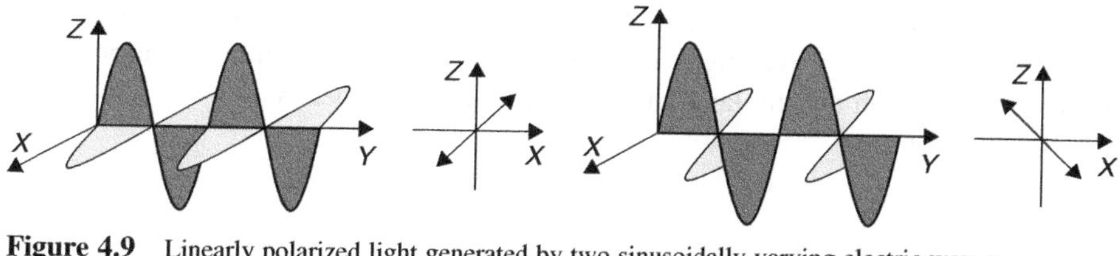

Figure 4.9 Linearly polarized light generated by two sinusoidally varying electric waves.

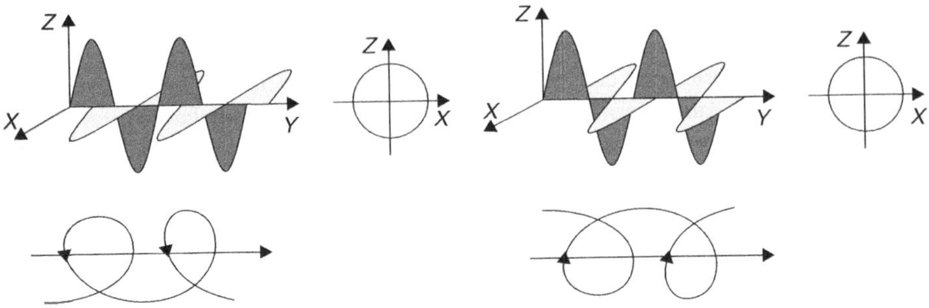

Figure 4.10 Circularly polarized light generated by two sinusoidally varying electric waves.

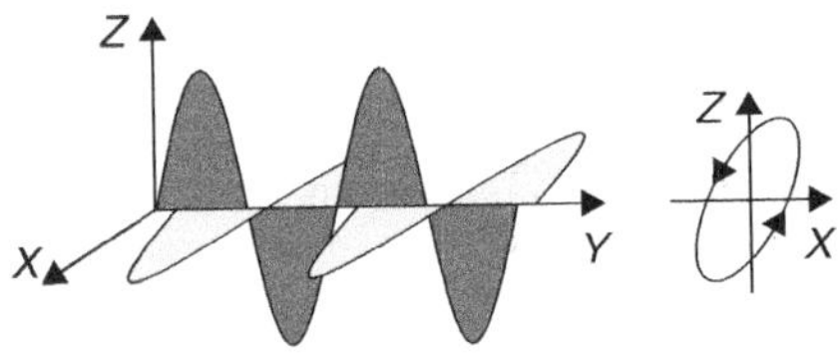

Figure 4.11 Elliptically polarized light generated by two sinusoidally varying electric waves.

field vectors for two such waves. Circularly polarized light forms a chiral left-handed or right-handed helix when it propagates in space.

The sum of the two aforementioned circularly polarized light waves will result in a linear polarized light wave. In other words, a linear polarized wave can be decomposed into two opposed circular polarized light waves.

In addition to the linear polarization and circular polarization mentioned above, there is another method, called elliptical polarization, in which the electric field vector traces out an ellipse when the phase difference is at an arbitrary angle (see Fig. 4.11).

ORD Spectroscopy

When a linearly polarized light passes through an optically active substance, the travel speed of left-hand circular and right-hand circular light composed of linearly polarized light is different because the optical substance has different absorbance indices for left and right circularly polarized light ($n_L \neq n_R$). This will result in a change of the polarization angle, which is wavelength-dependent. When the optical substance is irradiated with different wavelengths of light, different optical rotations or molar rotations will be obtained. This is called an ORD spectrum. The spectrum can be made with the wavelength as the X-axis and rotation or molar rotation as the Y-axis. The rotation and molar rotation are calculated as below.

Specific rotation:

$$[\alpha]_\lambda^T = 100\frac{\alpha}{cd}$$

α: rotation in degree

c: concentration in g/100 mL

d: path length in cm

T: temperature in Celsius

λ: wavelength in nm

Molar rotation:

$$[\Phi] = [\alpha]_{\lambda}^{T} \frac{M}{100}$$

M: the molecular weight of the substance.

CD Spectroscopy

CD spectroscopy is widely used to study chiral molecules containing one or more chiral chromophores. Professor Koji Nakanishi, a famous pioneer of natural product chemistry at Columbia University in New York, has determined many stereo-structures of complex natural compounds based on CD spectra. Now, CD has been coupled to LC for performance of online LC–CD analysis for identification of compounds in herbal extracts.[20]

In a CD experiment, equal amounts of left and right circularly polarized lights of a selected wavelength are alternately radiated into a chiral sample. Since the absorptions of a chiral molecule to left- and right-hand circularly polarized light are different, one of the two polarized light is more absorbed than the other one. The wavelength-dependent difference of absorption is recorded, yielding the CD spectrum of the sample. The spectrum can be created with wavelength as the X-axis and difference of absorbances or molar ellipticity as the Y-axis. The difference of absorbances or the molar ellipticity are calculated as below.

$$\Delta A = A_L - A_R$$

ΔA: the difference of absorbances of left and right circularly polarized light

A_L and A_R: the absorbances of left and right circularly polarized light at a given wavelength, respectively

By applying Beer's law, it can also be expressed as:

$$\Delta A = (\varepsilon_L - \varepsilon_R)cd = \Delta\varepsilon cd$$

ε_L and ε_R: the molar absorption coefficients for left and right circularly polarized light, respectively

$\Delta\varepsilon$: the difference of the two absorption coefficient

c: the molar concentration; d: the path length in centimeters (cm)

Although ΔA is usually measured, molar ellipticity ($[\theta]$) is more often used as Y-axis in CD spectra.

$$[\theta] = 3298\Delta\varepsilon$$

4.3 IDENTIFICATION OF COMPOUNDS BY HPLC AND TLC

Traditionally, identification of target compounds in a complex sample is usually carried out by comparison with their standards after HPLC or TLC separation. The principles of TLC and HPLC separation have been introduced in Chapter 3.

Identification of a target compound is generally performed first by comparing its retardation factor (Rf) in TLC and the retention time (t_R) in HPLC with that of the known standard. The Rf and t_R are characteristics for a specific compound in TLC and HPLC analysis, respectively. The compound can only be identified as the standard compound when it has the same Rf in TLC and t_R in HPLC as the standard. However, compounds with the same Rf and t_R at given chromatographic condition may not be identical in structure. An accurate identification of unknown compounds by comparison with standards on TLC or HPLC should be confirmed under at least three chromatographic conditions, for example, three different developing solvent systems for TLC and three different elution solvents for HPLC. The retention behavior of different compounds will be observed with the change of chromatographic conditions.

Identification of target compounds in comparison with standards by HPLC is usually performed by changing the mobile phase, and, if necessary, the stationary phase as well. The common mobile phases for reversed phase (RP) HPLC are methanol/water and acetonitrile/water. The t_R of a compound will change with different stationary phases, including different brand columns and/or different types of columns (ODS and CN), especially columns packed with different materials involving different separation mechanisms.

Identification of target compounds by comparing t_R with chromatography is less accurate than by comparing spectra, but spectra comparison is not suitable for identifying an unknown compound in a complex sample. Hence, the hyphenated technique combining the HPLC separation and identification of characteristic spectra has a great advantage by providing more accurate and convenient identification. DAD is a cheap and commonly used detector for HPLC. The detection range of DAD is 200 to 800 nm. UV-Vis spectra of every peak can be recorded by DAD, which permits identification by comparing spectra of an unknown compound with standards. If an unknown compound and a standard have the same t_R and same UV-Vis spectra, then the unknown compound can be identified.

Recently, the connection of HPLC-DAD to MS (HPLC-DAD/MS) has become more widely applied for unambiguous identification of natural compounds in herbal extracts by comparison with standards.[21–26] In addition to the value of t_R given by HPLC and UV-Vis spectra obtained from DAD, the MS detector provides the information of molecular weight and structural information by CID. Except for the case of chiral isomers, there is little chance of different compounds having the same retention parameters, UV-Vis spectra, molecular weight, and fragment ions in MS^2. With advantages in accuracy and rapid analysis, HPLC-DAD/MS serves as a very effective tool for identification of compounds in herbal extracts by providing varieties of information through one analysis and comparing with standards. In this section, identification of several main types of bioactive compounds in herbs by TLC and HPLC are introduced.

4.3.1 Identification of Flavonoids

Silica gel TLC is an effective method for separation and identification of flavonoids with weak polarities. The most common developing reagent for TLC separation of flavonoids is toluene-methylformate-formic acid (5:4:1), in which the proportion of toluene and formic acid can be adjusted based on the polarity of compounds. Bezene-acetone (9:1) and bezene-acetic acid (7.5:2.5) are often used to separate medium polar derivatives of flavonoids, such as methyl ethers and acetates of flavonoid.

Polyamides TLC is extensively used to separate flavonoide glycosides and flavones containing free phenolic hydroxyl groups. The developing reagents for separation of flavonoids on polyamides TLC are always very polar due to the strong affinity of flavonoids to polyamides. Ethanol, acids, and water are often used in the developers to destroy hydrogen bonding. The common developers are ethanol-water (3:2), water-ethanol-acetylacetone (4:2:1), water-ethanol-formic acid-acetylacetone (5:1.5:1:0.5), acetone-water (1:1), acetone-95% ethanol-water (2:1:2), and 95% ethanol-acetic acid (100:2).

Most separations of flavonoid by HPLC are performed on RP–HPLC columns. Chemically bonded alkyl stationary phase columns, especially ODS, are particularly suitable for the separation of flavonoids and flavonoid glycosides. The most commonly used mobile phase is a binary solvent of methanol/water or acetonitrile/water. Formic acid or acetic acid is often added into the mobile phase to improve the peak shape by inhibiting dissociation, since flavonoids are normally acidic compounds with multiple phenolic hydroxyl groups.

UV is the most widely used detector for flavonoids. Generally speaking, the maximum absorption wavelength (λ_{max}) of flavonoids is within the range of 220–280 nm produced by the A-ring (band II) and 300–400 nm produced by the C-ring (band I) (Fig. 4.12). The λ_{max} of flavone is within the range of 250–260 nm for band II and 330–350 nm for band I, while that for band I of flavonols shifts to 360–380 nm. Isoflavones and flavanones only have one λ_{max}, at about 260 nm, due to lack of a

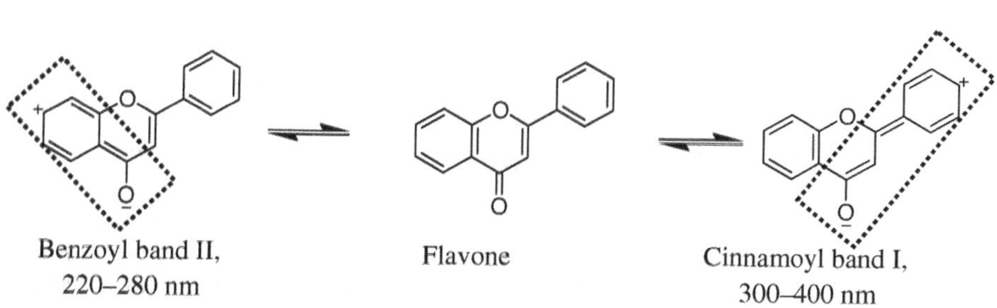

Figure 4.12 The absorption band of flavonoid compounds in UV.

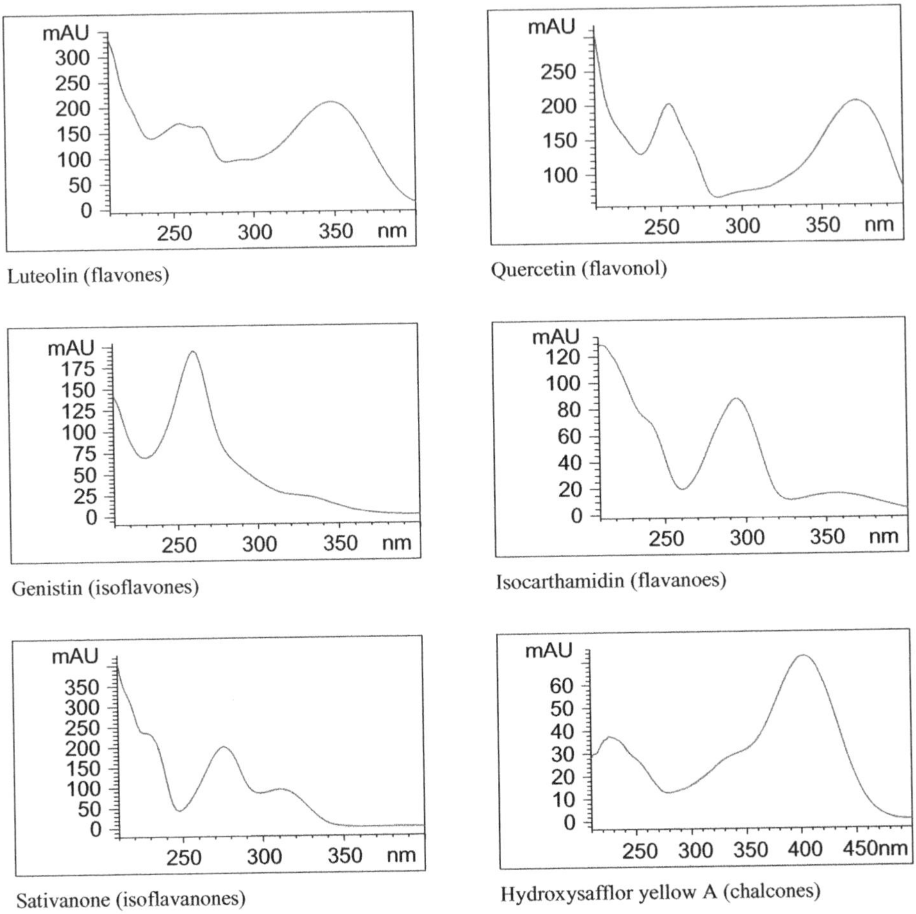

Figure 4.13 The typical UV-Vis spectra of flavonoids.

conjugated system in the C-ring. The λ_{max} of chalcone is at about 400 nm, which is a strong absorption of band I. Some examples are shown in Figure 4.13.

A recent report shows that flavonoids can also be separated with hydrophilic interaction liquid chromatography (HILIC).[27] HILIC is an alternative HPLC mode for the separation of polar compounds with polar stationary phases, as that in normal phase (NP) LC, but with aqueous mobile phases that are similar to those used in RP–LC mode.[28] In the HILIC mode, water is the strongest eluting solvent and its volume percentage in the mobile phase is usually between 5% and 40%, which is just contrary to that in RP–LC mode. Flavonoid glycosides can be separated at both RP-LC and HILIC modes, and the retention behaviors of flavonoid glycosides in these two separation modes are different. The good orthogonality between RP-LC and HILIC modes can be utilized to determine the purity of flavonoid glycosides.

4.3.2 Identification of Saponins

Silica gel TLC is a common method for identifying saponins. The most commonly used developers for saponins are $CHCl_3$–MeOH–H_2O or *n*–BuOH–HOAc–H_2O. The mixed developer saturated with water will greatly decrease the tailing phenomena. The most frequently used chromogenic reagents include Carr–Price reagent, Liebermann–Burchard reagent, phosphotungstic acid, 10% H_2SO_4 in EtOH, phenol–H_2SO_4, and 0.5% *p*-anisaldehyde and 1% H_2SO_4 in HOAc.[29]

The separations of saponins are usually performed on RP-HPLC with C_{18} columns. The mobile phase always consists of acetonitrile and water with addition of trifluoroacetic acid. Only a few saponins, for instance glycyrrhetinic acid and its glycosides, have relatively stronger absorptions in the UV range due to α, β-unsaturated ketone in their structures, and thus can be detected by a UV detector. Most saponins contain none or very few chromophore groups, and therefore have a poor response to a UV detector. The separation of saponins aglycones or their glycosides has to be traced at lower UV wavelengths, ranging from 200 to 210 nm. Some mobile phases, such as methanol, and additives, such as formic acid and acetic acid, are not suitable for the separation of saponins when a UV detector is used because their cutoff wavelengths are higher than the detection wavelength of saponins. Thus, detection is the main problem in HPLC/UV analysis of saponins.

The evaporative light scattering detector (ELSD) is an alternative detector for the determination of non-chromophoric compounds. Therefore, many qualitative and quantitative analyses of saponins are performed with ELSD.[30] When saponins are determined together with other kinds of compounds in a mixture, it is ideal to connect the UV and ELSD in series to obtain higher sensitivity for each analysis target.[31] MS is another effective tool to resolve the detection problem of saponins with higher sensitivity.[32]

4.3.3 Identification of Alkaloids

TLC has been a popular method for the identification of alkaloids. On TLC plates, some specific chromogenic reagents can react with different types of alkaloids to show different definitive colors for qualitative analysis by comparison with Rf of standards.[33] The most commonly used spray reagent is Dragendorff's reagent. Silica gel is most popularly used for TLC of alkaloids. The developers are usually composed of nonpolar solvents, such as chloroform, acetone, and acetyl acetate. Alcohols are often added as polarity adjusters, while diethylamine or ammonia is always added to reduce spot tailing. For example, acetone-water-25% ammonia (80:20:1) and bezene–diethylamine (1:1) can be used to separate the parent alkaloids and the dihydro bases of quinine and quindine.

Separation of alkaloids by HPLC is challenging work due to the interaction between the alkaline groups in alkaloids and residual silanols of the solid phase, which lead to tailing peaks, poor column efficiency, irreproducible t_R, and irreversible adsorption. Metal oxides and polymer are also used as solid phases for separation of alkaloids. However, alumina-based column shows poor selectivity and the polymer column shows low performance. Therefore, the silica-based columns are still the first choice for separations of alkaloids. Binding of shorter alkyl chain or embedding of polar groups give some shielding of the silanol groups. Rigorous endcapping for shielding the silanol group is one effective method to decrease the effect of free silanol group on the separation of alkaloid.

Peak shape of basic compounds can be improved by organic modifier. In many cases, acetonitrile is the first selected modifier at acidic pH for alkaloids separation. Tetrahydrofuran (THF) is another recommendation, which has higher inherent activity toward basic compounds.

The pH of the mobile phase is another important factor to get good peak shape and alter the selectivity of separation. Alkaloids can be separated on both acidic and basic conditions. The separations of alkaloid are mostly performed at acidic conditions. On many columns, a pH below 3 is used to obtain good peak shape because silanol groups are protonated; thus ion-exchange effects are probably limited. The common mobile phase additive is formic acid, acetic acid, and phosphoric acid at acid conditions. Buffers such as ammonium formate and ammonium acetate can be added to the mobile phase to obtain good peak shape by weakening the ion-exchange interactions. If the alkaloids are weak bases, the use of phosphate, citrate, or Tris can help to obtain good peak shape at neutral to medium basic conditions in some cases. Sometimes, ion pair reagents such as hexane-, heptane-, or octane-sulfonic acid are also used to avoid the ionic mechanism effectively. They are often used in concentrations up to 50 mM.

When the pH value is more than two units above the pK_a of alkaloid, the protonated alkaloid will be converted to their corresponding base so that the peak tailing can be resolved. But this method is impractical on the usual RP columns due to the instability of silica at basic condition. Now some silica-based stationary phases have been synthesized to bear high pH.

4.3.4 Identification of Anthraquinones

Silica gel TLC is mostly used for the isolation and identification of anthraquinones. The reported developers include hexane-acetone-tert-butanol (85:10:5),[34] benzene-ethyl acetate-methanol (5:3.5:1.5), petroleum ether-hexane-ethyl acetate-acetic acid-methanol (150:300:150:10:10), benzene-ethyl formate-formate acid-methanol-water (3:1:0.05:0.1:0.5), petroleum ether-ethyl acetate-acetic acid (87:6.4:6), and petroleum ether-ethyl formate-formate acid (15:5:1). Two-dimensional TLC method has been developed to analyze five kinds of hydroxyanthraquinones. The plate was first developed with ethyl acetate-methanol-H_2O (100:16.5:13.5) from one bottom where the sample was spotted on the corner, then developed by the second developer of petroleum ether-hexane-ethyl formate-formate acid (1:3:1.5:0.2) from the perpendicular direction where the developed spots are located as a raw at the bottom. Anthraquinone can be easily observed under UV-Vis light.

The separation of anthraquinone is mostly performed by RP-HPLC with C_{18} column. Sometimes C_8 and CN column are applied. Methanol-water and acetonitrile-water are commonly used as mobile phases for the separation of anthraquinone. Acid or buffer is often added to improve the peak shape by restraining the dissociation of phenolic hydroxyl on the anthraquinone. Anthraquinones can be easily detected with UV-Vis or DAD detector since they have several significant absorption peaks in UV-Vis spectrum (see Identification of anthraquinones by UV section).

4.3.5 Identification of Phenylpropanoids

Coumarins and lignans are two main kinds of phenylpropanoids. Silica gel TLC is commonly used to identify coumarins by observing their fluorescence spots at 365 nm under UV light. The common TLC developers are hexane-acetylacetate and chloroform-acetylacetate. Lignans are usually detected at 254 nm under UV light or by spraying with 5% sulfuric acid in ethanol.[35] The commonly used TLC developer for lignans include petrol-acetylacetate, petrol-ether, benzene-acetylacetate, and chloroform-methanol. The selection depends on the chemical structures of the targets.

Most coumarins and lignans have medium polarity, so their separation on HPLC is relatively easy. Separation of coumarins and lignans by HPLC is mostly performed using RP columns. Methanol or acetonitrile have been extensively used as organic solvent in the mobile phase. Acid is usually added in the mobile phase because of the acidity of the phenolic hydroxyl groups.

4.4 IDENTIFICATION OF COMPOUNDS BY SPECTRA

Unambiguous structure identification for a compound can be fulfilled by elucidating its spectra mainly using its UV, IR, MS, and NMR spectra. The systematic structure identification with these four spectra not only complementarily provides structural related information, but also verifies the structural elucidation from each other. More information about structure identification with these four spectra is available in many books about phytochemistry. This section will briefly introduce the main characteristics and functions of these four spectra in structure identification, and the characteristic spectra of main chemical compounds.

4.4.1 Identification of Flavonoids

Identification of Flavonoids by UV

The UV spectrum of a typical flavonoid is composed of two absorptions bands; band II (220–280 nm) is generated from the benzoyl part of the structure, while band I (300–400 nm) is generated from the cinnamoyl part. The change in position of the wavelength range of bands I and II in a compound is often used to differentiate the type of skeleton and the substituted position of the oxygen-containing group. Chemists usually use the red or blue shift to describe the movement of the absorption band in UV spectrum. A red shift refers to the moving of absorption band toward the right side of the spectrum, that is, an increase of the wavelength, while blue shift refers to moving toward the left side, that is, a decrease of the wavelength.

If aromatic hydrogen is replaced by a hydroxyl group, the red shift of band I or band II will be observed in the UV spectrum, and the value of the red shift will help to infer the substituted position of the hydroxyl group in the structure. The methylation and glycosidation of the hydroxyl group can result in a relevant blue shift of the UV absorption. On the contrary, the acetylation of the hydroxyl group can lead to a red shift of UV absorption.

A variety of diagnostic reagents, such as sodium methylate, sodium acetate, or aluminum chloride, are often added to determine the substituted position of hydroxyl groups in flavonoids based on the shift of UV absorption of band I and band II. The characteristic UV spectra of subclasses of flavonoids are listed in the following.

Flavones Band I locates between 304 and 350 nm and band II ranges from 240 to 280 nm. Substitution of hydroxyl to B-ring leads to red shift of band I, while substitution of hydroxyl to A-ring results in red shift of band II. 5-Hydroxyl substitute makes the absorption of band I 3–10 nm red shift and band II 6–17 nm red shift at the same time.

Flavonones Band II, located between 270 and 295 nm, is a strong peak due to the conjugated absorption of acetophenone derivatives. In comparison with the UV spectrum of flavone, the decreases of intensity and blue shift of band I makes it a shoulder peak of band II or a weak absorption peak.

Isoflavones Similar to flavonone, a strong absorption of band II between 245 and 270 nm is observed. When three oxygenic substituent groups connect to the A-ring, the range of band II moves to 265–270 nm. Band I often appears as a shoulder peak of band II or a low intensity absorption peak. However, band I becomes strong when positions 6 and 7 are all substituted. 5-hydroxyl substitute makes band II 7–17 nm red shift. On the contrary, methylation or glycosidation of 5-hydroxyl makes band II 5–10 nm blue shift.

Flavonols The substitution of 3-hydroxyl leads to a red shift of band I (352–385 nm).

Chalcones The wavelength and absorption intensity of band I (340–390 nm) all increase due to the opening of C-ring. At the same time, the intensity of band II (220–270 nm) weakens to become a shoulder peak or a low intensity peak. As the number of oxygen substitution increases in A- and B-ring, red shift occurs to both band I and band II; sometimes, they are divided into two sub-bands (I_a, I_b and II_a, II_b, respectively).

Identification of Flavonoids by IR

The IR spectra of flavonoid have no obvious characteristics in functional absorptions. Corresponding absorption of carbonyl, phenolic hydroxyl, benzene, and sugar can be observed. Carbonyl absorption in flavanone without substitution is between 1685 and 1680 cm^{-1}, in chalcone is at 1659 cm^{-1}, and in flavone is at 1649 cm^{-1}. Absorption of phenolic hydroxyl and hydroxyl on sugar are within 3650–3200 cm^{-1}, and the vibration of the benzene skeleton shows absorption at 1600, 1580, 1500, and 1450 cm^{-1}.

Identification of Flavonoids by MS

The main fragmentations of flavonoids aglycone occur at C-ring, resulting in structurally informative $^{i,j}A^+$ and $^{i,j}B^+$ ions. These fragment ions can be rationalized by retro-Diels-Alder (RDA) reactions and designated according to the nomenclature proposed by Ma and his colleagues[36] (see Fig. 4.14). For free aglycones, $^{i,j}A^+$ and $^{i,j}B^+$, labels refer to the fragments containing intact A-ring and B-ring, respectively, in which the superscripts i and j indicate the C-ring bonds that have been broken. For conjugated aglycones, an additional subscript 0 to the right of the letter is used to avoid confusion with the A_i^+ and B_i^+ ($i \geq 1$) labels that have been used to designate carbohydrate fragments containing a terminal (nonreducing) sugar unit. $^{i,j}A^+$ and $^{i,j}B^+$ are the most diagnostic fragments for flavonoid aglycone identification since they provide information on the number and type of substituents in A-ring and B-ring. In most classes of flavonoids, cleavage at positions 1/3 of the C-ring are the main fragmentation pathways in both positive and negative ion modes. The substitution pattern and the class of flavonoids also lead to characteristic fragmentation pathways at C-ring; for example, the additional hydroxyl group in position 3 of flavonols results in $^{0,2}A^+$ and $^{0,2}B^+$ as characteristic fragment ions in the positive ion mode, and $^{1,2}A^-$ and $^{1,2}B^-$ as characteristic fragment ions in the negative ion mode.[37] The cleavage at position 1/4 of C-ring ($^{1,4}A^+$) is the main fragmentation of flavanone in the positive ion mode. Moreover, the B-ring of flavanone is easy to cleave, and results in B^+ ion.[38]

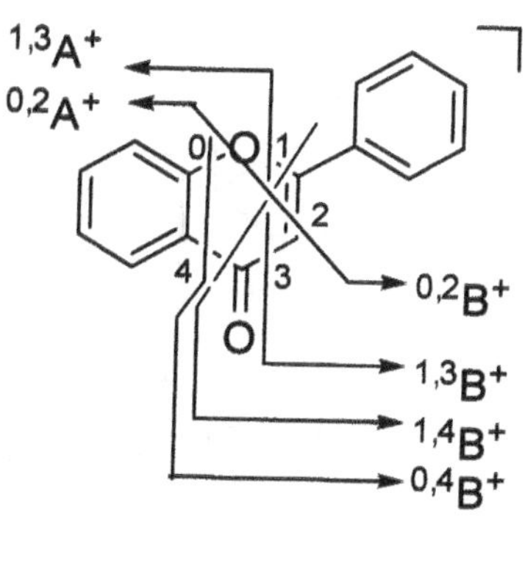

Figure 4.14 Characteristic MS fragmentations of flavonoid aglycones.

Neutral loss is also the main fragmentation of flavonoids aglycone. The most common neutral losses are 18 Da (H_2O), 28 Da (CO), 44 Da (CO_2), and 42 Da (C_2H_2O). Successive neutral losses are always observed, for example, 36 Da ($2H_2O$), 46 Da (H_2O+CO), and 56 Da (2CO). Losses of 56 Da (C_4H_8) may indicate the presence of a prenyl substituent.[39] Some specific neutral losses are helpful for judging the class of flavonoid aglycone; for example, a specific neutral loss of 30 Da can be observed in flavonols, which can reasonably be attributed to a formaldehyde (HCHO) rupturing carbonyl at C-3 position.[40] Neutral losses, such as CO, CO_2, and C_3O_2, as well as their successive neutral losses, can be observed in flavanone.

In addition to neutral loss, the radical cleavage can also be observed when the methoxy group is conjugated to the aglycone. Fragment ions losing 15 (CH_3), 29 (HCO), 30 ($2CH_3$), 33 (CH_3+H_2O), 43 (CH_3+CO), and 61 (CO+H_2O+CH_3) can be observed in polymethoxylated flavonoids.[41]

In the discussion of the fragmentation pathway of the glycosyl flavonoid, the nomenclature proposed by Domon and Costello for glycoconjugates was adopted to denote the product ions of flavonoid glycosides.[42] ${}^{k,l}X_j$, Y_j, Z_j represent the ions still containing the aglycone, in which j is the number of the interglycosidic bonds broken, counted from the aglycone, and k and l denote the cleavage within the carbohydrate rings (see Fig. 4.15).

The characteristic fragmentation of sugars in flavonoid O-glycosides is the neutral loss of sugar residue (Y type). The neutral loss of a glucose, a rhamnose, and a glucuronic acid will produce fragment ions losing 162 Da, 146 Da, and 176 Da, respectively. The neutral losses of sugar residue are used to judge the number of sugar and the category of sugars by the stepwise neutral loss of glycan. Fragmentation rules have been concluded to identify the sequence and/or the interglycosidic linkage effectively.[43] The (1→2) and (1→6) interglycosidic linkage are the most common linkage modes of glycan. In the negative ion mode, the cleavage of (1→6) interglycosidic linkage produces less intermediary ions with relevant abundance resulting from the rupture of the interglycosidic linkage.[44] The (1→2) interglycosidic linkage can produce more fragmentation ions among which the characterization of the (1→2) interglycosidic linkage was defined by the high relative abundance of the

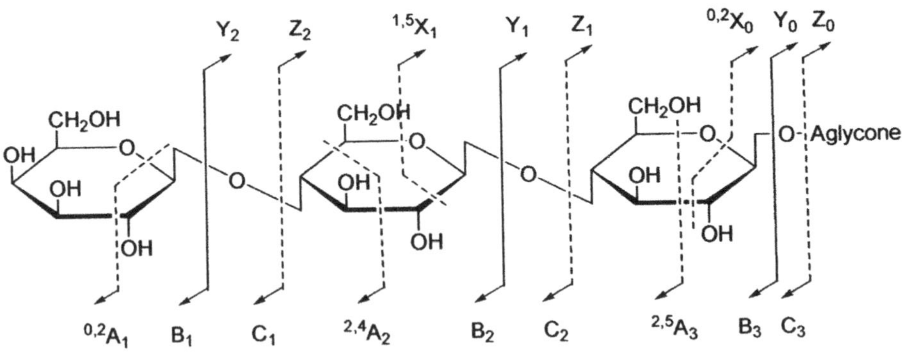

Figure 4.15 Carbohydrate ion nomenclature according to Domon and Costello.

Y^- and Z^- ions (>10%).[45] In the positive ion mode, the fragmentation of (1→6) interglycosidic linkage was unique. The rather irregular Y* ion corresponded to the loss of the inner glucose residue.[46]

Glycosylation can happen to hydroxyl at any position in flavonoids. It is difficult to determine the position of glycosylation by MS characterization. The susceptibility to cleavage of C–O bond of O-glycosyl flavonoids depends on the attachment position of sugar. To deprotonated flavonol with substitutes at 3–O and 7–O positions, the substitute at 7–O position is more easily lost than that at 3–O position in negative ion mode, while the protonated flavonol is just the contrary.[47] Under certain given condition, the relative intensity can be used to judge the glycosylation position. If intensities of fragment ions are used for elucidation of the glycosylation sites, it is highly instrument- (and parameter) dependent. Some known standard compounds should be analyzed first to get the pattern showing fragment ion ratios using the same instrument.

Great differences in fragmentation between C-glycosyl flavonoids and O-glycosyl flavonoids have been observed in MS^n, which make it easy to distinguish them. The C–C bond of C-glycosyl flavonoids is resistant to rupture, so the main cleavages happen on bonds in sugar.[48,49] The characteristic fragment ion is $^{0,2}X^+[M+H-120]^+$ in the positive ion mode. At the same time, the cleavage at the bonds in sugar produce a series of fragment ions $^{0,1}X^+$ $[M+H-150]^+$, $^{0,3}X^+$ $[M+H-90]^+$, $^{0,4}X^+[M+H-60]^+$, and $^{2,3}X^+$ $[M+H-30]^+$ (see Fig. 4.16). The successive losses of H_2O producing fragment ions $E_1^+[M+H\text{-}H_2O]^+$, $E_2^+[M+H\text{-}2H_2O]^+$, and $E_3^+[M+H\text{-}3H_2O]^+$ are also the characteristic fragmentation of C-glycosyl flavonoids. The fragmentation at bonds of sugars is always followed by successive losses of H_2O.

The fragmentation of C-glycosyl chalcone is different from other subclasses of flavonoid due to its open C-ring. In addition to the typical internal cleavage at sugar moiety, the same as other C-glycosyl flavoinds, there is also a fragmentation due to a special cleavage at carbon-carbon bond.[50] In positive ion mode, the cleavage of C-glycosyl chalcone results in $[M+H-162]^+$ ion by neutral loss, while in negative ion mode, the cleavage of C-glycosyl chalcone leads to $[M-H-163]^-$ ion by radical cleavage.[51] Another characteristic fragmentation of C-glycosyl chalcone

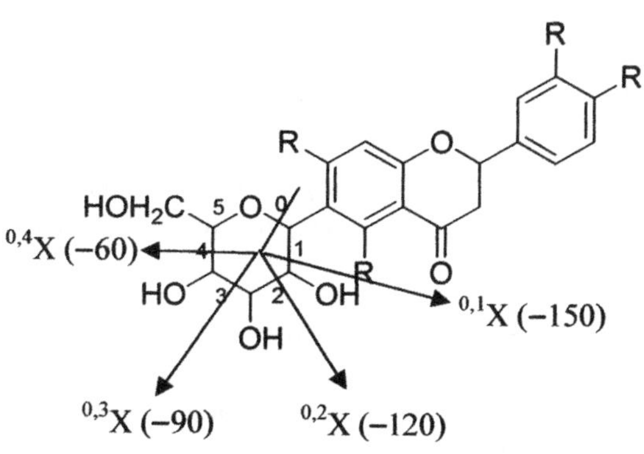

Figure 4.16 Characteristic MS fragmentations of C-glycosyl flavonoids.

is the cleavage from carbonyl group. This fragmentation produces fragment ions containing A-ring or B-ring, which will help to deduce the substituent groups on A- and B-ring.

Identification of Flavonoids by NMR

The common solvents used for determining the NMR of flavonoids include $CDCl_3$, C_5D_5N, $(CD_3)_2CO$, and DMSO-d_6, among which DMSO-d_6 is the most ideal solvent because of its good solubility for most flavonoids and reduced effect on signals of flavonoids. The ^{1}H-NMR and ^{13}C-NMR spectra of flavonoids are introduced as follows.

1. *^{1}H-NMR Spectrum of Flavonoids*

The chemical shifts of hydrogens at the C-ring are very helpful to differentiate the subclasses of flavonoids in ^{1}H-NMR spectrum. The data of some typical flavonoids are summarized in Table 4.6. The signal of H-3 in flavones and H-2 in isoflavones shows a sharp single peak in NMR. The former is at δ 6.30 ppm, while the latter is at δ 7.60–7.80 ppm. The signal of H-2 will move to low field at δ 8.50–8.70 ppm if

Table 4.6 Chemical Shifts of Protons in C-Ring of Flavonoids

		δ_{H-2} (ppm)	δ_{H-3} (ppm)
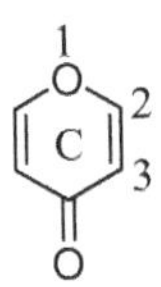	Flavone	—	6.30 s
	Flavonol	—	—
	Isoflavone	7.60–7.80 s	—
	Flavanone	5.00–5.50 dd	≈2.80 dd
	Flavanonol	4.80–5.00 d	4.10~4.30 d

Table 4.7 Chemical Shifts of Flavonoids with Hydroxyl at 5,7-position or 7-position

		δ_{H-5} (ppm)	δ_{H-6} (ppm)	δ_{H-8} (ppm)
Flavonoids with 5,7-dihydroxyl groups	Flavone, flavonol, isoflavanone	—	6.00–6.20 d	6.30–6.50 d
	7-O-glucosides of flavone, flavonol, and isoflavanone	—	6.20–6.40 d	6.50–6.90 d
	Flavanone, flavanonol	—	5.75–5.95 d	5.90–6.10 d
	7-O-glucosides of flavanone and flavanonol	—	5.90–6.10 d	6.10–6.40 d
Flavonoids with 7-hydroxyl group	Flavone, flavonol, isoflavanone,	7.90–8.20 d	6.70–7.10 dd	6.70–7.00 d
	Flavanone, flavanonol	7.70–7.90 d	6.40–6.50 dd	6.30–6.40 d

Table 4.8 Chemical Shifts of H-2′,6′ and H-3′,5′ of Flavonoids with Oxygen Substituent Group at 4′-postion

Flavonoids with –O– at $C_{4'}$	$\delta_{H\text{-}2',6'}$ (ppm)	$\delta_{H\text{-}3',5'}$ (ppm)
Flavanone	7.10–7.30 d	6.50–7.10 d
Flavanonol	7.20–7.40 d	6.50–7.10 d
Isoflavone	7.20–7.50 d	6.50–7.10 d
Chalcone	7.40–7.60 d	6.50–7.10 d
Aurone	7.60–7.80 d	6.50–7.10 d
Flavone	7.70–7.90 d	6.50–7.10 d
Flavonol	7.90–8.10 d	6.50–7.10 d

the solvent DMSO-d_6 is used. The signal of H-3 is easy to confuse with that of H-6 in flavonoids with 5,7,8-substituent groups or H-8 in flavonoids with 5,6,7-substituent groups. The signals of H-2 and H-3 of flavanones locate at 5.20 ppm (1H, dd, $J_{2,3\,trans} = 11.0$ Hz, $J_{2,3\,cis} = 5.0$ Hz) and 2.80 ppm (2H, dd, $J_{3,3} = \sim 17$ Hz, $J_{2,3\,trans} = 11.0$ Hz, $J_{2,3\,cis} = 5.0$ Hz), respectively. H-2 and H-3 in flavanonols are axial bonds in *trans* and present as signals of doublet (d, $J_{aa} = 11.0$ Hz) separately. The signals of H-α and H-β in chalcones (see structure in Fig. 4.1) also appear as doublet (d, $J = 17.0$ Hz) and their chemical shift (δ) at 6.70–7.40 ppm and 7.00–7.70 ppm, respectively.

The most common substituent group(s) at A-ring of flavonoid is hydroxyl at 7-position or 5,7-positions. Chemical shifts of H–6 and H–8 of a flavonoid having two hydroxyls at 5,7-positions are between δ 5.70 and 6.90 ppm, each being a doublet, and the signal of H–6 is at higher field relatively. When 7–OH is glycosidated, the signals of H–6 and H–8 will move to a lower field. When a flavonoid only has a hydroxyl at 7-position, the signal of H–5 appears as a doublet (d, $J = 8.0$ Hz) at δ 8.0 ppm because of the vicinal coupling with H–6. The signal of H–8 is also a doublet (d, $J = 2.0$ Hz) because of the coupling with H–6. The signal of H–6 is a doublet of doublet (dd, $J_{5,6} = 8.0$ Hz, $J_{6,8} = 2.0$ Hz) because of the coupling with both H–5 and H–8. The chemical shifts of H–6 and H–8 are between δ 6.30 and 7.10 ppm and the sequence of these two signals is not confirmed (Table 4.7).

The proton at B-ring can be divided into two groups, namely H–2′, H–6′ and H–3′, and H–5′. When a flavonoid has an oxygen group at 4′-position, signals of these two groups of protons show as doublets ($J = 8.0$ Hz) between δ 7.10 and 8.10 ppm (H–2′ and H–6′) and δ 6.50 and 7.10 ppm (H–3′ and H–5′). The oxidation of C-ring has great effect on the chemical shift of H–2′ and H–6′. More detailed information is shown in Table 4.8. For flavones and flavonol having substituent groups at 3′,4′-position, the signal of H–2′, H–5′, and H–6′ appears as doublet, doublet, and doublet of doublet, respectively. The chemical shift of H–5′ is at δ 6.70–7.10 ppm, and that of H–2′ and H–6′ are summarized in Table 4.9.

2. *^{13}C-NMR Spectrum of Flavonoids*

^{13}C-NMR is very useful in identifying the skeleton and the substituent group of flavonoids. The chemical shifts of carbon atoms in flavone without substituent group are shown in Fig. 4.17. The identification of the flavonoid skeleton is by virtue of the chemical shifts of carbons at C-ring (see Table 4.10), and the identification of the substituent position is by virtue of the chemical shift of carbons in rings A and B. When B-rings have a substituent group, the effect of the substituent group on carbons in the ring is similar to that of the simple derivative of benzene. Although the effect of substituent group on the chemical shift of carbons is usually limited in the respective ring, the substituent group at 5-position of A-ring will also affect the chemical shift of carbons in the C–ring. The signals of C–6 and C–8 of 5,7–OH flavonoid are between δ 90 and 100 ppm; $\delta_{C\text{-}6}$ is higher than $\delta_{C\text{-}8}$. The chemical shift of carbons in aglycone and glycan will change after glycosidation. The change of chemical shift will help to identify the position of the glycosidation. The common chemical shifts of carbons in aglycone of O-glycoside flavonoids are listed in Table 4.11.

Table 4.9 Chemical Shifts of H-2′ and H-6′ of Flavonoids with Oxygen Substituent Group at 3′, 4′-postion

Subclasses	$\delta_{H\text{-}2'}$ (ppm)	$\delta_{H\text{-}6'}$ (ppm)
Flavone (3′,4′-OH or 3′-OH, 4′-OCH_3)	7.20–7.30 d	7.30–7.50 dd
Flavonol (3′,4′-OH or 3′-OH, 4′-OCH_3)	7.50–7.70 d	7.60–7.90 dd
Flavonol (3′-OCH_3, 4′-OH)	7.60–7.80 d	7.40–7.60 dd
Flavonol (3′,4′-OH or 3′-O-glycoside)	7.20–7.50 d	7.30–7.70 dd

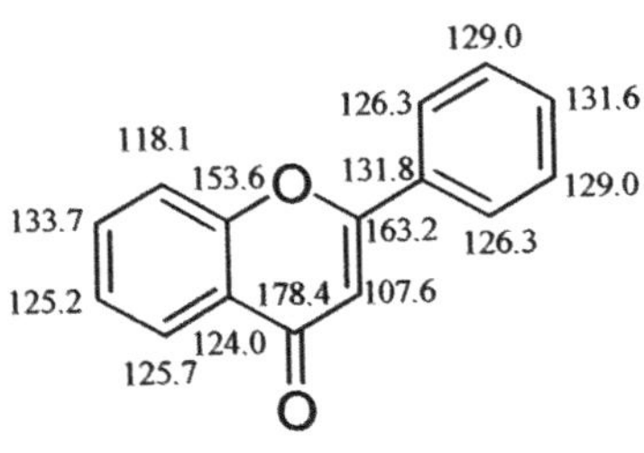

Figure 4.17 Attribution of ^{13}C-NMR of flavone without substituent groups.

Table 4.10 Chemical Shifts of Carbon Atoms in C-Ring of Flavonoids

	$\delta_{C=O}$ (ppm)	$\delta_{C\text{-}2}$ (ppm)	$\delta_{C\text{-}3}$ (ppm)
Flavone	176.3–184.0	160.0–165.0	103.0–111.8
Flavonol	172.0–177.0	145.0–150.0	136.0–139.0
Isoflavone	174.5–181.0	149.8–155.4	122.3–125.9
Flavanone	189.5–195.5	75.0–80.3	42.8–44.6
Flavanonol	188.0–197.0	82.7	71.2
Chalcone	188.6–194.6	136.9–145.4	116.6–128.1

Table 4.11 Average Changes of the Chemical Shifts of Aglycone of O-Glycoside Flavonoids

Δδ	Glycosidation position						
No.	7-O-sugar	7-O- rha	3-O-sugar	3-O- rha	5-O-glu	3′-O-glu	4′-O-glu
2			+9.2	+10.3	−2.8	−0.5	+0.1
3			−2.1	−1.1	+2.2	+0.4	
4			+1.5	+2.0	−6.0		+0.1
5			+0.4	+0.6	−2.7		
6	+0.8	+0.8			+4.4		
7	−1.4	−2.4			−3.0		
8	+1.1	+1.0			+3.2		
9					+1.4		
10	+1.7	+1.7	+1.0	+1.1	+4.3		
1′			−0.8		−1.3		+3.7
2′			+1.1		−1.2	+1.6	+0.4
3′			−0.3		−0.4	0	+2.0
4′			+0.7		−0.8	+1.4	−1.2
5′					−1.0	+0.4	+1.4
6′			+1.5		−1.2	+3.2	0

rha = rhamnose; glu = glucose.

An Example for Spectral Identification of Flavonoid

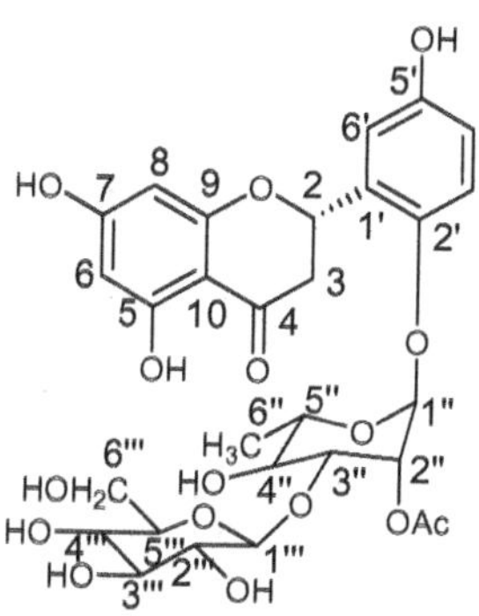

(2*S*)-5,7,5′-trihydroxyflavanone 2′-*O*-β-D-glucopyranosyl-(1→3)-α -L-2-*O*-acetylrhamnopyanoside

Compound of (2*S*)-5,7,5′-trihydroxyflavanone 2′-*O*-β-D-glucopyranosyl-(1→3)-α-L-2- *O*-acetylrhamnopyanoside is a flavonoid isolated from the rhizomes of *Cyclosorus acuminatus*.[52] Its structure was elucidated based on the following spectra.

UV (MeOH) λ_{max} nm (logε): 216 (4.12), 290(4.04), 330 (3.48)

IR (KBr) ν_{max} cm^{-1}: 3423, 2929, 1733, 1639, 1498, 1457, 810

Table 4.12 Attribution of ^{1}H- and ^{13}C-NMR Signals of (2*S*)-5,7,5′-trihydroxyflavanone 2′-*O*-β-D-glucopyranosyl-(1→3)-α-L-2-*O*-acetylrhamnopyanoside

Position	δ_H (ppm)	δ_C (ppm)
2	5.61 (dd , *J* = 13.0, 3.0)	77.3
3	3.08 (dd, *J* = 17.0, 13.0)	43.5
	2.81 (dd, *J* = 17.0, 3.0)	
4	—	198.3
5	—	166.1
6	5.92 (d, *J* = 2.2)	98.0
7	—	169.6
8	5.99 (d, *J* = 2.2)	97.1
9	—	165.4
10	—	103.8
1′	—	131.0
2′	—	148.2
3′	7.04 (d, *J* = 8.8)	118.5
4′	6.74 (dd, *J* = 8.8, 3.0)	117.4
5′	—	154.8
6′	6.99 (d, *J* = 3.0)	115.5
1″	5.31 (d, *J* = 1.7)	98.6
2″	5.40 (m)	74.3
3″	4.00 (m)	81.0
4″	3.60 (m)	73.1
5″	3.80 (m)	71.1
6″	1.30 (d, *J* = 7.4)	18.5
1‴	4.35 (d, *J* = 7.7)	106.9
2‴	3.19 (m)	75.8
3‴	3.31 (m)	78.1
4‴	3.30 (m)	71.2
5‴	3.09 (m)	78.1
6‴	3.62 (m)	62.5
	3.58 (m)	
2″-OCOCH$_3$	2.10 (s)	173.5
		21.3

ESI-MS *m/z*: 661.5 [M + Na]$^+$

HRESIMS (positive-ion mode) *m/z*: 661.1734 [M + Na]$^+$ (calcd. for $C_{29}H_{34}O_{16}Na$, 661.1745).

^{1}H-NMR (CD_3OD, 400 MHz) and ^{13}C-NMR (CD_3OD, 100 MHz): data are shown in Table 4.12.

4.4.2 Identification of Steroidal Saponins

Identification of Steroidal Saponins by UV

There is no UV absorption in saturated steroidal saponins. The presence of independent double bonds, carbonyl, α,β-unsaturated ketone, or conjugated double bonds in the structure will produce UV absorption. An absorption within 205–225 nm indicates a structure of a steroidal saponin with independent double bond(s). A weak absorption at 285 nm shows a carbonyl group in steroidal saponin. A characteristic absorption of 240 nm tells the existence of α,β-unsaturated ketone in steroidal saponin. An absorption of 235 nm suggests that the steroidal saponin has a conjugated diene.

Identification of Steroidal Saponins by IR

There are four specific absorption bands in the IR spectrum of steroidal saponin. They are at 980 cm^{-1} (A), 920 cm^{-1} (B), 900 cm^{-1} (C), and 860 cm^{-1} (D), respectively, in which band A shows the most strong absorption. The isomer in C_{25} can be distinguished by the ratio of band B and band C. Band B is stronger than band C in 25S saponin. On the contrary, band B is weaker than band C in 25R saponin.

Identification of Saponins by MS

The saponins easily produce a predominant $[M+Na]^+$ ion in the positive ion mode and $[M-H]^-$ ion in the negative ion mode. They are used to confirm the molecular mass of the saponins. The aglycones of saponins are relatively stable in the MS, so the characterization of saponins with MS mainly focuses on the interpretation of the nature, sequence, and linkage of multiple sugars. The nomenclature of fragment ions in sugar chains has been introduced in Fig. 4.15, where ions retaining the charge at the reducing terminus are termed Y and Z (glycosidic cleavages) and X (cross-ring cleavages), whereas those ions retaining the charge at the nonreducing terminus are termed B, C (glycosidic cleavages), and A (cross-ring cleavages). Cross-ring cleavage ions are designated by superscript numbers indicating the two bonds cleaved. An example to illustrate the procedure of characterization of sugar of saponin is given in Figure 4.18.[53]

In the positive ion mode, the metal ion adduct of saponins not only yields Y, Z, B, and C types of fragment ions, which can help to illustrate the aglycone, the terminal sugar, and the sequence and branching in terms of classes of monosaccharides in oligosaccahride chains, but also generates cross-ring A and X types of ions providing the information of linkages between sugar residues. In the above exampled structure, the $B_{3\alpha}$ and $C_{3\alpha}$ ions are produced from the cleavage of an ether bond between aglycone and glycan. The $B_{3\alpha}$ ion represents dehydration ion of a sodiated whole sugar chain, while $C_{3\alpha}$ is formed by glycosidic cleavages. The difference of $[M+Na]^+$ and $B_{3\alpha}$ is the mass of aglycone. This structure has two pentose terminal sugars and one deoxyhexose terminal sugar, so in the MS^2 spectra, it can be observed

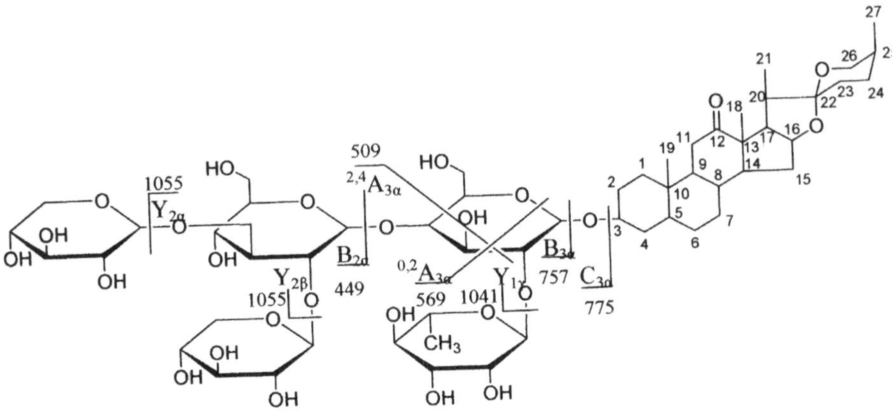

Figure 4.18 Characteristic MS fragmentations of saponins.

that the neutral losses of these terminal sugars produce Y type fragment ions ($Y_{2\alpha}$, $Y_{2\beta}$ and $Y_{1\gamma}$) with strong intensity. The successive fragmentation from $Y_{2\alpha}$, $Y_{2\beta}$, and $Y_{1\gamma}$ fragment ions will show low intensity due to more bond cleavages.

The tandem MS characterization of the B type ion can be used to elucidate the nature, sequence, and branching of sugar, for example, the CID spectrum of the whole sugar chain ($B_{3\alpha}$, *m/z* 757) produced the ion at *m/z* 625 by losing a pentose ($B_{3\alpha}$-132) and *m/z* 611 by losing a deoxyhexose ($B_{3\alpha}$-146); both are the neutral loss of terminal sugar. The CID spectrum of *m/z* 611 produced the ion at *m/z* 479 ($B_{3\alpha}$-146-132) and *m/z* 449 ($B_{3\alpha}$-146-162, $B_{2\alpha}$), and the CID spectrum of *m/z* 449 produced only one daughter ion at *m/z* 317 ($B_{2\alpha}$-132), which prove that two pentose residues connect with a hexose residue, respectively.

After the nature, sequence, and branching of the sugar chain are determined, the linkage of sugar is deduced by cross-ring A and X types of ions, for example, $^{0,2}A_{3\alpha}$ and $^{2,4}A_{3\alpha}$ prove that the deoxyhexose residue linked with the hexose residue at position 2, and another sugar residue linked with the hexose residue at position 3 or 4. The fragment of *m/z* of $^{0,3}A_{3\alpha}$ further confirms that the second sugar residue is linked with the hexose residue at position 4 rather than 3.

Adding a metal ion to a saponin molecule generates a new bond, which is an exothermic reaction. So the energy released is deposited into the internal degree of freedom and may excite more bond cleavages under CID conditions. In general, $[M+Na]^+$ ion gives a fair degree of fragmentation via several stages of CID, providing both sugar sequence and linkage information. Because of the inevitable presence of sodium ions during the process of sample preparation and the strong affinity of sugar to sodium ions in the gas phase, the fragmentation of $[M+Na]^+$ ion of saponins is usually used to identify the structure of saponins.

The fragmentation of metal ion adducts of saponins has been compared. Alkali metal ions (Li^+, Na^+, K^+) and other metal ions (Ag^+, Rb^+, Cs^+) affect the degree of fragmentation tremendously. With increasing atomic number of metal, the tendency of fragmentation from glycosidic cleavage and cross-ring cleavages decreases, for

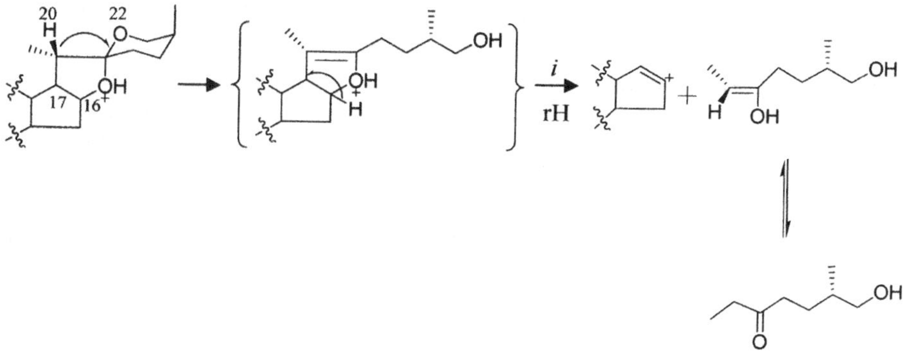

144 mass units

Figure 4.19 Characteristic MS fragmentations of the aglycones of steroidal saponins.

example, $[M+Li]^+$ ion can produce much more fragment ions, including cross-ring ions, than $[M+Na]^+$ ion, which may provide more detailed information to interpret the structure of saponins in some cases. $[M+Ag]^+$ ion is mainly fragmented through glycosidic cleavages. For adduct ions of $[M+Rb]^+$ and $[M+Cs]^+$, no fragmentation is observed.

Compared with the stable aglycone of triterpenoidic saponins, the E-ring of steroidal saponins can cleave to produce the characteristic fragment ion eliminating 144 Da. The proposed fragmentation pathway includes the rearrangement of the hydrogen atom as C-20 moves toward the oxygen atom in the F-ring, which results in the cleavage of the C-22-O bond. Then the transfer of the hydrogen atom from C-16 to C-20 accompanies the C-17-C-20 bond cleavage; meanwhile, the positive charge on the oxygen of the E-ring migrates to the D-ring to form a stable ion involving the loss of 144 Da (see Fig. 4.19).[54]

Identification of Steroid Saponins by NMR

In the ^{1}H-NMR spectrum of steroids, four characteristic signals of methyl protons at positions 18, 19, 21, and 27 (see structure in Fig. 4.2) can be observed in higher fields. The signals of methyls at 18-CH_3 and 19-CH_3 are singlet; the former is at a higher field. Two doublets are observed representing 21-CH_3 and 27-CH_3; the latter is at a higher field. If there is a hydroxyl group at C-25, the signal of 27-CH_3 as a singlet will shift down to the lower field. The shift of α-H-27 (equatorial bond, 25R) is at a higher field than that of β-H-27 (axial bond, 25S); thus the chemical shift of H-27 can be used to distinguish the isomer 25R from 25S.

If 27 carbon signals can be observed in ^{13}C-NMR spectrum, this compound might be a steroid saponin. The carbon signals of C-16 and C-22 are characteristic in ^{13}C-NMR spectra of steroid saponins. For spirostanols and isospirostanols, the chemical shifts of C-16 and C-22 are at δ 80 and 109 ppm, respectively. For pseudo-spirostanols whose F-ring is a furan ring, the carbon signal of C-22 is at δ 121 ppm and the carbon signal of C-25 is at δ 86 ppm. For furostanols, the carbon signal of

C-22 is at δ 90 ppm. When a hydroxyl group links to C-22, the carbon signal is observed at δ 111 ppm.

NMR is very useful for structure elucidation of glycoside compounds. Compared with MS, NMR is used to identify not only the nature of aglycone and glycan, the number and interlinkage of sugars, but also the substituent sites and configuration of glycosidic bonds.

The types of sugars can be determined based on chemical shifts in both ^{1}H-NMR and ^{13}C-NMR spectra, as well as adjacent coupling constants of protons in ^{1}H-NMR or 2D-NMR. As the anomeric proton resonances are usually well discerned, they commonly serve as a special window for identifying the number of sugars. It is easy to confirm the glycosidation position based on the change of shift by comparing the ^{13}C-NMR spectra of glycoside and aglycone. Glycosidation generally results in the chemical shifts of carbons in aglycone that directly link to the sugar (C-α) shifts to a lower field by 4–10 ppm and the adjacent carbons (C-β) shifts to a higher field by 1–4 ppm. The glycosidation position of aglycones and the linkage position of sugars can also be determined by HMBC.

NMR is also used to confirm the configuration (α- or β-) of glycosidic linkages based on different coupling constants of the end proton. For xylose, glucose, and galactose, when $J_{1,2}$ is ~8 Hz, it indicates a β-glucoside; when $J_{1,2}$ is ~4 Hz, it indicates an α-glucoside. However, it does not work for rhamnose and mannose due to their similarity in coupling constants.

An Example of Spectral Identification of Steroid Saponins

Filiasparoside D Filiasparoside D is a cytotoxic steroid saponin isolated from the roots of *Asparagus filicinus*.[55] The structure was elucidated based on the following spectra.

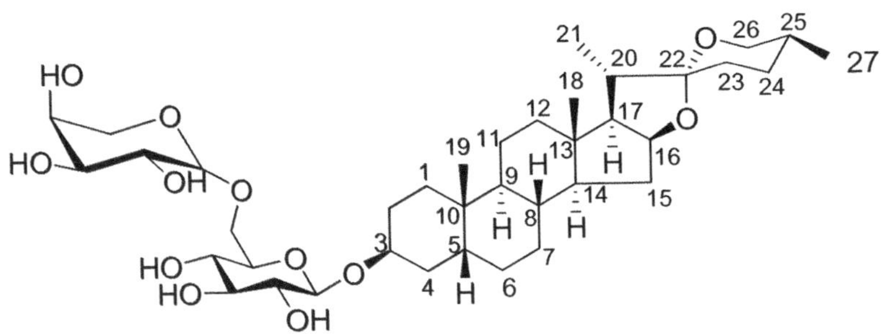

Filiasparoside D

IR (KBr) ν_{max} cm^{-1}: 3417 (OH), 2935 (CH), 1450 (OH), 1068 (OH), 986, 918, 897, 850

HRESIMS *m/z*: 709.4167 (calcd for $C_{38}H_{62}O_{12}$ [M–H]$^-$, 709.4163)

^{1}H-NMR (pyridine-*d5*, 400 MH$_Z$) and ^{13}C-NMR (pyridine-*d5*, 100 MH$_Z$): see Table 4.13.

Table 4.13 Attribution of ^{1}H- and ^{13}C-NMR Signals of Filiasparoside D

Position	δ_H (ppm)	δ_C (ppm)
1	1.68, 1.76 (m)	30.4
2	1.70, 1.98 (m)	26.8
3	4.38 (m)	74.3
4	1.48, 1.70 (m)	30.9
5	1.98 (m)	36.8
6	1.02, 1.05 (m)	26.8
7	0.93, 1.22 (m)	26.6
8	1.45 (m)	35.4
9	1.02 (m)	40.1
10		35.1
11	1.30, 1.26 (m)	21.0
12	1.24, 1.64 (m)	40.1
13		40.7
14	1.03 (m)	56.3
15	1.38, 2.00 (m)	32.0
16	4.57 (m)	81.2
17	1.82 (m)	62.8
18	0.78 (s)	16.5
19	0.80 (s)	23.7
20	1.99 (m)	42.3
21	1.14 (d, J = 7.0)	14.8
22		109.6
23	1.42, 1.88 (m)	26.2
24	1.33, 2.12 (m)	26.1
25	1.56 (m)	27.4
26	3.36 (d, J = 9.3) 4.04 (m)	64.9
27	1.04 (d, J = 7.0)	16.1

4.4.3 Identification of Triterpenes

Identification of Triterpenes by UV

Most of triterpenes have no conjugate system, so there is no characteristic UV absorption. If there are two or more unsaturated groups in the structure, UV spectra can help to determine whether they are conjugated.

Identification of Triterpenes by IR

The three types of triterpenes (oleanane, ursane, and tetracyclic triterpenoids) can be easily differentiated based on absorption in region A (1392–1355 cm^{-1}) and region

B (1330–1245 cm^{-1}). The oleanane type shows two absorption peaks in region A (1370–1355 cm^{-1} and 1392–1379 cm^{-1}) and three peaks in region B (1269–1250 cm^{-1}, 1306–1299 cm^{-1}, and 1330–1315 cm^{-1}). The ursane type shows three absorption peaks in region A (1364–1359 cm^{-1}, 1383–1370 cm^{-1}, and 1392–1386 cm^{-1}) and three peaks in region B (1250–1245 cm^{-1}, 1276–1270 cm^{-1}, and 1312–1308 cm^{-1}). The tetracyclic triterpenoid shows only one peak in regions A and B, respectively.

Identification of Triterpenes by NMR

^{1}H-NMR spectra of triterpenes are complex since there are many signals at high-field region. The characteristic of triterpene is that many signals as singlet of methyl appear between δ 0.5 and 2.0 ppm. The skeleton of triterpene can be determined based on the number and peak shape of methyl signals. More than three singlets of three-hydrogen in a high-field region can be found in ^{1}H-NMR spectrum of triterpene, but not in that of steroids. The signal of one-hydrogen is between δ 4.3 and 6.0 ppm. The chemical shift of proton of endo-double bond is usually more than 5 ppm, while the chemical shift of proton of exocyclic double bond is less than 5 ppm.

In ^{13}C-NMR spectrum, the chemical shifts of carbons are always under 60 ppm except for carbon linked with oxygen and olefinic carbon. Signals between δ 60 and 105 ppm represent carbons that are linked with oxygen in aglycones and sugars; signals between δ 109 and 160 ppm represent olefinic carbons; signals between δ 170 and 220 ppm represent carbonyl carbons. The characteristic that distinguishes triterpene from steroid is that most triterpenes produce 30 carbon signals, but the carbon signals of steroids are normally less than 30. According to the number of signal of quaternary carbon in ^{13}C-NMR spectrum, types of triterpene skeletons can be differentiated from each other. Oleanane triterpene contains six quaternary carbons (C-4, C-8, C-10, C-14, C-17, C-20) with chemical shifts between δ 37 and 42. Ursane triterpene and lupane triterpene have five quaternary carbons (C-4, C-8, C-10, C-14, C-17).

An Example of Spectral Identification of Triterpenes

Cycloart-24-en-3β,23(R)-diol 3-sulfate Cycloart-24-en-3β,23(*R*)-diol 3-sulfate is a triterpene isolated from the red Alga *Tricleocarpa fragilis*.[56] The structure was elucidated based on the following spectra.

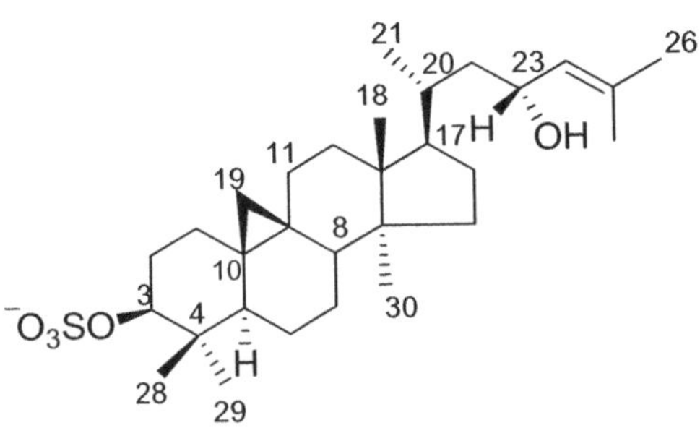

Cycloart-24-en-3*β*,23(*R*)-diol 3-sulfate

UV λ_{max} nm (in MeOH): 201

IR ν_{max} cm^{-1} (film on NaCl): 3400 br, 2930, 2869,1632 br, 1467, 1462, 1453, 1444, 1374, 1202 str/br, 1068, 1053,945, 841

FABMS (negative-mode) *m/z*: (rel int) 521 $[M]^-$

HRFABMS *m/z*: $[M]^-$ 521.3311 (Cald for $C_{30}H_{49}O_5S$, 521.3301)

^{1}H-NMR and ^{13}C-NMR: (DMSO-*d6*) data are shown in Table 4.14.

Table 4.14 Attribution of ^{1}H- and ^{13}C-NMR Signals of Cycloart-24-en-3β,23(*R*)-diol 3-sulfate

Position	δ_H (ppm)	δ_C (ppm)
1	1.42α (m), 1.18β (m)	31.3
2	2.06α (m), 1.43β (m)	27.4
3	3.67 (dd, *J* = 4.1,10.6)	82.2
4		39.6
5	1.26 (m)	47.3
6	1.50 (m)	20.6
	0.76 (m)	
7	1.03 (m)	25.5
	1.27 (m)	
8	1.48 (m)	47.3
9	—	19.5
10	—	25.8
11	1.91α (m), 1.13β (m)	25.9
12	1.58 (m)	32.5
13	—	44.9
14	—	48.4
15	1.24 (m)	35.0
16	1.82α (m), 1.25β (m)	27.7
17	1.51 (m)	52.3
18	0.93 (s)	17.9
19	0.30α (d, *J* = 3.9), 0.48β (d, *J* = 3.6)	29.1
20	1.58 (m)	31.9
21	0.86 (d, *J* = 5.9)	18.2
22	0.85 (m), 1.49 (m)	44.4
23	4.21 (m)	64.0
24	5.09 (dt, *J* = 8.2,1.3)	131.0
25	—	129.5
26	1.61 (d, *J* = 1.0)	25.4
27	1.57 (d, *J* = 1.0)	17.8
28	0.87 (s)	25.6
29	0.71 (s)	15.0
30	0.84 (s)	19.0
23-OH	4.29 (d, *J* = 4.9)	

4.4.4 Identification of Anthraquinones

Identification of Anthraquinones by UV

Anthraquinones are highly unsaturated; four absorption bands appear in their UV spectra (240–260 nm, 262–295 nm, 305–389 nm, >400 nm). Their intensity and wavelength depend on the nature, number, and arrangement of substituted groups. Anthraquinone, with only one phenolic hydroxyl, shows absorption at about 230 nm. With an increase of the number of phenolic hydroxyl groups, the absorption of anthraquinone is red shifted. Although UV spectra can provide structural related information to a certain extent, it is only considered as a complementary structural analysis tool due to many exceptions.

Identification of Anthraquinones by IR

IR spectra can provide very useful information in structural identification of anthraquanones, particular helping to determine the positions of substitutes. The characteristic IR absorption peaks of anthraquinones is at 1675–1653 cm^{-1}, 3600–3150 cm^{-1}, and 1600–1480 cm^{-1}, which represent the stretching vibration of carbonyl group, stretching vibration of hydroxyl, and benzene skeleton vibration, respectively. Attention should be paid to the effect of hydroxyl groups on the absorption of carbonyl group in anthraquinones. Absorptions of the two carbonyls in anthraquinone without any substitution are identical. The α-OH can associate with the carbonyl group and make the absorption of carbonyl group greatly decrease. Therefore, in the IR spectrum of anthraquinone with one α-OH, two absorption peaks of the carbonyls are observed (1675–1647 cm^{-1} and 1637–1621 cm^{-1}). For the same reason, 1,8-dihydroxyanthraquinone also has two peaks of the carbonyls, at 1678–1661 cm^{-1} and 1626–1616 cm^{-1}, while 1,4- or 1,5-dihydroxyanthraquinone shows only one absorption peak of the carbonyls, at 1645–1608 cm^{-1}. Anthraquinone with three or four α-OH has only one weak peak at 1616–1592 cm^{-1} and 1592–1572 cm^{-1}, respectively, which sometimes overlaps with the skeleton vibration of C=C.

Identification of Anthraquinones by MS

The skeleton of anthraquinones is relatively stable, so fragmentation of anthraquinones mainly occurs at its substituent groups. The anthraquinones aglycones can be ionized in the negative ion mode. CO elimination may originate from C-9 or C-10, but a hydroxyl group at C-1 and/or C-8 may protect the CO elimination of C-9 due to the hydrogen bonding. The common substituent groups of anthraquinones are hydroxyl, hydroxymethyl, methoxyl, and carboxyl. Various combination of these substituent groups results in many isomers of anthraquinones, which can be differentiated by the tandem MS. The anthraquinones with one hydroxyl can produce fragment ions by losing a H_2O. The anthraquinones with one hydroxymethyl are easy to occur radical rupture and produce $[M{-}H{-}CHO]^{-\bullet}$ fragment ion. Losing a methoxyl radical and a CO_2 is a characteristic of the anthraquinones with one methoxyl and carboxyl, respectively. Based on these characteristic fragmentations, the nature of substituent groups of anthrquinones can be confirmed, but the conjunctive sites are not easy to judge.[57] Anthraquinone glycoside can be detected with both the positive and negative ion modes. Glucose is the most common sugar in anthraquinone glycoside, so fragment ion losing 162 Da is the characteristic ion for anthraquinone glycoside.

Identification of Anthraquinones by NMR

The aromatic protons in anthraquinones are divided into α-H and β-H (see structure in Fig. 4.6). Due to the impact of the carbonyl group, the chemical shift of α–H in the deshielding region of the carbonyl group shifts to the lower field at δ 8.07 ppm in ^{1}H-NMR. However, the carbonyl group has little effect on β-H, so it appears at δ 6.67 ppm in the relatively higher field. To substituted anthraquinones, the isolated aromatic protons displays as signals of singlet, the two ortho-protons display as signals of doublet (d, $J = 6.0$–9.4 Hz), and two meta-protons display as signals of doublet (d, $J = 0.8$–3.1 Hz), respectively. The hydrogen bond between α-OH and the carbonyl group make the signal of α-OH shift to a lower field in ^{1}H NMR spectrum. When only one α-OH presents in the structure, its chemical shift is greater than 12 ppm. When two hydroxyl groups present at α-position of a carbonyl group, the signals appear at δ 11.6–12.1 ppm.

The aromatic carbons in anthraquinones are divided into four kinds: α-carbon, β-carbon, carbonyl carbon, and quaternary carbon. If there is no substituent, their signals are at δ 127, 134, 182, and 133 ppm in ^{13}C-NMR, respectively. When only one ring is substituted, the carbon signals of another ring show only little change, suggesting the substituents have little transannular effect. In words, the influence of the substituents at one ring on chemical shifts of another ring in ^{13}C-NMR spectrum could be taken out of consideration.

An Example of Spectral Identification of Anthraquinones

2-Methoxy-1,3,6-trihydroxyanthraquinone 2-Methoxy-1,3,6-trihydroxyanthraquinone is an anthraquinone isolated from the fruits of *Morinda citrifolia* (Noni).[58] The structure was elucidated based on the following spectra.

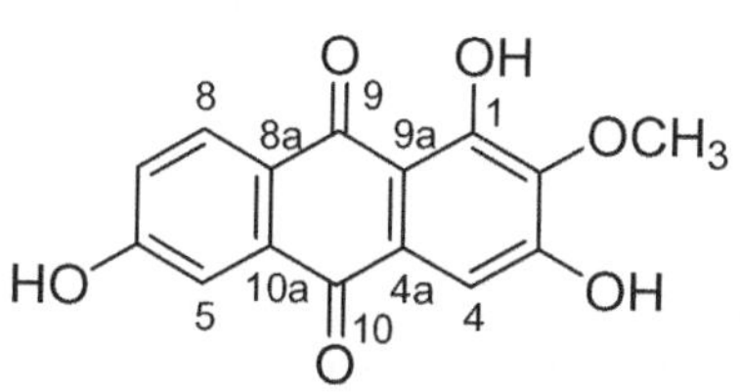

2-Methoxyl-1,3,6-trihydroxyanthraquinone

Table 4.15 Attribution of ^{1}H- and ^{13}C-NMR Signals of 2-methoxy-1,3,6-trihydroxyanthraquinone

Position	δ_H (ppm)	δ_C (ppm)
1	—	156.9
2	—	140.3
3	—	157.1
4	7.32(s)	109.5
4a	—	130.0
5	7.55(br s)	113.5
6	—	163.6
7	7.17(br d)	121.5
8	8.15(d, J = 8.4)	129.8
8a	—	126.0
9	—	187.1
9a	—	111.1
10	—	183.2
10a	—	136.0
OMe-2	4.02(s)	60.9

UV (MeOH) λ_{max} nm (logε): 212 (4.22), 280 (4.37), 301 (3.97, sh), 414 (3.57, br)

IR (dried film) ν_{max} cm^{-1}: 3391, 2926, 2853, 1731, 1595, 1455, 1393, 1365, 1290, 1130, 1092

LREIMS (70 eV) *m/z*: 286 $[M]^+$ (15), 201 (37), 103 (100)

HREIMS *m/z*: 309.0370 $[M + Na]^+$ (calcd for $C_{15}H_{10}O_6Na$, 309.0375)

^{1}H-NMR ($CDCl_3$, 400 MHz) and ^{13}C-NMR ($CDCl_3$, 100 MHz): data are shown in Table 4.15.

4.4.5 Identification of Alkaloid

Identification of Alkaloid by UV

It is difficult to summarize the UV characteristics of alkaloid due to its complex and numerous subclasses. UV spectrum can help to judge the aromatic nitrogen heterocyclic, such as pyridine (λ_{max} 195 nm, 257 nm, 270 nm), quinoline (λ_{max} 230 nm, 270 nm, 314 nm), isoquinoline (λ_{max} 238 nm, 279 nm, 320 nm), inode (λ_{max} 280~290 nm).

Identification of Alkaloid by IR

IR can help to distinguish between primary amine and secondary amine. The former shows two sharp absorptions at 3500–3150 cm^{-1}, while the latter shows absorption at 3400 cm^{-1}.

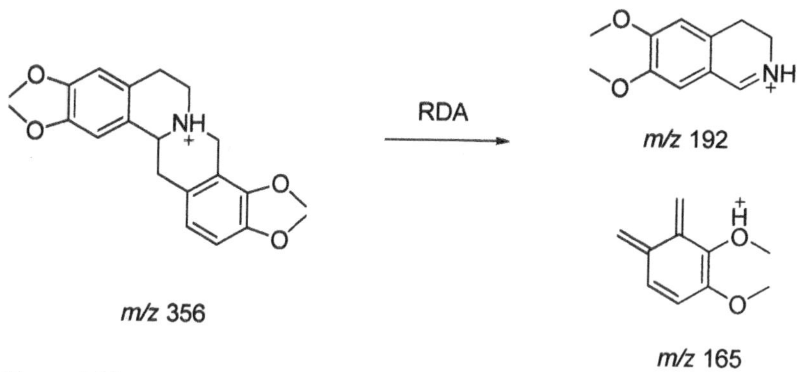

Figure 4.20 Proposed retro-Diels-Alder pathway for tetrahydropalmatine.

Identification of Alkaloid by MS

Alkaloids are usually detected in the positive ion mode because of the nitrogen atom in their structures. The number of nitrogen atoms can be deduced with the N rule in MS. The fragmentation at the nitrogen atom will lead to the cleavage of the alkaloid skeleton, such as RDA fragmentation. Due to the complexity of alkaloid structures, only characteristics of protoberberines and berberine types of alkaloids MS spectra are introduced in this chapter.

The fragment produced by RDA cleavage is the characteristic signal in the MS spectrum of protoberberines, a type of tertiary alkaloid. For example, the fragmentation of tetrahydropalmatine produces a predominant ion appearing at *m/z* 192, which corresponds to RDA reaction of C-ring opening. Two fragment ions produced from this fragmentation mode were obtained from the part of tetrahydroisowuinolin and the part of benzene-ring at *m/z* 192 and 165, respectively[59] (see Fig. 4.20).

Different from the tertiary alkaloids, it is difficult for quaternary alkaloids, such as berberine and palmatine, to open the ring due to the conjugated plane structure; thus, no fragments are produced by RDA reaction in MS^2 spectra. The fragment ions of quaternary alkaloids are produced by losing small molecular with neutral loss or radical rupture. For example, the predominant fragment ion appearing at *m/z* 337 in MS spectrum of palmatine (*m/z* 352) was produced by losing a methyl radical of methoxyl ($[M-15]^{+\bullet}$). Then the successive loss of an H radical and a CO neutral molecular produces the fragment ion at *m/z* 308 (see Fig. 4.21). Further neutral loss and radical rupture will lead to more fragmentation ions. For example, loss of an H radical from *m/z* 337 produced fragment ion at *m/z* 336, and successive loss of an H_2 produced *m/z* 334, and then a methyl radical rupture from *m/z* 334 led to the fragment ion at *m/z* 319.[60]

Identification of Alkaloid by NMR

Nitrogen atoms in alkaloids give a great impact to the atoms around them. In ^{1}H-NMR spectrum, chemical shifts of protons connected with different types of nitrogen

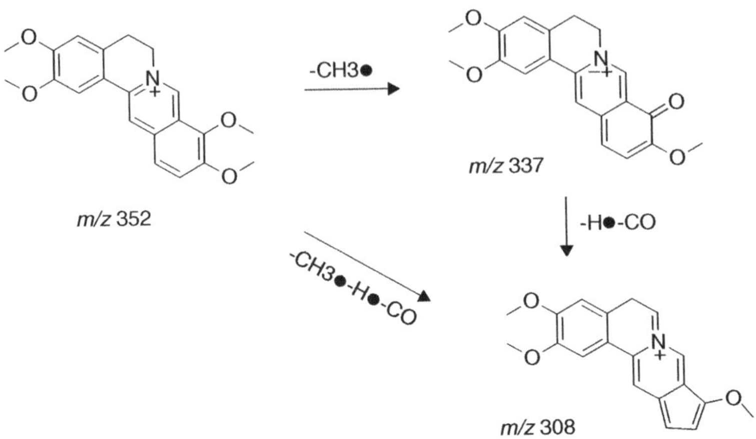

Figure 4.21 Proposed MS fragmentation pathway for palmatine.

atoms are different, for example, proton of aliphatic amine is at δ 0.3–2.2 ppm; proton of aromatic amines is at δ 2.6–5.0 ppm; and proton of amide is at δ 5.2–10.0 ppm. Chemical shifts of methyl connected to different types of nitrogen atoms are also different, for example, methyl connected with tertiary amine is at δ 1.97–2.56 ppm; methyl connected with secondary amine is at δ 2.3–2.5 ppm; methyl connected with quarter amine is at δ 2.7–3.5 ppm; methyl connected with aromatic tertiary and secondary amine and amide is at δ 2.6–3.1 ppm.

In ^{1}H-NMR spectrum, the spin–spin coupling of H–N–C–H is difficult to observe due to the rapid proton exchange of N–H. If there is a CH–NH–C= group in alkaloids, such as an enamine, aromatic amine, and amide, a splitting peak can often be observed. The coupling constant (*J*) can help to determine the configuration and conformation of the compound. For example, the $J_{H\text{-}N\text{-}C\text{-}H}$ is 5~6 Hz when the bond of H–N–C–H can freely rotate.

^{13}C-NMR spectrum is one of the most important methods to identify structures of alkaloids. With increasing electro-negativity of the nitrogen atom, the chemical shift of carbon atom around the nitrogen atom shifts to the lower field. The general rule is α-carbon > β-carbon > γ-carbon. The connected methyl group moves to the low field between δ 30–47 ppm.

An Example of Spectral Identification of Aalkaloid

14,15-Didehydro-10,11-dimethoxy-16-epivincamine The compound 14,15-Didehydro-10,11-dimethoxy-16-epivincamine is an alkaloid isolated from a plant in *Ervatamia* genus.[61] The structure was elucidated based on the following spectra.

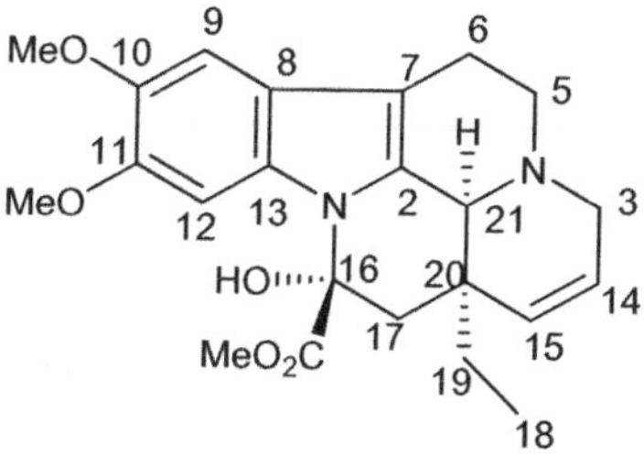

14,15-Didehydro-10,11-dimethoxy-16-epivincamine

UV (MeOH) λ_{max} nm (log ε): 303 (3.92), 299 (3.91), 275 (3.86), 228 (4.42)

IR (KBr) ν_{max} cm^{-1}: 3427, 2929, 2852, 1736, 1626,1481, 1443, 1365, 1265, 1230, 1205, 1163, 1107, 1032, 781, 719

EIMS *m*/*z*: 412 $[M]^+$ (100), 394(9), 383 (19), 365 (23), 353 (14), 344 (50), 326 (13), 310 (43), 309(33), 295 (10)

HREIMS *m*/*z*: 412.1971 (calcd for $C_{23}H_{28}N_2O_5$, 412.1998)

^{1}H-NMR and ^{13}C-NMR ($CDCl_3$): spectral data is shown in Table 4.16.

4.4.6 Identification of Iridoids

Iridoids are a type of monoterpenes derived from geraniol. Their characteristic is having the general form of cyclopentanopyran (see structure in Fig. 4.5). Cleavage of a bond in the cyclopentane ring gives rise to a subclass known as seco-iridoids. Iridoids are typically found in plants as glycosides; the sugar is mostly glucose.

Identification of Iridoids by UV

Iridoids with –COOH or –COOR substituent group at C_4 position show strong absorption at 230–240 nm in UV spectrum because of the α, β-unsaturated acid or ester.

Identification of Iridoids by IR

IR spectrum can be used not only to distinguish iridoid from secoiridoid, but also to determine the present of substitution of –COOR on C_4 and substitution of hydroxyl, carbonyl, double bond, and epoxy group in the part of cyclopentane.

- A characteristic strong absorption around 1640 cm^{-1} is caused by the stretching vibration of the vinyl ether double bond in the skeleton.
- A strong peak appearing at 1680 cm^{-1} is attributed to the carbonyl absorption of α, β-unsaturated ester. It can be used to distinguish substitution of –COOR on C_4 from the other substituent group such as $-CH_3$ and $-CH_2OH$ or non-substitution at C_4 position.

Table 4.16 Attribution of ^{1}H- and ^{13}C-NMR Signals of 14,15-didehydro-10,11-dimethoxy-16-epivincamine

Position	δ_H (ppm)	δ_C (ppm)
2	—	131.2
3	3.01 (br s)	43.7
5	3.38 (dd, J = 13.8, 6.9)	49.8
	3.24 (ddd, J = 13.8,11.0, 6.3)	
6	3.07 (m)	16.7
	2.49 (ddd, J = 16.0, 6.3, 1.8)	
7	—	106.3
8	—	121.6
9	6.88 (s)	100.1
10	—	145.2
11	—	146.1
12	7.06 (s)	96.9
13	—	131.0
14	5.48 (dt, J = 10.3, 3.1)	125.8
15	5.24 (br d, J = 10.3)	126.6
16	—	84.1
17	2.57 (d, J = 14.0)	45.9
	1.98 (d, J = 14.0)	
18	0.91 (t, J = 7.6)	8.4
19	1.76 (dq, J = 15.0, 7.6)	35.2
	1.41(dq, J = 15.0, 7.6)	
20	—	38.5
21	3.80 (br s)	57.1
10-OMe	3.91 (s)	56.2
11-OMe	3.85 (s)	56.4
CO_2Me	3.48 (s)	172.0
CO_2Me	—	52.6
16-OH	4.24 (br s)	—

- A strong peak appearing at 1740 cm^{-1} can be attributed to cyclic ketone in cyclopentane.
- If there is an epoxy group in the part of cyclopentane, two peaks at 1250 cm^{-1} and 890–830 cm^{-1} should be observed. In most case, the vinyl group in seco-iridoid can produce absorption peaks at 990 cm^{-1} and 910 cm^{-1}.

Identification of Iridoids by MS

The fragmentations of iridoid in both positive and negative ion modes are investigated. In positive ion mode, the iridoids easily form metal ion adducts, such as

$[M+Na]^+$. For this reason, the alkali metal adduct are used to investigate the fragmentation of iridoid. The iridoids having a C_7, C_8-epoxy ring can produce RDA fragmentation by high-energy CID when there is a double bond of C_3 and C_4, or the RDA fragmentation cannot occur.[62] The alkali metal adduct of iridoids having a saturated cyclopentane ring can produce fragmentation in the skeleton at low-energy collision. Some neutral losses produce fragment ions, losing 44 (CO_2), 18 (H_2O), and so on can also be observed.[63] The skeleton of iridoids having an unsaturated cyclopentene ring could not be cleaved in the positive ion mode, therefore, only cleavage of substituent group can be observed. Secoiridoid can produce $[M+Na]^+$, $[M+K]^+$, $[M+NH_4]^+$, and fragmentation in the skeleton can be observed.[64]

The fragmentation of iridoids in negative ion mode has been investigated in recent years. If there is a carboxyl at C_4 position, no acid adduct can be observed in MS spectrum, while fragment ion losing 44Da can be seen in the MS^2 spectrum. If the substituent group of C_4 is $COOCH_3$ or OCH_3, acid adduct can be observed in the MS spectrum. Different from the fragmentation in the positive ion mode, the fragmentation in the skeleton of iridoids having an unsaturated cyclopentene ring can be identified in the negative ion mode. Moreover, the isomer of iridoids can be distinguished by comparing the relative intensity of fragment ions in the negative ion mode. More detailed elucidation for fragmentation of iridoids in negative ion mode can be found in reference literature.[26,65]

Identification of Iridoids by NMR

In ^{1}H-NMR spectrum, the chemical shift of H-1 in iridoid is at δ 4.5–6.2 ppm. When $J_{1,9}$ is 0–3 Hz, H-1 can be identified at 1α position; when $J_{1,9}$ is 7–10 Hz, H-1 can be identified at 1β position. It is difficult to determine the relative configuration of C-6, C-7, and C-8 on cyclopentane of iridoid glycosides having no substituent. When $J_{5,6}$ is less than 1 Hz, the substituent group can be identified at 6β position.

In ^{13}C-NMR, if there is no methyl on C-4 position, the chemical shifts of C-3 and C-4 are at δ 139–143 ppm and δ 102–111 ppm, respectively. If there is a double bond at C-7 and C-8 position, but no methyl on C-8 position, their chemical shifts are at δ 134–136 ppm. If C-7 and C-8 form an epoxy ring, their chemical shift is at δ 56–60 ppm.

An Example of Spectral Identification of Iridoids

Gardaloside Gardaloside is an iridoid isolated from the fruits of *Gardenia jasminoides*.[66] The structure was elucidated based on the following spectra.

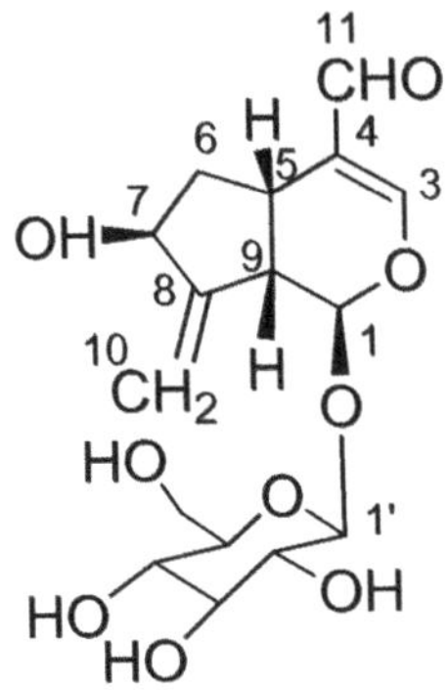

Gardaloside

UV (MeOH) λ_{max} nm (log ε) 244 (3.9)

IR (KBr) ν_{max} cm^{-1}: 3367 (OH), 2927, 1626 (CHO), 1410, 1242, 1157

FABMS (positive ion mode) *m/z*: 381 $[M + Na]^+$ (12), 358 $[M + H]^+$ (2), 329 (20), 289 (11), 176 (100)

HRFABMS *m/z*: 381.1167 $[M + Na]^+$ (calcd for $C_{16}H_{22}NaO_9$, 381.1161)

^{1}H-NMR (MeOH-d_4, 500 MHz) and ^{13}C-NMR (MeOH-d_4, 125 MHz): see Table 4.17.

Table 4.17 Attribution of ^{1}H- and ^{13}C-NMR Signals of Gardaloside

Position	δ_H (ppm)	δ_C (ppm)
1	5.65 (d, *J* = 3.8)	97.7
3	7.36 (s)	164.5
4		124.1
5	3.18[a]	29.5
6	2.16 (ddd, *J* = 12.6, 6.3, 4.4)	39.0
	1.86 (ddd, *J* = 12.6, 7.2, 7.2)	
7	4.31[a]	73.9
8		152.4
9	3.05[a]	44.8
10	5.35 (s)	112.8
11	9.19 (s)	193.0
1′	4.67 (d, *J* = 7.9)	100.1
2′	3.18[a]	74.7
3′	3.25[a]	78.0
4′	3.26[a]	71.6
5′	3.32[a]	78.5
6′	3.90 (dd, *J* = 11.7, 5.3)	62.8
	3.63[a]	

[a] Overlapped signals.

4.4.7 Identification of Coumarins

Identification of Coumarins by UV

Coumarins with no substituent groups show maximum absorption at 274 nm and 311 nm, resulting from the benzene ring and pyrone ring in its skeleton, respectively. The UV spectra of linear furan coumarins show four absorptions: 205–225 nm, 240–255 nm, 260–270 nm, and 298–316 nm, but angular furan coumarins have no absorption within 240–255 nm and 260–270 nm. It is easy to distinguish the UV spectra of coumarins and chromone, despite their difference, only in substituted position of carbonyl on pyrone ring. The absorption in 240–250 nm is strong in chromone, but it is the weakest in coumarins.

Identification of Coumarins by IR

- Coumarin has three strong absorption bands of C=C stretching vibration in the region of 1660–1600 cm^{-1}. It is easy to distinguish it from chromone because the absorption of chromone is much simpler.
- The carbonyl of coumarin shows a strong absorption band at 1750–1700 cm^{-1}. The carbonyl absorption of coumarin glycoside with sugar substitution at C_6–O– or C_7–O– is lower than 1700 cm^{-1}.
- Furanocoumarin has two or three absorption bands with weak to moderate intensity in the region of 3175–3025 cm^{-1}, which are attributed to the C–H stretching vibration of pyrone, benzene, and furan ring.

Identification of Coumarins by NMR

Most natural coumarins are 3,4-free substitute. In ^{1}H-NMR spectrum, H-3 and H-4 form an AB system, which is easy to identify by chemical shifts of H-3 and H-4 at δ 6.1- 6.5 ppm and *δ* 7.5- 8.2 ppm, respectively, with a characteristic coupling constant ($J_{3,4}$ = ~9.5Hz).

In ^{13}C-NMR spectrum, the skeleton of coumarins has nine carbon atoms, and the characteristic signals of the coumarin skeleton are summarized as follows: C-2 is carbonyl carbon, and C-7 is often connected with a hydroxyl or other oxygen-containing group and always effected with confederated carbonyl, so the chemical shifts of C-2 and C-7 are observed at a low field around δ 160 ppm. The chemical shifts of C-3 and C-4 with no substituent group are at δ 110–113 ppm and δ 143–145 ppm, respectively. The chemical shifts of the two quaternary carbons are δ 149–154 ppm (C-9) and δ 110–113 ppm (C-10).

An Example of Spectral Identification of Coumarins

3,8-Dimethyl-5-isopropyl-6-methoxycoumarin Compound 3,8-dimethyl-5-isopropyl-6-methoxycoumarin is a coumarin isolated from the heartwoods of

Table 4.18 The Attribution of ^{1}H- and ^{13}C-NMR Signals of 3,8-Dimethyl-5-isopropyl-6-methoxycoumarin

Position	δ_H (ppm)	δ_C (ppm)
2	—	162.1
3	—	124.6
4	7.90 (s)	136.7
4a	—	117.8
5	—	129.6
6	—	153.7
7	6.90 (s)	116.7
8	—	123.8
8a	—	146.6
9	3.56 (m)	26.6
10	1.38 (d, J = 7.3)	17.6 (2C)
11	2.23 (s)	21.4
12	2.42 (s)	15.7
13	3.83 (s)	56.2

Mansonia gagei Drumm.[67] Its structure was elucidated based on the following spectra.

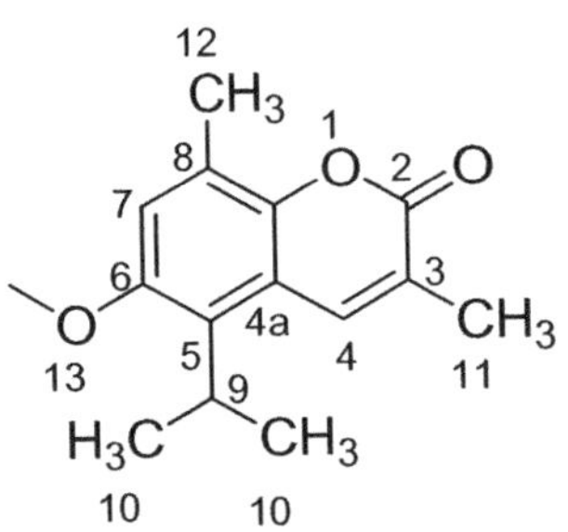

3,8-Dimethyl-5-isopropyl-6-methoxycoumarin

UV λ_{max} nm (logε): 232 (4.15), 292 (4.13), 344 (3.43)

IR ν_{max} cm^{-1}: 1711, 1600, 1450, 1300, 1200, 1035

EIMS 70 eV *m/z* (rel.int.%): 246 [M$^+$] (62), 231 (100), 203 (5). (Found: C, 72.98; H, 7.59. $C_{15}H_{18}O_3$ requires: C, 73.17; H, 7.31)

^{1}H-NMR ($CDCl_3$, 500 MHz) and ^{13}C-NMR ($CDCl_3$, 125 MHz): spectral data are shown in Table 4.18.

4.4.8 Identification of Lignans

Identification of Lignans by UV

UV spectrum is mainly used to determinate the presence of the benzene ring in lignan. Lignans have absorption around 230 nm and 280 nm, similar to phenolic derivatives. When double bonds on side chain conjugate with the benzene ring, absorptions at 235, 295, and 335 nm will be observed. When the side chain forms another benzene ring, the characteristic absorptions at 220, 260, 295, 310, and 350 nm, similar to naphthalene, will be observed.

Identification of Lignans by IR

Lignans show absorption of the benzene within 1600–1500 cm^{-1}. Lignans with an independent five-member lactone ring have absorption at 1780–1760 cm^{-1}. Lignans with the conjugated five-member lactone ring have absorption at 1760–1740 cm^{-1}. Lignans with the unsaturated five-member lactone ring have absorption at 1725 cm^{-1}. The absorption of the double bond on side chain is at 1625 cm^{-1}; the ketone group is at 1670 cm^{-1}; and the diketene is at 1640 cm^{-1}.

Identification of Lignans by MS

Lignans with a free phenolic group can be analyzed in the negative ion mode. Most of the lignans with lactone can give the characteristic fragment ion $[M-H-44]^-$ presumably by the loss of CO_2 from the lactone ring. Introduced here is MS fragmentation of dibenzylbutanediol lignans and furofurano lignans.

Dibenzylbutanediol lignans can produce $[M-H-48]^-$ fragment ion by loss of formaldehyde and water from the diol structure. It is easy to cleave at the β-position of dibenzylbutanediol[68] (Fig. 4.22). Furofurano lignans can cleave in the furofuran

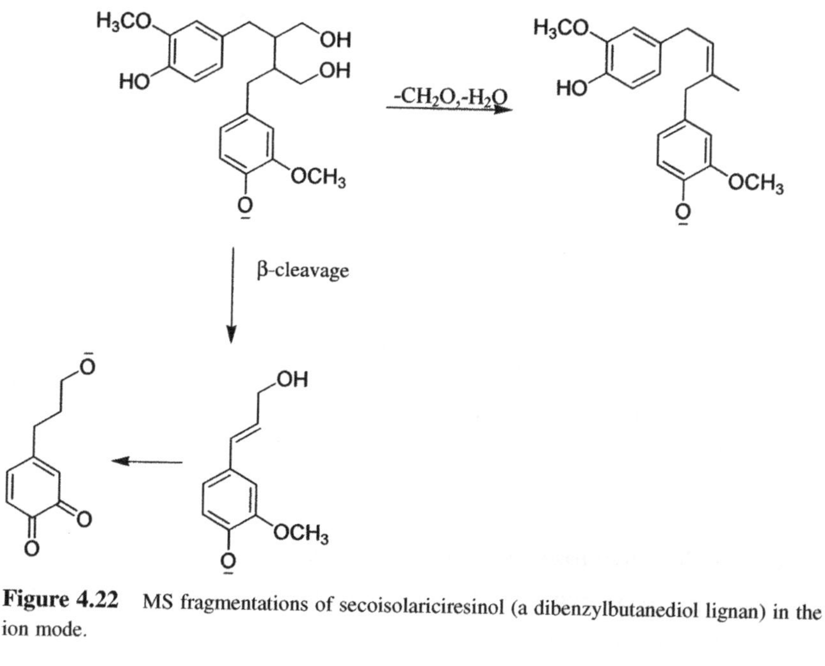

Figure 4.22 MS fragmentations of secoisolariciresinol (a dibenzylbutanediol lignan) in the negative ion mode.

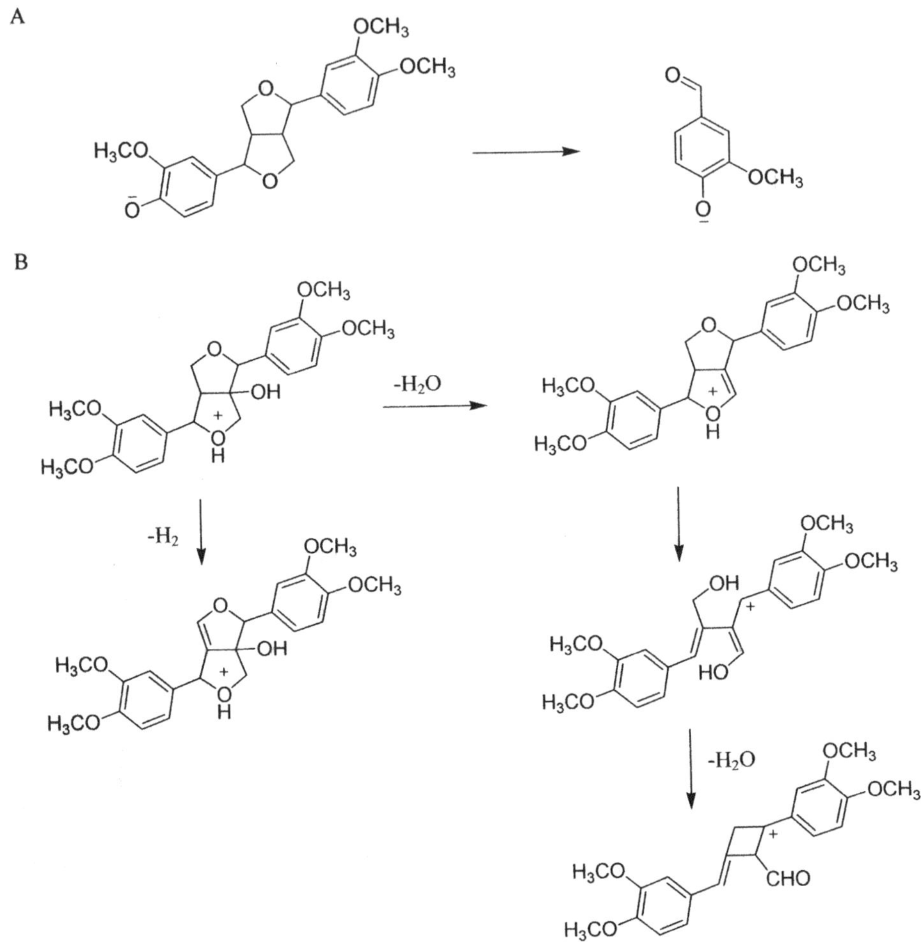

Figure 4.23 Characteristic MS fragmentations of phillygenin (a furofurano lignan). (A) In the negative ion mode; (B) in the positive ion mode.

ring, which is considered the most characteristic fragmentation for this class of lignans in the negative ion mode (Fig. 4.23A). In the positive ion mode, the consecutive losses of two H_2O molecules and H_2 molecules can take place from the furan rings of furofurano lignans[69] (Fig. 4.23B). Loss of H_2O and CH_3• are usually observed when there are hydroxyl and methoxyl substituent groups in the structure.

An Example of Spectral Identification of Lignans

(7S*,7′S*,7″R*,8R*,8′R*,8″S*)- 3′-methoxy-4,4″,9″-trihydroxy-4′,7″:7,9′:7′,9-triepoxy-5′,8″,8,8′-sesquineolignan

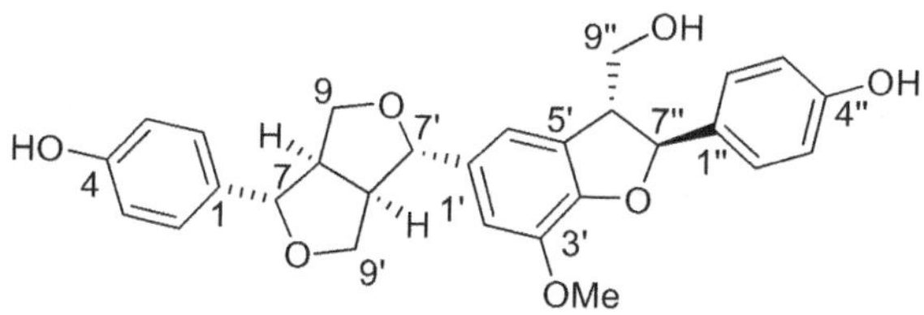

(7*S**,7′*S**,7″*R**,8*R**,8′*R**,8″*S**)- 3′-methoxy-4,4″,9″-trihydroxy-4′,7″:7,9′:7′,9-triepoxy-5′,8″,8,8′-sesquineolignan.

Compound (7*S**,7′*S**,7″*R**,8*R**,8′*R**,8″*S**)- 3′-methoxy-4,4″,9″-trihydroxy-4′,7″:7,9′:7′,9-triepoxy-5′,8″,8,8′-sesquineolignan was isolated from the fruits of *Broussonetia papyrifera*.[70] The structure was elucidated based on the following spectra.

UV (MeOH) λ_{max} nm (logε) 278 (3.80), 225 (4.45), 208 (4.70)

IR (film) ν_{max} cm^{-1}: 3420, 2936, 2874, 1614, 1516, 1451, 1370, 1331, 1217, 1171,1144, 1048, 831

FABMS (positive ion mode) *m/z*: 477$[M+H]^+$

HRESIMS (positive ion mode) *m/z*: 499.1741 $[M+Na]^+$ (calcd for $C_{28}H_{28}O_7Na$, 499.1732)

^{1}H-NMR (methanol-d_4, 500 MHz) and ^{13}C-NMR (methanol-d_4, 125 MHz): data are shown in the Table 4.19.

4.5 STRUCTURE ELUCIDATION OF UNKNOWN COMPOUNDS BY HYPHENATED TECHNIQUE

Traditionally, unknown compounds need first to be isolated and purified to a certain percent for structural identification. The purity is usually examined by TLC or HPLC. Most commonly, the pure compounds are then measured for their melting point, molecular weight by MS, skeleton and functional substitutes by UV and IR. The exact structures are normally elucidated by comprehensive analysis of chemical shifts of NMR and fragmentations of MS in combination with information from UV and IR. Compounds with chiral carbons often need CD or ORD spectrum for determination of the stereo-structures; occasionally, X-ray might be a necessary tool for structure elucidation. However, isolation of a target compound, especially a low-concentrated component, from a complex extract is usually time-consuming and tedious with traditional procedures, which usually include several chromatographic separations.

Hyphenated techniques have become powerful tools for rapid online qualitative analysis of unknown compounds in complex samples. As introduced previously, numerous structurally related compounds could be produced in plants by the second metabolism. The similar structures generally share the same or similar absorptions in UV-Vis spectra, fragmentation patterns in MS, and signals (chemical shift and coupling constant). With the popular application of LC-DAD-MS/MS^n

Table 4.19 Attribution of ^{1}H- and ^{13}C-NMR Signals of 3′-methoxy-4,4″,9″-trihydroxy- 4′,7″:7,9′:7′,9-triepoxy-5′,8″,8,8′-sesquineolignan

Position	δ_H (ppm)	δ_C (ppm)
1		132.7
2	7.20 (d, J = 8.3)	128.0
3	6.77 (d, J = 8.3)	116.0
4		158.0
5	6.77 (d, J = 8.3)	116.0
6	7.20 (d, J = 8.3)	128.0
7	4.71 (d, J = 5.2)	87.0
8	3.13 (m)	55.3
9	4.20 (m) 3.77 (m)	72.5
1′		135.6
2′	6.90 (br s)	111.7
3′		145.2
4′		148.8
5′		129.9
6′	6.90 (br s)	116.0
7′	4.70 (d, J = 4.8)	87.4
8′	3.13 (m)	55.0
9′	4.24 (m) 3.80 (m)	72.2
1″		133.7
2″	7.17 (d, J = 8.3)	128.5
3″	6.74 (d, J = 8.3)	115.8
4″		158.2
5″	6.74 (d, J = 8.3)	115.8
6″	7.17 (d, J = 8.3)	128.5
7″	5.51 (d, J = 6.1)	88.8
8″	3.47 (m)	54.9
9″	3.79 (m)	64.6
OMe	3.86 (s)	56.5

and availability of LC-NMR, phytochemists now are able to rapidly characterize unknown compounds online based on the spectra of unknown peak in HPLC and to elucidate its structure by comparing its spectra to these of known peak with similar spectra in the chromatogram. Such online structural elucidation eliminates the laborious and time-consuming purification process, and is therefore particularly suitable for identification of isomers and unstable compounds. In this section, we will introduce the online structural elucidation methods with these two hyphenated techniques.

4.5.1 Structure Elucidation of Unknown Compounds by HPLC–DAD-MS

The structure elucidation of unknown compounds by HPLC–DAD-MS/MS^n is becoming popular due to its fast speed, high sensitivity and selectivity, and ability to provide comprehensive information. This method makes it possible to investigate chemical components in herbal medicines in a wider field and at a deeper level. Sometimes, isolation of target compounds is required to verify the result of online structure elucidation.

Techniques for Structure Elucidation of Compounds by HPLC-DAD-MS

The most commonly used hyphenated technique is HPLC-DAD-ESI-MS^n. With such a setup, the UV-Vis spectrum of each isolated compound through HPLC is recorded by DAD and used to primarily speculate on the possible skeleton of this compound. The molecular weight is measured in MS, and the tandem mass with auto MS^n can simultaneously produce abundant fragment ions that help to deduce the structure based on the fragmentation rules introduced before. For the purpose of structure elucidation, ion trap MS is preferred in this system due to its function of MS^n. ESI is the most commonly used ion source. Selection of positive or negative ion mode mainly depends on the properties of the compounds. These two modes can be switched during the running of chromatography.

Many known compounds can be identified by comparing their MS with those in literature and further confirmed by comparing their t_R, UV spectra, molecular weights, and fragment ions with standards. For unknown compounds, MS experts or phytochemists who are good at MS analysis will capture important structural information from fragment ions.

The development of new techniques on chromatography and MS detectors contribute to not only the speedup, but also the accuracy of structure elucidation with LC-MS. To isolate compounds more effectively, stationary phases with a particle size of less than 2 μm have been recently developed. The advantages of ultra-performance liquid chromatography (UPLC) with high efficiency of separation and shorter analysis time prompt its hyphenation to the MS for application in structure elucidation. UPLC is designed to undertake the high pressure resulting from smaller particle size. According to the van Deemter curve, if a stationary phase of less than 2 μm is used, not only are higher efficiencies gained, but these efficiencies also no longer diminish with flow rate. This makes it possible to take full advantage of chromatographic principles to run separation using shorter column with higher flow rates, but without loss of superior resolution and sensitivity,[71] that is, to analyze a complex sample in a shorter time, specifically, an approximate three-fold reduction in the analysis time in comparison with conventional HPLC.[72]

Accurate structure elucidation of target compound depends on accurate mass measurement. The invention of high-resolution MS such as TOF is a significant improvement in MS for structure elucidation. Accurate mass measurement of molecular ions in TOF gives the elemental composition of a compound. The generation of accurate mass of the product ions by CID in the Q-TOF mode facilitates the identification of fragment ions and interpretation of the fragmentation pathways. Combining the advantages of UPLC in separation and Q-TOF in structure elucidation, UPLC-Q-TOF has become an effective tool for structure elucidation of unknown compounds in herbal medicines. Applications of UPLC-Q-TOF are described in literature.[73,74]

Thus far, characterization of herbal medicine with RP-LC-MS mostly focuses on the research of medium and weak polar compounds. The research on polar compounds is relatively less applied due to the lack of an effective method. With the development of HILIC for separation of polar compounds, characterization of polar compounds in herbs with HILIC-MS is becoming popular. A comprehensive 2-D HILIC-Q-TOF has been developed for analysis of complex samples of hydrophilic compounds.[75] A column-switching of HILIC and RP-LC was also developed to analyze both the polar compounds and medium or weak polar compounds systematically.[76] This will help scientists explore more new biological chemical compounds and reveal mechanisms of herbal medicines.

Screening Compounds by HPLC-DAD-MS Based on Structural Characteristics

With high selectivity and sensitivity, HPLC-DAD-MS system is very useful for screening and characterizing certain types of secondary metabolites in complex herbal samples on the basis of their spectral similarity. DAD can provide the UV spectrum that is helpful to primarily know the type of the compound, MS spectrometer can provide the MS spectrum for further structure identification or elucidation. Several types of MS, such as ion trap, Q, and TOF can be used for screening. Scan modes, such as full scan, MS^n, and MRM are used for scanning.

Full scan is usually used at the beginning of the analysis for primary scanning. Full scan acquisition provides a typical total ion chromatogram (TIC). The range of the scan depends on the samples. Because it takes time for the acquisition of a range of ions, when the scan range is increased, the times (or number) of scan on each ion decrease, resulting in lower sensitivity. Generally, information for very low concentrated compounds in a mixture might not be obtained from a full scan because their signals might be buried by those of major compounds having the same t_R in the chromatogram.

MS^n allows several steps of fragmentation and isolation of ions. Each subsequent generation of product ions exhibits unique cleavage patterns. MS^n can provide key information to structure identification of an unknown compound. However, both full scan and MS^n might miss some minor components because of their lower sensitivities and selectivities than MRM mode. In MRM mode, the chromatogram records a particular parent ion producing special daughter ions. Thus, scanning with this mode can greatly improve sensitivity and selectivity. High-resolution MS such as Q-TOF can give more accurate screening result.

For systematical screening, an MS screening table can be designed before the experiment. For example, a screening table of flavonoids was designed for analysis of total flavonoids content in plants and foodstuffs by HPLC/ESI-TOF.[77] In this study, molecular weights of skeletons and substituent groups were selected reasonably according to analytical requirement, the screening table was designed by adding the molecular weight of skeletons and substituent groups; no concrete structures were provided in this stage (Table 4.20).

In another example, an MS method was developed to characterize the flavonoid glycosides in *Carthamus tinctorius* L.[78] As a result, 77 flavonoid glycosides in total were found by this method and their structures were deduced by MS^n. The glycosylation modes of flavonoids glycosides, C-glycosylation, and O-glycosylation were first identified, and then the structure of their aglycones and glycans was identified. The advantage of tandem MS makes the nature and number of sugars, as well as the sequence and/or the interglycosidic linkages of the glycan portion, easily deducible. However, it is difficult to determine the glycosylation position with this method. To determine the connecting positions of aglycones and sugars for final elucidation of complete structures of these flavonoids glycoside, NMR spectra are necessary. This can be achieved either by preparative HPLC separation first, then NMR measurement, or by the LC-NMR system.

The HPLC-DAD-MS screening method for characterizing similar compounds is proven to be useful for systematic and rapid characterization of compounds in herbal extracts. Application of this method has two necessary conditions. First, researchers should be familiar enough with the structures of the target compounds and their similar compounds to make the screening table include more comprehensive information. To ensure the analysis is successfully performed, the corresponding screening table should be carefully designed, and an appropriate scan method should be selected. Second, the fragmentation rule of the target type of compounds should be mastered so that correct elucidation can be carried out. Many new compounds can be found through HPLC-DAD-MS, and their accurate structures are then identified by combined analysis with other spectra after preparative separation. The HPLC-DAD-MS analysis has become an important part of systematic research on chemical components in herbs. Its popular application will definitely facilitate the depth and speed of modernization of herbal medicine.

HPLC-MS Analysis of Flavonoid Glycoside by Neutral Loss Scan

If the fragmentation of a type of compound has been well known, the detection or identification of this group of compounds will be made easier by selecting an appropriate scan method and optimizing the MS condition for better sensitivity and selectivity. Neutral loss scan is useful in the selective identification of structural related compounds in a mixture. The mass spectrometric identification and quantification of glycosyl flavonoids with neutral loss scan mode was recently reported.[50] The analysis used HPLC-triple Q MS with neutral loss scan in the positive ion mode. With the following brief summary of this analysis, the key points of HPLC-MS analysis for compounds in herbal medicines with this method are introduced.

Table 4.20 Molecular ions in MS Used for Automatic Ion Chromatogram Extraction from LC/MS Data by Software[77]

Aglycone type ESI$^-$	Monoglycosides				Diglycosides				
	$[M-H]^-$	+ara/xyl	+rha	+glu/gal	+rha +xyl/ara	+rha +rha	+glu/gal +xyl/ara	+glu/gal +rha	+glu/gal +glu/gal
Flavanol+OH=flavone+2×OH	**269**	(401)	**415**	**431**	(547)	(561)	(563)	**577**	**593**
Flavanol+2×OH=flavone+3×OH=flavanone+2×OH+Me	**285**	(417)	(431)	(447)	(563)	(577)	(579)	(593)	(609)
Flavonol+3×OH=flavanone+3×OH+Me	**301**	(433)	**447**	**463**	(579)	(593)	(595)	**609**	**625**
Flavonol+4×OH	**317**	(449)	(463)	**479**	(595)	(609)	(611)	(625)	**641**
Flavanol+2×OH+Me	**299**	(431)	**445**	**461**	(577)	(591)	(593)	**607**	**623**
Flavanol+3×OH+Me	**315**	(447)	(461)	**477**	(593)	(607)	(609)	(623)	**639**
Flavanone+OH	**255**	**387**	**401**	**417**	**533**	(547)	**549**	**563**	**579**
Flavanone+2×OH	**271**	**403**	(417)	(433)	(549)	(563)	**565**	(579)	**(595)**
Flavanone+3×OH	**287**	**419**	**433**	**449**	(565)	(579)	**581**	**595**	**611**
Flavone+OH	**253**	**385**	**399**	(415)	**531**	(545)	**547**	**561**	(577)
Flavone+OH+ Me	**267**	(399)	**413**	**429**	**545**	**559**	(561)	**575**	**591**
Flavone+2×OH+ Me	**283**	(415)	**429**	**445**	(561)	(575)	(577)	(591)	(607)
Flavanone+5×OH (taxifolin)	**319**	**451**	**465**	**481**	**597**	(611)	**613**	**627**	**643**

Note: The ions shown in bold font were used to create the ion chromatogram, and the others (in brackets) have the same nominal molecular weights as some of those shown in bold. Ara, arabinose; xyl, xylose; rha, rhamnose; glu, glucose; gal, galactose.

1. *Fragmentation of flavonids glycoside:* The characteristic fragmentation of flavonoid O-glycosides is the neutral loss of sugar residue. The neutral loss of one glucose will produce fragment ions losing 162 Da. The main cleavages of C-glycosyl flavonoids are at the bonds in the sugar, and the main fragment ions are neutral loss of 120 and 150 Da. Hence, the neutral losses of 162, 150, and 120 Da were selected for screening glycosyl flavonoid.
2. *Optimization of collision energy:* The fragmentation depends on the collision energy and the nature of chemical structure. In other words, the fragment patterns in an MS spectrum of a compound vary with different collision energies. Optimization of collision energy is a key step for detection of a certain kind of chemical compound because an appropriate collision energy can help to detect compounds with high selectivity. In this report, 15 and 30 eV were selected for identification of glycosyl flavonoids by optimization.

If analysis needs to be performed in a wider range, energy-gradient neutral loss scan is a good choice.[79] With the energy linear increase in a certain range, more glycosides can be detected. Because of this, the ratio of peak intensity of a characteristic ion changes at different collision energy was used to distinguish the type of glycosyl flavonoids. As shown in Table 4.21, the ratio of peak intensity of 150 Da/120 Da at

Table 4.21 Ratios of Peak Intensity and Characteristic Ion Peaks in LC/MS Chromatogram for Identification of Glycosyl Flavonoids[50]

Type of glycosyl flavonoid	Characteristics in neutral loss scan			
	A	B	C	D
	−150(30eV)[a]/ −120(30eV)[b]	−120(15eV)[c]/ −120(30eV)[b]	−162(15eV)[d]	−308(15eV)[e]
8-C-Glycosyl flavonoid	<1	<1	n.d.	n.d.
8-C-Glycosyl flavonoid	>1	<1	n.d.	n.d.
C-Glycosyl dihydrochalcone	>1	>>1	+	n.d.
O-Glycosyl flavonoid (monosaccharide)	n.d.	n.d.	+	n.d.
O-Glycosyl flavonoid (disaccharide)	n.d.	n.d.	+	+

Note: This diagnostic table shows peak intensity in LC-MS chromatogram of neutral loss scan of [a] 150 Da at collision energy of 30 eV, [b] 120 Da at 30 eV, [c] 120 Da at 15 eV, [d] 162 Da at 15 eV, and [e] 308 Da at 15 eV. Ratio A is the ratio of peak intensity in LC-MS chromatogram of neutral loss scan of 150 Da at collision energy of 30 eV to that of 120 Da at 30 eV; Ratio B is the ratio of peak intensity in LC-MS chromatogram of neutral loss scan of 120 Da at 15 eV to that of 120 Da at eV; n.d., not detected.

collision energy of 30 eV and the ratio of peak intensity of 120 Da (15 eV)/120 Da (30 eV) of different types of glycosyl flavonoids in neutral loss scan MS were summarized for identification.

3. *Identifying glycosyl flavonoids in a complex sample:* After the scan mode was selected, the MS condition was optimized, and the selection criterion was determined, the glycosyl flavonoids in Rooibos tea were analyzed in this report. Due to better selectivity, the chromatogram of neutral loss scan is much simpler than that of TIC.
4. *Structure elucidation:* The aglycones and sugars have been determined based on their fragmentation during the screening procedure. Further structure elucidation to the target compounds is necessary to identify the intact structure.

The target screening with a selected scan method can obtain structural related information for interesting compounds with high selectivity and sensitivity. Fragmentation rule and optimal MS conditions are key factors for developing this method. The application of such a method for screening other kinds of natural compounds should be based on mastery of characteristic fragmentation rule and the characteristics of the individual sample.

4.5.2 Structure Elucidation by LC-NMR

NMR is one of the most powerful techniques for elucidating the structure of organic compounds. Although LC-MS has gradually become one of the most powerful hyphenated techniques currently available to researchers in phytochemistry to analyze mixtures without prior isolation of the components, in many cases, NMR is still needed for an unambiguous structure identification. The requirement of qualitative analysis of complex samples has accelerated the development of LC-NMR. Recent advances in NMR technology have allowed for the practical application of LC-NMR, which combines the separation power of HPLC with the superior structure elucidation of NMR. Similar to LC-MS, it provides information of molecular structures of components in an extract or fraction with a single online experiment.

General Introduction to LC-NMR

The working mode of LC-NMR can be classified into on-flow and stop-flow, according to the status of the sample during the measurement.

On-flow LC-NMR operation is to let LC eluent directly enter the NMR flow probe. The NMR spectra are acquired when a chromatographic separation is continuous, so the sample is analyzed under a dynamic condition.

In the stop-flow LC-NMR system, NMR spectra of the HPLC isolated compounds in a sample can be measured under static conditions. Depending on how a sample is transferred into the NMR detection cell, the stop-flow method can be classified into direct stop-flow and loop storage or loop transfer. In the direct stop-flow method, chromatographic separation must be interrupted to provide a static condition when the sample arrives in the NMR detection cell. In the loop storage or loop transfer method, samples separated from the chromatography are temporarily stored in the loop, and then transferred into NMR detection cell after the chromatographic separation is completed.

As a summary, the on-flow mode is able to acquire NMR spectra of all NMR active compounds but has less sensitivity due to the limited NMR data acquisition time; therefore, only ^{1}H and ^{19}F NMR can be acquired in this mode. On the contrary, there is no restriction for scanning time in static measurement of stop-flow; therefore, the sensitivity is relatively high, and 2-D correlation experiments are also allowed. Stop-flow is a widely used method for detailed structure elucidation of low-concentrated compounds in samples. The disadvantage is that only selected compounds (peaks) are measured in NMR.

Inevitably, there are disadvantages to these working modes. Gradient elution is commonly used in HPLC analysis of a complex sample. Because the chemical shift of solvent and sample resonances depend on solvent composition, a change in the composition of eluent in gradient elution will lead to changes in chemical shifts during the acquisition process in the on-flow mode. Moreover, the mobile phase used in regular chromatographic system is usually not an ideal matrix for the acquisition of NMR spectra.

In direct stop-flow mode, possible peak dispersion can decrease the chromatographic resolution. Another concern of this mode is the carry-over (memory) effects. Compared with the peak volumes and other void volumes in the chromatographic system, the volume of the NMR detection cell is relatively large (60–240 μL). It takes a certain amount of time to wash a peak out of the flow cell completely; thus, highly concentrated compounds may pollute the NMR detection cell and make measurement of the lower concentrated compounds following these peaks difficult. In the loop storage mode, such a disadvantage is avoided because there is no interference with separation and arbitrary analytical sequence in NMR. However, degradation or isomerization might occur to some unstable compounds during the long storage time in the loop storage mode. Another disadvantage of stop-flow is that only selected compounds (peaks) are measured in NMR.

In on-flow mode, neither special instrument nor software is required because the eluent directly flows to the NMR detection cell. But the scanning rate per increment and the LC resolution should be compatible. In stop-flow mode, an additional device is required to control the temporary LC flow stop or for the storage and transfer of an LC fraction to the temporary sample storage unit. In general, UV or fluorescence detectors can be used to control the flow stop via either manual or automatic operation. When compounds have none or only weak UV or fluorescence response, other detectors such as MS can be coupled to LC-NMR. MS control is more sensitive, versatile, and suitable for selection of the interest fractions. More detailed introduction about LC-NMR is available from books and review articles.[80,81]

The Key Concerns of LC-NMR

Two key concerns must be addressed when LC-NMR is applied for structure elucidation. One is how to improve the sensitivity of NMR, and the other is how to resolve the compatibility of solvent between LC and NMR. Great effort has been made to improve the sensitivity of LC-NMR, including the use of miniature flow cells, a novel flow probe, and their combination. Solving this concern has focused on the design of the flow probe. The most significant advance in the NMR probe achieved is the introduction of CryoProbes, which is able to increase the sensitivity three- to four-fold in high-resolution NMR compared with conventional probes.

Because the sensitivity of NMR is primarily limited by thermal noise of the detection components, the electronic components are cryogenically cooled down to –20 K in these NMR probes. By operating the electronic components at such low temperature, while the sample remains at an ambient temperature, the electronic noise is greatly reduced.[82] Recently, a novelty cryoflow probe was built in LC-NMR.[83] It combines the advantages of cryogenic NMR technology with on-flow configuration. A solenoid probe is designed by wounding a solenoid rf coil directly on the separation capillary[84,85] for the application to capillary LC-NMR system.

The mobile phase of LC is usually acetonitrile/water or methanol/water. Because the intensity of protons in these solvents is so high, the characteristic NMR signals of samples are buried. The use of deuterated solvents as the mobile phase of LC can eliminate the interfering solvent signal, but they are too expensive for routine LC analysis. The solvent signal interfering problem will not be resolved until the invention of solvent suppression techniques. There are three ways to perform solvent signal suppression, namely pre-saturation (NOESY pre-saturation), soft-pulse multiple irradiation, and water suppression enhanced through T1 effects (WET) pre-saturation.

In the pre-saturation method, a highly selective low-power pulse irradiates the selected solvent signals for 0.5 to 2 s prior to data acquisition, leading to saturation of the solvent signal frequency. As a consequence, no irradiation occurs during data acquisition. NOESY-type pre-saturation uses a pulse sequence consisting of three 90° pulses. In the soft-pulse multiple irradiation method, the pre-saturation is performed with the use of a shaped pulse. This method is therefore more suitable for the suppression of multiple solvents. The WET sequence uses four solvent selective pulses of variable lengths, thus providing fast and highly efficient saturation of multiple solvent frequencies.

Solvent signal suppression is necessary in order to achieve a reduction of the NMR signals of solvents entering the receiver, so that small signals of analytes can be observed. But this method can also lead to the loss of signals of analytes. An alternative method is to construct an LC-solid phase extraction-NMR (LC-SPE-NMR) system by introducing a post-column SPE between LC column and NMR flow probe. One function of SPE is to trap compounds eluted from the LC column and concentrate them. Therefore, its application significantly improves the sensitivity. Another advantage of online LC-SPE-NMR is that the SPE unit allows the use of normal solvents for LC to separate compounds, and then the deuterated solvents to wash out the concentrated compounds trapped on SPE. By doing so, the expense of high-purity deuterated solvents is no longer an issue, and the procedure of solvent suppression becomes unnecessary.

Development of LC-NMR

The LC-NMR-MS and the miniaturized LC-NMR instrument are two aspects of the development of LC-NMR. The LC-NMR-MS system is now commercially available, and this setup allows the acquisition of MS and NMR data simultaneously in a single LC run. The introduction of MS to the LC-NMR aims to make up the shortage of NMR at low sensitivity and slow speed, and acquire more structural information at the same time.

The complementary information of MS and NMR data make the structure elucidation easier and more accurate. MS cannot only provide the data of molecular weight, but also structural related information for all kinds of compounds, including those that cannot be measured by NMR, such as sulfate. It can also be used to justify, if the unstable compounds degrade and peak disperses during the delay for NMR measurement, the acquisition of more information of low-concentrated compounds with its advantage in high sensitivity, and to help control the stop-flow or collection of loop storage so that the compounds of interest can be accurately analyzed by NMR.

The common way of coupling LC with NMR and MS is to place them in parallel. The LC eluent is split at a suitable ratio. The major portion of fraction is directed to NMR flow probe and the minor portion of fraction is directed to the MS detector. This connection mode enables two detectors to perform under their respective optimal conditions. In addition, parallel configuration avoids the concern of back-pressure produced by the interface of LC-MS that could damage the NMR flow probe. The split-flows can be adjusted easily to satisfy the need of experiments being conducted.

As mentioned above, MS detection can help to sensitively control the stop-flow NMR experiments for particular compounds of interest. This is one reason that the parallel connection mode is widely adopted. In this mode, the detection of MS is adjusted earlier than that of NMR by connecting LC to MS interface with a shorter capillary tubing. An accurate delay time should be set up to give an accurate trigger signal in the software. The SPE is also introduced in the LC-NMR-MS setup. LC-SPE-NMR-MS will show greater advantage in structure elucidation when its setup and software become more compatible.

The development of chromatography also prompts the improvement of LC-NMR. Over the last years, there has been a trend of hyphenating miniaturized separation techniques, which has great impact in the field of LC-NMR. Capillary LC (capLC)-NMR is one of the important miniaturized techniques. Because of the small amount of solvents needed, when capLC is performed, the deuterated solvents can be used as the mobile phase without increasing cost. CapLC also needs smaller samples, which is much more suitable for mass/volume limited samples such as biomedicine, metabolite, and high-throughput production of natural product libraries. Because capLC can provide higher concentrated analyte from the elution due to its high separation efficiency, the detecting sensitivity will be improved; thus this technique is more applicable than standard LC for the analysis of low concentrated components in natural products. In order to couple with the capLC, the high-field NMR equipped with miniaturized micro/nano liter NMR probes is required. For a microcoil NMR probe, the diameter of its typically solenoidal RF coils is always less than 1 mm, and the observed volume ranges from 5 nl to 1 μL. With the development of miniaturized LC-NMR instrument, the structure elucidation can be performed with smaller samples, shorter acquisition periods, and more 2-D correlation information.

Application of LC-NMR in Structure Elucidation for Herbal Medicines

Analysis of natural products is one of the most important applications for LC-NMR. Some compounds in herbal medicines, such as iridoids, are unstable. Traditional structure elucidation for natural compounds requires certain purity, meaning compounds must be isolated and purified with multiple procedures first. High-temperature concentration and air exposure of unstable compounds during the procedure would lead to changes in their structures. LC-NMR, especially the on-flow working mode, is well suited for identification of unstable compounds in natural products because it can avoid the tedious isolation steps and the possibility of structure change.

LC-MS is unable to distinguish between the different geometrical and optical stereoisomers. LC-NMR can make up for this shortage. One of the advantages of LC-NMR in structure elucidation lies in its ability to determine isomers. The direct coupling of LC-NMR is one of the most useful tools for the determination of isomers using ^{1}H, ^{13}C, and 2-D correlation spectra.

Stop-flow mode and WET are most commonly applied in the LC-NMR system. In comparison with the limitation of LC-NMR, LC-SPE-NMR has great advantage in structure elucidation of natural products. When analytes are eluted from the LC column, the organic solvent used in the mobile phase of HPLC would lead to the washout of analytes from the SPE. Therefore, water is commonly added with optimal speed by a makeup pump prior to the concentration procedure on SPE cartridge to ensure the effective enrichment of analytes by SPE. Sometimes, it may be necessary to repeat this trapping procedure on SPE several times to acquire a 2-D correlation spectrum. SPE behaves as an interface between chromatography and NMR; the solvent change from non-deuterated eluent to deuterated solvent makes the LC-NMR analysis easier and inexpensive, and the solvent suppressing is no longer necessary in principle. In addition, the effective concentration by SPE also makes it possible to acquire 2-D correlation spectra, which greatly facilitates the structure elucidation. But in some cases, obtaining a ^{1}H–NMR spectrum requires using a pulse sequence based on the ^{1}D NOESY sequence, with double pre-saturation to suppress signals of residual water and acetonitrile.

Online screening for antioxidant compounds by LC-NMR application combining radical scavenging technique has been recently reported.[86,87] The antioxidants in natural products are detected and identified by means of LC-UV-NMR/DPPH$^{\bullet}$

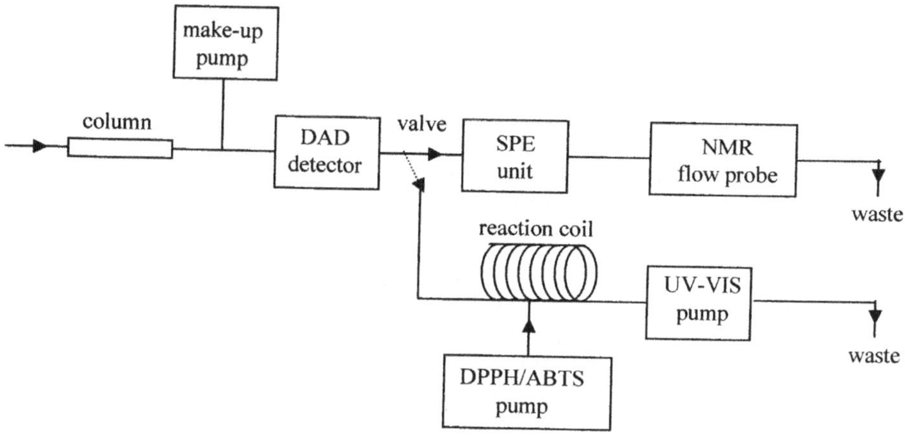

Figure 4.24 LC-UV-DPPH$^{\bullet}$/ABTS$^{\bullet+}$ and LC-UV-SPE-NMR instrumental setup.

(2,2-Diphenyl-1-picrylhydrazyl radical) or ABTS$^{\bullet+}$ [2,2-azinobis-(3-ethylbenzothiazoline-6-sulfonate)]. The design of LC-UV-NMR/DPPH or ABTS setup is shown in Figure 4.24.[87] The screening is parallel with LC-UV-NMR. The screening for antioxidant components in natural products is performed by adding DPPH$^{\bullet}$ and ABTS$^{\bullet+}$ after the LC separation. Following the screening, the antioxidant compounds were effectively identified by means of LC-NMR or combining LC-MS or LC-UV. By the combination of UV, NMR, and MS, antioxidant compounds can be unambiguously identified without isolation and use of standards.

REFERENCES

1. Friebolin, H. (1998) *Basic One- and Two-Dimensional NMR Spectroscopy*. Weinheim, Wiley-VCH Verlag GmbH.
2. Xiao, C.N., et al. (2008) Revealing the metabonomic variation of rosemary extracts using H-1 NMR spectroscopy and multivariate data analysis. *Journal of Agricultural and Food Chemistry* 56(21):10142–10153.
3. Dass, C. (2007) *Fundamentals of Contemporary Mass Spectrometry*. Hoboken, NJ, John Wiley & Sons.
4. Xing, C.F., et al. (2007) Recent applications of liquid chromatography-mass spectrometry in natural products bioanalysis. *Journal of Pharmaceutical and Biomedical Analysis* 44(2):368–378.
5. March, R.E. (1997) An introduction to quadrupole ion trap mass spectrometry. *Journal of Mass Spectrometry* 32:351–369.
6. Schwartz, J.C., et al. (2002) A two-dimensional quadrupole ion trap mass spectrometer. *Journal of the American Society for Mass Spectrometry* 13(6):659–669.
7. Hopfgartner, G., et al. (2004) Triple quadrupole linear ion trap mass spectrometer for the analysis of small molecules and macromolecules. *Journal of Mass Spectrometry* 39:845–855.
8. Xia, L., et al. (2008) Rapid and sensitive analysis of multiple bioactive constituents in compound danshen preparations using LC-ESI-TOF-MS. *Journal of Separation Science* 31(18):3156–3169.
9. Gross, M. and Rempel, D. (1984) Fourier transform mass spectrometry. *Science* 226(4672): 261–268.

10. Comisarow, M.B. and Marshall, A.G. (1974) Fourier transform ion cyclotron resonance spectroscopy. *Chemical Physics Letters* 25(2):282–283.
11. Dienes, T. (1996) Fourier transform mass spectrometry—advancing years (1992 mid 1996). *Mass Spectrometry Reviews* 15(3):163–211.
12. Hu, Q.Z. (2005) Special feature: perspective the orbitrap: a new mass spectrometer. *Journal of Mass Spectrometry* 40(4):430–443.
13. Perry, R.H. (2009) Rephasing ion packets in the orbitrap mass analyzer to improve resolution and peak shape. *Journal of the American Society for Mass Spectrometry* 20(8):1397–1404.
14. Makarov, A. (2000) Electrostatic axially harmonic orbital trapping: a high-performance technique of mass analysis. *Analytical Chemistry* 72(6):1156–1162.
15. Robert, J., et al. (1985) Coupling a quadrupole mass spectrometer and a fourier transform mass spectrometer. *International Journal of Mass Spectrometry and Ion Processes* 64(1):66–67.
16. Patrie, S.M., et al. (2004) Construction of a hybrid quadrupole/fourier transform ion cyclotron resonance mass spectrometer for versatile MS/MS above 10 Kda. *Journal of the American Society for Mass Spectrometry* 15(7):1099–1108.
17. Wang, Y., et al. (2000) Mass-selective ion accumulation and fragmentation in a linear octopole ion trap external to a fourier transform ion cyclotron resonance mass spectrometer. *International Journal of Mass Spectrometry* 198(1–2):113–120.
18. Syka, J.E.P., et al. (2004) Novel linear quadrupole ion trap/FT mass spectrometer: performance characterization and use in the comparative analysis of histone H3 post-translational modifications. *Journal of Proteome Research* 3(3):621–626.
19. Makarov, A., et al. (2006) Dynamic range of mass accuracy in LTQ orbitrap hybrid mass spectrometer. *Journal of the American Society for Mass Spectrometry* 17(7):977–982.
20. Iwasa, K., et al. (2008) Online structural elucidation of alkaloids and other constituents in crude extracts and cultured cells of nandina domestica by combination of LC-MS/MS, LC-NMR, and LC-CD analyses. *Journal of Natural Products* 71(8):1376–1385.
21. Liu, R.X., et al. (2005) Liquid chromatography/electrospray ionization mass spectrometry for the characterization of twenty-three flavonoids in the extract of Dalbergia odorifera. *Rapid Communications in Mass Spectrometry* 19(11):1557–1565.
22. Horvath, C.R., et al. (2005) Identification and quantification of eight flavones in root and shoot tissues of the medicinal plant huang-qin (Scutellaria baicalensis Georgi) using high-performance liquid chromatography with diode array and mass spectrometric detection. *Journal of Chromatography A* 1062(2):199–207.
23. Zhou, Y., et al. (2006) Analyses of stemona alkaloids in Stemona tuberosa by liquid chromatography/tandem mass spectrometry. *Rapid Communications in Mass Spectrometry* 20(6):1030–1038.
24. Shi, P.Y., et al. (2007) Characterization and identification of isomeric flavonoid O-diglycosides from genus Citrus in negative electrospray ionization by ion trap mass spectrometry and time-of-flight mass spectrometry. *Analytica Chimica Acta* 598(1):110–118.
25. Zhou, Y., et al. (2007) Multistage electrospray ionization mass spectrometric analyses of sulfur-containing iridoid glucosides in Paederia scandens. *Rapid Communications in Mass Spectrometry* 21(8):1375–1385.
26. Ren, L., et al. (2007) Studies of iridoid glycosides using liquid chromatography/electrospray ionization tandem mass spectrometry. *Rapid Communications in Mass Spectrometry* 21(18): 3039–3050.
27. Zhang, H., et al. (2008) HILIC for separation of co-eluted flavonoids under RP-HPLC mode. *Journal of Separation Science* 31(9):1623–1627.
28. Alpert, A.J. (1990) Hydrophilic-interaction chromatography for the separation of peptides, nucleic acids and other polar compounds. *Journal of Chromatography A* 499:177–196.
29. Oleszek, W.A. (2002) Chromatographic determination of plant saponins. *Journal of Chromatography A* 967(1):147–162.
30. Chai, X.Y., et al. (2005) Quality evaluation of Flos Lonicerae through a simultaneous determination of seven saponins by HPLC with ELSD. *Journal of Chromatography A* 1070(1–2):43–48.
31. Qi, L.W., et al. (2006) Quality evaluation of Radix Astragali through a simultaneous determination of six major active isoflavonoids and four main saponins by high-performance liquid chromatography

coupled with diode array and evaporative light scattering detectors. *Journal of Chromatography A* 1134(1–2):162–169.

32. Oleszek, W. and Bialy, Z. (2006) Chromatographic determination of plant saponins – an update (2002–2005). *Journal of Chromatography A* 1112(1–2):78–91.
33. McCalley, D.V. (2002) Analysis of the cinchona alkaloids by high-performance liquid chromatography and other separation techniques. *Journal of Chromatography A* 967(1):1–19.
34. Danielsen, K. and Francis, G.W. (1994) An alternative solvent system for the separation of anthraquinone aglycones from rhubarb on silica thin-layers. *Chromatographia* 38(7–8):520–520.
35. Willfor, M., et al. (2006) Chromatographic analysis of lignans. *Journal of Chromatography A* 1112(1–2):64–77.
36. Ma, Y.L., et al. (1997) Characterization of flavone and flavonol aglycones by collision-induced dissociation tandem mass spectrometry. 15th Annual Informal Meeting on Mass Spectrometry (IMMS), Smolenice, Slovakia, pp. 1357–1364.
37. Fabre, N., et al. (2001) Determination of flavone, flavonol, and flavanone aglycones by negative ion liquid chromatography electrospray ion trap mass spectrometry. *Journal of the American Society for Mass Spectrometry* 12(6):707–715.
38. Zhou, D.Y., et al. (2007) Rapid qualitative and quantitative analyses of flavanone aglycones in Fructus Aurantii HPLC ion-trap MS. *Journal of Separation Science* 30(6):858–867.
39. Takayama, M., et al. (1992) Mass-spectrometry of prenylated flavonoids. *Heterocycles* 33(1): 405–434.
40. Ye, M., et al. (2005) Characterization of phenolic compounds in the Chinese herbal drug Tu-Si-Zi by liquid chromatography coupled to electrospray ionization mass spectrometry. *Rapid Communications in Mass Spectrometry* 19(11):1469–1484.
41. Justesen, U. (2001) Collision-induced fragmentation of deprotonated methoxylated flavonoids, obtained by electrospray ionization mass spectrometry. *Journal of Mass Spectrometry* 36(2): 169–178.
42. Domon, B. and Costello, C.E. (1988) A systematic nomenclature for carbohydrate fragmentation in FAB-MS/MS spectra of glyconjugates. *Glycoconjugate Journal* 5:397–405.
43. Ma, Y.L., et al. (2001) Mass spectrometric methods for the characterisation and differentiation of isomeric O-diglycosyl flavonoids. *Phytochemical Analysis* 12(3):159–165.
44. Cuyckens, F., et al. (2001) Structure characterization of flavonoid O-diglycosides by positive and negative nano-electrospray ionization ion trap mass spectrometry. *Journal of Mass Spectrometry* 36(11):1203–1210.
45. Ferreres, F., et al. (2004) Characterization of the interglycosidic linkage in di-, tri, tetra- and pentaglycosylated flavonoids and differentiation of positional isomers by liquid chromatography/electrospray ionization tandem mass spectrometry. *Journal of Mass Spectrometry* 39(3):312–321.
46. Kovacik, V., et al. (1995) Oligosaccharide characterization using collision-induced dissociation fast-atom-bombardment mass-spectrometry-evidence for internal monosaccharide residue loss. *Journal of Mass Spectrometry* 30(7):949–958.
47. Shahat, A.A., et al. (2005) Structural characterization of flavonol di-O-glycosides from Farsetia Aegyptia by electrospray ionization and collision-induced dissociation mass spectrometry. *Rapid Communications in Mass Spectrometry* 19(15):2172–2178.
48. Waridel, P., et al. (2001) Evaluation of quadrupole time-of-flight tandem mass spectrometry and ion-trap multiple-stage mass spectrometry for the differentiation of C-glycosidic flavonoid isomers. *Journal of Chromatography A* 926(1):29–41.
49. Kite, G.C., et al. (2006) Data-directed scan sequence for the general assignment of C-glycosylflavone O-glycosides in plant extracts by liquid chromatography-ion trap mass spectrometry. *Journal of Chromatography A* 1104(1–2):123–131.
50. Kazuno, S., et al. (2005) Mass spectrometric identification and quantification of glycosyl flavonoids, including dihydrochalcones with neutral loss scan mode. *Analytical Biochemistry* 347(2):182–192.
51. Jin, Y., et al. (2008) Characterization of C-glycosyl quinochalcones in Carthamus tinctorius L. by ultraperformance liquid chromatography coupled with quadrupole-time-of-flight mass spectrometry. *Rapid Communications in Mass Spectrometry* 22(8):1275–1287.

52. Fang, W., et al. (2006) Acetylated flavanone glycosides from the Rhizomes of Cyclosorus acuminatus. *Journal of Natural Products* 69(11):1641–1644.
53. Liu, S.Y., et al. (2004) Structural analysis of saponins from medicinal herbs using electrospray ionization tandem mass spectrometry. *Journal of the American Society for Mass Spectrometry* 15(2):133–141.
54. Liang, F. (2002) Structural characterization of steroidal saponins by electrospray ionization and fast-atom bombardment tandem mass spectrometry. *Rapid Communications in Mass Spectrometry* 16(12):1168–1173.
55. Zhou, L.B., et al. (2007) Filiasparosides a-D, cytotoxic steroidal saponins from the roots of Asparagus Filicinus. *Journal of Natural Products* 70(8):1263–1267.
56. Horgen, F.D., et al. (2000) New triterpenoid sulfates from the red alga Tricleocarpa Fragilis. *Journal of Natural Products* 63(2):210–216.
57. Ye, M., et al. (2007) Analysis of phenolic compounds in rhubarbs using liquid chromatography coupled with electrospray ionization mass spectrometry. *Journal of the American Society for Mass Spectrometry* 18(1):82–91.
58. Pawlus, A.D., et al. (2005) An anthraquinone with potent quinone reductase-inducing activity and other constituents of the fruits of Morinda citrifolia (Noni). *Journal of Natural Products* 68(12):1720–1722.
59. Ding, B., et al. (2007) Qualitative and quantitative determination of ten alkaloids in traditional Chinese medicine Corydalis yanhusuo Wt Wang by LC-MS/MS and LC-DAD. *Journal of Pharmaceutical and Biomedical Analysis* 45(2):219–226.
60. Wang, D.W., et al. (2004) Structural elucidation and identification of alkaloids in Rhizoma Coptidis by electrospray ionization tandem mass spectrometry. *Journal of Mass Spectrometry* 39(11):1356–1365.
61. Zhang, H., et al. (2007) Indole alkaloids from three species of the Ervatamia genus: E. Officinalis, E. Divaricata, and E. Divaricata Gouyahua. *Journal of Natural Products* 70(1):54–59.
62. Es-Safi, N.E., et al. (2007) Fragmentation study of iridoid glucosides through positive and negative electrospray ionization, collision-induced dissociation and tandem mass spectrometry. *Rapid Communications in Mass Spectrometry* 21(7):1165–1175.
63. Madhusudanan, K.P., et al. (2000) Effect of alkali metal cationization and multiple alkali metal exchange on the collision-induced dissociation of loganic acid studied by electrospray ionization tandem mass spectrometry. *Rapid Communications in Mass Spectrometry* 14(10):885–896.
64. Suryawanshi, S., et al. (2006) Liquid chromatography/tandem mass spectrometric study and analysis of xanthone and secoiridoid glycoside composition of Swertia chirata, a potent antidiabetic. *Rapid Communications in Mass Spectrometry* 20(24):3761–3768.
65. Li, C.M., et al. (2008) Structural characterization of iridoid glucosides by ultra-performance liquid chromatography/electrospray ionization quadrupole time-of-flight tandem mass spectrometry. *Rapid Communications in Mass Spectrometry* 22(12):1941–1954.
66. Chang, W.L., et al. (2005) Immunosuppressive iridoids from the fruits of Gardenia jasminoides. *Journal of Natural Products* 68(11):1683–1685.
67. Tiew, P., et al. (2002) Coumarins from the heartwoods of Mansonia gagei Drumm. *Phytochemistry* 60(8):773–776.
68. Eklund, P.C., et al. (2008) Identification of lignans by liquid chromatography-electrospray ionization ion-trap mass spectrometry. *Journal of Mass Spectrometry* 43(1):97–107.
69. Yan, Q., et al. (2007) Electrospray ionization ion-trap time-of-flight tandem mass spectrometry of two furofurans: sesamin and gmelinol. *Rapid Communications in Mass Spectrometry* 21(22):3613–3620.
70. Mei, R.Q., et al. (2009) Antioxidant lignans from the fruits of Broussonetia papyrifera. *Journal of Natural Products* 72(4):621–625.
71. Wren, S.A.C. (2005) Peak capacity in gradient ultra performance liquid chromatography (UPLC). *Journal of Pharmaceutical and Biomedical Analysis* 38(2):337–343.
72. Liu, M., et al. (2007) Extraction and ultra-performance liquid chromatography of hydrophilic an lipophilic bioactive components in a Chinese herb Radix Salviae Miltiorrhizae. *Journal of Chromatography A* 1157(1–2):51–55.

73. Chen, J.H., et al. (2008) Analysis of alkaloids in Coptis chinensis Franch by accelerated solvent extraction combined with ultra performance liquid chromatographic analysis with photodiode array and tandem mass spectrometry detections. *Analytica Chimica Acta* 613(2):184–195.
74. Zheng, X.T., et al. (2008) Rapid analysis of a Chinese herbal prescription by liquid chromatography-time-of-flight tandem mass spectrometry. *Journal of Chromatography A* 1206(2):140–146.
75. Wang, Y., et al. (2008) Development of a comprehensive two-dimensional hydrophilic interaction chromatography/quadrupole time-of-flight mass spectrometry system and its application in separation and identification of saponins from Quillaja saponaria. *Journal of Chromatography A* 1181(1–2):51–59.
76. Wang, Y., et al. (2008) Novel, fully automatic hydrophilic interaction/reversed-phase column-switching high-performance liquid chromatographic system for the complementary analysis of polar and apolar compounds in complex samples. *Journal of Chromatography A* 1204(1):28–34.
77. Tolonen, A. and Uusitalo, J. (2004) Fast screening method for the analysis of total flavonoid content in plants and foodstuffs by high-performance liquid chromatography/electrospray ionization time-of-flight mass spectrometry with polarity switching. *Rapid Communications in Mass Spectrometry* 18(24):3113–3122.
78. Jin, Y., et al. (2008) Systematic screening and characterization of flavonoid glycosides in Carthamus tinctorius L. by liquid chromatography/UV diode-array detection/electrospray ionization tandem mass spectrometry. *Journal of Pharmaceutical and Biomedical Analysis* 46(3):418–430.
79. Qu, J., et al. (2004) Screening and identification of glycosides in biological samples using energy-gradient neutral loss scan and liquid chromatography tandem mass spectrometry. *Analytical Chemistry* 76(8):2239–2247.
80. Albert, K. (2002) *On-Line LC-NMR and Related Techniques*. Hoboken, NJ, John Wiley & Sons, Inc.
81. Yang, Z. (2006) Online hyphenated liquid chromatography-nuclear magnetic resonance spectroscopy-mass spectrometry for drug metabolite and nature product analysis. *Journal of Pharmaceutical and Biomedical Analysis* 40(3):516–527.
82. Serber, Z., et al. (2000) New carbon-detected protein NMR experiments using cryoprobes. *Journal of the American Chemical Society* 122(14):3554–3555.
83. Spraul, M., et al. (2003) Advancing NMR sensitivity for LC-NMR-MS using a cryoflow probe: application to the analysis of acetaminophen metabolites in urine. *Analytical Chemistry* 75(6):1536–1541.
84. Wu, N.A., et al. (1994) ^{1}H–NMR spectroscopy on the nonoliter scale for static and online measurements. *Analytical Chemistry* 66(22):3849–3857.
85. Olson, D.L., et al. (1995) High-resolution microcoil ^{1}H–NMR for mass-limited, nanoliter-volume samples. *Science* 270(5244):1967–1970.
86. Pukalskas, A., et al. (2005) Development of a triple hyphenated HPLC-radical scavenging detection-DAD-SPE-NMR system for the rapid identification of antioxidants in complex plant extracts. *Journal of Chromatography A* 1074(1–2):81–88.
87. Exarchou, V., et al. (2006) Hyphenated chromatographic techniques for the rapid screening and identification of antioxidants in methanolic extracts of pharmaceutically used plants. *Journal of Chromatography A* 1112(1–2):293–302.

Chapter 5

Bioassays for Screening and Functional Elucidation of Herbal Medicines

Willow J.H. Liu

Bioactive compound screening and functional elucidation of herbal medicines are two different concepts. Bioactive compound screening uses an established bioassay to screen out bioactive extracts or fractions to isolate the biological compounds responsible for this particular assay. Samples for screening may be a synthetic compound, an extract from a plant collected from a forest, an animal from an ocean, or a microbial from earth. Bioactive compound screening is the main method for finding new drugs in pharmaceutical companies. The functional elucidation or mechanism study of an herbal medicine is the attempt to use different *in vitro* or *in vivo*, occasionally *in situ*, biological and pharmacological methods to explain why an herb has a known therapeutic effect. Bioassays are performed for both purposes.

In practice, functional elucidation is more complicated and challenging. With a well set up bioassay, screening of an extract or a fraction is usually terminated if the result is negative. However, for functional evaluation of an herbal medicine with a known function, the study should not be stopped if a negative result is obtained from one assay. Other bioassays with different mechanisms should continue to be tried out for the test sample. For example, red clover and black cohosh have been used by women with menopause to relieve their hot flash symptoms, so their methanol extracts were both studied on estrogenic activity beginning with the estrogenic receptor (ER) binding assay. However, unlike the red clover extract binding to the ERs directly to exhibit estrogenic activity, bioassay results showed that the extract of black cohosh neither bound to the ERs, nor showed any estrogenic activities in other assays measuring the estrogen-related enzyme or gene expression.[1] But the extract was found to be bound to serotonin receptors 5 HT_{1A} and 5-HT_7, indicating

Traditional Herbal Medicine Research Methods, Edited by Willow J.H. Liu

that this herbal medicine may take effect through regulating the nerve system.[2] The details of this example will be given later in this chapter.

Enzyme or receptor binding studies are usually the first step of a series of bioassays for new drug development or mechanism study of herbal medicines. Once a positive result is confirmed, regardless of whether it is for drug development or for mechanism study, further study with other assays should be followed up. For example, when the ligand binds to a G-protein coupled receptor, it will trigger the intracellular response, including enzyme activation; when the ligand binds to a steroid hormone receptor within the nucleus, it will trigger the gene expression, causing cascade reactions to build up various proteins to meet the needs of cell growth and reproduction. Thus, *in vitro* assays for detecting the intracellular changes by measuring enzyme activity, gene expression, or other reactions need to be continued.

There are many bioassays available for screening new drug candidates or for mechanism studies of herbal medicines. Bioassays can be categorized into different groups based on the experimental objectives used, such as enzyme or receptor binding and gene expression, or according to their actions, such as antioxidant, anti-inflammation, antiaggregation, anticancer, antibacteria, antivirus, or antifungus. The targets of screening have been extended from plants to marine organisms and microorganisms. However, with broad screening of drug candidates through decades all over the world, the likelihood of finding new potential targets with commonly applied assays is decreasing. Mainly because of this, natural product departments in many big pharmaceutical companies were cut off in the past two decades. New high-throughput assays with low cost are always expected for screening. In fact, new assays come out every year.

It is important for researchers to learn new technologies all the time. However, the purpose of this chapter is not to simply list all of the available bioassays and detailed procedures for each method. Such information is easily obtained by searching reference books, review articles, research papers, and web sites. The details of each method are usually given within the literature or are available from companies that either sell the assay kits or provide experimental materials such as receptors, enzymes, genes, or cells. If you purchase enzymes, receptors, or test kits from a commercial company in the United States, the company usually provides not only the experimental procedure, but also technical support if you have any questions or trouble during the experiment. In fact, no matter what method is performed, the researchers should know the principle and have enough knowledge about the properties of both the test sample and material used in the assay, such as receptor, enzyme, or cell. This chapter will focus on introducing basic knowledge and methodology of the use of bioassay for screening new drug candidates and mechanism studies of herbal medicine.

5.1 HISTORY OF SCREENING COMPOUNDS FROM NATURAL PRODUCTS FOR DRUG DEVELOPMENT

For many decades, scientists in universities, research institutes, and pharmaceutical companies have been trying to screen biological active components from synthesized compounds to extracts from plants, microbes, and marine organisms from all over the world, for new drug development. The history of screening natural products leads down the path from traditionally long and tedious bioassay-guided fractionation to recent, fast, and direct online high-throughput biological screens of isolated compounds, fractions, or extracts. The speed of screening has been dependent on the technologies of isolation and structure identification of natural products.

About half a century ago, macromolecules such as proteins and polysaccharides were discarded as wastes during extraction. Now, many proteins and polysaccharides have been demonstrated to be bioactive compounds with the potency to enhance immunity, with anti-inflammatory and antiviral properties, and so on. Thus, the width and depth of recognition of natural products have closely depended on the development of science in chemistry, biochemistry, biology, and other disciplines.

A survey by scientists from the National Cancer Institute (NCI) in the United States indicated that 61% of the 877 new small-molecule chemical entities introduced as drugs worldwide during 1981–2002 can be traced to or were inspired by natural products. These include natural products (6%), natural product derivatives (27%), synthetic compounds with natural product-derived pharmacophores (5%), and synthetic compounds designed on the basis of knowledge gained from a natural product (i.e., a natural product mimic; 23%).[3]

However, many pharmaceutical companies have cut down their natural product departments within the last 15 years. Maureen Rouhi, the deputy editor-in-chief of *Chemical & Engineering News* (*C&EN*) wrote a series of reports in the *C&EN* journal in 2003 about the trend of natural products' drug discovery. She cited a sentence by a researcher from a pharmaceutical company for explanation: "Natural products fell out of favor because they did not fit into the current in-house efforts of drug companies."[4] But she also wrote: "Cast aside for years, natural products drug discovery appears to be reclaiming attention and on the verge of a comeback." A recent article published in *Science* again raised the question, "Drug discovery and natural products: end of an era or an endless frontier?"[5] No doubt, with continuously emerging high-impact technologies for natural product screening,[6] natural products will continue to play an important role in drug discovery.[7]

The view the author puts forward here is that, besides screening natural products from the untapped biological resources with high throughput[8] and other advanced technology, functional mechanism evaluation of well-known traditional herbs and study of the theories of traditional medicines behind the treatment may inspire scientists to find new directions for the treatment of diseases.

5.2 BRIEF INTRODUCTION OF ENZYMES, RECEPTORS, CELLS, AND GENE EXPRESSION

Screening usually measures the interaction of extracts, fractions, or pure compounds with an enzyme, a receptor, a gene, or other objectives. The materials used for a bioassay may be isolated enzymes, receptors, or cell lines that contain the targeting enzymes, receptors, or genes. Many bioassay kits are now commercially available.

If there are not many samples to be tested, or the test is just applied for the mechanism study of one or a few herbs, it is easier to begin with a commercial kit and follow the instruction provided with the kit. However, it is necessary for performers to know the properties of the protein, gene, or cell line used for the assay and the mechanism behind the assay, so that the reason behind an unexpected result can be analyzed and the cause found.

Receptors are proteins on the cell membrane or within the cytoplasm or cell nucleus that a specific ligand can bind to. Binding of a ligand to a receptor initiates changes in behavior of receptor proteins, resulting in a cellular response to carry out physiological changes. The ligands are usually small molecular compounds, such as monoamine or peptide neurotransmitters, steroid hormones, or other substances such as drugs or toxins. Many drugs take action through binding to receptors. Receptors are very sensitive and selective to ligands, like locks to keys.

Enzymes are proteins that catalyze chemical reactions by converting the bound substrates into products. Almost all metabolic processes in cells are carried out by enzymes at significant rates. Enzymes are extremely selective for their substrates as receptors to ligands.

The most notable common feature between receptor and enzyme is their specificity. Keep in mind that the bioactivity of a protein is temperature and pH sensitivity; different proteins require different buffers to keep their bioactivities.

Cells are the smallest structural and functional unit of a living organism. The functions of a cell include cell growth, metabolism, division, and protein synthesis. Among these functions, ligand binding to receptor and enzyme catalysis play very important roles. For example, the binding of a ligand to a receptor on a cell membrane initiates an intracellular response such as metabolism, of which enzyme catalysis may also be involved; binding of a ligand to a receptor within the nucleus initiates gene expression and protein synthesis. Cell culture is very important in a cell bioassay. The temperature, humidity, and concentration of carbon dioxide of the incubator, and the composition of media used are all matters for cell growth and survival. An antibody is often added into the cell cultural media to prevent contamination. The selection of antibody depends on the cell line.

Genes are the units of heredity that carry inherited information. Each human gene consists of long strands of DNA that contain a promoter and a coding sequence. The promoter controls the activity of a gene, while the coding sequence determines what the gene produces. When a gene is active, the coding sequence is copied in a process called transcription, producing an RNA copy of the gene's information. This RNA can then direct the synthesis of proteins via the genetic code. The process of producing a biologically functional molecule of either RNA or protein is called *gene expression*, and the resulting molecule itself is called a gene product.

5.3 SELECTION OF BIOASSAY

With the development of modern science and biologically related technology, more and more assays are now available for screening and functional evaluation of herbal medicines, from the traditional enzyme and receptor binding assays to high-throughput microarrays. It is important to determine which assay is (or assays are) suitable for a specific research project and what the best way is to reach the project's goal. Ideally, bioassays for screening should be simple, accurate, reproducible, rapid, and affordable. Of course, the selection of bioassays should mainly depend on your research need. For mechanism study, the newest method is not necessarily the best for your project. Be sure that the assay is reliable and appropriate to your research project. The budget and your lab condition are also important considerations. When making the decision, the time schedule, cost, and lab condition should all be considered. Once the assay is selected, there are three ways to carry out your research.

1. *Have a service lab or find a partner to work with your assay.*

If only a few samples will be tested as part of your research project, and your lab is not a bioassay lab and lacks the basic testing conditions or technically skilled person for the assay, it is advised to have a service lab do it for you or to find a partner in another research group who has experience with performing the assay you selected.

Establishing a new assay may sometimes be time-consuming. To benefit both the research group and the lab, many service labs are available for testing many kinds of bioassays now. Most research labs in universities are willing to collaborate with others.

2. *Buy commercially available test kits for your assay.*

If your lab has the conditions for the assay, but only few samples need to be tested, buying the commercial kit may save the time of optimizing the experimental condition.

3. *Establish the whole assay system in your lab.*

If several dozens or hundreds of samples are to be tested, or the bioassay is the core of your research project, then it is necessary to set up the assay. When establishing a new assay, it is better to have an experienced person working on it because many factors will impact the result. The conditions need to be optimized for better sensitivity and reproducibility.

5.4 EVALUATION OF BIOASSAY RESULTS OF HERBAL SAMPLES

Screening extracts from plants is more difficult than from pure isolated or synthetic compounds because of the complex composition of extracts. Many factors may affect the result. Scientists need to be very careful when evaluating the test results.

5.4.1 Carefully Assessing the Negative Results

When a negative result of a test sample is obtained, one needs to carefully analyze the reason before giving a conclusion of "inactive." Three factors must be carefully investigated.

1. *Concentration of the test sample*

One needs to know the significant differences between screening isolated or synthesized compounds and crude herbal extracts. Crude herbal extracts are mixtures containing hundreds or even thousands of compounds. The bioactive components in an herbal extract may only be a small percentage or even a few thousands. Thus, the concentration of an herbal extract used for an assay should be much higher than that of an isolated or synthesized compound. If the concentration of pure standard for an assay used is in ng/mL, the concentration of the crude extract may need to be in μg/mL. It is best to prepare samples of extracts at several concentrations of different magnitudes at initial test.

When I first joined a bioassay lab, a graduate student showed me his test results from the ER binding assay for a few herbal extracts. All the results from his experiment were negative, and the test had been repeated several times. I noticed that the concentration of sample used in his test was 1 μg/mL, and thus suggested to him that he increase it to 10 μg/mL. This time, three of the extracts tested (red clover, chaste berry, and hops) turned into positive results. See the example of estrogenic screening later in this chapter.

2. *Extraction method*

The bioactive compounds may be nonpolar small molecules that are easily dissolved in chloroform, such as the estrogenic isoflavones from red clover,[1] but may also be polar compounds that are more easily dissolved in alcohol, such as serotonergic N_{ω}-methylserotonin from black cohosh,[9] or water-soluble big molecules such as immunomodulatory glycoproteins from *Echinacea pallida*.[10] Therefore, it is important to search the literature before conducting experiments to gain a rough idea of the possible types of bioactive compounds for the assay. To be safe, for initial tests, it is better to extract samples with a series of solvents having different polarities, starting from the least polar one, then the medium polar, and polar solvents, and end with water.

I personally witnessed a research group screening hundreds of plant extracts for anti-HIV activities for 2 years without any promising results. Out of curiosity, I talked to the person who was performing the bioassay. Through conversation, I hypothesized that the problem might be the wrong extracts being provided because it had been reported at that time that aqueous extracts of some Chinese herbs were found as anti-HIV inhibitors,[11] but the samples they tested were all extracted with chloroform.

To prove my point, I gave this person 10 Chinese herbal extracts prepared with different solvents from China. The bioassay showed that two of the water extracts presented positive results. The reason I share this story is to warn people of the importance of the extraction method used for an assay.

If the chemist for the extraction knew the mechanism of the bioassay, or the biologist for the bioassay knew the structure–efficacy relation, such a waste of time and money would have not occurred. This story also demonstrates the importance of exhaustively searching for literature before beginning the experiment.

It is impossible for one to know everything, but for a scientist undertaking a new project, it is very critical to update his or her knowledge by researching enough literature.

3. *Toxicity to cells*

When the assay is cell-based, one needs to be sure that the negative or positive result is not due to the death of cells killed by tested samples. Therefore, toxicity should be measured when an assay is performed using a cell line. For example, when we test the estrogenic activity, the extract of hops showed toxicity at the concentration being tested (the example is given later in this chapter).

5.4.2 Factors That May Affect Bioassay Results

As a summary, the following factors have to be considered before beginning the bioassay.

1. *Interference of chlorophyll, fatty acids, and tannins in assays*

In my experience, linolic acid has displayed positive results in both enzyme and receptor binding assays. For example, linoleic acid was screened out as the most active cyclooxygenase (COX) inhibitor from *Angelica Pubescens* (Duhuo)[12] and the most active estrogenic compound from *Vitex agnus-castus* (chaste berry).[13]

2. *Solvents and methods used for extraction*

As mentioned above, some solvents may be unable to extract the bioactive compound. Samples extracted with a different solvent or method may significantly affect the yields of the bioactive compounds in it.

3. *Concentration of extracts prepared for assay*

If the concentration of test sample is too low, the false negative result will be observed; if it is too high, the toxic components in the sample may kill cells in a cell-based assay.

4. *Solvent used for preparation of samples*

Searching literature carefully ensures that the solvent used for preparing samples will not affect the test result. Some solvents may affect the activity of proteins or function of cells.

5. *Some of the toxic ingredients in the extract may kill the cells*

When a cell line is used in a bioassay, be sure to check whether the cells could be killed by the extract.

6. *Experimental conditions*

Both receptors and enzymes are proteins. The affinity of a particular ligand–receptor or substrate–enzyme complex (dissociation constant) can be altered significantly with solution conditions, for example, temperature, pH, and concentration of the buffer. Therefore, pH and concentrations of the reagents prepared and the temperature at which it is stored will all impact the results.

If a cell line is used, great care needs to be given to possible contamination. Usually, an antibiotic needs to be added to the cultural media. If a new cell line is used, one should consult the provider or look for literature about the detailed condition for its growth. When establishing a new assay, one needs to carefully search for the literature to know enough about the assay.

7. *Other factors*

Sometimes, even the brand or grade of a reagent, or the materials of the plate used for an assay, may affect the result. For example, when the detection of a test is the absorption at certain wavelengths of the reacted products, the ultraviolet (UV) absorption of the plate should be checked to make sure it is not close to the measuring wavelength.

5.4.3 Factors That May Affect Reproducibility of Results

It is commonplace to hear researchers complain that their bioassay results could not be repeated. Many factors will impact the reproducibility of results, such as changes of buffer pH, room temperature, temperature or CO_2 concentration in incubator, concentration of reagents, skills of the operators, different cell batches, and expired or contaminated reagents used. When an assay is performed to compare the bioactivity of different fractions from one extract or potency of different herbal extracts, it is ideal to test them using the same batch of receptors, enzymes, or cells in one experiment. This will make the variation of the result as small as possible.

It is necessary to mention to phytochemists that the reproducibility of a sample in a bioassay cannot be as good as that of the retention time (Rt) of a compound in the chromatogram of high-performance liquid chromatography (HPLC). When I was a graduate student of natural product chemistry, I was mainly trained in the chemical field and did not have enough experience with evaluating bioassay data at the beginning. I isolated compounds and gave them to another lab for anti-platelet aggregation tests. When I first received the test results, I was not satisfied with the data received because the results of these samples from the three tests varied too much from what I expected as a phytochemist. Coincidently, when I later worked at a bioassay lab of a university in the United States, a similar thing happened to our collaborated phytochemists with results of their samples from a receptor binding assay. For this reason, one sample with different codes was given to us several times to test the reproducibility of the assay. The testing game finally ended when I explained that due to many manual steps involved in the assay, and the receptors being prepared from membranes of different cell batches each time, the reproducibility of the sample in the assay could not be as good as that of the Rt in HPLC.

5.5 ENZYME BINDING ASSAY

Although enzymes and receptors are both proteins with specific selectivity, their working mechanisms are different. The receptor–ligand complex initiates cell responses. However, the receptor does not change the structure of a ligand. On the other hand, enzymes change the substrate into products by its catalyzing reaction (see Fig. 5.1). Enzyme assays are usually performed to determine the interference of the test extract or compound to the enzymatic activity, that is, the inhibitory ability of the sample to the enzyme activity, by measuring either the consumption of the known substrate or its reaction product over time. Once a sample is screened out as a potential inhibitor to an enzyme for the first time, it is necessary to find out whether this compound works as a competitive or noncompetitive inhibitor by determining its kinetics curve.

5.5.1 Enzyme Kinetics

Enzyme kinetics is the study of the rate at which an enzyme takes effect. Enzyme kinetics is examined as a function of the concentration of substrate available to the enzyme. Measuring enzyme kinetics will help explain how enzymes work. For drug screening, it will tell us whether a drug candidate works as a competitive or non-competitive inhibitor.

The Michaelis–Menten equation describes how the reaction rate v depends on the substrate binding equilibrium and the rate constant K.

$$V_0 = \frac{V_{max}[S]}{Km + [S]}$$

This equation indicates that the initial reaction rate V_0 relates to the substrate concentration [S]. As mentioned above, enzyme-catalyzed reactions are saturable. At that point, the rate of catalysis does not show a linear response to increasing substrate. If the initial rate of the reaction is measured over a range of substrate concentrations ([S], molar), the reaction rate (V) increases as the substrate concentration [S] increases, as shown in Figure 5.2. However, as the substrate concentration [S] gets higher, the enzyme becomes saturated with substrate and the rate reaches V_{max}, the enzyme's maximum rate.

Figure 5.2 showed Michaelis–Menten kinetics of a single-substrate reaction. K_m, the Michaelis constant, is defined as the concentration at which the rate of the

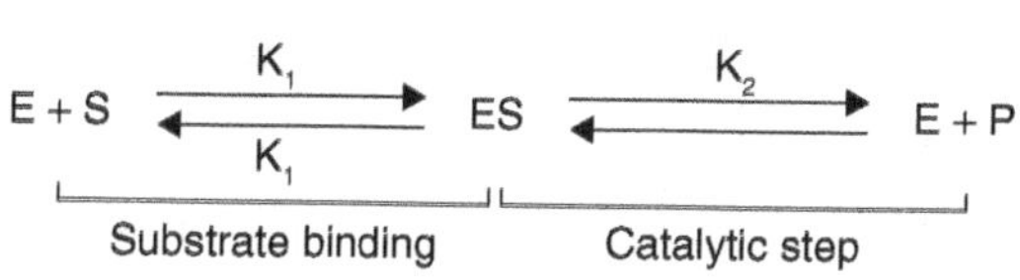

Figure 5.1 Single-substrate mechanism for an enzyme reaction (K_1 and K_2 are the rate constants for each individual steps).

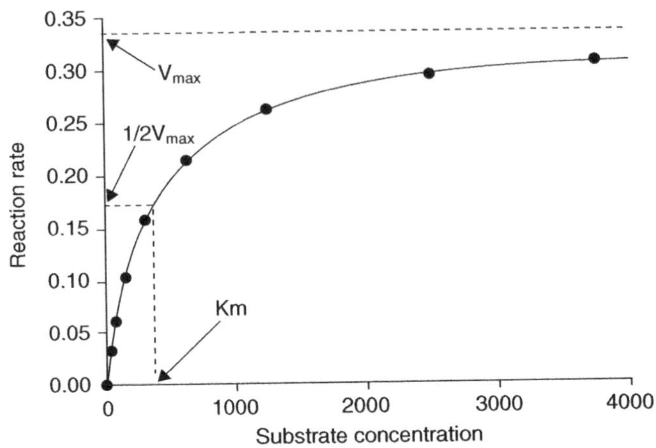

Figure 5.2 Saturation curve for an enzyme showing the relation between the concentration of substrate and rate (Michaelis–Menten kinetics curve). V_{max}, maximum velocity; K_m, substrate concentration when reaction rate is one-half V_{max} (Michaelis-Menton constant).

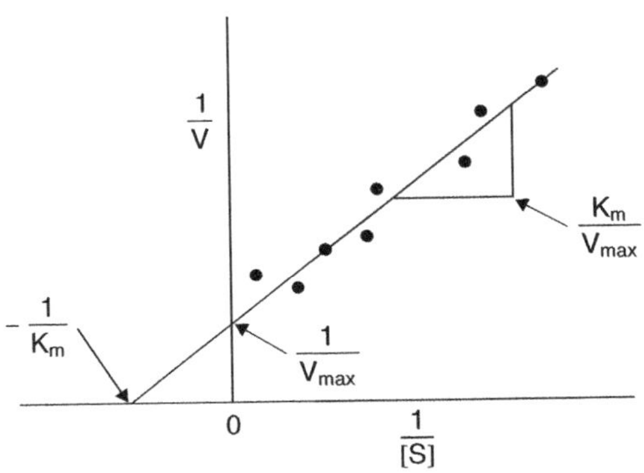

Figure 5.3 Lineweaver–Burk plot of kinetic data, indicating the significance of the axis intercepts and gradient.

enzyme reaction is half V_{max}. It indicates the affinity of binding between the enzyme and its substrate. The lower the K_m, the greater the affinity.

The Lineweaver–Burk plot, also called double-reciprocal plot, was widely used to determine K_m and V_{max} before computers and nonlinear regression software were made widely available.

From Figure 5.3, it is known that the *y*-intercept of such a graph is equivalent to the inverse of V_{max}; the *x*-intercept of the graph represents $-1/K_m$. K_m equals V_{max} times the slope of line. This is easily determined from the intercept on the X-axis. The Lineweaver–Burk plot is also used to determine the type of the inhibitor.

5.5.2 Binding Characteristics of Enzyme Inhibitors

Enzyme inhibitors are molecules that reduce or abolish enzyme activity. Many drugs act as inhibitors of enzymes to carry out pharmacological effect. Based on the binding characteristics, they can be divided into reversible or irreversible, competitive or noncompetitive inhibitors. The particular type of inhibitor can be discerned by studying the enzyme kinetics as a function of the inhibitor. More clearly, the distinction can be determined by plotting enzyme activity with and without the inhibitor present.

1. *Reversible and irreversible inhibitors*

Reversible inhibitors are not only able to bind to an enzyme, but also are dissociated from the enzyme when another compound with higher affinity competes with it. Removal of the reversible enzyme inhibitors will restore enzyme activity. However, irreversible inhibitors bind to enzymes covalently, and thus will permanently inactivate the enzyme, usually by modifying active site residues.

2. *Types of reversible inhibitors*

Reversible enzyme inhibitors can be classified as competitive, noncompetitive, or mixed, according to their binding sites (see Fig. 5.4) and effects on K_m and V_{max}. In special cases, some substances work as a partially competitive or an uncompetitive inhibitor.[14]

Competitive inhibitors compete with the endogenous substrate to bind to the free enzyme on the same active site as the substrate, preventing the substrate from binding, but not the enzyme-substrate (ES) complex. It does not hamper catalysis. Competitive inhibition increases K_m, but does not affect V_{max}. In other words, in the presence of a competitive inhibitor, it takes a higher substrate concentration to achieve the same velocities that were reached in its absence. So while V_{max} can still be reached if sufficient substrate is available, one-half V_{max} requires a higher [S] than before and thus K_m is larger.

Uncompetitive inhibitors bind only to the ES complex, but not to the free enzyme. The enzyme-inhibitor-substrate (EIS) complex is enzymatically inactive. This type of inhibition is rare. Uncompetitive inhibition causes a decrease in both V_{max} and K_m.

Noncompetitive inhibitors bind to both enzyme (E) and ES and have identical affinities for E and ES. They do not affect substrate binding, but reduce the catalytic efficiency of the enzyme. Both the enzyme-inhibitor (EI) and EIS complexes are enzymatically inactive. Noncompetitive inhibition does not change K_m, but decreases V_{max}. In other words, with noncompetitive inhibition, enzyme rate (velocity) is reduced for all values of [S], including V_{max} and one-half V_{max}, but K_m remains unchanged because the active site of those enzyme molecules that have not been inhibited is unchanged.

Mixed inhibitors bind to both E and ES, but their affinities for these two forms of the enzyme are different. They interfere with substrate binding and hamper catalysis. This type of inhibition resembles the noncompetitive type, except that the EIS complex has residual enzymatic activity. Mixed-type inhibitors increase K_m, but decrease V_{max}.

Partially competitive inhibitors have a mechanism similar to noncompetitive inhibitors. But partially competitive inhibition decreases catalytic activity (if the catalytic activity is increased, it is called partially competitive activation). Partially competitive inhibition typically displays a lower V_{max}, but has no effect on K_m.

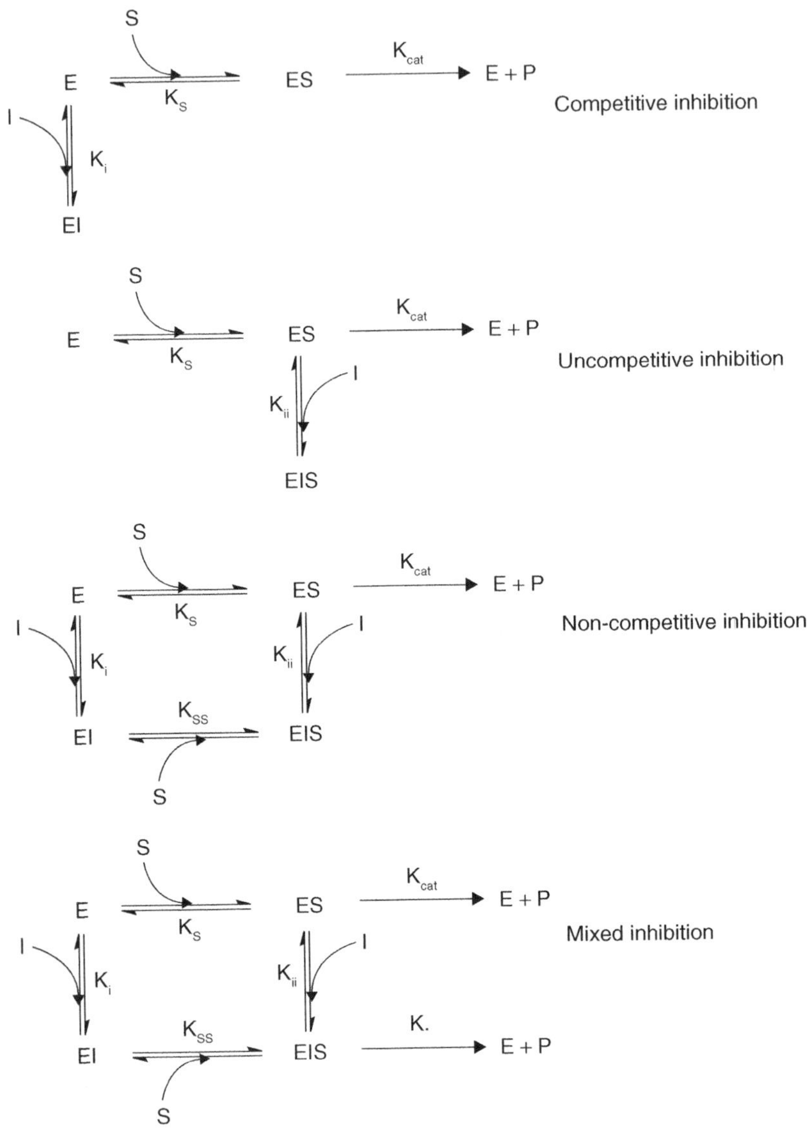

Figure 5.4 Types of reversible enzyme inhibitors.

3. *Determining enzyme inhibitor types*

The distinction between reversible inhibitor types can be determined by plotting enzyme activity, for example, enzyme kinetics with and without the inhibitor present.[15]

The Lineweaver–Burk plots between competitive, noncompetitive, and uncompetitive inhibitors can be clearly distinguished (see Fig. 5.5). Competitive inhibitors

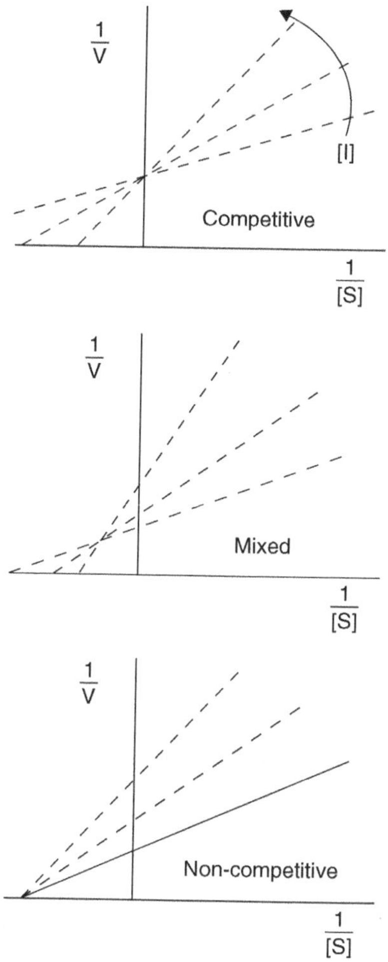

Figure 5.5 Lineweaver–Burk plots of different types of reversible enzyme inhibitors. The arrow shows the effect of increasing concentrations of inhibitor.

have the same *y*-intercept as uninhibited enzyme (since V_{max} is unaffected by competitive inhibitors, the inverse of V_{max} also does not change), but there are different slopes and *x*-intercepts between the two data sets. Meanwhile, noncompetitive inhibition produces plots with the same *x*-intercept as uninhibited enzyme (since K_m is unaffected) but different slopes and *y*-intercepts. Uncompetitive inhibition causes different intercepts on both the y- and x-axes but the same slope.

5.5.3 Methods of Enzyme Binding Assay

There are several methods to measure results, that is, an enzyme can be assayed in several different ways. The concentrations of substrates or products can be measured with continuous assays, meaning the assay gives a continuous reading of activity, or with discontinuous assays, meaning the concentration of substrates/products is determined after the reaction is stopped. In discontinuous assays, such as radiometric assay and chromatographic assay, when results are taken from an enzyme reaction at intervals, the amount of reaction production or substrate consumption is measured in the reaction mixture. But continuous assays are most convenient, with one assay giving the rate of reaction with no further work necessary. Now many different kinds of detectors are available for the measurement, such as spectrometric assay (colorimetric assay), fluorimetric assay, calorimetric assay, and chemiluminescence assay. HPLC-MS has also been used for this purpose. The principles of these mentioned assays are briefly introduced here, together with Enzyme-linked immunosorbent assay (ELISA) or enzyme immunoassay (EIA). Details of methods for each specific enzyme can be easily found by searching literature.

1. *Radiometric assay*

In a radiometric assay, the radioactive isotopes labeled substrates or its metabolites in a complex mixture is measured when cells are lysed after the reaction is catalyzed by an enzyme. The most frequently used isotopes in bioassays are ^{14}C, ^{32}P, ^{35}S, and ^{125}I. Radiometric assay is extremely sensitive and specific, and frequently used in bioassay labs. Radioactivity is traditionally measured using a scintillation counter. Now HPLC coupled with radioactivity monitor is also applied in many labs.

2. *Chromatographic assay*

In chromatographic assays, the product that formed in an enzymatic reaction is measured after separation by chromatography, usually by HPLC. The separation of the analyzing target from other compounds avoids the inference from the latter. Sensitivity can be increased by labeling the substrates/products with a radioactive or fluorescent tag. Examples are the anti-inflammatory effect of Chinese herbal extract *A. pubescens* in COX and 5-lipooxygenase (5-LO) assays. 5-LO assay was performed using intact leucocytes, while COX assay was performed with a preparation of microsomal COX from ram seminal vesicles. Both enzymes were incubated with 1-[^{14}C]-arachidonic acid whose radiolabeled metabolites were separated and determined with reversed phase HPLC using a radioactivity monitor[12]). Assay sensitivity can also be increased by switching protocols to improved chromatographic instruments (e.g., ultrahigh-pressure liquid chromatography) that operate at pump pressure a few folds higher than HPLC instruments.

3. *Spectrometric assay (colorimetric assay)*

In spectrometric assays, the change of the light absorbence in the reacting solution during the course of reaction is measured. Most of the absorption is within the UV region. If this light is in the visible region, in which people can actually see a change in the color of the assay, it is called colorimetric assay. An example of such an assay is the MTT assay, a redox assay for measuring the activity of enzymes that reduce MTT [3-(4,5-dimethylthiazol-2-yl)-2,5-diphenyltetrazolium bromide] or MTS [3-(4,5-dimethylthiazol-2-yl)-5-(3-carboxymethoxyphenyl)-2-(4-sulfophenyl)-2H-tetrazolium] + PMS (phenazine methosulfate) to formazan, giving a purple color. The results can be read on a multiwell scanning spectrometer (ELISA reader). The main advantages of the colorimetric assay are its rapidity and precision, and the lack of any radioisotope.[16]

By using a coupled assay, the spectrometric assay can also be used when the enzyme reaction does not result in a change in the absorbance of light. In such an assay, the product of one reaction is used as the substrate of another, easily detectable reaction. For example, the enzyme hexokinase catalyzes glucose, converting into glucose 6-phosphate; neither the substrate nor the product has UV absorption. However, the consequent reaction involves NADPH, which is the reduced form of $NADP^+$ (nicotinamide adenine dinucleotide phosphate) and absorbs UV light. When the hexokinase assay is coupled with another assay using glucose-6-phosphate dehydrogenase, by which glucose 6-phosphate can be further made into 6-phosphogluconate accompanied by $NADP^+$ into NADPH, the increase in UV absorbance can be measured at 340 nm.

4. *Fluorimetric assay*

Fluorometric assay utilizes the difference of emitted fluorescence between substrate and product to measure the enzyme reaction. This assay is relatively more sensitive than spectrometric assay, but the result can be easily interfered with by impurities; meanwhile, many fluorescent compounds are unstable when exposed to light.

NADH (the reduced form of nicotinamide adenine dinucleotide, NAD^+) and NADPH are also used in the fluorometric assay, in which the reduced forms are fluorescent and the oxidized forms nonfluorescent. Therefore, a decrease of fluorescence will be observed in an oxidation reaction, while an increase of fluorescence will be observed in reduction reactions.[17]

5. *Calorimetric assay*

In a calorimetry assay, the heat released or absorbed by enzymatic reactions is measured. Since many reactions involve some change in heat, with the use of a microcalorimeter, the assay needs only few enzymes and substrates. This assay is used when no other assays are available.[18]

6. *Chemiluminescence assay*

Chemiluminescence refers to the emission of light by a chemical reaction. Some enzyme reactions also generate light, which can be measured to detect the product formation. The generated light can be captured by photographic film over days or weeks, so this assay is very sensitive. However, since it is impossible to detect all the light released by a reaction, it is hard to quantify.

The detection of horseradish peroxidase by enhanced chemiluminescence (ECL) is a sensitive and commonly used method for detecting antibodies in Western blotting.[19] Another commonly used enzyme is luciferase, which is found in fireflies and naturally emits light from its substrate luciferin. Luciferases have been used for real-time, low-light imaging of gene expression in cell cultures, individual cells, whole organisms, and transgenic organisms over the last two decades. Luciferase imaging has also been used to trace bacterial and viral infection *in vivo* and to visualize the proliferation of tumor cells in animal models.[20]

7. *ELISA or EIA*

ELISA, also called EIA, is a biochemical technique used mainly in immunology to detect the presence of an antibody or an antigen in a sample. Briefly, an antigen is affixed to a solid support (usually a polystyrene microtiter plate), and then a specific antibody is washed over the surface so that it can bind to the antigen, forming an antigen–antibody complex. This antibody can be covalently linked to an enzyme, or can itself be detected by a secondary antibody that is linked to an enzyme through bioconjugation. Between each step, the plate is typically washed with a mild detergent solution to remove any proteins or antibodies that are not specifically bound. In the final step, the plate is developed by adding an enzymatic substrate to produce a visible signal, such as color or fluorescence for measurement by ELISA reader. Older ELISAs use chromogenic substrates, while newer assays employ fluorogenic substrates, enabling much higher sensitivity.

The ELISA has been used not only as a diagnostic tool in clinic labs for detection of antigens or antibodies, but also as a bioassay in the lab for mechanism study of herbal medicine, plant pathology, as well as a quality control check in various industries. An example is the intracellular cyclic adenosine monophosphate (cAMP) assay for mechanism study of black cohosh using the Trevigen's HT Direct cAMP EIA Kit. This kit is a high-throughput competitive immunoassay for the quantitative determination of cAMP in samples treated with 0.1M HCl. The kit uses a polyclonal antibody to cAMP to bind, in a competitive manner, the cAMP in the standards or sample. Details of the assay are given in Section 5.11.

5.6 RECEPTOR BINDING ASSAY

Substances that are able to bind to receptors are usually called ligands. There are several types of receptor families. The cellular responses of a ligand binding to a receptor differ between receptor families with different mechanisms. Researchers need to know the mechanism of each type. Therefore, major receptor families are briefly introduced below.

The first step of screening for bioactive ligands to receptors is the binding assay. However, merely knowing whether the sample has the ability to bind to a receptor is not enough. It is necessary to know whether the ligand binding to a receptor is an agonist, a partial agonist, or an antagonist, and whether the binding is competitive, mixed competitive, or noncompetitive. This section provides only a brief introduction about receptor binding. Details of methods for measurement can be found in books about receptor binding.[21] Examples of screening and mechanism study of herbal extracts used for menopause symptoms of red clover and black cohosh are given in Sections 5.10 and 5.11.

5.6.1 Major Receptor Families

Based on the location of the receptors, they can be divided into membrane receptors (also called transmembrane receptors) and intracellular receptors. The binding of hormones, neurotransmitters, cytokines, and other signaling compounds to membrane receptors from outside a cell will trigger changes of the cellular function. Iontropic receptors, G-protein-coupled receptors, and enzyme-linked receptors are membrane receptors. Some receptors locate within the cells, for example, inositol trisphosphate (IP_3) receptor on the endoplasmic reticulum, and steroid hormone receptors in the cell nucleus.

1. *Ionotropic receptors:* also called ligand-gated ion channels. They are a group of intrinsic transmembrane ion channels that are opened or closed in response to binding of a chemical messenger. Examples are cholinergic receptors, γ-aminobutyric acid (GABA) receptor, and 5-HT_3 serotonin receptor, and glutamate receptor.
2. *G-protein-coupled receptors:* also known as seven-transmembrane domain receptors or heptahelical receptors. This is a large family of transmembrane receptors. The activation of this group of receptors and their consequent cellular responses involve three parts: the ligand, the G-protein-coupled receptors, and the G-protein. When a ligand binds to a G-protein-coupled receptor, it creates a conformational change in the receptor, causing activation of a G-protein. Further effect depends on the type of G-protein. The receptors of this group transduce signals derived from odor, light, and numerous neurotransmitters, including most norepinephrine receptor, dopamine receptor, serotonin receptor (see Fig. 5.6), and acetylcholine receptor.
3. *Enzyme-linked receptors:* binding of a ligand to an extracellular domain of such receptors activates or inhibits a cytosolic enzyme activity. Insulin receptor belongs to this family.
4. *Steroid hormone receptor:* when the extracellular steroid hormones, such as androgen, estrogen, progesterone, and glucocorticoid, enter into the cell nucleus, they bind to their corresponding receptors and result in modulation of gene expression to synthesize proteins (Fig. 5.7).

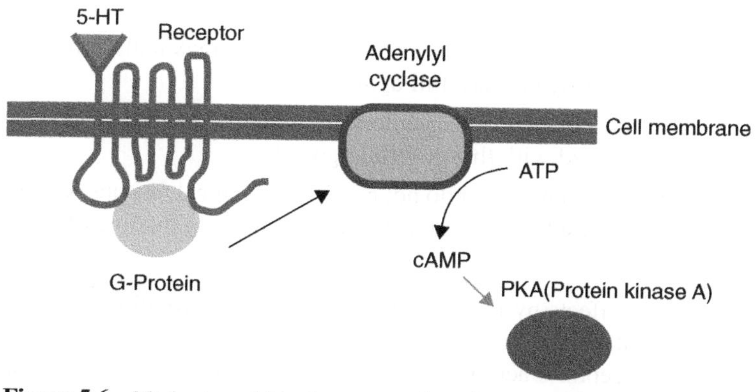

Figure 5.6 Mechanism of 5-hydroxytryptamine (5-HT) binding to serotonin receptor.

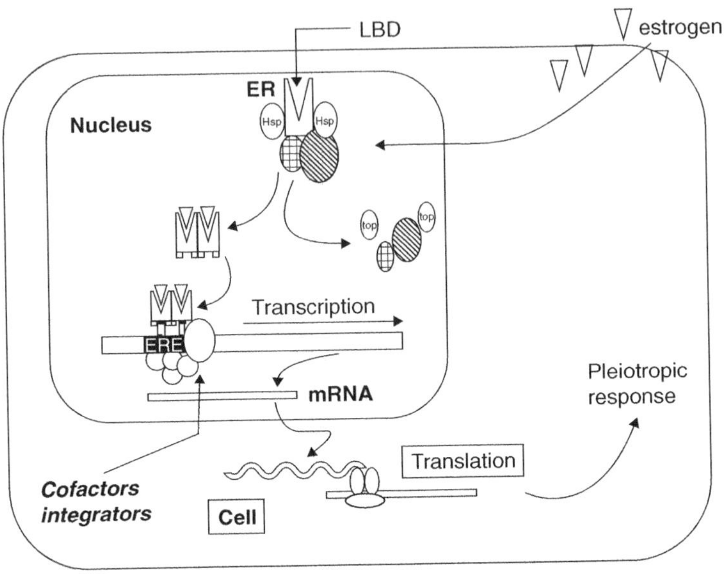

Figure 5.7 Mechanism of 17β-estradiol (E_2) binding to estrogen receptor. LBD, ligand binding domain; ERE, estrogen receptor element.

5. *Immune receptor (or immunologic receptor):* when a substance, such as a cytokine, binds to an immune receptor, it will cause a response in the immune system. Antigens (short for antibody generators) are defined as exogenous substances that bind to specific immune receptors and elicit an immune response. They are usually on a cell membrane.

5.6.2 Agonist versus Antagonist

Based on biological response of ligands binding to receptors, ligands can be divided into the following categories:

1. *Agonists:* are able to activate receptors and result in a maximal biological response. Most natural ligands are full agonists.
2. *Partial agonists:* do not activate receptors thoroughly, causing responses that are partial compared with those of full agonists.
3. *Antagonists:* bind to receptors but do not provoke biological responses upon binding. The binding results in receptor blockage, inhibiting binding of other agonists.

Antagonists mediate their effects by binding to the active site or to allosteric sites on receptors, or interact at unique binding sites that are not normally involved in biological regulation of receptor's activity. Antagonist activity can be *reversible* or *irreversible* depending on the nature of antagonist receptor binding, the antagonist–

receptor complex. Irreversible binding usually involves a covalent binding between ligand and receptor. The majority of drug antagonists achieve their potency by competing with endogenous ligands at structurally defined binding sites on receptors.

5.6.3 Competitive versus Noncompetitive Ligands

Ligands can be divided into competitive and noncompetitive.

- *Competitive ligand:* binds to a receptor at the same site as an endogenous agonist, thus, compete for the binding of the agonist. A competitive antagonist shifts the dose–response curve to the right (less potent).
- *Noncompetitive ligand:* binds to a site other than where an agonist binds. It either prevents the binding of an agonist or prevents an agonist from activating the receptor. A noncompetitive antagonist decreases the maximal response (less effective).

5.6.4 The Efficacy and Potency of Binding

The formation of a ligand–receptor complex leads to a biological response, and the magnitude of the response is proportional to the number of ligand–receptor complexes.

$$\text{Ligand} + \text{Receptor} \rightarrow \text{Ligand–Receptor Complex} \rightarrow \text{Effect}$$

Ligand binding to receptor is an equilibrium process. A ligand binds to a receptor and dissociates from it according to the law of mass action.

The affinity of a ligand to a receptor is an important factor in determining potency of the binding. It describes how tightly a ligand binds to a particular receptor. It can be influenced by non-covalent intermolecular interactions between the two molecules such as hydrogen bonding, electrostatic interactions, hydrophobic, and Van der Waals forces.

The dissociation constant is commonly used to indicate the affinity of a ligand to a receptor.

$Kd = [R][L]/[C]$, where [R], [L], and [C] represent the concentrations of receptor, ligand, and complex, respectively.

Binding affinity is inversely related to the dissociation constant *Kd*. A ligand with a lower *Kd* has a higher affinity to the receptor.

The biological response is achieved only when a significant number of receptors are activated. *Potency* is a measurement of how many ligands are required to elicit a given response. It is also termed as effective dose concentration. It is most often expressed as the concentration of a ligand when it achieves 50% of the maximal response, or EC_{50}. A ligand with a lower EC_{50} is more potent than one with a larger EC_{50}.

Efficacy is the maximal response produced by binding of a ligand to a receptor. It depends on the number of ligand–receptor complexes formed and the efficiency with which the activated receptor produces a cellular action. Efficacy is analogous to the maximal velocity of enzymes catalyzing reaction.

An antagonist binds to a receptor but does not elicit a response, thus having affinity but no efficacy. The binding will disrupt the interaction and inhibit the function of an agonist. In this case, it is said to have zero efficacy. Efficacy is more important than potency since it focuses on the effectiveness of a ligand.

5.6.5 Methods of Receptor Binding Assay

Radiolabel receptor binding is an easy, traditional, and the most popularly applied method to measure the ligand binding to receptor. The materials can be either isolated receptors or cell membranes (if it is a membrane receptor). Most of the receptors are commercially available; they can also be isolated in the lab. Methods for estrogen and serotonin receptor binding assays are given in Sections 5.10 and 5.11 as examples. In addition, fluorescence polarization, immunocytochemical and "rosetting" methods, and a complementary approach making use of radiolabel-binding and immunohistological protocols in tandem are also used for investigation of the transferrin receptors.[22]

Recently, liquid chromatography-mass spectrum (LC-MS) has been applied for detection of receptor binding capability of a ligand in a mixture such as an herbal extract by comparing the intensity changes of peaks in the HPLC chromatograms before and after the sample is incubated with the receptors and followed by a dissociation and an filtration procedures (examples are available in Sections 5.10 and 5.11).

5.6.6 Determination of Binding Characteristics

When a ligand displays binding to a receptor, the characteristics of the binding should be determined. Radiolabel receptor binding can be used to determine the ligand is competitive, noncompetitive, or mixed. The method of determination is similar with that for an enzyme assay. An example of black cohosh acts as a mixed competitive ligand is available in Section 5.11.

A cellular assay that measures change of enzyme activity, or gene expression and others induced by the receptor binding, is usually needed to follow up to evaluate the efficacy of the ligand. If the result from the cellular assay shows no response to the test ligand but response to the positive control, it tells that this ligand works as an antagonist with zero efficacy; however, if the result shows no response from both the test ligand and positive control, a reason needs to be found. An assay for induction of alkaline phosphatase (AP) in the Ishikawa cell line for evaluation of the binding of red clover and other herbal extracts to the ER, and an assay for cAMP elevations study in HEK293 cell line for evaluation of the binding of black cohosh extract to 5-HT_7 receptors are available in Sections 5.10 and 5.11.

It should be noted that if a cell line used contains only the recombinant receptor on its cell membrane, you can only use the cell membrane for a binding assay, but will not see any biological response from the cells.

5.7 GENE EXPRESSION ASSAYS

Gene expression implemented by a cell is very dynamic, responding rapidly to external stimuli. Therefore, analysis of gene expression becomes necessary for providing clues about regulatory mechanisms, biochemical pathways, and broader cellular function. Many herbal extracts or isolated compounds have been found to be able to regulate gene expression. Examples for the regulation of *ER*, progestin receptors (*PR*), and presenelin-2 (*pS2*) gene expressions in different cell lines by red clover and chaste berry extracts[1,13] are available in Sections 5.10 and 5.11.

Now many experimental techniques are available for measurement of gene expression, ranging from expression vector, reporter gene, Northern blot, and fluorescent *in situ* hybridization, to reverse transcriptase-polymerase chain reaction (RT-PCR), real-time PCR, multiplex PCR, rapid amplification of cDNA ends (RACE), serial analysis of gene expression (SAGE), DNA microarray, tiling array, ChIP-Seq, mRNA-Seq. Here we only introduce a few of them.

To measure the up- or down-regulation of gene expression by an herbal extract, fraction, or isolated compound, the test samples are usually fed into cells that contain the target gene(s). The cells are then homogenized, and the mRNA is extracted for quantification by means of Northern blotting, RT-PCT, DNA microarray, or other methods. An example of regulation of *PR* and *pS2* gene expression by herbal extracts is given later in this chapter.

- *Northern blotting* is a process in which a sample of RNA is separated on an agarose gel and hybridized to a radiolabeled RNA probe that is complementary to the target sequence. Northern blotting uses radioactive reagents, but its benefit is allowing the discrimination of alternately spliced transcripts.
- *RT-PCR* is a method measuring band strength in an image of a gel for quantification of mRNA. Real-time PCR, which is also known as quantitative RT-PCR or quantitative PCR (qPCR), measures mRNA abundance with a carefully constructed standard curve qPCR that can produce an absolute measurement, such as the number of copies of mRNA per nanoliter of homogenized tissue. The lower level of noise in data obtained via qPCR often makes this the method of choice, but the price of the required equipment and reagents can be prohibitive.
- *DNA microarray* technology or "tag-based" technologies like SAGE or the more advanced version SuperSAGE provide a relative measure of the cellular concentration of different mRNAs. Recent advances in microarray technology allow for the quantification, on a single array, of transcript levels for every known gene in the human genome. The great advantage of tag-based methods is the "open architecture," allowing for the exact measurement of any

transcript, known or unknown. In particular, SuperSAGE recommends itself, therefore, for studying organisms with unknown genomes.

Gene expression assays originally measure expression of only one gene, or a select group of genes. Recent technological advances have made it possible to analyze the expression of the entire genome in a single experiment.[23] Since gene expression triggers a cascade reaction to build up various proteins, measurement of protein levels using Western blotting or other techniques is another way to determine the regulation of gene expression.

5.8 NEW TECHNOLOGIES AND OTHER BIOASSAYS FOR SCREENING AND MECHANISM STUDY

With traditional bioassay, the only way to identify the bioactive compounds in an extracted or fractioned mixture is continuous bioassay-guided isolation until the compound is purified for identification. With the combined application of LC/MS or LC/NMR today, the bioactive compounds can be easily detected in a mixture by comparing the relative intensity of the peaks of HPLC chromatograms before and after the mixture was incubated with the receptor or enzyme buffer followed by a disassociation. If a known compound has bound to a receptor or enzyme, its relative intensity in HPLC chromatogram will increase after the receptor or enzyme binding. By comparing its MS spectrum with the standard in the MS spectrum library, its structure should be determined. For a complex unknown compound, LC/NMR might be necessary. If LC/NMR is not available, a column chromatographic or HPLC separation might be needed for structural elucidation.

Such techniques allow researchers to distinguish between known compounds and new molecules directly from crude plant extracts. Thus, the tedious isolation of known compounds can be avoided, and only a targeted isolation of constituents presenting novel or unusual spectroscopic features needs to be performed. This not only saves time, but also makes it possible to find trace known bioactive components in a mixture. See examples of estrogenic and serotonergic compounds identification, respectively, from red clover and black cohosh by means of combination of receptor binding with LC/MS in Sections 5.10 and 5.11.

Many herbs have different preventative and therapeutic functions because herbs are usually composed of several types of compounds that have different pharmacological activities with different mechanisms. Therefore, functional elucidation of one herb may need several types of bioassays. Such work may need scientists from several groups working together. With the microarray available, a sample can be tested simultaneously on many receptors and enzymes; this greatly accelerates the speed of the mechanism study of herbal medicine.

As mentioned before, it is impossible to list all of the bioassays in this chapter or even in one book. In addition to the aforementioned assays using cells, proteins, or genes from human beings and animals, it is necessary to mention the small Drosophila. Study with this little creature has advantages that other whole animal models cannot compete with, which include small individual space occupation, low costs, high propagation speed, short life-cycle (50 days), obvious age-related degradation of the neurons, and small sample consumption (5–50 mg). Drosophila has become an ideal model for studying age-related neuron degenerated diseases (e.g., Alzheimer's disease) and for screening new drugs. New iridoid types of compounds isolated from Chinese herbs Gardenia Fruit and Scrophularia screened using Drosophila have been patented for anti-Alzheimer's (Chinese Patent CPCH0860106N-B0128). While there are methods for detection of the psychostimulant and analeptic properties of an extract obtained from the plant *Acorus calamus*, using mutant Drosophila melanogaster Sh^5 eag^1 has also been applied for patent (U.S. Patent 6617491).

High-throughput screening (HTS) has been well known in pharmaceutical companies and universities for drug discovery. In comparison with traditional bioassays in labs, using robotics, computers with fast data processing and control software, liquid handling devices, and sensitive detectors, HTS allows researchers to simultaneously conduct thousands or even millions of biochemical, genetic, or pharmacological tests. Modern microplates for HTS generally have either 384, 1536, 3456 wells, or even more. These are all multiples of 96, reflecting the original 96 well microplate. Through this process one can rapidly evaluate the activity of samples and identify active compounds, antibodies, genes, or extracts from nature that modulate a particular biomolecular pathway. The results of these experiments provide starting points for drug development.[6,24–26] HTS is a relatively recent innovation. It is still expensive to run an HTS. Small to moderately sized research institutions can use the services of an existing HTS facility rather than set up one by themselves.

5.9 KEYS TO FUNCTIONAL MECHANISM STUDY OF HERBAL MEDICINES

As mentioned before, the difference of mechanism study of herbal medicine from general screening for new drug development is that the negative result of an herbal extract from one assay does not indicate that this extract has no bioactivity. Other assays must be tried out until its functional mechanism is explored based on the knowledge about the diseases and functions of the test herbal extraction.

Although herbal medicines have been used on human beings for a long time, most of the mechanisms of their therapeutic functions are too complicated to be well explained so far.

The unknown mechanisms for known treatments are not only due to the complicated composition of herbs, but also due to the fact that physiological knowledge about the human body is still limited and the etiology and pathology of many diseases, that is, the causes of many diseases, are not yet known.

It is known that cells are the smallest metabolically functional unit of all living organisms, including plants, animals, and the human body. Cells contain proteins such as enzymes and receptors that are coded by genes. The cell functions are controlled by nerves and hormones. Enzymes, receptors, and genes play important roles in cell growth and division. It is also known that cells form the organs, muscles, blood, nerves, and bones in our body. The nutrients are provided through blood vessels to the whole body to nourish cells. However, the disorders of many diseases are not well known at the level of cells and molecules. For instance, more receptors and enzymes and their subtypes are being explored every year, and the impact of their imbalance to the functions of our body still remains unclear. Therefore, the mechanisms of herbal treatment for diseases currently cannot be elucidated.

With deeper understanding of the human body and broader knowledge about etiology and pathology, as well as the advances in techniques and equipment since the last century, we can now at least try to use the available information to explore the mechanism of the therapeutic and preventative functions of traditional herbs on our body. In other words, it is possible for scientists to elucidate the preventative and therapeutic functions of herbs with a combination of chemistry, biochemistry, biology, and pharmacology.

Thanks to emerging advanced technologies and equipment available for isolation and structure identification, as well as biological methods of the past decades, the elucidation of functional mechanisms of herbs has become not only possible, but also sped up. Many different test kits are commercially available with clear, detailed instructions for operation.

The key work to successfully get a positive result is in choosing the right testing targets, meaning the right enzyme, receptor, or gene, and extracting herbal materials with the right extraction methods, using proper solvents. To reach this goal, the researchers in the team need to have solid foundations in biology, biochemistry, pathology, and pharmacology.

Mechanism studies of traditional medicine may give Western medicine inspirations to find new directions in pathological study.

An herb may not directly act on a targeted receptor as expected, but instead through regulation of the functions of the central nervous system or the release of the hormones involved. For example, when we studied herbs used for menopause symptoms, estrogenic activity was tested at the beginning for several herbs that were clinically used for menopausal women. As a result, black cohosh, which folks use often for hot flashes or other symptoms of menopause, did not show any positive estrogenic activity in assays using ERs. But it turns out that its extract could significantly bind to 5HT1 and 5HT7 serotonin receptors, indicating that the action is through the regulation of the central nervous system. See the details of this example in the following sections.

5.10 EXAMPLE 1. SCREENING AND EVALUATION OF ESTROGENIC ACTIVITY OF HERBAL MEDICINES

Estrogen regulates gene expression by binding to intracellular ERs α (ER_α) and β (ER_β), which influence the growth, differentiation, and functioning of many target tissues. When estrogens bind to an ER, structural change results in receptor dimerization, allowing the dimer to bind to an estrogen-responsive element (ERE) in the DNA. Consequently, the ER complex activates genetic expression, such as the *ER*,

PR, and *pS2* genes, and ultimately stimulates cell growth and differentiation (see Fig. 5.7). High levels of ER_α mRNA have been reported in the uterus, and smaller quantities were detected in the ovaries, testes, skin, and gut. In contrast, fetal ovaries, testes, adrenal glands, and the spleen contain high amounts of ER_β mRNA. ER_α and ER_β are also found in the central nervous system, breast, cardiovascular tissue, and bone.

When estrogen decreases at menopausal age, women start to feel symptoms such as hot flashes, depression, mood swings, sleeping disorders, vaginal dryness, and joint pain. Meanwhile, the risk for osteoporosis, cardiovascular disease, dementia from Alzheimer's, and certain types of cancers increases. Hormone replacement therapy (HRT) is able to relieve these menopause symptoms and reduce the above risks. However, side effects such as increase in the risk of developing breast and endometrial cancer due to estrogen replacement therapy have been reported. Recently, experts have found that nearly 6 million American women could be adversely affected by side effects of HRT and estrogen replacement therapy for menopause symptoms. The nation's largest study in the United States, involving 16,000 women, came to an abrupt end when new data showed that the use of HRT and estrogen replacement therapy significantly increases the risks of breast cancer, heart attacks, strokes, and blood clots.

As a result, women are increasingly using herbal remedies as an alternative therapy. For example, epidemiological data have shown that a diet rich in phytoestrogens, such as those found in soybeans, reduces the number of hot flashes and the incidence of cancer in Japanese women. To evaluate the estrogenic activity of herbal medicine widely used for the improvement of women's health concerns, the UIC/NIH Center for Botanical Dietary Supplement Research in Women's Health, College of Pharmacy, University of Illinois at Chicago (UIC), has performed the following studies in the past few years. All of the results have been published in journals;[1,13,27] some of them are presented below. Literature references in these papers are not given here. They can be found from the original papers listed.

Eight botanicals were tested *in vitro* for estrogenic activity to investigate their mechanisms of action. Black cohosh (*Cimicifuga racemosa*, CR), red clover (*Trifilium pratense*, TP), hops (*Humulus lupulus*, HL), and chaste berries (*V. agnus-castus*, VA) are the most frequently used herbs in Western countries for menopause symptoms or PMS. Dong-quai (*Angelica sinensis*, AS) and licorice (*Glycyrrhiza glabra,* GG) are common Chinese herbs used for women's health concerns. Several case reports implicate ginseng (*Panax ginseng*, PG) as a possible candidate for the treatment of menopause. Therefore, these herbs, together with American ginseng (*Panax quinquefolius*, PQ), were selected for the purpose of studying their estrogen properties.

First, ER binding assay was performed as guidance to isolate the active compounds; then induction assay of AP in Ishikawa cells followed to investigate the estrogenic/antiestrogenic activity of the extract and compounds. The expressions of the estrogen-responsive *pS2* gene in S-30 cells, *PR* gene in Ishikawa cells, and *ER*β gene in T47D:A18 cells were also performed to confirm the results from the two above assays.

5.10.1 Experimental Methods

Extraction and Fractionation Each powder of plant (100 g) was soaked in 600 mL MeOH overnight. After filtering, the plant material was re-extracted twice with 600 mL MeOH with gentle heating (<45°C, 10 min). The extract solution was filtered and the solvent was evaporated.

Cell Culture Conditions Ishikawa cells were maintained in Dulbecco's modified Eagle's medium (DMEM)/F12 media with 10% heat inactive fetal bovine serum (FBS), sodium pyruvate (1%), penicillin-streptomycin (1%), and Glutamax (1%). The S-30 cells were maintained in phenol-free minimum essential medium (MEM) supplemented with 1% penicillin-streptomycin, 6 mg/L insulin, 500 μg/mL G418, 1% glutamax, and 5% charcoal-dextran-treated FBS. The T47D:A18 cells were grown in RPMI-1640 media with heat-inactivated FBS (10%), penicillin-streptomycin (1%), nonessential amino acids (1%), L-glutamine (1%), and bovine insulin (6 ng/mL). At least 1 day prior to the experiment, the medium was replaced with estrogen-free (phenol red-free) DMEM/F12 for Ishikawa and S-30 cells, and estrogen-free (phenol red-free) RPMI-1640 for T47D:A18 cells, both containing charcoal/dextran-twice stripped FBS (10%).

ER Competitive Binding Assay with ^{3}H-estradiol Briefly, 24 h before the assay, 50% v/v hydoxylapatite (HAP) slurry was prepared using 10 g HAP/60 mL of TE buffer (50 mM Tris-Cl, 1 mM EDTA, pH 7.4) and stored at 4°C. ER binding buffer (10 mM Tris, 10% glycerol, 2 mM dithiothrietol, 1 mg/mL bovine serum albumin, pH 7.5), ERα (40 mM Tris, 100 mM KCl, 1 mM EDTA, 1 mM EGTA, pH 7.5), and ERβ (40 mM Tris, pH 7.5) wash buffers were subsequently prepared. The reaction mixture consisted of 5 μL of test samples in dimethyl sulfoxide (DMSO), 5 μL of pure human recombinant diluted ERα or ERβ (0.5 *p*mol) in ER binding buffer, 5 μL of "Hot Mix" (400 nM, prepared fresh using 3.2 μL of 25 μM, 83 Ci/mM ^{3}H-estradiol, 98.4 μL of ethanol, and 98.4 μL of ER binding buffer), and 85 μL ER binding buffer. The incubations were performed at room temperature for 2 h, then 100 μL of 50% HAP slurry was added and the tubes were incubated on ice for 15 min with vortexing every 5 min. Each ER wash buffer was added (1 mL), the tubes were vortexed, and centrifuged at 2000 × *g* for 5 min. The supernatant was discarded and this wash step was repeated three times. The HAP pellet containing ligand–receptor complexes was then resuspended in 200 μL ethanol and transferred to scintillation vials. Cytoscint™ (4 mL/vial) was added and the tubes were counted using a Beckman LS 5801 liquid scintillation counter (Beckman Coulter Inc., Schaumburg, IL). The percent inhibition of ^{3}H-estradiol binding to each ER was determined as $[(dpm_{sample} - dpm_{blank})/(dpm_{DMSO} - dpm_{blank}) - 1] \times 100$. The competitive binding capability (%) of the sample was calculated in comparison with the inhibition of estradiol (50 nM, 100%). The data represent the average ± S.D. of triplicate determinations.

Induction of AP in Ishikawa Cell Line Briefly, Ishikawa cells (5×10^4/well) were incubated overnight with estrogen-free media in 96-well plates. Test samples in DMSO were added and the cells in a total volume of 200 μL media/well were incubated at 37°C for 4 days. For the determination of antiestrogenic activity, 2×10^{-8} M estradiol was added to the media. Enzyme activity was measured by reading the liberation of *p*-nitrophenol at 340 nm every 15 s for 16–20 readings with an ELISA reader. The maximum slope of the lines generated by the kinetic reading was calculated using a computer program. The percent induction for determination of estrogenic activity was calculated as $[(slope_{sample} - slope_{cells})/(slope_{estrogen} - slope_{cells})] \times 100$. For antiestrogenic activity, the percent induction was determined as $[(slope_{sample} - slope_{cells})/(slope_{DMSO} - slope_{cells})] \times 100$. The data represent the average ± SD of triplicate determinations.

Cytotoxic Assay Ishikawa (15000 cells/well) and S-30 cells (4000 cells/well) were incubated in 96-well plates overnight in estrogen-free media. The Ishikawa cells were incubated with test samples for 4 days, and S-30 cells were incubated for 1 day. As an indication of cell viability, absorbance was measured at 515 nm on a microtiter plate reader after the cell was fixed with 20% trichloroacetic acid (TCA) and stained with 0.4% sulforhodamine B (SRB), and the bound dye was liberated with 0.1 M TRIS. The data represent the average ± S.D. of triplicate determinations.

RT-PCR Analysis of PR, ER, and pS2 mRNA Expressions in Ishikawa, T47D:A18 and S-30 Cell Lines Ishikawa cells (2×10^5/well) were pre-incubated overnight with estrogen-free media in a 6-well plate. Test samples dissolved in DMSO were added and incubated at 37°C for 4 days. T47D:A18 cells (1×10^5/well) were pre-incubated overnight in estrogen-free media in 6-well plates. Test samples dissolved in DMSO were added and incubated at 37°C for 2 days. S-30 cells (4×10^4/well) were pre-incubated overnight with estrogen-free media in a 24-well plate. Test samples dissolved in DMSO were added and incubated at 37°C for 24 h. Total mRNA from each cell line was extracted with Trizol, and RT-PCR was performed with SuperScript one-step RT-PCR system using a DNA thermal cycler 480. The following sense and antisense primers were used, respectively, for PR expression: 5′-CCATGTGGCAGATCCCACAGGAGTT-3′ and 5′-TGGAA ATTCAACACTCAGTGCCCGG-3′; for *ER*β expression: 5′-GTCCA TCGCCAGTTATCACATC-3′ and 5′-GCCTTACATCCTTCACACGA-3′; for *pS2* expression: 5′-CATGGAGAACAAGGTG-ATCTG-3′ and 5′-CAGAAGCGTG-TCTGAGGTGTC-3′; and for β-actin expression: 5′-ACACT GTGCCCATCTACGAGG-3′, 5′-AGGGGCCGGACTCGTCATACT-3′. The PCR products (5 μl) of PR (271 bp), ERβ (242 bp), *pS2* (365 bp), and β-actin (621 bp) were separated by electrophoresis in 2% agarose gel and visualized by staining with ethidium bromide.

Detection of ER Ligands in Red Clover Extract using Ultrafiltration and LC-MS Human recombinant ERβ (50 *p*mol or 100 *p*mol) was mixed with the test sample in binding buffer containing 50 mM Tris (pH 7.5), 10% glycerol, 50 mM KCl, and 1 mM EDTA in a total volume of 150 μL. After a 2-h incubation at room temperature, the reaction mixture was filtered through a Microcon® YM-30 centrifugal ultrafiltration filter (Millipore) containing a regenerated cellulose ultrafiltration membrane with a 30,000 molecular weight (MW) cutoff by centrifugation at 10,000 rpm for 7 min at 4°C. The filter was washed with NH_4Ac buffer (pH 7.5) at 4°C by centrifugation (3 × 150 μL). Unbound compounds were removed by washing the filter three times by centrifugation with 150 μL aliquots of ammonium acetate buffer at pH 7.5 at 4°C. To disrupt the ligand–receptor complexes and release the bound ligands, 400 μL of MeOH/H_2O (90:10) was added, followed by centrifugation at 10,000 rpm for 10 min. The solvent in the ultrafiltrate was removed under vacuum, and the ligands were redissolved in 60 μL of H_2O/MeOH (80:20, 60 mL). Aliquots (10 μL) of this solution were analyzed by using LC-MS, which consisted of a Waters 2690 LC system coupled to a Micromass Quattro II electrospray triple quadrupole mass spectrometer. HPLC separation were carried out using a Micra (Micra Scientific, Inc., Northbrook, IL) C_{18} HPLC column, 4.6 × 21 mm, containing 1.5 μm nonporous silica. The mobile phase consisted of H_2O/MeOH (95:5, v/v) containing 0.01% acetic acid (A) and MeOH containing 0.01% acetic acid (v/v) (B), using linear gradients of 5–98 % B (v/v) in 20 min. The electrospray source was operated at 155°C in negative ion mode. Nitrogen was used as both nebulizing gas and drying gas at flow rates of 20 and 450 L/h, respectively. A control was used to correct for nonspecific binding of the sample, in which ERβ was absent from the incubation solution.

5.10.2 Results

Summary Methanol extracts of red clover, chaste berry, and hops showed significant competitive binding to ERα and ERβ. In Ishikawa cells, red clover and hops exhibited estrogenic activity as indicated by induction of AP activity and upregulation of *PR* mRNA. Meanwhile, hops acted as a cytotoxic agent. Chaste berry stimulated *PR* expression in Ishikawa cells and *ERβ* expression in T47D:A18 cells, but no induction of AP activity was observed. In S-30 cells, *pS2* was expressed in the presence of red clover, hops, chaste berry, Asian ginseng, and American ginseng extracts. Dang guai and licorice showed only weak ER binding and *PR* and *pS2* mRNA induction. Black cohosh demonstrated no activity in the three assays. Bioassay-guided isolation utilizing ER competitive binding as monitor and screening using ultrafiltration LC/MS revealed that genistein was the most active component of red clover. Linoleic acid was isolated from chaste berry as the most estrogenic compound.

Relative Affinity of Plant Extracts, Isoflavones, and Fatty Acids for ERα and ERβ Among the eight tested methanol extracts, hops, red clover, and chaste berry showed significant binding affinities to ERα and ERβ on the basis of their 50% inhibitory (IC_{50}) values (Table 5.1). The order of binding potency was red clover > hops > chaste berry, and their affinities to ERα and ERβ were not significantly different. Dong quai and licorice showed weak binding affinity (IC_{50} > 50 μg/mL), whereas Asian ginseng, North American ginseng, and black cohosh displayed no binding (<20% at a concentration of 200 μg/mL).

Table 5.1 ER Binding, AP Induction, and Cytotoxicity of Methanol Extracts

Extract	ERα binding	ERβ binding	AP induction, Ishikawa cells	Toxicity, Ishikawa cells
	IC_{50}, μg/mL	IC_{50}, μg/mL	IC_{50}, μg/mL	ED_{50}, μg/mL
Dong quai	NA*	NA	NA	>20
Black cohosh	NA	NA	NA	>20
Licorice	NA	NA	NA	>20
Hops	30 ± 0.4	27 ± 2.8	13.1 ± 6.1	<2.5
Asian ginseng	NA	NA	NA	>20
American ginseng	NA	NA	NA	>20
Red clover	5.6 ± 2.1	2.5 ± 0.6	1.0 ± 0.2	>20
Chaste berry	46 ± 3	64 ± 4	NA	>20

* NA, not active (IC_{50} > 50 μg/mL for ER binding; IC_{50} > 20 μg/mL for AP induction).

Table 5.2 ER Binding, AP Induction, and Cytotoxicity of Phytoestrogens in Red Clover

Compound	ERα binding	ERβ binding	AP induction, Ishikawa cells	Toxicity, Ishikawa cells
	IC_{50}, μM	IC_{50}, μM	IC_{50}, μM	ED_{50}, μM
Genistein	0.3 ± 0.01	0.018 ± 0.002	0.51 ± 0.1	>5000
Daidzein	17 ± 2.5	1.2 ± 0.0	1.2 ± 0.6	>5000
Biochanin A	35 ± 1.4	4.1 ± 0.8	5.1 ± 0.4	47 ± 6.0
Formononetin	104 ± 8.2	60 ± 7.1	12 ± 3.0	>100
Estradiol	0.0065 ± 0.00058	0.0024 ± 0.00014	0.00014 ± 0.000014	>0.005

Standards of isoflavones, known as corresponding estrogenic compounds presented in red clover, were tested with both ER receptor subtypes and exhibited competitive binding potency following the order genistein > daidzein > biochanin A > formononetein, based on their IC_{50} values (Table 5.2). These four isoflavones displayed higher affinity with ERβ compared with ERα.

Linoleic acid, a fatty acid that is ubiquitous in nature, was isolated from chaste berry as the most estrogenic compound under the guidance of ER receptor binding with an IC_{50} value of 27 ± 2 μM and 30 ± 6 μM, respectively, to ERα and ERβ. It has been reported that linoleic acid also demonstrated the ability to interact with the opiod receptor, the nucleoside transport protein, and COX. Although researchers screening for new drug candidates are not interested in fatty acids, such findings help to elucidate the functional mechanism of some herbal medicines.

Detection of ER Ligands in Red Clover and Chaste Berry by Combination of Ultrafiltration and LC-MS Since red clover and chaste berry showed higher ER binding affinity without cytotoxicity among the crude extracts tested, the two extracts were isolated to find the active compounds.

The chloroform fraction of red clover displayed the greatest active potency compared with the other fractions. Therefore, this fraction was further isolated with column chromatography into fourteen fractions for screening with ER binding assay. The four fractions showing the highest competitive binding ability were analyzed using affinity ultrafiltration LC-MS for the detection of ERβ ligands in the mixture. As a result, genistein, daidzein, and biochanin A were identified as corresponding ERβ ligands in the red clover chloroform fraction on the basis of molecular weight, tandem MS, and HPLC retention time in comparison with authentic standard compounds (Fig. 5.8).

ER assay guided isolation of chaste berry extract resulted in the identification of linoleic acid as the corresponding ER ligand. The identification of linoleic acid was based on ^{1}H-NMR, MS, and HPLC comparison with the reference standard. An affinity ultrafiltration LC-MS method was also used to confirm linoleic acid as an ER ligand. The affinity of linoleic acid for the ER was confirmed by the enhancement of the LC-MS peak at the retention time of 11.9 min following affinity ultrafiltration compared with the control sample, which did not contain any receptors (Fig. 5.9). The peak at 12.5 min is an isomer of linoleic acid that does not show a large enhancement in the LC trace following ultrafiltration, compared with the control, indicating it is not as specific for the ER.

AP Induction in Ishikawa Cells by Herbal Extracts and Phytoestrogens Ishikawa is an ER-positive endomentrial adenocarcinoma cell line derived from a glandular epithelial cell line. This cell responds to estrogens and antiestrogens at a concentration approximating physiological levels. Induction of AP activity in Ishikawa cells indicates an estrogenic response, whereas inhibition represents an antiestrogenic effect. Data in Table 5.1 indicated that, in Ishikawa cells, red clover displayed the strongest estrogenic induction ability, and hops exhibited estrogenic activity. While also showing cytotoxicity to Ishikawa cell line, chaste berry only displayed weak estrogenic activity (40%) at a concentration of 20 μg/mL, while the others did not show activity. The chloroform fraction of red clover also exhibited strong estrogenic activity as compared with the three other fractions. Subfractions Fr-6, Fr-7, and Fr-8 showing the strongest ER binding affinity and demonstrated genistein as the active principle, displaying higher induction of AP in the Ishikawa cells, while Fr-2, containing biochanin A as the binding component, showed mainly toxicity to Ishikawa cells. Genistein, daidzein, biochanin A, and formononetin exhibited estrogenic activity in the same order as in ER binding assay according to their IC_{50} values (Table 5.2). None of the extracts or isoflavones exhibited antiestrogenic activity (data not shown).

Stimulation of pS2 mRNA Expression in S-30 Cells by Herbal Extracts and Phytoestrogens The stimulation of *pS2* expression in the ER-positive breast cancer cell MCF-7 has been reported; thus, induction of *pS2* expression was first used to confirm the results from the AP induction assay. However, the expression of *pS2* gene in MCF-7 was constitutive in our experimental condition; despite a change in estrogen-free media for 4 days, even the control group gave positive

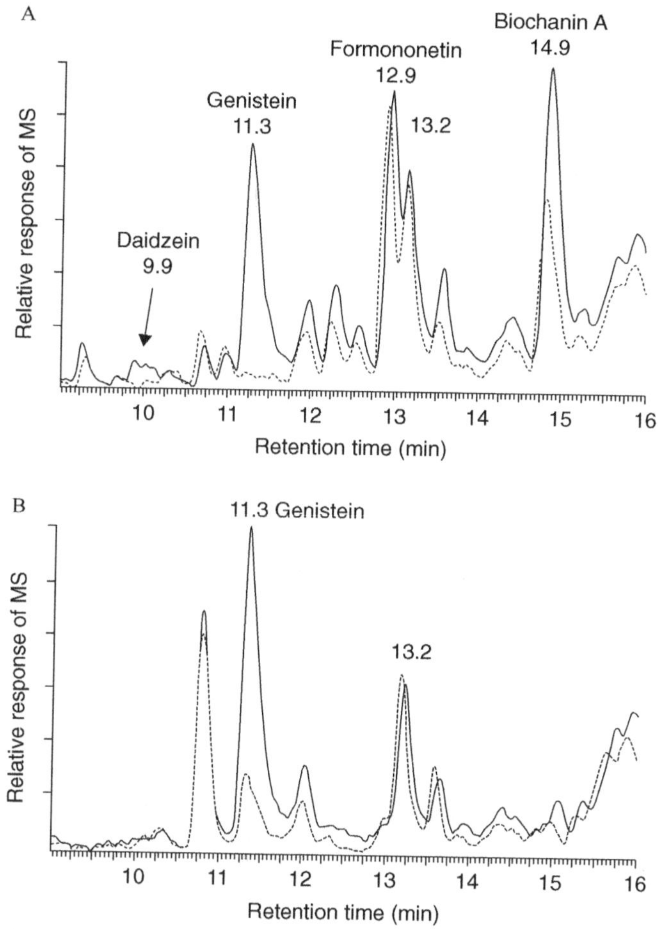

Figure 5.8 Overlaid total ion chromatograms showing affinity ultrafiltration and LC-MS screening results of (A) red clover chloroform extract (10 μg/mL) and (B) one of its bioactive subfractions (fraction 7, 20 μg/mL). The solid line represents the experiment with ERβ (0.667 μM), and the dashed line indicates the control experiment without the receptor. By a combination of ultrafiltration and LC-MS, the enhanced peaks of genistein (11.3 min), daidzein (9.9 min), and biochanin A (14.9 min) were identified and confirmed as active ligands in the chloroform extract. Genistein was the active ligand in subfraction 7. The peak at 13.2 min is due to an impurity in the water mobile phase collected during the equilibration of the LC column. It serves as an internal standard for normalizing the experiment with the control, because the equilibration time of the LC column was the same in the LC sequence before sample injection for both samples.

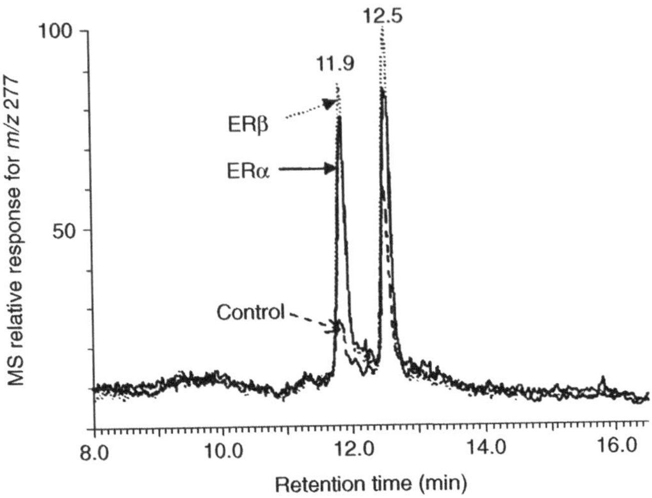

Figure 5.9 Negative ion electrospray mass chromatogram of *m/z* 277 for the ultrafiltrate of a *Vitex agnus castus* incubation with ER_{α} and ER_{β} ligand binding domain. The control incubation contained no ER. The peak at 11.9 min. corresponds to the deprotonated molecule of linoleic acid, which showed specific binding to ER_{α} and ER_{β}. The unidentified peak at 12.5 min probably represents an isomeric fatty acid that showed less specific binding. (---), control; (—), estrogen receptor α; (.....), estrogen receptor β.

expression (data not shown). S-30 is a subclone of the ER-negative MDA-MB-231 breast cancer cell line that is stably transfected with ERα. This cell line was selected for *pS2* expression because it was responsive to estradiol. Results showed that all extracts except that of black cohosh induced *pS2* expression in S-30 cells (see Fig. 5.10) at a concentration of 20 μg/mL. It should be noticed that Asian ginseng and North American ginseng did not show activity in the above assays.

The chloroform fraction of red clover showed stronger *pS2* expression than the petroleum ether, BuOH, and H_2O extracts at 20 μg/mL (data not shown). Expression induced by genistein and daidzein was significantly stronger than that of biochanin A and formononetin at a concentration of 0.1 μM (see Fig. 5.10).

Stimulation of PR mRNA Expression in Ishikawa Cells by Herbal Extracts, Phytoestrogens, and Linoleic Acid The Ishikawa cell line was used to investigate the estrogenic or antiestrogenic effect of the test samples by measuring their ability of AP induction. In our experiments, estradiol-mediated *PR* expression was not observed in S-30 cells; meanwhile, Ishikawa cells did not show *pS2* expression in incubations with estradiol either (data not shown). Because Asian ginseng and North American ginseng show induction of *pS2* expression in S-30 cells but negative in AP induction in Ishikawa cells, *PR* expression was carried out with this cell line to further confirm the results of AP induction.

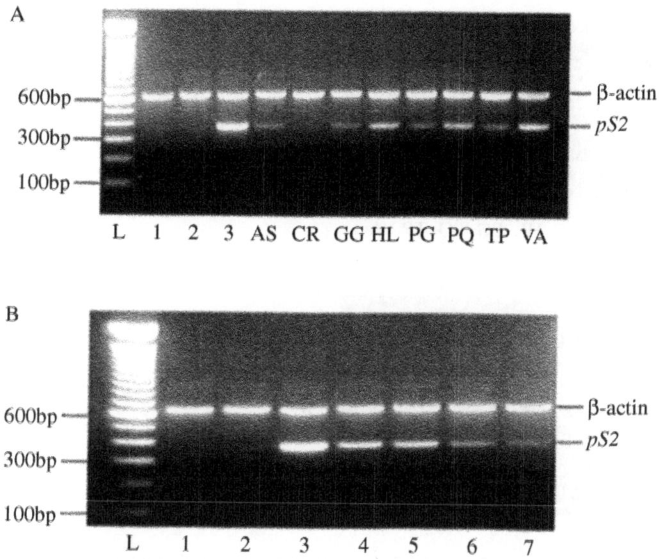

Figure 5.10 Induction of *pS2* mRNA expression in S-30 cells: (A) methanol extracts (20 μg/mL) (1, control; 2, DMSO; 3, estradiol; AS, *A. sinensis* [dong quai]; CR, *C. racemosa* [black cohosh]; GG, *G. glabra* [licorice]; HL, *H. lupulus* [hops]; PG, *P. ginseng* [Asian ginseng]; PQ, *P. quinquefolius* [North American ginseng]; TP, *T. pratense* [red clover]; VA, *V. agnuscastus* [chaste berry]); (B) phytoestrogens (0.1 μM) (1, control; 2, DMSO; 3, estradiol; 4, genistein; 5, daidzein; 6, biochanin A; 7, formononetin).

As measured by RT-PCR, *PR* expression was significantly up-regulated by red clover, hops, and chaste berry extracts at a concentration of 20 μg/mL (Fig. 5.11). Dong quai and licorice exhibited weak stimulation of *PR* expression at this concentration, while extracts of black cohosh and the two ginseng species did not show activity. Standards of the four isoflavones identified in red clover and linoleic acid isolated from chaste berry also induced *PR* expression at a concentration of 5 nM and 3.6 μM, respectively (see Figs. 5.11 and 5.12A). These data are well consistent with that from the ER binding and AP induction assays.

Stimulation of ER mRNA Expression by Chaste Berry and Linoleic Acid in T47D:A18 Cells T47:A18 breast cancer cell line had previously served as a model of ER regulation by estrogens and antiestrogens. Since chaste berry only displayed weak estrogenic activity (40%) at a concentration of 20 μg/mL, it is necessary to discern whether chaste berry extracts and pure linoleic acid bound not only to the ER but also up-regulated *ER* gene expression. Therefore, an *ER* mRNA experiment was carried out. The result showed that *ERα* mRNA expression in T47D:A18 cell line was constitutive using RT-PCR (data not shown); however, expression of *ERβ* mRNA was inducible in the presence of ER ligands. Both the chaste berry extract (10 μg/mL) and linoleic acid (3.6 μM) enhanced the expression of *ERβ* mRNA in T47D:A18 cells (see Fig. 5.12B).

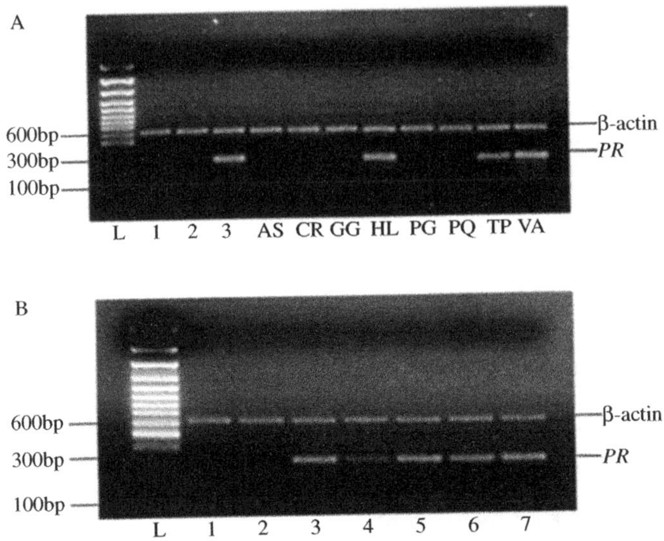

Figure 5.11 Induction of *PR* mRNA expression in Ishikawa cells: (A) methanol extracts (20 μg/mL) (1, control; 2, DMSO; 3, estradiol; AS, *A. sinensis* [dong quai]; CR, *C. racemosa* [black cohosh]; GG, *G. glabra* [licorice]; HL, *H. lupulus* [hops]; PG, *P. ginseng* [Asian ginseng]; PQ, *P. quinquefolius* [North American ginseng]; TP, *T. pratense* [red clover]; VA, *V. agnuscastus* [chaste berry]); (B) phytoestrogens (5 nM) (1, control; 2, DMSO; 3, estradiol; 4, genistein; 5, daidzein; 6, biochanin A; 7, formononetin).

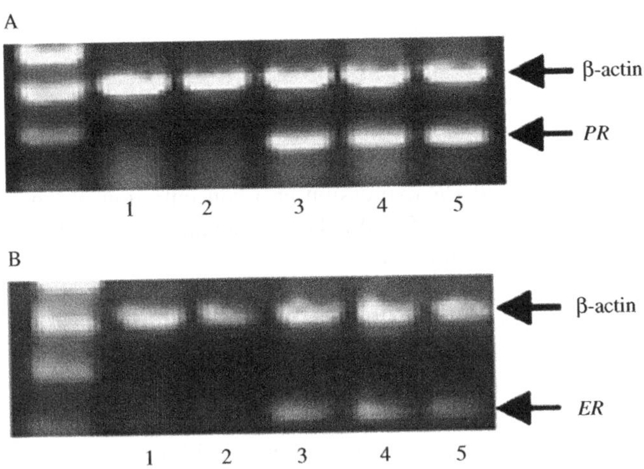

Figure 5.12 Agarose gel visualizing RT-PCR products. (A) Induction of *PR* mRNA expression in Ishikawa cells; (B) induction of *ER*β mRNA expression in T47D:A18 cells. 1, control; 2, DMSO; 3, estradiol, 1 nM; 4, chaste berry MeOH extract, 10 μg/mL; 5, linoleic acid, 3.6 μM)

5.10.3 Brief Discussion

Different cell lines express different estrogen related genes. Therefore, three cell lines were used to confirm the results of ER binding and AP induction assays for these extracts. Ginseng and American ginseng displayed *pS2* expression, but not ER binding and induction of AP. Whether *pS2* expression in S-30 is ER-specific needs to be further investigated. Another explanation of this result is that the constituents of ginseng probably work on one or more elements involved in ER function instead of ER itself.

Red clover is a well-known isoflavone-containing botanical dietary supplement like that found in soy (*Glycine max*). The combined utilization of ER binding guided isolation with ^{3}H-estradiol and the screening by using centrifugal ultrafiltration and LC-MS demonstrated that genistein plays the most important role in terms of the phytestrogenic effects of red clover.

Although chaste berry did not significantly stimulate the AP in Ishikawa cells, its extract consistently exhibited ER binding affinity and up-regulation of *ER* expression in T47:A18 cells, *PR* expression in Ishikawa cells, and *pS2* expression in S-30 cells. The combination of ER binding-guided isolation with ultrafiltration LC-MS confirmed that linoleic acid corresponds to the estrogenic effect of chaste berry.

Hops presented strong cytotoxicity to Ishikawa cells as well as significant estrogenic activity. Both the toxic and estrogenic compounds need to be explored.

Extract of black cohosh displayed no estrogenic activity in all of assays presented here. The data agree with the reports on this herb so far. This indicates that black cohosh may alleviate menopause symptoms by actions discrete from ER regulation.

5.11 EXAMPLE 2. FUNCTIONAL ELUCIDATION OF BLACK COHOSH FOR MENOPAUSE SYMPTOMS

C. racemosa L. (Nutt) (syn. *Actaea racemosa* L., black cohosh in English) (Ranunculaceae) is a popular and well-documented medicinal plant, native to eastern North America. Extracts of the black cohosh rhizome have been used since the 1950s and tested successfully in several clinical trials for mitigating hot flashes. The mechanism of action for black cohosh has been presumed to involve hormonal signaling through the ER or related targets since this was the pathway through which other HRT acted to reverse menopause symptoms. Unfortunately, as results showed in the above example, extracts of black cohosh do not exhibit estrogenic effect. Neither test of black cohosh extract on ovariectomized (OVX) Sprague–Dawley rats administered by gavage for 2 weeks with or without estradiol (50 μg/[kg·d]) caused any increase in uterine weight or vaginal cellular cornification, indicating black cohosh extract had no estrogenic or antiestrogenic properties. These data support the theory that black cohosh might relieve menopausal hot flashes by a mechanism different from HRT; however, no such pathway has been confirmed.

Since black cohosh had previously been reported to reduce luteinizing hormone (LH) levels, mechanisms for potential LH reduction were considered as possible pathways through which black cohosh might reduce hot flashes. The theory that LH may be involved in the etiology of the hot flash began in the 1970s, and although this gonadotropin is no longer considered the trigger for hot flashes, the connection has been documented. Serotonin receptors are known to at least partially control hot flashes and are present in the hypothalamus, the area of the brain modulated by estrogen in a negative inhibition pathway to reduce LH secretion from the pituitary. Serotonin selective reuptake inhibitors (SSRI) have also received attention for reducing hot flashes in menopausal women for whom HRT is contraindicated and in men who received androgen deprivation therapy. The 5-HT_7 and 5-HT_{1A} serotonin receptors have been known to be involved in thermoregulation, which suggests that agonists for these receptors might be beneficial for the alleviation of hot flash. For these reasons, serotonin receptors were investigated as potential targets for black cohosh. To evaluate other potential pathways by which black cohosh might reduce menopausal hot flashes, serotonin activity was assessed by the inhibition of radioligand binding to cell membrane preparations containing recombinant human serotonin receptor (5-HT) subtypes.

Binding of several hydroalcoholic extracts of black cohosh to several subtypes of 5-HT receptors was performed. Binding of a ligand to this G-protein will increase the release of intracellular cAMP, so the induction of cAMP in 5-HT7-transferred HEK293 cells, and the ability to block the reuptake of serotonin into hSERT-transferred HEK293 cells were also measured. cAMP is one of the most important "secondary messengers" involved as a modulator of physiological processes, including regulating neuronal, glandular, cardiovascular, immune, and other functions and actions. Measurement of its level change after adding samples of black cohosh will help to determine the sample act as an agonist or antagonist of the receptor. As a result, a new biological target of black cohosh that is independent of the ER has been confirmed, and the bioactive compounds have been identified.

Black cohosh is one of the herbal medicines that having been evaluated for its effect on menopause symptoms in the UIC/NIH Center for Botanical Dietary Supplement Research in Women's Health, College of Pharmacy, University of Illinois at Chicago (UIC), as mentioned in the first example. Results from this example have all been published in journals.[2,9,28]

5.11.1 Experimental Methods

Extraction of Back Cohosh The air-dried, milled roots/rhizomes of *C. racemosa* were exhaustively extracted by percolation with 100% methanol, 75% ethanol, or 40% isopropanol, and then dried to powder form.

Cell Culture Conditions The human 5-HT_7-transfected CHO (Chinese hamster ovary) cell line was cultured with Ham's F-12 medium, containing FBS (10%), 1 mM of MEM sodium pyruvate, 50 mg/mL gentamycin, and 50 U/mL of penicillin/

streptomycin. The rat 5-HT_{1A} transfected LZD_7 (transfectant of the Ltk mouse fibroblast) cells were cultured with MEM with nonessential amino acids containing FBS (10%) and antibiotic-antimytotic (1%). The 5-HT_7 transfected HEK293 (human epithelial kidney) cells were cultured in DMEM supplemented with puromycin (10 µg/mL), antibiotic-antimycotic (1%), and FBS (10%). Four days before cAMP assays were performed, the culture media were replaced with Cellgro Complete Serum-Free Media. hSERT-transfected-HEK293 cells were maintained in DMEM supplemented with 10% dialyzed FBS, 1% penicillin-streptomycin, 1% glutamax, and 1.2% geneticin (50 mg/mL). Cells were grown in at 37°C.

Membrane Preparation Cells (human 5-HT_7 CHO or rat 5-HT_{1a} Ltk mouse fibroblast) were plated into dishes (150 mm × 10 mm) and cultured to confluence in order to collect membranes as previously described. A hypotonic buffer (15 mM Tris, 1.25 mM $MgCl_2$, and 1 mM EDTA, pH 7.4) was added to the dishes, incubated at 4°C for 15 min, the cells were scraped from the dishes, and the lysate was centrifuged. The hypotonic buffer was removed and the membrane pellet was resuspended in TEM buffer (75 mM Tris, 1 mM EDTA, 12.5 mM of $MgCl_2$, pH 7.4). The cell membranes were homogenized and centrifuged twice at 12,000 × g for 20 min. The pellets were dissolved in TEM buffer and were stored at −70°C. Protein concentrations were determined by the Lowry method using bovine serum albumin as the standard.

Serotonin Receptor Binding Assays Initial radioligand binding studies were performed by Panlabs for serotonin receptors 1A, 1B, 1D, 2A, 2B, 2C, 3, 5A, 6, and 7. For the 5-HT_7 receptor, additional assays were performed with minor modifications using human recombinant CHO cell membrane and [^{3}H]lysergic acid diethylamide (LSD) (5 nM), in an incubation buffer (75 mM Tris-HCl, 1.25 mM $MgCl_2$, 1 mM EDTA, pH 7.4). Assays with the rat 5-HT_{1A} were performed with recombinant LZD_7 cell membranes using [^{3}H] 8-OH-DPAT (5 nM) incubated at 37°C for 60 min in incubation buffer (75 mM Tris-HCl, 1.25 mM $MgCl_2$, 1 mM EDTA, 0.1% l-ascorbic acid, 10 µM pargyline, pH 7.4). After a 1-h incubation at 37°C, the mixtures for both receptors were filtered over 934-AH Whatman filters that had been presoaked in 0.5% polyethylenimine (PEI) and washed two times in ice-cold 50 mM Tris buffer (pH 7.4) using a 96-well Tomtec-Harvester. Each filter was dried, suspended in Wallac microbeta plate scintillation fluid, and counted with a Wallac 1450 Microbeta Liquid Scintillation Counter. 5-Hydroxytryptamine (serotonin, 5-HT) (250 nM) was used to define nonspecific binding, which accounted for less than 10% of total binding. The percent inhibition of [^{3}H]-ligand bound to each 5-HT receptor was determined as $[1 - (dpm_{sample} - dpm_{blank})/(dpm_{DMSO} - dpm_{blank})] \times 100$. The data represent the average ± S.D. of at least triplicate determinations.

Bioassay-Guided Isolation and Fractionation Isolation and fractionation of rhizomes/roots of *C. racemosa* (black cohosh) extract were guided by serotonin receptor binding assays. Briefly, black cohosh was extracted with methanol. The methanol extract was first partitioned with water-ethyl acetate. The water partition was then subject to column chromatography consisting of Amberlite XAD-2, yielding water and methanol fractions. The methanol fraction was further isolated by fast centrifugal partition chromatography (FCPC) and yielded 7 fractions. N_ω-methylserotonin was identified by MS by means of accurate mass measurements in the XAD-MeOH fraction 7 after it was separated on a TOSOH TSK-Gel Aide-80 column. The identification was confirmed by comparison with the HPLC retention time and MS fragmentation pattern of authentic N_ω-methylserotonin.

Receptor Kinetics The mechanism of receptor binding was characterized using the methods described above with the modification that concentrations of [^{3}H]LSD were varied from 0.5 to 6 nM at fixed, variable concentrations of the black cohosh methanol extract that ranged from 0 to 20 µg/mL. The K_d and K_i values were determined using the equations for a single substrate single inhibitor model and the software available in Sigmaplot Enzyme Kinetics module. Direct binding and reciprocal plots were analyzed to determine the receptor ligand interaction. Full and partial competitive, uncompetitive, and noncompetitive inhibition models were all evaluated and ranked according to the best fit based on their R^2 and Akaike Information Criterion (AIC) values.

Intracellular cAMP Assays HEK293 human transfected 5-HT_7 cells were grown for 4 d in serum-free media and then plated in poly-d-lysine coated 12-well plates (100×10^4 cells/well) and left overnight. The following day, the cells were washed twice with 2 mL of incubation buffer (150 nM NaCl, 5 mM KCl, 1 mM $MgSO_4$, 2 mM $CaCl_2$, 10 mM glucose, 10 mM HEPES, 500 µM isobutylmethylxanthine, 1 µM ascorbic acid, 10 µM pargyline, pH 7.4) and then incubated for 20 min at 37°C. The compounds and extracts were added at concentrations 1000-fold higher than their final concentration in DMSO and incubated at 37°C for 10 min. The reactions were terminated by aspirating the buffer and adding 1 mL of boiling water to lyse the cells. The cells were then subjected to four rapid freeze/thaw cycles using a methanol and dry ice bath and a 37°C water bath. An aliquot of cell lysate (50 µL) was then transferred into the cAMP binding assay using the Amersham Kit (TRK 432). In the separation step, Whatman filters were used instead of charcoal as described in the binding assay protocol.

SSRI Assay Briefly, 200 µL of poly-D-lysine solution (20 µg/mL) was added to each well of 24-well plates. After the plate had been subjected to 5 min of shaking, the solution was aspirated, and the plates were dried for 1 h. hSERT-HEK293 cells (12×10^4 cells/well) were plated in the pre-coated 24-well culture plates. After 24 h, the plates were washed with KRH buffer (120 mM NaCl, 4.7 mM KCl, 2.2 mM $CaCl_2$, 10 mM Heps, 1.2 mM $MgSO_4$, pH 7.4), incubated with the extracts at 37°C in KRH buffer containing D-glucose (1.8 g/L), LASCORBIC ACID (100 µM), and pargyline (100 µM) for 10 min, and treated with [^{3}H]-5HT (20 nM). Fluoxetine (100 µM) was used as a positive control to validate the method. Cells were incubated for an additional 10 min. Uptake of [^{3}H]-5HT was terminated by washing the plates three times with KRH buffer. After the addition of 200 µL of 1% SDS to each well, the plates were incubated at room temperature for 30 min with gentle shaking. The cell solution was transferred to scintillation vials with 4 mL of Cytoscint. The tubes were counted using a Beckman LS 5801 liquid scintillation counter. The percent inhibition was determined as: % sample binding = $[1 - (cpm_{sample} - cpm_{blank})/(cpm_{DMSO} - cpm_{blank})] \times 100$. All black cohosh extracts were subjected to SSRI assays, at 40 µg/mL, and were considered to be active if inhibition was ≥50%.

5.11.2 Results

Summary First, a 40% isopropanol extract of black cohosh was tested against 10 subtypes of the serotonin receptor, revealing the presence of compounds with strong binding to the 5-HT_{1A}, 5-HT_{1D}, and 5-HT_7 subtypes. Subsequent binding studies were carried out using 5-HT_{1A} and 5-HT_7 receptors because of their association with the hypothalamus, which has been implicated in the generation of hot flashes. The black cohosh 40% isopropanol extract inhibited [^{3}H]LSD binding to the human 5-HT_7 receptor ($IC_{50} = 2.4 \pm 0.4$ µg/mL) with greater potency than binding of [^{3}H] 8-hydroxy-2-(di-N-propylamino) tetralin to the rat 5-HT_{1A} receptor ($IC_{50} = 13.9 \pm 0.6$ µg/mL).

Analysis of ligand binding data indicated that components of a black cohosh in methanol extract functioned as a mixed competitive ligand of the 5-HT_7 receptor. In addition, a black cohosh methanol extract elevated cAMP in 293T-5-HT_7-transfected HEK cells, suggesting the extract acted as a partial agonist at the receptor. The elevation in cAMP mediated by the black cohosh extract could be reversed in the presence of the antagonist methiothepin indicating a receptor mediated process. N_ω-methylserotonin was identified as the possible corresponding compound for the serotonergic activity of black cohosh by showing 5-HT_7 receptor binding, inducing cAMP, and blocked serotonin reuptaking (see data below). Triterpenoids and phenolic acids only bound weakly to the 5-HT_7 receptor with no cAMP or SSRI activity.

Binding of Black Cohosh Extract to Different Subtypes of the Serotonin Receptor Since black cohosh extracts did not demonstrate any discernable estrogenic behavior in any of the *in vitro* assays or in the OVX rat model, other biological targets were studied. Serotonin receptors were investigated on the basis of their neuroendocrine links to menopausal hot flashes and previous literature reported that *Cimicifuga foetida* bound to the 5-HT_{1A} receptor. Serotonin receptor binding was evaluated to determine if black cohosh displayed other pharmacological properties.

A 40% isopropanol extract of black cohosh was screened by Panlabs to determine if the extract contained any potential ligand(s) of several serotonin receptors including 1A, 1B, 1D, 2A, 2B, 2C, 3, 5A, 6, and 7. The extract (250 µg/mL) produced the greatest inhibition of receptor binding for the 1A, 1D, and 7 receptors (>95%). A black cohosh isopropanol extract bound most strongly (>50%) to all subtypes except for 5-HT_{2A} and 5-HT_3 (data not shown). The black cohosh isopropanol extract did not significantly alter radioligand binding to the serotonin transporter (data not shown).

Since 5-HT_{1A} and 5-HT_7 are both located in the hypothalamus where they might be involved in hormonal regulation, and since they both have similar drug sensitivities, dose-response studies were conducted to further distinguish the relative binding potency of black cohosh extract for only these two receptors. In order to validate the method, serotonin (5HT) was incubated as an inhibitor of [^{3}H]LSD binding for the 5-HT_7 receptor and [^{3}H] 8-OH-DPAT as an inhibitor for the 5-HT_{1A} receptor binding. Experiments generated K_i values of 1.5 nM for 5-HT_7 and 2.8 nM for 5-HT_{1a} that were consistent with the literature. Three hydroalcoholic extracts were compared to determine an extraction procedure that resulted in the greatest percent inhibition of binding of a radiolabeled compound to the serotonin receptors including a 100% methanol extract, a 40% isopropanol extract, and a 75% ethanol extract. The results showed that the methanol extract displaced radioligands from the 5-HT_{1A} ($IC_{50} = 2.5 \pm 0.6\,\mu g/mL$) and 5-$HT_7$ ($2.2 \pm 0.2\,\mu g/mL$) receptors equally well. However, the 75% ethanol and the 40% isopropanol extracts inhibited binding of 8-OH-DPAT to the 5-HT_{1A} ($IC_{50} = 13 \pm 2$; $14 \pm 1\,\mu g/mL$) receptor less effectively than binding of LSD to the 5-HT_7 receptor ($IC_{50} = 3.1 \pm 0.5$; $2.4 \pm 0.4\,\mu g/mL$). Since the black cohosh methanol extract exhibited the highest binding potency for both receptor studies, only this extract was used for further analyses.

Receptor Binding Mechanism To determine if a black cohosh methanol extract functioned as a competitive, noncompetitive, uncompetitive, or mixed inhibitor(s) of LSD binding to the 5-HT_7 receptor, LSD binding experiments were performed in the absence and presence of 10 and 20 µg/mL of the extract. Figure 5.13A represents a direct binding isotherm plot of triplicate data for each concentration of black cohosh and [^{3}H]LSD. Saturable binding with increasing concentrations of [^{3}H]LSD cannot be achieved in the presence of black cohosh methanol extract, indicating that the ligand(s) contained in the extract did not act as a full competitor(s) of LSD for the receptor binding site. When reciprocal plots (Fig. 5.13B) were analyzed, the data best fit to a model for mixed inhibition in which the ligand(s) present in black cohosh could bind to the receptor binding site alone but could also bind in the presence of the known ligand, [^{3}H]LSD. The K_d of LSD in this study was found to be 0.9 ± 0.3 nM. The model for how the ligand(s) in black cohosh methanol extract binds to the receptor in the presence and absence of LSD is shown schematically in Figure 5.14.

cAMP Elevations Studies 5-HT_7 receptors are coupled to stimulatory G-proteins. Upon agonist occupancy of the receptor, the G_α subunit of the G-protein dissociates and activates adenylate cyclase to catalyze the formation of cAMP. The HEK293 stably transfected 5-HT_7 cells produced cAMP in the presence of serotonin (Fig. 5.6). This stimulation could be eradicated by co-incubation of serotonin with the receptor antagonist methiothepin. HEK/5-HT_7 cells were studied to determine if compounds in the black cohosh methanol extract acted as agonists or antagonists of the receptor. Incubations of the black cohosh methanol extract with HEK/5-HT_7 cells demonstrated that the extract contained serotonin receptor agonist(s) that increased intracellular levels of cAMP. Black cohosh extract was also co-incubated with 5-HT

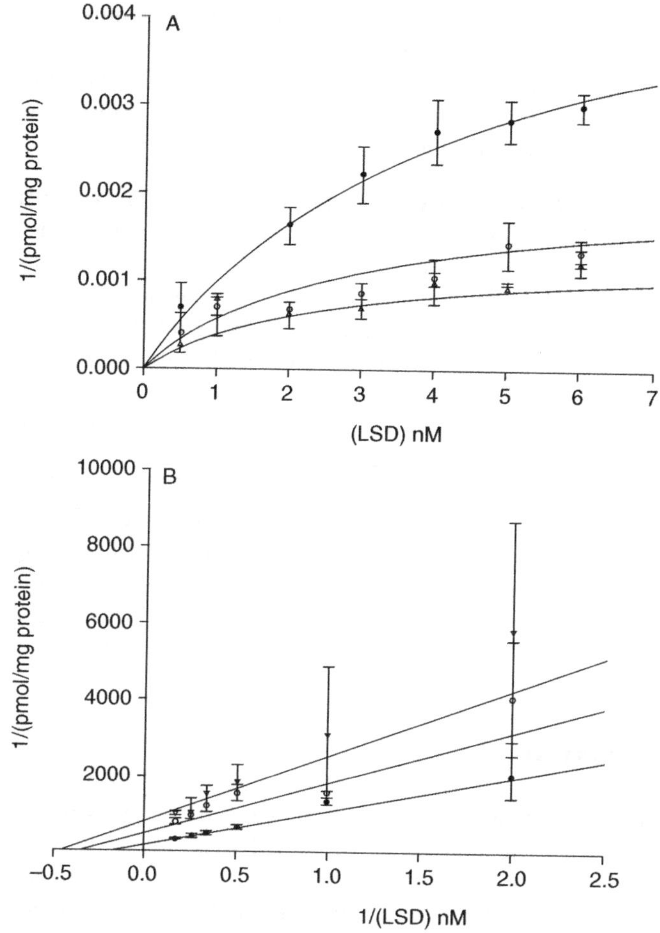

Figure 5.13 Binding plot displaying the effect of black cohosh on the binding of [^{3}H]LSD to the 5-HT$_7$ receptor. (A) Direct binding plots are shown for increasing concentrations of [^{3}H]LSD incubated with increasing concentrations of black cohosh. [^{3}H]LSD was not able to bind to all of the receptor binding sites in the presence of black cohosh, indicating a mixed full inhibition. (●), no black cohosh; (○), 10 μg/mL black cohosh; (▼), 20 μg/mL black cohosh. (B) Reciprocal plots of the data in (A) illustrate that black cohosh methanol extract acts as a mixed competitor of the 5-HT$_7$ receptor in the presence of [^{3}H]LSD. The plot reveals that a black cohosh methanol extract contains a mixed ligand of the serotonin receptor such that both ligands can be simultaneously bound. The *Kd* for [^{3}H]LSD was 0.9 nM; the *Ki* of black cohosh was 6.1 μg/mL; and the R factor was 2.9. (●), no black cohosh; (○), 10 μg/mL black cohosh; (▼), 20 μg/mL black cohosh.

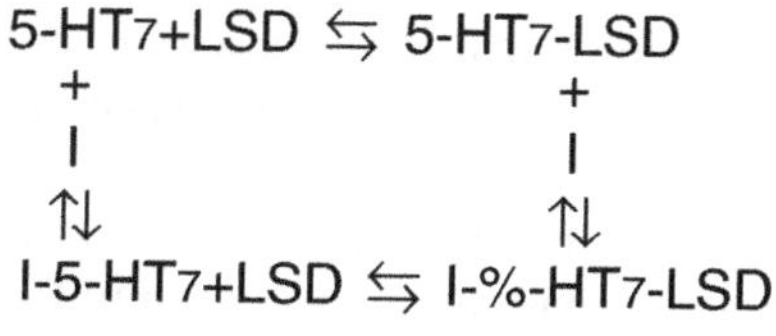

Figure 5.14 Scheme depicting the mixed competitive inhibition model for the interaction of black cohosh methanol extract and the inhibitor (i) with the serotonin receptor 5-HT$_7$ in the presence of [^{3}H]LSD.

to determine if any antagonist(s) were found within the plant material, but no attenuation of the 5-HT stimulated production of cAMP was observed at doses of 40 µg/mL (data not shown). Finally, the black cohosh methanol extract was incubated in the presence of the antagonist, methiothepin, and the formation of cAMP was reduced, indicating this process is receptor-mediated.

Identification and Serotonergic and SSRI Activity of Nω-methylserotonin in Black Cohosh Extract N_{ω}-methylserotonin was identified as the corresponding bioactive compound from the most serotonergic fraction by LC-MS by means of accurate mass measurements in combination with CrossFire Commander database search, and comparison with an authentic sample of the compound. It was also screened against all known serotonin receptor subfamilies. It showed the highest affinity to the 5-HT_1, 5-HT_6, and 5-HT_7 particularly 5-HT_{1A} and 5-HT_7 receptors, and the lowest to the 5-HT_3 and 5-HT_4, which do not appear to be associated with thermoregulation. The addition of the methyl group to the omega nitrogen of serotonin appears to increase its selectivity for 5-HT_6 (IC_{50} = 55 nM) and 5-HT_7 (IC_{50} = 0.02 nM) serotonin receptors when compared with serotonin itself (5-HT_6: IC_{50} = 340 nM; 5-HT_7:IC_{50} = 0.86 nM). N_{ω}-methylserotonin demonstrated a strong ability to increase intracellular cAMP level (EC_{50} = 22 nM) in the cAMP induction assay and strong SSRI activity (IC_{50} = 490 nM) in the SSRI assay. cAMP formation could be reversed by co-incubation with the 5-HT_7 antagonist, SB269970, indicating the process was receptor mediated.

5.11.3 Brief Discussion

Since all of our data indicated that black cohosh functioned in an estrogen-independent manner, another biological target was selected based on the following rationale. Although HRT alleviates hot flashes, a decline in estrogen itself is not the trigger of hot flashes since levels do not differ significantly between symptomatic and asymptomatic women. Also, levels of estrogen remain low for the remainder of a woman's life, but hot flashes eventually cease. Two studies conducted with black cohosh suggest that the extract might act through the serotonin pathway. One study reported a water extract of *C. foetida* L. bound to the serotonin receptor 5-HT_{1A}; the other study found that a methanol extract of Cimicifuga rhizomes inhibited 5-HT induced diarrhea in mice. In addition to this evidence, there are a number of documented connections between the estrogenic and serotonergic systems.

First, when estrogen levels fall, the amount of tryptophan hydroxylase in the body responsible for converting tryptophan into serotonin is also lowered, limiting the formation of serotonin. Second, the level of serotonin in the bloodstream of menopausal women declines when estrogen declines. HRT is capable of augmenting this diminished serotonergic activity. Clinical trials using SSRIs have shown success in reducing menopausal hot flashes in both healthy women and cancer patients taking selective ER modulators.

The above results and others indicated that black cohosh extracts contained serotonin receptor ligands. The black cohosh ligand(s) had the highest affinity for the 5-HT_{1A}, 5-HT_{1D}, and 5-HT_7 receptors at a high dose (250 μg/mL). Little is known about the 5-HT_{1D} receptor, but it has been suggested that it has pharmacological properties similar to the 5-HT_{1B} receptor. The 5-HT_{1B} receptor physiological functions are not clearly understood either because there are no selective compounds for this receptor. The 5-HT_{1A} and 5-HT_7 receptors were further examined because they are both found in the hypothalamus, a key area for thermoregulation. Indeed, the administration of a 5-HT_{1A} agonist into the anterior hypothalamus of a rat induces hypothermia. The results showed that the extract was somewhat selective for the 5-HT_7 receptor because all three extracts tested (100% methanol, 40% isopropanol, and 75% ethanol) bound to the receptor with an IC_{50} values of 3.12 μg/mL or less.

Serotonin has been shown to inhibit secretion of LH from the hypothalamus through the 5-HT_{1A} receptor, and black cohosh has previously been reported to reduce LH concentrations. Serotonin is known to control the pulsatile release of LH by acting on the hypothalamus and, using an OVX rat model, serotonin receptors have been found that terminate directly onto LH-releasing hormone (LhRh) neurons and inhibit the secretion of LH from the pituitary. This is of interest since the selective agonist, 8-OH-DPAT, used to determine that 5-HT_{1A} receptors were involved in LH reduction is also a partial agonist of the 5-HT_7 receptor, the most recently identified serotonin receptor, and the data at least partially implicate both receptors as being involved in the regulation of LH release by controlling the hypothalamus. Also, selective 5-HT_7 receptor ligands have not been identified and, therefore, black cohosh may contain lead compound(s) that could be used to develop more selective ligands for the receptor.

Black cohosh extract was shown to interact with the 5-HT_7 receptor by a mixed competitive model. The extract was capable of binding to the receptor in the presence and absence of [^{3}H]LSD. Determination of the cAMP levels showed that a black cohosh methanol extract acted as a 5-HT_7 agonist and it stimulated the G-protein coupled activation of adenylate cylase. This process is receptor-mediated as demonstrated by the co-incubation of black cohosh methanol extract with the antagonist methiothepin. These data are consistent with the idea that a treatment capable of increasing serotonergic activity in the body may help to reduce hot flashes, as shown by the clinical trials performed with menopausal women using SSRIs. Since the ligand(s) in black cohosh was able to increase the amount of cAMP inside the cell, this supports the idea that the ligand(s), despite its mixed interaction with the receptor, has the same downstream effects as other competitive ligands.

SSRIs are currently used for the treatment of depression. An animal study using black cohosh extract used the tail suspension test in rats as a marker for antidepressant activity and found that black cohosh extract can produce behavior indicative of antidepressant activity. The positive control, imipramine, used in the animal study was a mixed inhibitor of the serotonin and norepinephrine transporters. The data are consistent with the theory that black cohosh may be stimulating serotonin receptors. In addition, the data demonstrate that black cohosh extract is capable of producing *in vivo* behavioral changes consistent with those seen by other serotonergic compounds.

The data indicate that black cohosh acts on serotonin receptors but not on ERs. Despite the lack of estrogenic effects, many clinical trials have found that black cohosh effectively reduces menopausal hot flashes, indicating that another target is possibly involved in the alleviation of hot flashes. Nonhormonal treatments may reduce menopause symptoms without increasing a woman's risk for developing breast cancer. Although the results presented do not show how black cohosh extracts alleviate hot flashes, another biological target has been identified for black cohosh providing a possible alternative mechanism of action. Active compounds in the extract must be identified and characterized for their receptor binding and agonistic properties. Future studies will be conducted to determine if black cohosh actually reaches serotonin receptors in the brain and has effects on hot flashes in women.

The UIC/NIH center for botanical and dietary supplements research also performed antioxidant study on black cohosh. The results indicated that methanol extracts of black cohosh effectively scavenged 2,2-diphenyl-1-picrylhydrazyl (DPPH) free radicals and exhibited dose-dependent decreases in DNA single-strand breaks and oxidized bases induced by the quinone menadione using the comet (single-cell gel electrophoresis assay) and fragment length associated repair enzyme (FLARE) assays, respectively. Bioassay-directed fractionation of the methanolic extracts using the DPPH assay as a monitor led to the isolation of nine antioxidant active compounds: caffeic acid, methyl caffeate, ferulic acid, isoferulic acid, fukinolic acid, cimicifugic acid A, cimicifugic acid B, cimicifugic acid F, cimiciracemate A, and cimiciracemate B. Six of these antioxidants were found to reduce menadione-induced DNA damage in cultured S30 breast cancer cells with the following order of potency: methyl caffeate > caffeic acid > ferulic acid > cimiracemate A > cimiracemate B > fukinolic acid. These data suggest that black cohosh can protect against cellular DNA damage caused by reactive oxygen species and this may be of benefit for women using this product for the relief of menopause symptoms.[25]

A group in Germany did a study on black cohosh extract for its effect on bone and fat tissue in OVX rats. In this study, bone mineral density (BMD) of the tibia was measured and fat depots were quantified by computer-assisted tomography (CT) scans; bone turnover (osteocalcin, crosslaps) and lipocyte activity (leptin) were also determined. Meanwhile, uterine weights were measured and gene expression of estrogen-regulated uterine genes (IGF-1, ERbeta) was determined by RT-PCR. Results showed that treatment of the OVX rats over a period of three months with the extract showed osteoprotective effects by significantly reducing the loss of BMD in tibia. Serum osteocalcin levels, a paratibial fat deposit, and serum leptin concentration were significantly reduced, but not in the uterus of OVX rats.[29]

REFERENCES

1. Liu, J.H., et al. (2001) Evaluation of estrogenic activities of plant extracts on the potential treatment of menopausal sympotoms. *Journal of Agricultural and Food Chemistry* 49(5):2472–2479.
2. Burdette, J.E., et al. (2003) Black cohosh acts as a mixed competitive ligand and partial agonist of the serotonin receptor. *Journal of Agricultural and Food Chemistry* 51(19):5661–5670.
3. Newman, D.J., et al. (2003) Natural products as sources of new drugs over the period 1981–2002. *Journal of Natural Products* 66(7):1022–1037.
4. Rouhi, A.M. (2003) Rediscovering Natural Products. *Chemical & Engineering News* 81(41):77–78, 82–83, 86, 88–91.

5. Li, J.W. and Vederas, J.C. (2009) Drug discovery and natural products: end of an era or an endless frontier? *Science* 325(5937):161–165.
6. Koehn, F.E. (2008) High impact technologies for natural products screening. *Progress in Drug Research* 65(175):177–210.
7. Harvey, A.L. (2008) Natural products in drug discovery. *Drug Discovery Today* 13(19–20): 894–901.
8. Mishra, K.P., et al. (2008) A review of high throughput technology for the screening of natural products. *Biomedicine & Pharmacotherapy* 62(2):94–98.
9. Powell, S.L., et al. (2008) In vitro serotonergic activity of black cohosh and identification of N_{ω}-methylserotonin as a potential active constituent. *Journal of Agricultural and Food Chemistry* 56(24):1718–1726.
10. Roxas, M.N.D. and Jurenka, J.M.T. (2007) Colds and influenza: a review of diagnosis and conventional, botanical, and nutritional considerations. *Alternative Medicine Review* 12(1):25–48.
11. Collins, A., et al. (1997) A comparison of human immunodeficiency virus type 1 inhibition by partially purified aqueous extracts of Chinese medicinal herbs. *Life Sciences* 60(23):345–351.
12. Liu, J.H., et al. (1998) Comparison of duhuo with its substitutes in terms of constituents and inhibitory effect on 5-LO and COX. *Phramaceutical Biology* 36(3):207–216.
13. Liu, J.H., et al. (2004) Isolation of linoleic acid as an estrogenic compound from the fruits of *Vitex agnus-castus* L. (chaste berry). *Phytochemistry* 11(1):18–23.
14. Segel, I.H. (1993) *Enzyme Kinetics: Behavior and Analysis of Rapid Equilibrium and Steady-State Enzyme Systems*. Hoboken, NJ, Wiley.
15. Cleland, W.W. (1963) The kinetics of enzyme-catalyzed reactions with two or more substrates or products. II. Inhibition: nomenclature and theory. *Biochimica et Biophysica Acta* 67:173–187.
16. Mosmann, T. (1983) Rapid colorimetric assay for cellular growth and survival: application to proliferation and cytotoxicity assays. *Journal of Immunological Methods* 65(1–2):55–63.
17. Passonneau, J.V. and Lowry, O.H. (1993) *Enzymatic Analysis, A Practical Guide*. Totowa, NJ, Humana Press.
18. Todd, M.J. and Gomez, J. (2001) Enzyme kinetics determined using calorimetry: a general assay for enzyme activity? *Analytical Biochemistry* 296(2):179–187.
19. Stanley, P.E. and Kricka, L.J. (2002) *Bioluminescence & Chemiluminescence: Progress & Current Application*. Singapore, World Scientific Publishing.
20. Greer, L.F. and Szalay, A.A. (2002) Imaging of light emission from the expression of luciferases in living cells and organisms: a review. *Luminescence* 17(1):43–74.
21. William, M. and Wood, P. (1986) *Neuromethods, Vol. 4—Receptor Binding in Drug Discovery and Development*, Totowa, NJ, Humana Press.
22. Galbraith, G.M.P. and Galbraith, R.M. (1981) Combined radiolabel-binding and immunocytochemical evaluation of receptor-ligand interactions—studies of transferrin receptors on activated lymphocytes. *Biochemical Journal* 200:173–176.
23. Kirby, J., et al. (2007) Gene expression assays. *Advances in Clinical Chemistry* 44:247–292.
24. Michael, S., et al. (2008) A robotic platform for quantitative high-throughput screening. *Assay and Drug Development Technologies* 6(5):637–657.
25. Wunder, F., et al. (2008) Functional cell-based assays in microliter volumes for ultra-high throughput screening. *Combinational Chemistry & High Throughput Screen* 11(7):495–504.
26. Yasgar, A., et al. (2008) Compound management for quantitative high-throughput screening. *JALA, Charlottesville, VA* 13(2):79–89.
27. Burdette, J.E., et al. (2002) *Trifolium pratense* (red clover) exhibits estrogenic effects in vivo in ovariectomized Sprague-Dawley rats. *The Journal of Nutrition* 132(1):27–30.
28. Burdette, J.E., et al. (2003) Black cohosh (*Cimicifuga raceosa* L.) protects against menadione-induced DNA damage through scavenging of reactive oxygen species: bioassay-directed isolation and characterization of active principles. *Journal of Agricultural and Food Chemistry* 50(24): 7022–7028.
29. Seidlová-Wuttke, D., et al. (2003) Pharmacology of Cimicifuga racemosa extract BNO 1055 in rats: bone, fat and uterus. *Maturitas* 14(44 Suppl. 1):S39–S50.

Chapter 6

Functional Evaluation of Herbal Medicines by Animal Experiments

Chun-fu Wu, Fang Wang, and Chun-li Li

According to Dorland's Medical Dictionary for Health Consumers, *pharmacology* is the science that deals with the origin, nature, chemistry, effects, and uses of drugs; it includes pharmacognosy, pharmocokinetics, pharmacodynamics, pharmacotherapeutics, and toxicology. *Pharmacognosy* is the study of medicines derived from natural sources. The American Society of Pharmacognosy defines pharmacognosy as "the study of the physical, chemical, biochemical and biological properties of drugs, drug substances or potential drugs or drug substances of natural origin as well as the search for new drugs from natural sources." It is the oldest and broadest of the five branches. Contents in all the chapters in this book are under the scope of pharmacognosy study. *Pharmacodynamics* is the study of the biochemical and physiological effects of drugs and the mechanisms of their actions, including the correlation of their actions and effects with their chemical structure. *Pharmocokinetics* studies how a body reacts to the presence of a drug, nutrient, or other foreign compound from its introduction to its final elimination from the body. *Pharmacotherapeutics* is the study of the therapeutic uses and effects of drugs. *Toxicology* studies the effects of, antidotes to, and detection of poisons. This chapter mainly introduces the *in vivo* pharmacodynamics study of herbal medicine. Pharmocokinetics study of herbal medicine is not covered here because its methods are the same as those for synthetic drug study, with an exception of the complex chemical composition of herbs, whose identification and qualitative and quantitative analytical methods will be introduced in Chapters 4 and 9. Toxicology will be introduced in Chapter 7.

In fact, many modern drugs are compounds originally isolated from plants or derivatively made from components in plants. In the past century, pharmacological

Traditional Herbal Medicine Research Methods, Edited by Willow J.H. Liu

research has been applied worldwide to traditional herbal medicine. Animals are the subjects used for pharmacological studies.

The use of animals in experimental medicine, pharmacology, pharmaceutical development, safety assessment, and toxicological evaluation has become a well-established and essential practice. Whether serving as a source of isolated organelles, cells, or tissues; a disease model; or a prediction for drug or other xenobiotic action or transformation in human, experiments on animals have provided the necessary building blocks that have permitted the explosive growth of medical and biological knowledge in the second half of the twentieth century and into the twenty-first century. Animal-based research has significantly contributed to the advancement of scientific knowledge in general, and to biomedical progress specifically.

Animals have been used as models for centuries to predict how chemicals and environmental factors affect humans. According to the book *Animal Models in Toxicology,* written by C. G. Shayne,[1] the earliest clearly described use of animals was for a study of the effects of environmental agents. Soon after, the first application of animals for systematic screening of a wide variety of poisons and drugs was reported. This report included dosing test animals with known quantities of agents and detailed recording of the experimental results such as signs and gross necropsy observations of animals. Since then, the use of animals as predictors of potential effects of substances on the human body has grown.

Pharmacological and toxicological studies of substances on animals provide information on changes in their biology and physiology, which are closely relevant to that of human beings. Animal models of human diseases or disorders provide invaluable information in many aspects of the pathophysiology, treatment, and diagnosis of human diseases. Without exception, research on traditional herbs is also heavily dependent on animals in experiments *in vivo, in situ*, and *in vitro*.

6.1 PURPOSES AND SIGNIFICANCE OF PHARMACOLOGICAL RESEARCH FOR HERBAL MEDICINES

According to the definition in *Research Guidelines for Evaluating the Safety and Efficacy of Herbal Medicines* of the World Health Organization (WHO), issued in 1994, herbal medicine is a plant-derived material or preparation with therapeutic or other human health benefits, which contains either raw or processed ingredients from one or more plants. Although inorganic and animal materials are also used for treatment of diseases in some countries, herbs always dominate in traditional medicines.

There are many purposes to carrying out pharmacological research for traditional herbal medicines:

1. To scientifically evaluate the safety and efficacy of traditional herbs and elucidate their mechanisms by a modern medicine research system.
2. To validate the efficacy of a new combination of herbal medicines or a new herbal product and to establish their therapeutic doses, or to evaluate a new indication, or a new administration route for existing herbal products.

3. To evaluate the pharmacological effect of purified or semi purified compounds isolated or derived from herbal medicine.
4. To discover new pharmacologically active plant materials.

Although traditional herbs have been applied with experience for protection, restoration, and improvement of health over hundreds and thousands of years, only a few herbs have been scientifically evaluated, and most of them need to be experimentally studied.

Pharmacological research plays an important role in the modernization of traditional herbal medicines, because the experimental method is the most fundamental method of modern science. Information that cannot be obtained from human beings can be obtained through animal experiments. Pharmacological research of herbal medicine will lay a foundation for clinical study.

An overview of methods for functional evaluation of herbal medicines by *in vivo* pharmacological experiments will be presented here.

6.2 CHARACTERISTICS OF PHARMACOLOGICAL RESEARCH OF HERBAL MEDICINES

6.2.1 Characteristics of Traditional Herbal Medicines and Their Pharmacological Studies

Compared with well-defined synthetic drugs that are developed based on modern pharmaceutical research system, herbal medicines exhibit the following markedly different characteristics:

1. Most traditional herbs have been empirically used in folks with a long history; therefore, their therapeutic and side effects are well known by practitioners.
2. Their mechanisms need to be elucidated with modern pharmacology and biology.
3. Very few herbs have proven their efficacy and safety through well-controlled double-blind clinical and toxicological studies.
4. Most of the traditional herbs have a wide range of therapeutic uses. They generally work on different targets in the body simultaneously, but show effect more slowly than modern drugs.
5. The undesirable side effects of herbs are presumed to be less and more rare, but well-controlled randomized clinical trials have revealed that they certainly exist.
6. The herbal extracts are composed of diverse chemical compounds and the active principles are mostly unknown.
7. Standardization, stability, and quality control of herbal medicine are feasible, but more difficult because of their chemical complexity.

8. The availability and quality of raw materials for pharmacological study are sometimes problematic because species growing in different regions or collected at different times may vary significantly in concentration and composition because they differ with environment and season.
9. They usually cost less than synthetic drugs.

Pharmacological research methods for modern drug development are applied for pharmacological study on herbal medicines. Many advanced technologies and research methods for modern drugs have been applied to herbal study. However, one must keep in mind that research for traditional herbs should not simply copy the steps of these methods for modern drug development. In connection with the above characteristics, the following are critical when pharmacological study is performed for herbal medicine.

1. It is critical to ensure the bioactive components are extracted out.
2. It is critical to ensure the materials for the study are collected from an appropriate place at an appropriate time, so as to contain reasonably concentrated bioactive components.
3. The extract or product for the study should be chemically standardized and controlled, and its stability should be analyzed to ensure the experimental result could be repeated.
4. The dosages of the extracts given to animals are usually much higher than the positive controlled Western drugs. Sometimes, the difference may be even a thousand times or more, depending on the relative concentration of the bioactive compounds in the extracts. Take the experiment of red clover on ovariectomized (OVX) Sprague–Dawley rats for estrogenic effect as an example: the dose of the positive control of 17β-estradiol was 50 μg/(kg × d), but doses of red clover to exhibit significant estrogenic activity was 500 and 750 mg/(kg × d).[2]
5. The *in vivo* experiment, that is, the treatment with test samples of herbal medicine on animals, usually takes longer time to see the positive result in comparison with the same one for Western drugs. For example, the experiment on OVX Sprague–Dawley rats for estrogenic effect of a Western drug usually needs 1 week (17β-estradiol exhibits significant estrogenic activity on the next day of the treatment), but the above-mentioned experiment for red clover extract did not show estrogenic activity until the third week.[3]
6. If the experimental result for an herbal extract shows negative without any doubt, other mechanisms of this herb should be considered, because it may work indirectly. An example is that black cohosh has been known to relieve menopause symptoms. But *in vivo* and *in vitro* studies showed that the extract of black cohosh does not directly regulate estrogen release by binding to estrogen receptor; it may act through the regulation of serotonin via the central nervous system.[3–5]

7. A special model of an animal may need to be established for pharmacological study of some traditional medicine.

6.2.2 Herbal Pharmacological Study Should Refer to Theory of Traditional Medicine

Many herbs have been used empirically through generations, with theories of traditional medicine in some countries. The current modern research of traditional herbal medicine has been divided in two directions. One is the westernized research that aims to isolate the active substances from traditional herbs and then evaluate their efficacy for new drug development. Successful stories include taxol from Yew tree, vincristine from periwinkle, and artemisinin from wormwood. Another is the research with special features by using modern experimental methods but combining the direction of theory of traditional medicine to elucidate the mechanism of the therapeutic efficacy. An example is the study of traditional Chinese herbs with modern technology and methodology for evaluation of their functional mechanisms.

Traditional Chinese medicine (TCM), as one of the oldest surviving traditional medicines, is an integral system of healing that started to develop in China about 3000 years ago. It has been constantly replenished and improved during the past millennia and has maintained its own identity up to the present day. With thousands of years of experience in treating diseases with natural materials, TCM still plays an important role currently in the health-care system of China and is officially recognized not only in China, but also in some other eastern and southeastern Asian countries with similar cultural traditions.

Because TCM was developed before modern medicine, there are many differences between TCM and conventional modern medicine. In order for TCM or other traditional medicine to be modernized and accepted by people worldwide, particularly medical doctors and scientists, simple, word by word translation of the meaning is far from enough. For example, righteous *qi* (antipathogenic *qi*) is used to describe the ability of humans, including immune ability, to fight disease, while evil *qi* (pathogenic *qi*) is used to prescribe the pathogens, including bacteria and viruses. Evaluation of the modern pharmacological efficacy of Chinese and other herbal medicine with modern research methodology is necessary to explain their therapeutic functions with modern scientific terminology. But to do this, the design of the study methods must be based on the theories that guide the clinical application of these herbal medicines. Otherwise, the study can only be used to screen out drug candidates.

The theoretical basis of TCM originates from the ancient Chinese philosophy of yin–yang and the five elements. TCM theory covers *qi*, blood, body fluids, essence, shen, the Zang–Fu internal organ theory, etiology and pathogenesis, as well as the principles of the prevention and treatment of diseases (see detailed introduction in Chapter 10).

Unlike modern medicine, which seeks the mechanism at cellular or molecular level by exploring whether a drug acts as an agonist or antagonistic to the targets, TCM explains the pathology and pharmacology with its own terminology. In addition to infection (exogenous *qi*) and damage, emotional impacts, and improper diet and overstrain or lack of physical exercise, TCM believes that the stagnant blood and phlegm fluid may cause many types of diseases. Another saying is "blocked *qi* and blood cause a hundred types of diseases." Corresponding to the causes, the symptoms and signs of diseases are prescribed and differentiated into different types of "*zheng*" (see explanation in Section 6.2.5), such as excessive heat or yang, deficient *qi* or yin, or stagnation of *qi* and blood.

The principle of TCM treatment is to balance the body by regulating yin, yang, *qi*, and blood. The treatment of disease is either to help the body enhance the righteous *qi* to fight the evil *qi* (pathogenic *qi*) or to improve the blocked circulation of the *qi* or blood. Chinese medicine, mainly composed of herbs, is used to restore and maintain balance between yin and yang, *qi* and blood, as well as organs and tissues. For example, symptoms of excessive heat should be treated by herbs with cool or cold properties; symptoms of Yin deficiency should be treated with yin tonified herbs; symptoms of *qi* and blood stagnation should be treated with herbs that can improve the circulation of *qi* and blood. The main functions of these most commonly used Chinese herbs have been pharmacologically studied with animal tests.[6,7] Brief introductions about pharmacological effects of different types of Chinese herbs are also available in Chapter 10.

It should be addressed that, as an integral part of TCM, the characteristic application of Chinese herbal medicine is the combined use of several herbs in one formula. A typical prescription of a formula is composed of the principal herb(s), the associate herb(s), the adjuvant drug(s), and the messenger drug(s). They are often described as herbs of emperor, minister, assistant, and envoy, respectively, according to their roles in the formula. The strategy of such a combination is to increase the therapeutic effectiveness of the major ingredients, reduce complex symptoms simultaneously, minimize toxic or side effects, alter the unwanted actions, increase the solubility, or correct the unpleasant tastes of others.[8] Therefore, the pharmacological action of a Chinese formula is not a simple addition of the action of an individual herb in the formula. Knowing these characteristics of Chinese herbal medicine will no doubt help to direct scientists studying Chinese herbs; it will help them not only save time and money, but also obtain the right conclusion.

6.2.3 Herbal Pharmacological Research Should be Combined with Clinical Practice

The ultimate purpose of pharmacological research of herbal medicine is to provide scientific evidence by elucidating the mechanism of the therapeutic functions of herbal medicines for treating diseases of human beings. There is a big difference between Western new drug development and the research of traditional herbs. For a new drug development in Western medicine, the clinical trial is performed after its animal experiments indicate significant effect(s) in animal tests. Many of these new drug candidates may be abandoned through clinical trial due to their insignificant efficacies on humans or due to severe side effects. But with traditional herbal medicine, although sometimes lacking confirmation through modern clinical trials, the efficacies of most herbs are well known in the clinic. The results of pharmacological studies in the research of traditional herbal medicine are mainly used to support clinical efficacy with modern medical exam indicators and by means of statistical methods, thus elucidating the mechanism with modern medicine. The negative results of an herbal medicine in a pharmacological study on animals should not summarily be used to negate its clinical effect. Negative results from animal tests may not only be due to the species difference between humans and animals, but also to the use of inappropriate animal models attributed to the unknown functional mechanism of the herbal medicine.

First, make sure that the design of the animal experiment is scientifically correct, that all the procedures in the operation have no mistakes, and especially, that the herb is properly extracted and the dosage and period of administration are appropriate. If a negative result is obtained, the problem may possibly be the wrong animal model, because the functional mechanism of the test herbal medicine is unknown. Pharmacologists should try their best to set up a new model with different mechanisms based on the traditional application. An example is the aforementioned black cohosh for menopause symptoms through regulating serotonin rather than estrogen release.[5,6] Because the causes of many diseases are still unknown thus far, scientists may even find a new mechanism for treatment of disease through such a pharmacological study.

6.2.4 Experiments *In Vivo* for Herbal Mechanism Study Should be Combined with Experiments *In Vitro*

Once the therapeutic effects of an herb are confirmed with pharmacological experiments, scientists should try to find the corresponding bioactive compounds by *in vitro* bioassay-guided screening to further evaluate the mechanism on the cell or molecular level. Knowing bioactive compounds in an herb could also provide more accurate standardization and quality control of the product.

On the other hand, *Research Guidelines for Evaluating the Safety and Efficacy of Herbal Medicines*, issued from the WHO (1994), emphasizes that research on animals must be carried out with respect to their welfare, and encourages that one should consider reducing experimentation on intact animals by using *in vitro* methods. Experiments *in vitro* usually need smaller amounts of test samples than *in vivo* testing and are able to provide information at the cellular or molecular levels.

The process of selection of experiments *in vivo,* experiments *in vitro,* or their combination should consider economic and ethical issues; more importantly, it should be based on the purpose of the research projects.

Experiments *in vivo* and *in vitro* can be mutually complementary to explore the action and mechanism of traditional herbal medicine from different views and levels. Results obtained from *in vivo* experiments are closer to clinical application. Some of the results can even be used to direct clinical practice. However, compared with *in vitro* experiments, *in vivo* experiments are sometimes unable to clearly explore the mechanism of action of herbal medicine and explain the causes of all changes; the results are easily altered by various complex factors. To explore the action mechanism of herbal medicine, it should be combined with *in vitro* experiments, from which results can be obtained accurately without disturbance from various complex factors that happen in *in vivo* experiments.

Of course, experiments *in vitro* also have limitations and disadvantages. Results obtained from *in vitro* tests are still far from clinical application. Efficacy of the testing sample is easily altered by various factors when being taken into the body. Impact of absorption, distribution, and metabolism within the body could not be reflected through *in vitro* testing. On the other hand, experiments *in vitro* would not give a positive result if the herbal medicine shows its pharmacological activity only when its components are metabolized within the body.

6.2.5 Establishment of Animal Models to Mimic "*Zheng*" in TCM for Pharmacological Study on Chinese Herbal Medicine

The meaning of "*zheng*" in TCM is similar to symptoms or syndrome in modern medicine, but with a different expression. TCM diagnosis is mainly performed using the four diagnostic methods of observation: hearing and smelling, enquiry and palpation (which cover the symptoms), signs, and patient's history. There are many types of *zhengs* with TCM special terminology according to TCM diagnosis, such as superficial or internal, excessive or deficient, hot or cold, yin or yang, dampness, and phlegm.

For example, the *zheng* of excessive heat could be fever, redness and swelling of skin, or nose bleeding caused by dryness and heat. Fatigue is considered as "*qi* deficiency;" stroke by blockage of blood can be diagnosed as "blood stagnation;" night sweating is thought to be a "yin deficiency;" and headache due to hypertension is caused by "yang excess." Detailed introduction of *zheng* is available in English books about TCM diagnosis.[9,10]

It is necessary to first figure out the physiological and biological changes and pathogenesis in each *zheng* with the terminology of modern medicine, then to establish animal models of *zheng* by using modern experimental methods according to the descriptions in TCM diagnosis. Take *qi* deficiency as an example. There are many different types of *qi* deficiency, such as heart *qi* deficiency, lung *qi* deficiency, kidney *qi* deficiency, and spleen *qi* deficiency. Different types of *qi* deficiency reflect the degradation of different functional activities in the nervous system, respiratory system, circulatory system, digestive system, endocrine system, immune system, and so on. In particular, decline in cellular immune function, reduced plasma estrodiol (E_2), and decreased ratios of estrodiol to testosterone (E_2/T) and cAMP/cGMP in plasma and platelet are found to be related to the *zheng* of heart *qi* deficiency. Animal models of most of the *zheng* in TCM have been established by Chinese scientists during the past half century.[8]

To scientifically evaluate the functions of Chinese herbs and their formulae, animal models with proper *zhengs* need to be established according to TCM theory. Scientists who are willing to dedicate themselves to the pharmacological study of Chinese medicine with success have to be equipped with knowledge of both TCM and modern medicine. Otherwise, they risk wasting time and money, as well as having their research results doubted due to incomplete understanding and application of either scientific or TCM theory.

6.3 DESIGN OF PHARMACOLOGICAL STUDY ON HERBAL MEDICINES

Pharmacological experiments must be carefully designed, efficiently executed, precisely analyzed, clearly presented, correctly interpreted, and ethically accepted. Many factors, such as animal model, control group, quality of samples, dosage and administration route, period of the experiment, measurement of the result, and data processing, can affect the experimental results. This is why inconsistent results of an herbal medicine are often found in different literature.

6.3.1 Basic Principles of Study Design

In the design and conduct of pharmaceutical research on herbal medicine, the basic principles of experimental design including control, reproducibility, and randomization still apply.

1. *Control*

It is essential to have a negative (vehicle only) control group and a positive control group of animals in all pharmacological studies of herbal medicine.

A *negative control group* is subjected to conditions under which a negative result is expected. A *positive control group* is subjected to conditions under which a positive result is expected, indicating the experimental system is working.

In pharmacological study, a negative control group can be a blank control group, model control group, and/or vehicle control group (or excipient control group). *Blank control group* refers to a group in which the animals do not receive any treatment during the experiment. *Vehicle control group* (or *excipient control group*) refers to a group in which the animals only receive vehicle (or excipient) during the experiment. In *model control group*, animal models with a specific disease or syndrome are used. Animals in the *positive control group* are given a drug with a known effect that is the same or similar to that of the test sample and expected to positively respond to the given drug to indicate the experiment is correctly performed. Both the negative and positive controls are used to be compared with the sample group(s) in order to quantify the effect of the test sample. All experimental results should be statistically processed.

2. *Randomization*

Randomization is essential for all pharmacological experiments as it minimizes the chance of a biased result. A randomized controlled trial is the most rigorous method for objectively evaluating the efficacy. This must be done throughout the whole experiment, rather than just randomizing the animals (or other experimental subjects) in the treatment groups.

3. *Reproducibility*

Only when the results can be repeated under the same or similar conditions can the experiment design be considered reliable. Besides verifying reliability, repeated experiment also can be used to understand the variance.

6.3.2 Selection of Experiment Method

1. *Basic principle*

Researchers must clearly know the purpose and significance of the research before selecting the experimental method.

It is known that herbal medicines have special characteristics different from synthetic medicine. Herbal medicine has various efficacies and plays its role through many ways.

The pharmacological experimental method must be selected according to the indications of herbal medicine to be tested. It is very important to choose an appropriate and characteristic experimental method and proper animal model for evaluating the efficacy of herbal medicine. In order to evaluate directly the main pharmacological action, a core experiment must be carefully designed, and several other experiments using different animal models, methods, or administration techniques can be selected as auxiliary experiments to indirectly confirm the main pharmacological method. The auxiliary experiments may also be used to confirm secondary pharmacological activity, if necessary.

One must be aware that most herbal medicines, especially extracts or fractions, are usually composed of complex components. The quality of the test sample should be well controlled, if possible, not only by the quantity of main components, but also by documenting the fingerprint of high-performance liquid chromatography (HPLC) or other chromatogram so that the test results can be repeated. Ideally, the same batch of sample should be used when the test needs to be repeated. The experiment *in vivo* should be taken as the main experiment method, combined with experiment *in vitro* as the subsidiary method.

Herbal medicines are considered drugs in many Asian and European countries. Each country has its own requirements for new drug application. When the pharmacological experiments are performed for new drug application, the method design has to meet the requirements.

2. *Application of recent development of pharmacological experimental methods*

The complex composition of herbal medicine greatly hampers its research progress. Standard pharmacological methods are usually employed in the conduct of nonclinical research on herbal medicines. However, the application of novel technologies and methods resulting from scientific progress should be encouraged. In the recent past decades, research methods of herbal medicine have been widely developed with the advanced development of modern science and technology. Some new methods have played a big role in promoting the research of herbal medicine. For example, gene chip technique, proteomic technology, and bioinformation have been applied to pharmacodynamic studies of herbal medicine. The detection by comet assay of herbal medicine protecting from DNA damage in animals has been proven to be more sensitive than that by other classical methods.

To find the active component from the complicated extracts of herbal medicine, scientists should try to establish new methods at the cell and molecular levels by using latest spatiotemporal and real-time monitoring technology, both *in situ* such as micro-array, instantaneous, online detection of single cell and *in vivo* such as microdialysis combined HPLC, ultramicroelectrode, molecular radar, and other technologies to accelerate the screening rate.

To reveal the complex multi-pathway, multi-linked, and multi-targeted characteristic of herbal medicine, it is necessary to boldly explore various new thoughts and methods. With the supplement and modification of previous research methods and the development of more new research methods, it is possible for research of herbal medicine to reach a newfound level.

6.3.3 Selection of Animals and Animal Models for Herbal Study

Selecting the right animal for study is critical to obtaining a convincing scientific result. Animals used for pharmacological studies must be a standard species. Mice, rats, guinea pigs, rabbits, cats, and dogs are often used for research of traditional herbal medicine. Detailed introductions for animals commonly used in pharmacological study, including characteristics of the animals such as strain, sex, age, and holding conditions, are available from literature.[1,11,12]

In addition to normal animals, animal models of disease are required in many pharmacological studies. Such animal models may have an existing, inbred, or induced disease or injury that is similar to a human condition. Sometimes, *in situ* organs or tissues are also used, depending on the purposes of the experiment. Considering the difference in observation between animal and clinical study, it is necessary to test one sample on several different animal species. The species of animal, number of animals in each test group, and number of groups needed mainly depend on the purpose and method index. In general, for small animals such as mouse and rat, the number of animals in each group is generally 8-12; for larger animals such as rabbit or dog, the minimum number in each group is 5-8.

6.3.4 Selection of Measurement Index or Target

Special attention should be given when selecting the measurement index (target). It is important to ensure that the measurement will provide a general accepted sensitivity and reproducibility for quantization or semiquantization. Each kind of index has its own advantages and disadvantages. Therefore, one index may not exactly and completely reveal the efficacy of an herbal medicine. Multiple indexes are often used comprehensively to supplement each other for evaluation of the effects of herbal samples.

A thorough search of literature will provide the most useful information and help to select an ideal test index (target) that is expected to most possibly predict clinical results. With the popularity of HPLC and mass spectrum (MS), and the availability of many other quantitative equipment, it is usually not difficult to find a proper measurement for the test index (target).

6.3.5 Requirement of Herbal Samples

Compared with modern drugs, which are composed of only a single active compound, herbal preparations, especially Chinese herbal formulae, usually consist of several different herbs. It is known that even a single herb contains many constituents. Therefore, it is difficult to predict whether a combination of several herbs may give rise to interactions between compounds in these herbs. Meanwhile, most modern herbal preparations are formulated in various forms, such as granular, tablet, cataplasm, or capsules. Therefore, quality control of herbal medicines and their preparations is much more difficult.

As mentioned in the Section 6.2, an herbal species collected from different regions or at different times may vary in concentration and composition of chemical compounds because they change with the environment and season. The extraction methods, including extraction time, temperature, and volume of solvent used, may affect the extraction efficacy. Therefore, herbal materials for pharmacological research must be standardized for quality control to guarantee the reproducibility of the experimental results. Establishment of standardization of herbal medicines should include species identification (name, origin, description, and purity; extraction procedures; and chemical analysis of main and bioactive compounds, heavy metals, pesticides, ash, moisture, and storage stability) (see details in Chapter 9).

6.3.6 Selection of Doses and Administration Route

1. *Selection of dosage*

For each testing sample, at least three dosages that mostly reflect the dose-effect and/or time-effect probability are required. When using big experimental animals such as a monkey or canine, or under special situations, the number of dosages can be decreased accordingly. Dosage selection should be within a reasonable range. A sample of an insufficient dosage may fail to produce pharmacological effects, while an excess dosage may cause toxicity. Both will lead to a wrong result. The following methods may be referenced when making dosage selection.

- Dosage calculation according to LD_{50}

For samples that are able to obtain LD_{50}, select doses that are around 1/10, 1/20, 1/30, or 1/40 of LD_{50} for the high-, middle-, and low-dosage groups of a pharmacodynamic experiment.

- Dosage calculation according to clinical dosage

For samples that have been clinically used for a long period, the dose that is several to several 10-folds of clinical dose can be used in animal experiments. The relative dose in animals can be converted from the dose used in humans according to body weight. The cursory isoeffect multiple are 1 (human), 3 (canine, monkey), 5 (cat, rabbit), 7 (rat, guinea pig), and 10–11 (mouse). For example, the dose used in rats should be 3-7 times (10-15 times for mouse) the clinical dosage; meanwhile, the highest dosage should be lower than the lowest dosage employed in the long-term toxicity study.

- Dosage estimation according to documents

The dosage of reported medicines that possess similar formulae and/or extraction methods to test the sample can be employed as a referee to estimate the reasonable dosage range of the test reagent.

- *Dosage calculation according to surface area*

For drugs with low safety coefficients, such as anticancer drugs or cardiac glycoside, the dosage in animal experiments can be calculated according to surface area. This is the so-called clinical "isoeffect" dosage, which means the calculated dosage of equivalent surface area unit (m^2/cm^2) according to a conversion algorithm of surface area. This calculation method is complex, and shows only relative rationality because of the disparity of tolerance and response of animal and human.

Whatever method is used to set sample dosages, a pretest should be performed to find an appropriate dosage range at which the pharmacological effect emerges. The formal experimental dosages can be decided based on the pretest result.

2. *Administration route*

There are many modes of administration for traditional herbal medicine: oral, nasal (smoking, snuffing, or steaming), topical, rectal, bathing, subcutaneous, or intramuscular injections.

Since oral dosage forms of herbal medicines are usually used in clinic, the oral route of administration is ordinarily the most suitable for test animals. Additional routes may be used to approximate the intended route of administration in humans. If it is very difficult to choose the same route of administration, a similar administration route can be selected.

It is necessary to clarify that two administration routes are generally adopted in pharmacological experiment, and one of them should be the same as the clinical application. If the sample is difficult to administer through a required route, another route can be employed to carry out experiments, with an explanation. For a sample with good solubility, it can be administered by injection (be careful to exclude nonspecific reactions). For a crude extract or preparation with low solubility, it is permitted to employ only one administration route.

6.3.7 Selection of Positive Control

"Control" is one of the basic principles in experimental design. There always should be a positive control group in pharmacological experiments.

For studies on herbal medicine, a positive control of Western drug is usually used to evaluate the reliability of the experiment system. In addition, if there is any available well-known positive control of herbal medicine, it is also important to add this in for comparison with the efficacy of the testing sample.

The selection of the positive drug control should follow the basic principles:

1. Comparability: The indication, efficacy, dosage form, and administration route of a positive control drug should be similar to that of the testing sample as much as possible. However, sometimes it is difficult to find the appropriate positive control drug that meets the requirements. In this case, one that differs slightly from the testing sample may be flexibly selected as a positive control. But the indications and administration route must be the same as the testing sample.
2. Legitimacy: The drug or herbal medicine used as a positive control must be a legal one that has been marketed and recorded in pharmacopeia. Please note that herbal medicines are collected in the Pharmacopeia in China, many Asian countries, and some European countries.
3. Optimization: The ideal candidate is a representative and generally accepted drug without severe side effects.

6.3.8 Data Process and Statistics

All the data obtained from the experiment have to be documented, organized, and analyzed statistically in order to give correct conclusion.

6.4 EXAMPLES OF *IN VIVO* EXPERIMENTS FOR HERBAL EXTRACTS

6.4.1 Effects of Saponins in Ginseng Stem and Leaf on Extracellular 3,4-Dihydroxyphenylacetic acid (DOPAC), Homovanillic Acid (HVA), and 5-Hydroxyindoleacetic Acid (5-HIAA) in Striatum of Rat

Introduction

Ginsenosides, saponins in ginseng, are the main active components in *Panax ginseng* C.A. Meyer, with various biological activities. The root of ginseng has been used as a Chinese herb for a long time. Pharmacological experiments proved its anti-inflammatory activity, antioxidant nature, antitumor effects, morphine-induced analgesia tolerance, morphine-induced hyperactivity, rewarding effects, and so on. In the recent decades, ginsenosides were also found in the stem and leaf of ginseng. In order to make full use of the resources of ginseng, the following studies were performed to evaluate the functions and their mechanisms of ginsenosides in ginseng stem and leaf.

The dopaminergic system is one of the important central neurotransmitter systems. It has been well known that DOPAC and HVA are two metabolites of the neurotransmitter dopamine, and 5-HIAA is the metabolite of the neurotransmitter serotonin. Therefore, changes of DOPAC, HVA, and 5-HIAA in striatum can indirectly reflect the functions of dopaminergic and serotonin systems, respectively.

Microdialysis is a technique for sampling the extracellular fluid of living tissue by means of an implanted dialysis fiber based on the dialysis principle. With this method, a tubular membrane that is permeable to water and small molecules can be inserted into body cavities to collect samples by diffusion. The extracellular fluid in almost any brain region can be reproducibly sampled using a hollow fiber with dialysis principle by means of stereotaxic surgery. The dynamics and regulation of neurotransmission can be monitored in the conscious brain by using this technique in combination with sensitive analytical chemical equipment such as HPLC.

The effects of ginsenosides in ginseng stem and leaf on extracellular DOPAC, HVA, and 5-HIAA in striatum of rat were studied by microdialysis with HPLC coupled with electrochemical detector (HPLC-ECD) as shown in the following.

Materials

Animal Male Wistar rats weighting 200–250 g.

Apparatus Stereotaxic frame; HPLC-ECD; microinjection pump; hollow fiber; dental cement; micro-amount collecting duct.

Chemicals and Extract Chloral hydrate; ginseng stem and leaf saponins (GSLS) containing 84.5% ginsenosides.

Methods

1. *Preparation of hollow fiber:* The internal diameter of the probe was 200 μm and the external diameter of the probe was 310 μm.
2. *Surgery:* The rats (divided into two groups, seven in each group) were anesthetized with chloral hydrate 350 mg/kg (*i.p.*) and mounted in a stereotaxic frame. The head skin was cut and temporal muscles were retracted from the bone and folded away. Holes were drilled bilaterally in the temporal bone. The dialysis probe was inserted transversally through the striatum (coordinates: A: +1.5 mm from bregma, V: –5.3 mm from occipital bone).

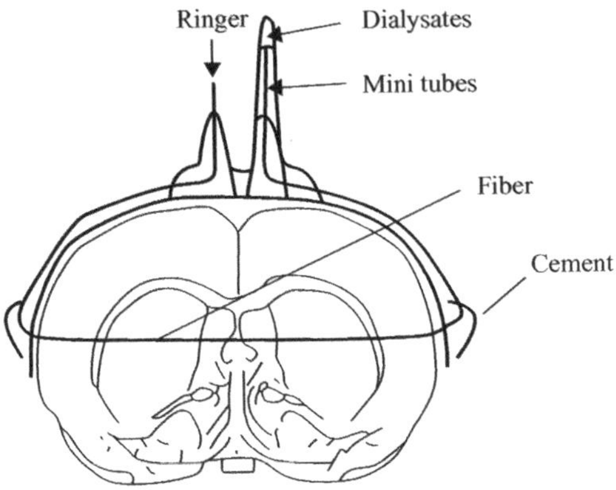

Figure 6.1 Schematic drawing of a dialysis probe implanted through the striatum of rat.

Do not fold the hollow fiber in the dialysis probe to acute angle, which can induce damage and blockage. After surgery, the animals were individually housed in a plastic cage and left to recover for about 24 h before being used for the experiment (see Fig. 6.1).

3. *Perfusion and detection:* On the second day after surgery, Ringer's solution was perfused through the dialysis fiber at a constant rate of 1–2 μL/min. The dialysis during the first 40–60 min was discarded and then the sample was collected every 20 min. Saline, GSLS of 200 mg/kg were administered when the output of serotonin (5-HT) became stable in the last three collected samples with variance of less than 10%.
4. *HPLC condition:* An aliquot of 20 μL dialysis samples was injected directly into the HPLC system. A reversed phase column (150 × 4 mm ID, 10 μm) was used with the mobile phase composed of 100 mM sodium acetate, 85 mM citric acid, 0.2 mM EDTA-Na_2, 1.8 mM sodium 1-octanesulfonate, and pH 3.7, 18% methanol (v/v). The mobile phase was pumped with LC-10A pump at a flow rate of 1.0 mL/min. The detector equipped with a glassy carbon working electrode and an Ag/AgCl reference electrode was set at +0.7 V. The current produced was monitored by a chromatography workshop. After the experiment, the rat was sacrificed and histological verification of probe placement was made via frozen or paraffin coronal sections (40 μm in thickness).
5. *Data analysis:* DOPAC, HVA, and 5-HIAA in the dialysis samples were measured by HPLC-ECD and the results were expressed as the percentage change compared with the respective basal value that was the mean of three consecutive samples before drug administration within a variance of 10%. Each point was expressed as mean ± S.E.M.

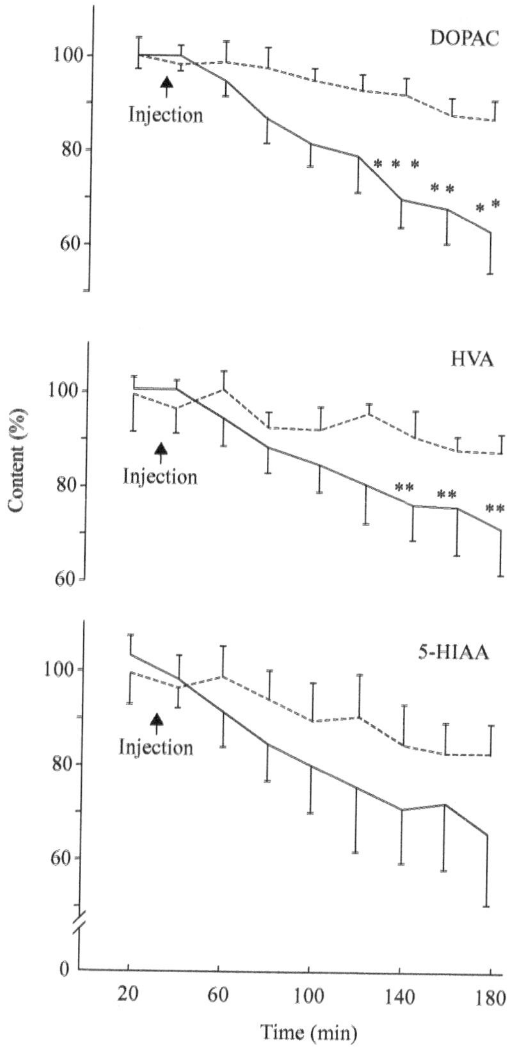

Figure 6.2 Effects of saponins in ginseng stem and leaf on extracellular DOPAC, HVA, and 5-HIAA in striatum of rat (---, saline ; —, GSLS 200 mg/kg).

Results

Basal contents of DOPAC, HVA, and 5-HIAA per dialysis were 7.3 ± 1.5 ng/20 μL, 7.9 ± 2.2 ng/20 μL, and 2.5 ± 0.4 ng/20 μL. GSLS at the dosage of 200 mg/kg significantly decreased the basal extracellular DOPAC and HVA at 80–120 min after administration; the concentration changes of DOPAC and HVA were both around −30%. Ginsenoside had no effect on the concentration of 5-HIAA (see Fig. 6.2). This result indicates that the mechanism of central action of GSLS might be through the dopamine system, rather than the 5-HT system.

Refer to details of the study in the literature.[13]

6.4.2 Blockage of Morphine Tolerance in Mice by Saponins in Ginseng Stem and Leaf

Introduction

Many experiments have demonstrated that ginseng extracts antagonized the morphine antinociception. The total ginseng saponins and ginsenoside-Rg1,-Rb1 have been reported separately to inhibit hyperactivity and conditioned place-preference induced by morphine. These findings suggest that ginseng may have potential application for the prevention of adverse actions of morphine abuse. Because the composition of saponins in ginseng stem and leaf is similar to the saponins in gingseng root, effects of saponins in ginseng stem and leaf on the development of morphine tolerance in mice were studied. The morphine tolerance tests include locomotor activity and analgesia tolerance.

Materials

Animal Male and female Swiss-Kunming mice weighing 18–22 g.

Apparatus Locomotor monitoring cage; clamps

Chemicals and Extract Morphine hydrochloride; GSLS

Methods

1. *Effects of GSLS on the development of morphine-induced locomotor sensitization*

Mice were randomly divided into six groups. Before the beginning of the experiments, mice were habituated to the activity cages without any injection for 60 min for 2 days. The six groups of mice (11 mice in each group) were treated with the following drug pairs, respectively, for 7 days: saline + saline (20 mL/kg), saline + GSLS (50 mg/kg), saline + GSLS (200 mg/kg), morphine (10 mg/kg) + GSLS (50 mg/kg), morphine (10 mg/kg) + GSLS (200 mg/kg), morphine (10 mg/kg) + saline (20 mL/kg). The saline and morphine were given by subcutaneous injection, while GSLS was administrated by intestinal gavage 60 min prior to the injection of morphine daily. Every day after morphine administration, the mice were immediately put into the test chambers and counts of motor activity were recorded for 30 min.

2. *Effects of GSLS on the tolerance to morphine analgesia as assessed by tail-pinch method in mice*

In this test, animal grouping and drug administration were the same as described in the Method section. All animals (19–23 mice in each group) were continually treated for 9 days. The inhibitory effect of GSLS on morphine-induced tolerance development was evidenced by the maintenance of analgesic response to a challenge dose of morphine (5 mg/kg) given 24 h after the final injection of morphine, when the clamp was put on the mouse tail, 1 cm from tail root. The animal's turning of its head was used as the index. The latency of biting response in the tail-pinch test was estimated at 30, 60, 90, and 120 min after administration of morphine.

3. *Effects of GSLS on the weight of mice*

Animals (10–11 mice in each group) were treated for 10 days and the weight was measured every day.

4. *Data analysis*

All data were expressed as mean ± S.E.M. Statistical analysis was performed using a mixed model and survival analysis. *p*-Values less than 0.05 were considered statistically significant.

Results

1. *Effects of GSLS on the development of morphine-induced locomotor sensitization*

GSLS, at the dosage of 50 mg/kg, significantly inhibited morphine-induced locomotor sensitization in mice. However, GSLS at the same dosage had no effect on locomotor activity (see Fig. 6.3).

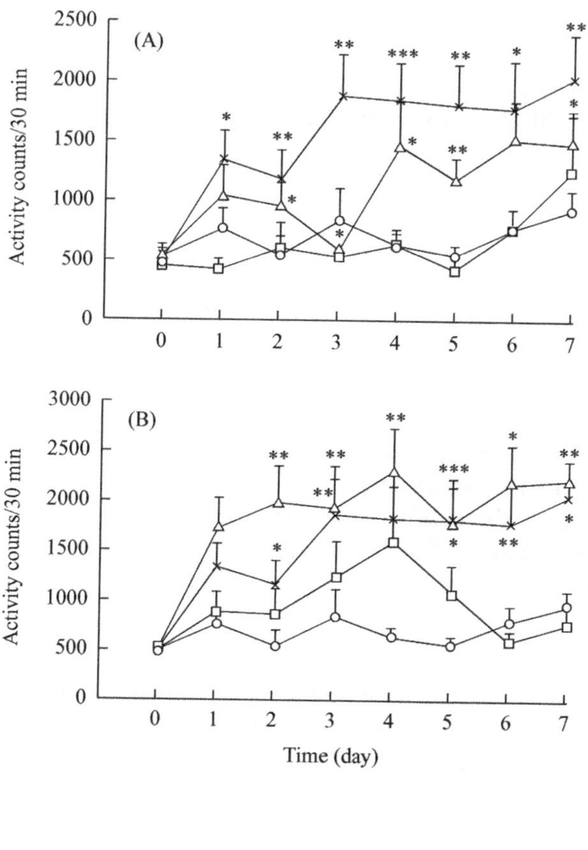

Figure 6.3 Effects of GSLS on the development of morphine-induced locomotor sensitization in mice. (A) GSLS 50 mg/kg; (B) GSLS 200 mg/kg. ○, saline; □, GSLS + saline; △, GSLS + morphine; ×, saline + morphine. All data are expressed as mean ± S.E.M. (* $p < 0.05$, ** $p < 0.01$, *** $p < 0.001$ vs. saline group; # $p < 0.05$ vs. morphine group).

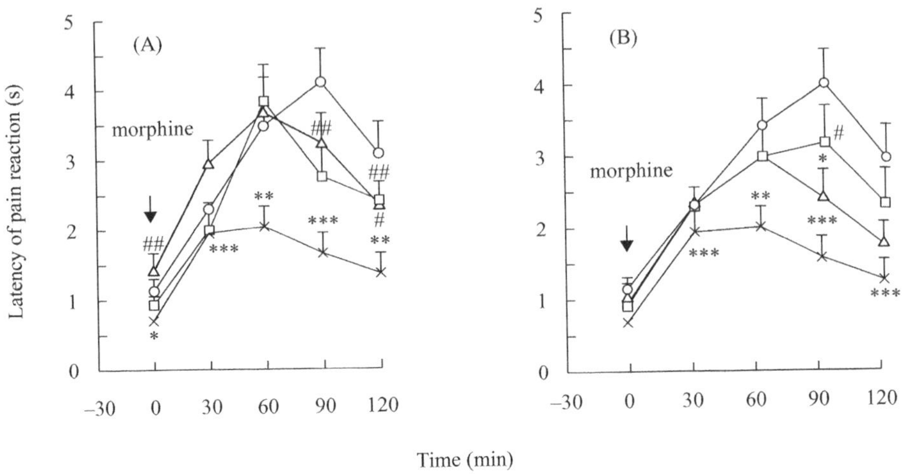

Figure 6.4 Effect of GSLS on morphine-induced analgesia tolerance in mice. (A) GSLS 50 mg/kg; (B) GSLS 200 mg/kg. ○, saline; □, GSLS + saline; △, GSLS + morphine; ×, saline + morphine. All data are expressed as mean ± S.E.M. (*$p < 0.05$, ** $p < 0.01$, *** $p < 0.001$ vs. saline group;. # $p < 0.05$, ## $p < 0.01$ vs. morphine group).

2. *Effect of GSLS on morphine-induced analgesia tolerance*

The morphine-induced analgesic tolerance was significantly attenuated by administration of GSLS at 50 mg/kg ($p < 0.01$) or 200 mg/kg ($p < 0.05$) (see Fig. 6.4).

3. *Effect of GSLS and morphine on the gain of body weight in mice*

GSLS, at the dosage of 50 mg/kg, had no effect on the body weight of mice. However, body weight in mice was significantly attenuated by DSLS at 200 mg/kg. GSLS and morphine had synergistic inhibitory effect on body weight gain ($p < 0.05$) (see Fig. 6.5).

Refer to the literature for the details of the study.[14]

6.4.3 Protective Effects of Total Saponins from Stem and Leaf of *P. ginseng* against Cyclophosphamide (CP)-Induced Genotoxicity in Mouse Bone Marrow Cells and Peripheral Lymphocyte Cells

Introduction

Clinically, prescriptions that contain ginseng have been frequently used in combination with chemotherapy on cancer patients to reduce the side effects of anticancer drugs. The main active components responsible for various biological activities of

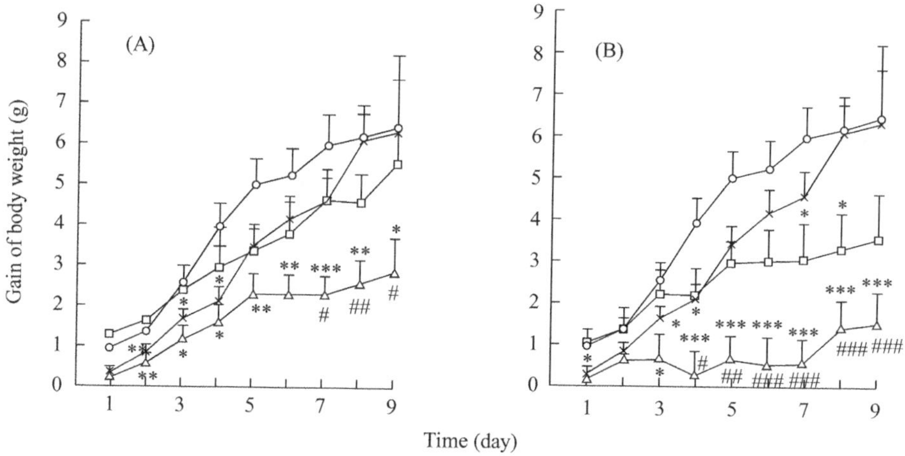

Figure 6.5 Effects of GSLS and morphine on the gain of body weight in mice. (A) GSLS 50 mg/kg; (B) GSLS 200 mg/kg. ○, saline; □, GSLS + saline; △, GSLS + morphine; ×, saline + morphine. All data are expressed as mean ± S.E.M. (* $p < 0.05$, ** $p < 0.01$, *** $p < 0.001$ vs. saline group; # $p < 0.05$, ## $p < 0.01$, ### $p < 0.001$ vs. morphine group).

P. ginseng are ginsenosides. In the following study, the protective effects of total saponins from stem and leaf of ginseng against CP-induced DNA damage were studied by the alkaline comet assay.

The comet assay, also called single cell gel electrophoresis (SCGE), is a sensitive and rapid technique for quantifying and analyzing DNA damage in individual cells. The resulting image that is obtained resembles a "comet" with a distinct head and tail. The head is composed of intact DNA, while the tail consists of damaged (single-strand or double-strand breaks) or broken pieces of DNA. Individual cells are embedded in a thin agarose gel on a microscope slide. All cellular proteins are then removed from the cells by lysis. The DNA is allowed to unwind under alkaline/neutral conditions. Following its unwinding, the DNA undergoes electrophoresis, allowing the broken DNA fragments or damaged DNA to migrate away from the nucleus. After staining with a DNA-specific fluorescent dye such as ethidium bromide or propidium iodide, the gel is read for amount of fluorescence in head and tail and length of tail. The extent of DNA liberated from the head of the comet is directly proportional to the amount of DNA damage.

Materials

Animal Male Swiss-Kunming mice with body weight 18–22 g.

Apparatus Electrophoresis apparatus; electrophoresis bath; specifically modified slides; fluorescence microscope; Laboratory Universal Computer Image Analysis (LUCIA) comet assay analysis software.

Chemicals and Extract CP; low-melting agarose; normal-melting agarose; total saponins from stem and leaf of *P. ginseng* C. A. Meyer (TSPG) (purity > 70%).

Methods

1. *Time-course and dose-response relationships of CP on peripheral lymphocyte DNA in mice*

For the time-course study, the mice were divided into five groups: saline control group and four CP groups (six in each group). Each mouse in CP group was given an intraperitoneal injection of CP with a dose of 140 mg/kg. Blood samples were collected from mice by puncture of the retro-orbital sinus at 30, 60, 120, and 180 min after drug administration.

For the dose-response study, the mice were divided into six groups: saline control group and five CP groups (six in each group) with the doses of 60, 80, 100, 120, and 140 mg/kg, respectively, by intraperitoneal injections. Mouse blood was drawn 2 h after CP administration.

2. *Protocol of TSPG administration in comet assay*

The mice were divided into five groups: control group, model group, low-dose group (25 mg/kg), middle-dose group (50 mg/kg), and high-dose group (100 mg/kg). Three doses of TSPG dissolved in sterile distilled water were orally administered to the animals for 3 consecutive days. The mice in control and model groups received the same volume of sterile distilled water. On the third day, CP was dissolved in saline and given intraperitoneally at a single dose of 100 mg/kg to mice 1 h after TSPG administration except for those in the control group. Two hours after the CP administration, blood samples were collected from mice by puncture of the retro-orbital sinus. Other mice were sacrificed by cervical dislocation following the above protocol after the CP administration. The bone marrow from the femurs was flushed in the form of a fine suspension into a centrifuge tube with phosphate buffer saline (PBS). The cells were dispersed by gentle pipetting and collected by centrifuge at 1500 rpm for 10 min at 4°C. Cell pellets were resuspended with PBS and adjusted until the density of cells reached 1×10^6/mL.

3. *Alkaline SCGE (comet assay)*

A fully frosted microscopic slide glass was coated with 1% (w/v) normal melting point agarose, and then covered by the coverslip and stored at 4°C until solidified. The cover slip was gently removed; subsequently, a mixture of 100 µL bone marrow cell suspension or 100 µL whole blood mixed with 100 µL of 0.5% (w/v) low melting point agarose was rapidly overlaid to solidify at 4°C for 10 min. After removal of the coverslip, the slides were immersed immediately in a chilled lysine solution (pH 10) of 2.5 M NaCl, 100 mM Na_4EDTA, 10% DMSO, and 1% Triton X-100 and kept at 4°C in the dark for 120 min. The slides were then placed on a horizontal gel-electrophoresis platform and covered with a chilled alkaline solution of 300 mM NaOH and 1 mM Na_2EDTA (pH 13). They were left in the solution in the dark at

4°C for 40 min, and then electrophoresed at 4°C in the dark for 30 min at 18 V and approximately 300 mA. The slide glass was neutralized by washing three times with 0.4 M Tris–HCl buffer, pH 7.5, stained with ethidium bromide, and covered with a coverslip. Fifty cells on one slide from each animal were examined and photographed through a fluorescence microscope (Olympus, at 200× magnification). All the protocol of the comet assay must be under slender light in a dark room. For the migration, Olive moment, which is one of the comet parameters, was measured with LUCIA comet analysis software.

4. *Statistical analysis*

For the experiments of the cell DNA damage induced by CP, the data were analyzed by one-way ANOVA followed by Fisher's least significant difference (LSD) with SPSS 12.0 software. For the experiments of the protective effects of TSPG against genotoxicity induced by CP and on antioxidative enzyme activities, two-way ANOVA followed by LSD was used to evaluate the interaction between drug treatment and CP with SPSS 12.0 software. $p < 0.05$ was considered statistically significant. Values were expressed as mean ± standard error.

Result

1. *Time-course and dose-response relationships of CP on peripheral lymphocyte DNA damage*

The time-course results showed that there were marked differences between the saline control group and CP groups at the time points of 120 and 180 min in the Olive moment (see Fig. 6.6). The dose-response study showed that there were significant differences between the saline control group and CP groups at the doses of 60, 80, 100, 120, and 140 mg/kg (see Fig. 6.7). These results demonstrated that CP could damage peripheral lymphocyte DNA in the time- and dose-dependent manners.

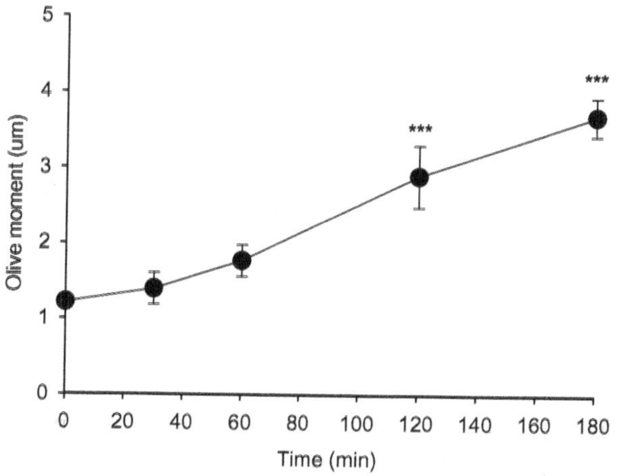

Figure 6.6 Time-course of DNA damage induced by CP in peripheral lymphocyte cells in mice. Each mouse, except those in the control group, was given an intraperitoneal injection of CP at the same dose of 140 mg/kg. Blood samples were collected at various time points from 30 to180 min. Data represented the mean ± S.E.M. *** $p < 0.001$ compared with control group.

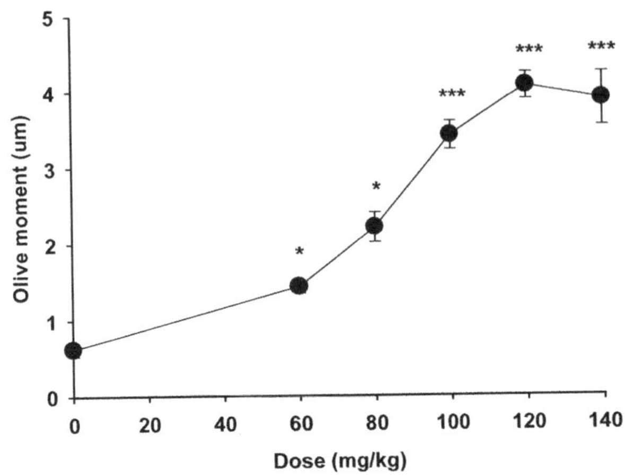

Figure 6.7 Dose-response of DNA damage induced by CP in peripheral lymphocyte cell in mice. Each mouse, except those in the control group, was given an intraperitoneal injection of CP at various doses from 60 to 140 mg/kg. Data represented the mean ± S.E.M. * $p < 0.05$ and *** $p < 0.001$ compared with control group.

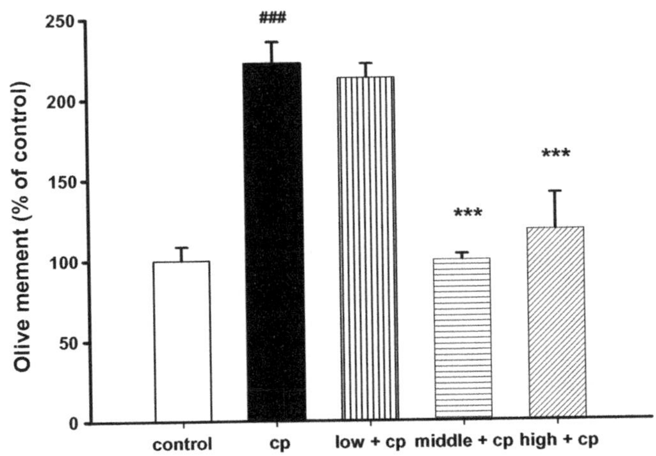

Figure 6.8 Effects of TSPG against CP-induced genotoxicity in mouse peripheral lymphocyte cells. The concentration of CP is 100 mg/kg; the low, middle, and high concentrations of TSPG are 25, 50, and 100 mg/kg respectively. (### $p < 0.001$ with respect to normal control group; *** $p < 0.001$ compared with the group treated with CP alone).

Therefore, in the following studies CP was taken as a model agent at the dose of 100 mg/kg, *i.p.*, and the samples were taken 120 min after its administration.

2. *The inhibitory effect of TSPG on DNA damage induced by CP treatment in mouse peripheral lymphocyte cells*

The data of saline control group were taken as 100% and the percentage changes were calculated for TSPG groups and CP group.

The effects of TSPG against DNA damage in peripheral lymphocyte cells are shown in Figure 6.8. Mice were treated with 25, 50, and 100 mg/kg of TSPG 1 h

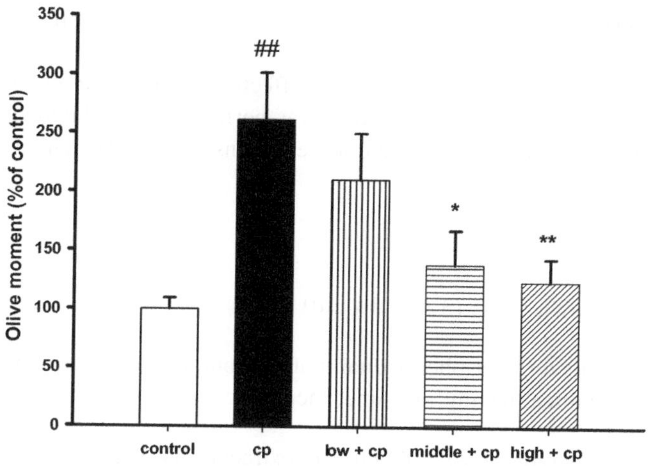

Figure 6.9 Effects of TSPG against CP-induced genotoxicity in mouse marrow cells. The concentrations of CP are 100 mg/kg; the low, middle, and high concentrations of TSPG are 25, 50, and 100 mg/kg, respectively. (## $p < 0.01$ with respect to normal control group; * $p < 0.05$; ** $p < 0.01$ compared with the group treated with CP alone).

prior to CP, *i.p.* The extent of DNA damage in the Olive moment was significantly decreased ($p < 0.001$) in mice orally administered TSPG at the doses of 50 and 100 mg/kg and CP, compared with those treated with only CP. TSPG at all three doses did not induce any genotoxicity in peripheral lymphocyte cells (data not shown).

3. *The inhibitory effect of TSPG on DNA damage induced by CP treatment in mouse marrow cells*

Figure 6.9 shows the effects of TSPG against CP-induced DNA damage in bone marrow cells. TSPG, at the dose of 50 and 100 mg/kg, significantly reduced DNA damage induced by CP in bone marrow cells as observed in the Olive moment ($p < 0.05, p < 0.01$). Again, TSPG at all three doses failed to induce any genotoxicity in bone marrow cells (data not shown).

Refer to the details of the study in the literature.[15]

6.4.4 Effect of *Sambucus williamsii* on Bone Mass and Bone Strength in OVX Rats

Introduction

Estrogen replacement therapy (ERT) was once a popular method for the prevention and treatment of postmenopausal osteoporosis. However, recent evidence suggests that ERT is associated with increased risk of development of breast, ovarian, and endometrial cancers. It has been reported that extract of the stem of *Sambucus sieboldiana* (SS) inhibited bone resorption in organ culture. Stem and ramulus of

S. williamsii HANCE (SWH) have been used in China for centuries for the treatment of inflammation, bone fractures, and joint diseases.

In the present study, investigation of the *in vivo* effects of SWH extract on postmenopausal osteoporosis, using OVX rat model, was performed. Biochemical markers of bone turnover, bone strength, and bone mineral density (BMD) changes were determined.

Materials

Animal Female Sprague-Dawley rats (3 months old)

Apparatus Automatic analyzer; peripheral quantitative computed tomography (pQCT) system; electromechanical testing machine; metabolic cage

Chemicals and Extract Estradiol (E_2); 60% ethanol extract of SWH extract (4% yield)

Methods

1. *Surgery*

Rats were bilaterally OVX using the dorsal approach. The surgery should be under asepsis conditions. A single longitudinal skin incision was made on the dorsal midline at the level of the kidneys. The ovaries were exposed and removed together with the surrounding fat, oviduct, and a small portion of the uterus. Animals of the control group underwent sham operations, during which the ovary was exposed but left intact. The surgery was carried out under anesthesia using ether. Rats were housed in cages under a 12/12 light/dark cycle at 22°C. Deionized water was provided to the animals *ad libitum.* During the study, OVX rats were pair-fed with a normal diet based on the average weekly food consumption of the sham control group. Husbandry of the animals was based on *Guide for Care and Use of Laboratory Animals.*

2. *SWH treatment*

The animals were either sham-operated ($n = 8$) or OVX ($n = 38$). OVX rats were randomly divided into four groups: OVX-control group ($n = 8$); estrogen group (E_2, $n = 10$, 2 mg/kg body weight/d); two SWH extract groups SWH30 ($n = 10$, 30 mg/100 g body weight/d); and SWH60 ($n = 10$, 60 mg/100 g body weight/d). The administration of E_2 and SWH by intestinal gavage started 4 weeks after the surgery and lasted for 3 months. The rats were then sacrificed for measurement of the parameters interested.

3. *Sample collection*

Rats were placed in the metabolic cages before sacrifice and allowed to have 24 h to accommodate the new environment. Urine samples were then collected continuously for 24 h and acidified with 2 mL of 1 M HCl, centrifuged at 1015 *g* for 5–10 min at 4°C to remove contaminating sediments, and aliquots were stored at −20°C until they were assayed. After sacrifice, blood samples were taken from the abdominal aorta and serum obtained was stored at −70°C before biochemical analysis. The uterus was collected and its weight was recorded. Left femora and tibias were removed and wrapped in gauzes saturated with physiological saline and stored at −20°C before analysis.

4. Biochemical Assay of Serum and Urine Samples

Serum and urine samples were analyzed as follows. Calcium (Ca) concentration in both serum and urine samples were measured by standard colorimetric methods using commercial kits and analyzed by an automatic analyzer ALCYON 300i. Urinary creatinine was determined by picric acid method. Serum alkaline phosphatase (ALP) activity was determined by using commercial ALP kit. Serum osteocalcin was assayed using a rat osteocalcin ELISA kit. Urine deoxypyridinoline (DPD) level was analyzed using a rat DPD ELISA kit. Urinary Ca excretion rate was expressed as the ratio of urine Ca to creatinine (Cr) level (Ca/Cr) while urinary DPD levels were expressed as urine DPD to creatinine level (DPD/Cr).

5. *Determination of trabecular bone mineral density*

Left tibiae were scanned with a pQCT system. The metaphyseal region of the left proximal tibia was scanned at a site of 2.5 mm from the proximal articular surface. Scans were analyzed using a threshold of 280 mg/ccm for delineation of the marrow cavity and an area of cortical and trabecular regions. Trabecular bone densities were measured using pQCT.

6. *Three-point bending test*

An electromechanical (universal) testing machine was used in conjunction with Instron Merlin software (version 4.03) for all tests. The left femur was cleaned to remove its surrounding soft tissue and stored at −20°C until analysis. The outer two supporting points were fixed 20 mm apart with a single central point positioned at the midpoint of the specimen. All loading points were 3 mm in diameter. The central loading point was displaced, and the load and displacement were recorded until the specimen was broken. Bending stiffness was then derived from the slope of the linear region of the resulting load versus displacement curve.

7. *Statistical analysis*

Data were reported as mean ±S.E.M. Significance of differences between group means was determined by one-way ANOVA. p-values <0.05 were considered statistically significant. Dunnett's procedure was used to determine the individual group difference at $\alpha = 0.05$.

Results

1. *Biochemical parameters of serum and urine samples*

Calcium Level in Serum and Urine Effects of SWH treatment on serum and urine Ca levels are shown in Figure 6.10. Serum Ca levels decreased in OVX rats

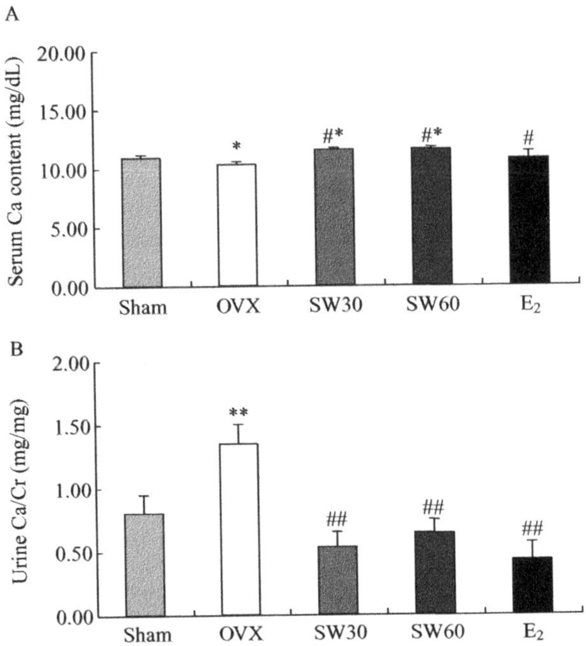

Figure 6.10 Effects of SWH extract on serum (A) and urinary calcium (B) level in OVX rats after 3 months of treatment. SW30 and SW60: respectively represent that OVX rats treated with 30 or 60 mg/100 g body weight/d of SWH extract. (* $p < 0.05$; ** $p < 0.01$, vs. sham control; # $p < 0.05$; ## $p < 0.01$; vs. OVX group).

($p < 0.05$, vs. sham). Such a decrease could be restored by treatment with E_2 ($p < 0.05$, vs. OVX) or SWH ($p < 0.05$, vs. OVX) (see Fig. 6.10A). On the other hand, the urine Ca level was significantly increased in OVX rats ($p < 0.01$, vs. sham, see Fig. 6.10B). Nevertheless, SWH treatment could significantly reduce the increase of urine Ca level in OVX rats in a manner similar to that by E_2 ($p < 0.01$, vs. OVX, see Fig. 6.10B).

Serum ALP Activity OVX treatment significantly increased serum ALP activity ($p < 0.05$, vs. sham, see Fig. 6.11A). SWH treatment at both dosages (30 mg/100 g body weight/d and 60 mg/100 g body weight/d) significantly reduced serum ALP activities in OVX rats to a level similar to that of the sham-operated group ($p < 0.05$). E_2 treatment at a dosage of 2 mg/kg body weight/d significantly reduced serum ALP activity in OVX rats ($p < 0.01$) to a level much lower than the sham-operated group ($p < 0.05$) (see Fig. 6.11A).

Serum Osteocalcin and Urinary DPD OVX treatment significantly increased serum osteocalcin (see Fig. 6.11B) and urinary DPD/creatinine levels (see Fig. 11C) ($p < 0.05$, vs. sham), indicating the induction of high bone turnover rate in rats by OVX. Treatment of OVX rats with SWH extracts resulted in significant reduction in serum osteocalcin level in the 30 and 60 mg/100 g body weight/d treated groups by 20% ($p < 0.05$) and 27% ($p < 0.05$), respectively. Similarly, SWH treatment significantly reduced urinary DPD excretion rate in OVX rats by 28% ($p < 0.05$)

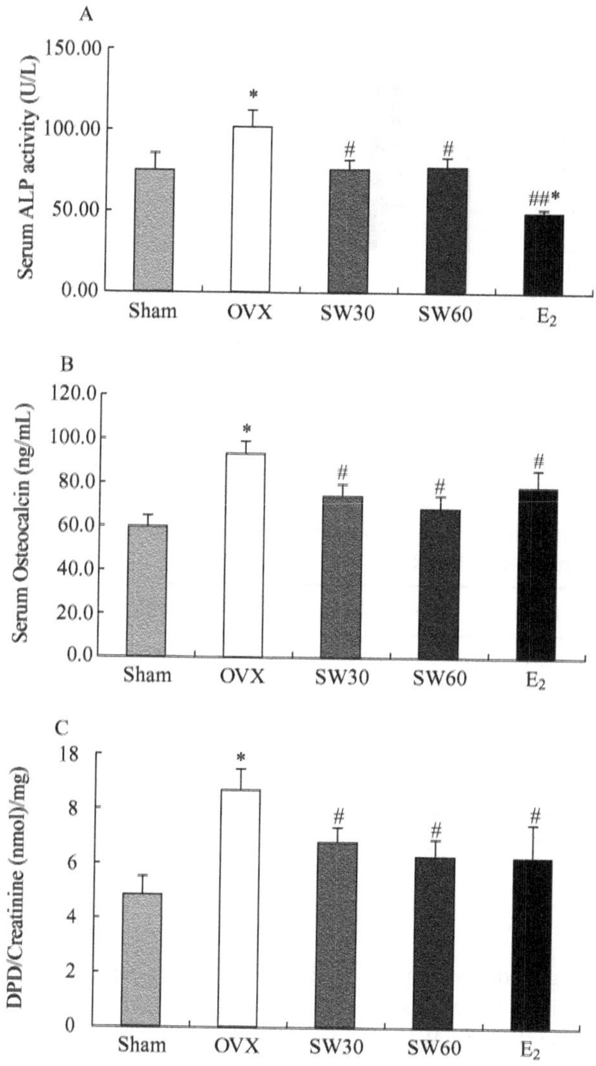

Figure 6.11 Effects of SWH extract on serum ALP activity (A) aerum osteocalcin; (B) urinary deoxypyridinoline/creatinine ratio; (C) OVX rats. SW30 and SW60: respectively represent that OVX rats treated with 30 and 60 mg/100 g body weight/d of SWH extract. (* $p < 0.05$ vs. sham control; # $p < 0.05$; ## $p < 0.01$ vs. OVX group).

and 22% ($p < 0.05$), respectively. Effects of SWH on serum osteocalcin (see Fig. 6.11B) and urinary DPD (see Fig. 6.11C) levels in OVX rats were similar to that of E_2, in which the increase in bone turnover rate by OVX were suppressed.

2. *pQCT analysis of left tibia*

pQCT analysis of proximal tibia was performed to determine if SWH treatment could increase BMD to protect against bone loss induced by OVX. As shown in Figure 6.12, there was a 33% reduction in trabecular BMD in OVX control group as compared with sham-operated group at 4 months postsurgery ($p < 0.05$). Treatment

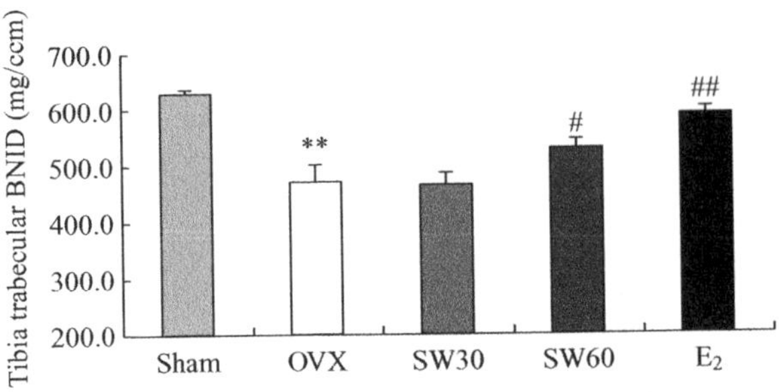

Figure 6.12 Measurement of bone mineral density of rat tibia by peripheral quantitative computed tomography (pQCT) after 3 months of treatment. SW30 and SW60: respectively represent that OVX rats treated with 30 and 60 mg/100 g body weight/d of SWH extract. (** $p < 0.01$ vs. sham control; # $p < 0.05$; ## $p < 0.01$ vs. OVX group).

Table 6.1 Effect of SWH Extract on Biomechanical Properties of Left Femur in Ovariectomized (OVX) Female Sprague-Dawley Rats

Group	*n*	Maximum load (N)	Energy absorption (J)	Stiffness (N/mm)
Sham	8	158.1 ± 3.3	88.0 ± 5.9	453.7 ± 33.8
OVX	8	145.0 ± 5.6*	58.1 ± 4.7*	412.8 ± 22.4*
SW30	10	140.6 ± 2.5	62.6 ± 5.4	502.6 ± 28.2##
SW60	10	148.1 ± 6.0	74.5 ± 8.0	465.8 ± 21.5#
E_2	10	153.3 ± 6.0	73.8 ± 5.8	532.2 ± 32.2##

* $p < 0.05$ vs. sham control, # $p < 0.05$, ## $p < 0.01$ vs. OVX group.

with SWH extracts (60 mg/100 g body weight/d) or E_2 for 4 months significantly increased trabecular BMD of tibia of OVX rats ($p < 0.05$ or $p < 0.01$). However, no obvious change in BMD was observed in SWH (30 mg/100 g body weight/d) treated OVX group ($p < 0.05$), suggesting that the protective effect of SWH on BMD in OVX rats was dose-dependent.

3. *Biomechanical testing of bone*

To determine if SWH treatment could improve bone strength, a three-point bending test of femur was performed. As shown in Table 6.1, OVX treatment significantly reduced the maximum load, energy, and stiffness of femur by 8%, 34%, and 10%, respectively ($p < 0.05$ vs. sham). Treatment with SWH or E_2 significantly prevented the reduction in bone stiffness induced by OVX. SWH or E_2 treatment appeared to demonstrate positive effect on the maximum load and energy of femur in OVX rats; however, the effects did not reach significant differences (Table 6.1).

Refer to the details of the study in the literature.[16]

REFERENCES

1. Shayne, C.G. (2007) *Animal Models in Toxicology* (2nd ed.), Boca Raton, FL, CRC Press.
2. Burdette, J.E., et al. (2002) *Trifolium pratense* (red clover) exhibits estrogenic effects in vivo in ovariectomized Sprague-Dawley rats. *The Journal of Nutrition* 132(1):27–30.
3. Liu, J.H., et al. (2001) Evaluation of estrogenic activities of plant extracts on the potential treatment of menopausal symptoms. *Journal of Agricultural and Food Chemistry* 49(5):2472–2479.
4. Burdette, J.E., et al. (2003) Black cohosh acts as a mixed competitive ligand and partial agonist of the serotonin receptor. *Journal of Agricultural and Food Chemistry* 51(19):5661–5670.
5. Powell, S.L., et al. (2008) In vitro serotonergic activity of black cohosh and identification of N_ω-methylserotonin as a potential active constituent. *Journal of Agricultural and Food Chemistry* 56(24):11718–11726.
6. Shen, Y.J. and Chen, C.X. (2008) *Traditional Chinese Pharmacology* (5th ed.), Shanghai, Shanghai Science and Technology Publisher.
7. Chen, Q. (2006) *Pharmacological Research Methodology of Chinese Medicine* (2nd ed.), Beijing, People's Health Publishing House.
8. Bensky, D. and Barolet, R. (1990) *Chinese Herbal Medicine: Formula & Strategies.* Seattle, WA, Eastland Press.
9. Maciocia, G. (2004) *Diagnosis in Chinese Medicine: A Comprehensive Guide.* Leithwalk, UK, Churchill Livingstone.
10. Deng, T.T. (1999) *Practical Diagnosis in Traditional Chinese Medicine.* Leithwalk, UK, Churchill Livingstone.
11. Hau, J. and Van Hooser, G.L. (2003) *Handbook of Laboratory Animal Science. Vol. I. Essential Principles and Practices* (2nd ed.), Boca Raton, FL, CRC Press.
12. Svendsen, P. and Hau, J. (1994) *Handbook of Laboratory Animal Science. Vol. II. Animal Models.* Boca Raton, FL, CRC Press.
13. Zhang, Y.Y., et al. (1991) Effecet of GSLS and its monomers on DOPAC, HVA and 5-HIAA in rat striatum in vivo. *Pharmacology and Clinics of Chinese Materia Medica* 7(1):9–12.
14. Wu, C.F., et al. (1998) Blockage by ginseng stem and leaf saponins on the development of morphine tolerance in mice. *Chinese Journal of Pain Medicine* 4(1):42–47.
15. Zhang, Q.H., et al. (2008) Protective effects of total saponins from stem and leaf of Panax Ginseng against cyclophosphamide-induced genotoxicity and apoptosis in mouse bone marrow cells and peripheral lymphocyte cells. *Food and Chemical Toxicology* 46(1):293–302.
16. Xie, F., et al. (2005) Increase in bone mass and bone strength by Sambucus Williamsii HANCE in ovariectomized rats. *Biological & Pharmaceutical Bulletin* 28(10):1879–1885.

Chapter 7

Safety Pharmacology and Toxicity Study of Herbal Medicines

Jing-yu Yang and Li-hui Wang

Herbal remedies are widely used for the treatment and prevention of various diseases and have become more popular over the past decades. Due to fear of side effects from Western medicine, more and more people every year turn to herbal medicine because it is considered to be generally effective and safe.

Not many herbal products have been investigated for their chronic toxicities, especially in countries where herbal products are used as dietary supplements, and there is no strict regulation for their quality and safety. Many people mistakenly believe that all herbal medicines are generally safe and free from toxicity. In fact, "natural" does not necessarily mean safe. A well-known example includes some plants, such as belladonna and tobacco, which contain poisonous alkaloids in the Solanaceae family, informally known as the nightshade family.

In fact, most herbal practitioners know herbal safety very well. Take Chinese herbal medicine as an example. Two thousand years ago, traditional Chinese herbs were divided into three categories based on clinical experiences in the first Chinese herbal book, *Shen Nong Ben Cao Jing* (*The Divine Farmer's Materia Medica*): herbs with no toxicity, minor toxicity, and heavy toxicity. Herbs with no toxicity are less likely to cause side effects if taken in the recommended dose range, and are thus safe in nature and most frequently used, for example, licorice and goji berry. Herbs with minor toxicity may cause mild side effects, particularly if overdosed or used with long tenure, for example, apricot seeds and ginkgo nut, causing mild breathing difficulties, abnormal heartbeat, or dizziness. Herbs with heavy toxicity may easily induce serious toxic effects, particularly if improperly used, for example, unprocessed croton fruits and monkshood, causing coma, nausea, and numbness of mouth and limbs. Since the potential side effects or toxicities of each herb have been clearly

Traditional Herbal Medicine Research Methods, Edited by Willow J.H. Liu

stated in traditional Chinese pharmacological books, most traditional Chinese herbs are safe if applied according to syndrome differentiation and with the recommended dosage for an appropriate length of time.

Residing in a country with a long history of application of herbal medicine, Chinese scientists have done a great job investigating all aspects of Chinese herbs and their formula, including pharmacology and toxicology. Unlike in the United States and other Western countries where herbs are considered dietary supplements, most Chinese herbal preparations are prescribed as drugs in China and other Asian countries. Beginning in the 1980s after the first drug administration law and official procedures for approval of new drug were promulgated in China, if an herbal preparation is applied as a new drug, safety and toxicity study results have to be presented together with the pharmacological and clinical data for registration.

Reports on toxicities of herbal medicine have been associated not only with the liver and kidney, but also with the nervous system and cardiovascular system, as well as with mutagenicity and carcinogenicity.[1–6] Numerous advanced biological experimental techniques have been used as standard safety tests along with the efficacy studies.[7,8]

Toxicological studies on herbal medicine mainly use the research methods of Western medicine. Because natural products are mostly composed of complex compounds, it is important to address again that, just as for primary pharmacological studies, researchers should make sure when performing toxicity studies on herbal products that the test samples have been chemically analyzed and characterized so that the result of the sample can be reproduced. Ideally, the same batch of sample should be used for all *in vitro* bioassay, *in vivo* pharmacology and toxicity studies, and clinical trials.

This chapter briefly introduces safety pharmacology, and methods for testing acute and chronic toxicity, as well as special toxicity such as genotoxicology, reproductive toxicity, and carcinogenicity for study on safety and toxicity of herbal medicine.

7.1 SAFETY PHARMACOLOGY

The criteria of safety for herbal medicine should be the same as for Western drugs. Therefore, it is necessary to give a brief introduction about the guidance and regulation of the safety of drugs.

The first attempt at defining safety pharmacology is contained in the guidelines of the Japanese Ministry of Health and Welfare (1995), which provides considerable detail on the types of pharmacology studies considered essential before an agent is administered to a human body. In July 2001, the International Conference on Harmonization of Technical Requirements for Registration of Pharmaceuticals for Human Use (ICH) issued a new *Guidance for Industry—Safety Pharmacology Studies for Human Pharmaceuticals* (S7A). S7A provides guidelines of safety pharmacology studies with the purpose of establishing a core battery to investigate the effects of the test substances on vital functions, which usually include cardiovascular, respiratory, and central nervous system (CNS).

The ICH S7A guideline has been adopted by the regulatory authorities of Europe, the United States, and Japan, and referenced by the relevant departments of many other countries. Described in this section are a number of important issues relating to safety pharmacology as it is currently defined by regulatory agencies.

According to this guideline, pharmacological studies can be divided into three categories: primary pharmacodynamic studies, secondary pharmacodynamic studies, and safety pharmacology studies.

Pharmacodynamic study is performed to investigate the mode of action and/or effects of a substance in relation to its desired therapeutic target. *Secondary pharmacodynamic study* is performed to investigate the mode of action and/or effects of a substance not related to its desired therapeutic target. *Safety pharmacology study* is to investigate the potential undesirable pharmacodynamic effects of a substance on physiological functions in relation to exposure in the therapeutic range and above.[9]

A recent review article based on the lecture of Dr. Tim Hammond, a recipient of the Distinguished Service Award of the Safety Pharmacology Society given in 2007 in Edinburgh, summarized the rationale behind the need for optimal nonclinical safety and secondary pharmacology testing, the evolution of safety and secondary pharmacology over the last decade, its impact on drug discovery and development, the value of adopting an integrated risk assessment approach, the translation of nonclinical findings to humans, and finally, the future challenges and opportunities facing these disciplines.[10]

The undesired effects of a substance may present a hazard, particularly in individuals who already have one or more compromised or limited organ functions.[11] Unlike other nonclinical evaluations of a drug's safety, these evaluations are usually conducted at doses close to the intended clinical dose.

To ensure the reliability of nonclinical safety pharmacology studies, not only the core battery but also follow-up and supplemental studies should be conducted in compliance with Good Laboratory Practice (GLP) with the most feasible extent.[9] Safety pharmacology investigations can be part of toxicology studies.

7.1.1 Objective

The specific objectives of safety pharmacology studies include:

- to identify undesirable pharmacodynamic properties of a substance that may have relevance to human safety
- to evaluate adverse pharmacodynamic and/or pathophysiological effects of a substance observed in toxicology and/or clinical studies
- to investigate the mechanism of the adverse pharmacodynamic effects observed and/or suspected.

7.1.2 Safety Pharmacology versus Efficacy Pharmacology and Toxicology

Safety pharmacological studies have been considered an important and independent discipline situated between traditional toxicology and efficacy/discovery pharmacology. Safety pharmacology is, however, a pharmacological rather than a toxicological discipline since it concerns the study of drug actions on physiological function rather than on physical/anatomical structure. Although using identical methods, safety pharmacology differs from efficacy pharmacology in that the former evaluates the potentially adverse effects of test substances on normal function, whereas the latter is aimed at establishing therapeutic potential. Both provide information critical for drug discovery and development.

Although structural and functional changes are sometimes clearly related, they are not always linked in terms of cause and effect. Changes in physiological function may occur in the absence of changes in organ structure and frequently occur at doses lower than those necessary to induce a structural change. On the other hand, not all structural changes are clearly associated with a detectable change in function. Thus, safety pharmacology and classical toxicology are complementary, both providing information important for determining the safety of a new drug. While regulatory authorities may place more emphasis on formal toxicology studies, clinical pharmacologists may find safety pharmacology data more reassuring when designing clinical trials.[12]

7.1.3 Reasons for Safety Pharmacology Studies on Traditional Herbal Medicine

With recent increasing interest in alternative or herbal medicine for the prevention and treatment of various illnesses, there is increasing concern about the safety of medicinal plants. There are general and herb-specific concerns regarding herbs and their potential to produce adverse effects. Accidental herbal adverse effects may occur as a result of collecting wrong raw materials and inappropriate preparation due to a lack of knowledge on active and toxic components in the materials and pharmaceutical quality control, or overdosed or over-lasting administration due to the mistaken belief that herbal remedies are harmless. Unfortunately, many countries have no official regulations for quality control on the manufacturing or labeling claims of herbal remedies and dietary supplements.

Not many people realize that when herbs are extracted and purified, the toxicity might be increased along with the increase in therapeutic efficacy. Here, take the rhizome of *Angelica pubescens f. biserrata,* a Chinese herb used for arthritis with anti-inflammatory activity. When this herb was extracted and then fractionated with petroleum benzene, chloroform, ethyl acetate and, *n*-butenol, respectively, into several fractions and given to a mouse with ear swelling induced by mixed inflammatory solution for pharmacological study in our lab, the petroleum benzene fraction showed not only the highest potential inhibition on ear swelling, but also the highest toxicity. Three out of 10 mice died after administration. Another example is ephedra, a Chinese herb used to induce sweating to disperse cold and wheezing and promote urination and reduce edema, used in traditional Chinese medicine (TCM). Under the guidance of a Chinese doctor, ephedra is a safe herb used in Chinese herbal formulae, but Western people extracted the alkaloids from it and sold the product as a stimulant, leading to the deaths of several people by improper administration of the product.

When performing safety pharmacological studies on herbal materials, the following two issues must be considered.

- The concentration of active or toxic compounds and other chemicals varies in different parts of the plant, for different harvest seasons, and when extracted with different methods. The stabilities of compounds depend on their structures.
- Geography, soil composition and its contaminants, and year-to-year variations in soil acidity, water, weather conditions, and other growth factors all have significant effects on the therapeutic properties and safety of the medicinal plants.

Currently, although adverse effects are often reported in patients to whom traditional herbal medicine was administered, there is no guidance for safety pharmacology of herbal medicine worldwide. The safety pharmacology guidance of traditional herbal medicine, including herbal extracts and preparations, is mainly followed for its chemical entities.

7.1.4 Study Designs

Safety pharmacology studies are normally conducted as a starting step under GLP. At the same time, unlike other safety assessment studies, safety pharmacology studies do not use doses that vastly exceed the intended therapeutic doses in order to reveal signs of toxicity.[11] Considerations in the selection and design of the studies should refer to guidance in S7A as stated in the following:

- Effects related to the therapeutic class of the test substance. The mechanism of action of some drugs may suggest specific adverse effects (e.g., proarrhythmia is a common feature of antiarrhythmic agents).
- Adverse effects associated with members of the chemical or therapeutic class, but independent of the primary pharmacodynamic effects (e.g., antipsychotics and QT prolongation [QT: Q wave and T wave in ECG]).
- Ligand binding or enzyme assay data suggesting a potential for adverse effects.
- Results from previous safety pharmacology studies, from secondary pharmacodynamic studies, from toxicology studies, or from human use that warrant further investigation to establish and characterize the relevance of these findings to potential adverse effects in humans.

Animal Species and Model

Generally, either rodent (such as mouse and rat) or non-rodent (such as dog) species can be used in safety pharmacology studies, depending on the individual experiment.

Although the S7A suggests the use of conscious, unrestrained, telemeterized, or trained animals, this may not be practical in all instances. Some cardiovascular and respiratory measures may require or may be best served by using animals that are restrained but acclimated.

Additionally, the following two issues should be addressed:

- whether the animals employed in *in vivo* studies need to be native or whether they could be used repeatedly for evaluations of different compounds
- whether sensitive, functionally compromised, or diseased model animals should be employed for some of the safety pharmacology applications, particularly those of the respiratory and renal systems where the presence of significant functional reserves can serve to reduce the ability of test methods to detect changes.

Group Size and Controls

The size of the groups for safety pharmacological study should be sufficient to allow meaningful scientific interpretation of the data generated. This means the statistical analysis of the data should indicate a significant difference between control and sample group and give the variability of the results.

In general, 10 rodents or 6 non-rodents per data point in each group should be adequate, and both male and female animals should be included. An exception is in the case of a freestanding Irwin screen or functional observatory battery (FOB) where many end points are assessed in the same animal with high variability associated with individual end points. Animals of one sex would be acceptable depending on the individual studies. For some studies, one sex of animal should be selected. For example, when estrogenic activity of a sample is investigated, the ovariectomized female rat has to be used.

Appropriate negative (vehicle or placebo) and positive control groups should be included in the experimental design. In a well-characterized *in vivo* study, positive controls may not be necessary. The exclusion of controls from studies should be justified.

Route, Dosage, and Duration of Administration

Normally, the expected clinical route of administration should be used when feasible. When the test substance is intended for clinical use by more than one route of administration (e.g., oral and peritoneal), assessment of effects by more than one route may be appropriate.

Whether for primary or safety pharmacological studies, herbs generally need larger dose and longer duration to see effect in comparison with conventional drugs.

For *in vivo* safety pharmacology studies, doses should meet and exceed the primary pharmacodynamic or therapeutic range. If the adverse effects are not observed within the first set test dosage range, the dosage should be increased enough to produce moderate adverse effects. Dose-response relationship of the adverse effect should be obtained.

The time course should be optimized. Determination of the course length should refer to clinical application and that used in primary pharmacodynamic study of the test herbs.

Some effects in the toxic range, for example, tremors during electrocardiogram (ECG) recording, may confound the interpretation of safety pharmacology effects and may also limit dose levels.

For *in vitro* safety pharmacology studies, concentrations should include a higher set to detect an adverse effect in the test system. The upper limit of the concentration range may be influenced by physicochemical properties of the test substance and other assay-specific factors. In the absence of an adverse effect, the range of concentrations selected should be justified. Just as for *in vivo* studies, the concentration-effect relationship should be obtained.

Safety pharmacology studies are generally performed by single-dose administration. In general, the herbal medicine needs a longer time to see their effect as mentioned above. The duration of the safety pharmacology studies should rationally base not only on results of the test sample from pharmacodynamics and pharmacokinetics studies, but also on results from repeat dose in nonclinical studies, as well as results from use in humans that give rise to concerns about safety pharmacological effects.

Statistical Design

Since identification of risk is the chief aim of a safety pharmacology test, it is essential that positive results not be overlooked. In other words, the generation of false negative results should be kept to a minimum.[13] False positive results, or the erroneous detections of possible risk, although bothersome, are less serious and can usually be corrected with supplementary testing. Thus, the risk of false negatives should be decreased as much as possible, even if there is increasing risk of obtaining false positives.

A test substance that has been found to pose no significant safety risks based on preclinical studies, even after the use of oversensitive statistics, is more likely to be truly devoid of risk. The generation of false negative results should be kept to a minimum. Safety pharmacology, in contrast to efficacy pharmacology, should employ statistical procedures possessing maximal sensitivity for detecting possible effects on a dose-by-dose basis, at the acknowledged risk of increasing the number of false positives.[14] When assessing drug-induced QT interval prolongation, a major concern in new drug candidate development, by the QT shift method, scientists found that the QT shift method was more sensitive than the QT correction method in some conditions.[15]

7.1.5 Core Battery and Supplemental Studies

The cardiovascular, respiratory, and CNSs are usually considered the vital organ systems, and therefore should be studied in the core battery. According to the ICH S7A guidance, the core battery should be supplemented or may need to be implemented in some instances based on scientific rationale.

Cardiovascular System

The cardiovascular system is a primary vital function that must be examined during safety pharmacology studies. Cardiovascular system functioning is maintained by cardiac electrical activity and by pump-muscle function, which contribute to hemodynamic efficacy. The aim of cardiovascular safety pharmacology is to evaluate the effects of test substances on the most pertinent organs of this system in order to detect potentially undesirable effects before engaging in clinical trials.[16,17]

In the basic program, a detailed hemodynamic evaluation is carried out in an anesthetized dog. It is completed by cardiac and/or cellular electrophysiology investigations in order to assess the arrhythmogenic risk. The basic program can be preceded by rapid and simple testing procedures during the early drug discovery stage. It should be completed, if necessary, by specific supplementary studies, depending on the data obtained during the early clinical trials. *The Non-Clinical Evaluation of the Potential for Delayed Ventricular Repolarization (QT interval prolongation) by Human Pharmaceuticals* (S7B) was issued by ICH in 2005 to emphasize the assessment of potential for QT prolongation in new drug development. The index of cardiovascular system safety pharmacology evaluations is given in Table 7.1.

CNS

The CNS is the body's control center and communication network. The CNS consists of the brain and spinal cord. It regulates vital bodily functions by sensing changes within the body and in the outside environment. The purpose of CNS safety pharmacology is to evaluate the effects of test substances on the most pertinent

Table 7.1 Cardiovascular System Safety Pharmacology Evaluations

Core	Hemodynamics (blood pressure, heart rate)
	Autonomic function (cardiovascular challenge)
	Electrophysiology (ECG in dog)
QT prolongation:	Cardiac action potential *in vitro*
(noncore)	ECG (QT measurements) in a cardiovascular study which would be covered in the core battery
	hERG channel interactions (human ether-à-go-go related gene [hERG] expressed in HEK 293 cells)

organs of this system in order to detect potentially undesirable effects before engaging in clinical trials.[18]

One characteristic that distinguishes CNS safety pharmacology from efficacy pharmacology is that CNS safety pharmacology is generally conducted in normal animals. The aim is to see whether the test substance induces adverse changes in normal function, rather than to observe potential therapeutic effects on abnormal function. Another characteristic of CNS safety pharmacology is that studies are almost exclusively carried out *in vivo,* using conscious animals. CNS safety study seldom use *in vitro* assay because the functional mechanisms of CNS are too complicated. The index of CNS safety pharmacology evaluations is given in Table 7.2.

Respiratory System

As early as 1964, it was well known that β-adrenergic blockers could lead to bronchoconstriction (and possible death) in asthmatics. Since then, many similar adverse effects of β-blockers or other drugs have been identified. These side effects of drugs

Table 7.2 Central Nervous System Safety Pharmacology Evaluations

Irwin test
General assessment of effects on gross behavior and physiological state*
Locomotor activity
Specific test for sedative, excitatory effects of compounds
Neuromuscular function
Assessment of grip strength
Rotarod
Test of motor coordination
Anesthetic interactions
Test for central interaction with barbiturates
Anti/proconvulsant activity
Potentiation or inhibition of effects of pentylenetetrazole
Tail flick
Tests for modulation of nociception (also hot plate, Randall Selitto, tail pinch)
Body temperature
Measurement of effects on thermoregulation
Autonomic function
Interaction with autonomic neurotransmitters *in vitro* or *in vivo*
Drug dependency
Test for physical dependence, tolerance, and substitution potential
Learning and memory
Measurement of learning ability and cognitive function in rats

* A functional observational battery (FOB) is usually integrated into a rodent (rat) repeat dose toxicity study to meet this requirement.

Table 7.3 Respiratory System Safety Pharmacology Evaluations

Respiratory functions
Measurement of rate and relative tidal volume in conscious animals
Pulmonary function
Measurement of rate, tidal volume, and lung resistance and compliance in anesthetized animals

from a variety of pharmacologic/therapeutic classes on the respiratory system support the need for conducting respiratory evaluations in safety pharmacology.

The objective of such studies is to evaluate the potential of drugs to cause secondary pharmacologic or toxic effects that influence respiratory function. Changes in respiratory function can result either from alterations in the pumping apparatus that controls the pattern of pulmonary ventilation or from changes in the mechanical properties of the lung that determine the transpulmonary pressures (work) required for lung inflation and deflation.[10,19]

Defects in the pumping apparatus are classified as hypo- or hyperventilation syndromes and are evaluated by examination of ventilatory parameters in a conscious animal model. Defects in the mechanical properties of the lung are classified as obstructive or restrictive disorders and are evaluated in anesthetized animal models by performing flow-volume and pressure-volume maneuvers, respectively (see Table 7.3).

The species used in safety pharmacological studies are the same as those used in toxicological studies, since pharmacokinetic and toxicologic/pathologic data are available in these species. These data can be used to help select test measurement intervals and doses and to aid in the interpretation of functional change. The techniques and procedures for measuring respiratory function parameters are well established in guinea pigs, rats, and dogs.[20] The index for respiratory system safety pharmacology evaluations is given in Table 7.3.

Supplemental Safety Pharmacology Studies

Supplemental studies are meant to evaluate potential adverse pharmacodynamic effects on organ system functions not addressed by the core battery or repeated dose toxicity studies when there is a cause for concern. The parameters that should be considered in supplemental safety studies are given in Table 7.4.

7.2 ACUTE TOXICITY STUDY

Acute toxicity testing describes the adverse effects of a substance that result either from a single exposure or from multiple exposures in a short space of time (within 14 days of the administration of the substance). Acute toxicity is involved in the investigation of half-lethal dose (LD_{50}), the dose at which it has been proven to be

Table 7.4 Secondary Organ Safety Pharmacology Evaluations

Renal system
Renal function: Measurement of effects on urine excretion in saline loaded rats
Renal dynamics: Measurement of renal blood flow, glomerular filtration rate (GFR), and clearance
Blood chemistry: Measurement of blood urea nitrogen, creatinine, and plasma proteins
Gastrointestinal (GI) system
GI function: Measurement of gastric emptying, intestinal transit, and ileal contraction (*in vitro*)
Secretion: Measurement of gastric acid secretion, bile secretion, and gastric pH
GI irritation: Assessment of potential irritancy to the gastric mucosa
Emesis: nausea, vomiting
Autonomic nervous system
Measurement of effects on the autonomic nervous system. For example, binding to receptors relevant for the autonomic nervous system, functional responses to agonists or antagonists *in vivo* or *in vitro*, direct stimulation of autonomic nerves and measurement of cardiovascular responses, baroreflex testing, and heart rate variability can be used.
Immune system
Passive cutaneous anaphylaxis (PCA)—test for potential antigenicity of compounds
Other
Blood coagulation
In vitro platelet aggregation
In vitro hemolysis

lethal—causing death to 50% of the tested group of animals. Acute toxicity study in animals is necessary for all pharmaceutical substances intended for human use. It will help choose doses for chronic toxicity study, provide preliminary identification of target organs of toxicity, and occasionally reveal delayed toxicity. It may also aid in the selection of starting doses for Phase 1 clinic study and provide information relevant to acute overdosing in humans.

Acute toxicity testing should be designed under the principle of "random, control, and repeat," and performed according to GLP. Methods for acute toxicity study and determination of median LD (LD_{50}) have recently been reviewed.[21]

Some of the following contents reference the *Guideline of Single Dose Toxicity* (3BS1a) from the European Medical Agent (EMEA, 1987).

7.2.1 Test Sample

The active substance should have the same pattern of impurities as the product to be marketed, when possible. Should the final dosage form be shown to have impurities significantly different either in quantity or quality from those in the test batch, then further steps should be taken to ascertain their possible toxicity. Consideration should be given to the physical characteristics of the substances in relation to the route of administration. Test product can be the crude drug because of the limitation of administration volume or administration method (EMEA, 1987).

For herbal extracts or products, the test sample should be qualified with fingerprints by TLC or HPLC, and quantified for the main or active compounds in the sample.

7.2.2 Animals and Groups

Single-dose toxicity tests must use equal numbers of both sexes. Rodents such as the mouse, rat, and hamster are suitable for the qualitative study of toxic signs and the quantitative determination of the approximate LD. If no difference in response is observed between the animals of the two sexes of the first rodent species, then only animals of one sex need be used in the other acute toxicity studies. Toxic signs should also be observed and recorded in detail for each animal used in the case of other mammals.

Whatever species or strain of animals are selected, it is essential that the following information should be provided: age, sex, weight, origin, the time in the laboratory before test, whether or not the animals are classified as being free of specific pathogens, and whether or not the animals have been vaccinated or submitted to any other procedure. Details of housing and environmental conditions should be given. Access to and the nature of the diet and the availability of water should be stated. All the above factors are known to affect the acute toxicity of substances (EMEA, 1987).

Healthy adult animals are usually used in single-dose toxicity tests. But young animals are recommended if the test sample will be administrated to children.

A control (and/or negative) group should be set except for groups of various test product dosage.

7.2.3 Route and Dosage of Administration

In the case of rodents in general, two routes should be used, and when possible should include those routes proposed for humans. At least one should ensure full access of unchanged substance into circulation. Ensure the animals have an empty stomach when intragastric administration is applied (EMEA, 1987).

The initial sample weight should be no more or less than 20% of the average body weight of the test animals. The volume of the sample solution should not surpass 20 mL/kg for rat and 40 mL/kg for mouse when intragastric administration is applied. The volume of sample solution for other animals and other routes of administration can be decided according to relative documents and practice conditions.

7.2.4 Observation Period and Index

Animals should be observed at regular intervals for 14 days after administration. All signs of toxicity and the time of their first occurrence and their severity, duration, and progression should be recorded. The time and mode of any death should be documented. The observation period should be extended until signs of toxicity become apparent if the toxic reaction appears slowly. Reasons must be presented if the observation period is less than 14 days.

All reactions of the animals should be recorded in the first 4 hours after administration. Beyond that time frame, record any changes every morning and afternoon.

The observation indexes include the progressive loss of weight, inhibition of growth, disorders of drinking and eating, as well as abnormal appearance, behavior, excretion, and excreta. Any signs of toxicity should be presented separately for each animal. Check gross anatomy on the dead animals or the animals on the brink of death and perform histopathological inspection if there are changes of volume, color, or texture on organs.

7.2.5 Data Processing and Analysis

The results from which any calculations have been made should be given in detail; the methods of calculation used should be stated. The toxic effects, including assessment of morbidity, should be described in each species and for each route of administration at all dose levels. The investigator should draw all relevant conclusions from the data obtained in these studies. Any significant deviations from these guidelines should be justified (EMEA, 1987).

7.2.6 Method for Acute Toxicity Testing

Lethal Dose (LD) Determination

LD includes minimal LD (LD_5), median LD (LD_{50}), and maximum LD (LD_{95}), representing the dose that will cause death to 5%, 50%, and 95% of animals, respectively, in the tested group. Among them, the LD_{50} determination is most frequently applied.

There are many methods for LD_{50} determinination.[21] Bliss method is the most commonly applied one. It involves first determining the extent of the lethal dose, then setting several groups within the dose range by geometric proportion; the dose ratio of adjacent groups should be 0.65–0.85, observing the general states of animals and recording the number of death animals during 7–14 days, and calculating the LD_{50} according to mortality. Due to the complexity of the calculation process, a computer program is usually used in practice.[22]

Maximum Tolerance Dose Experiment

Many herbs have no significant toxicity, thus making it impossible to measure their LD_{50}s. In this case, the maximum tolerance dose test should be measured. The maximum tolerance dose test measures the highest dose of a test sample at which no animals will die in a single dose. If it is impossible to measure the maximum tolerance dose by one single administration, multiple (e.g., two to three times) administrations within 24 hours should be measured. The test should apply the route that is used for patients in clinic with the highest concentration and largest volume that the animal can endure.

Maximum Dose Experiment

Maximum dose of administration refers to the maximum dose of a single administration or multiple administrations during 24h of tested substances. Maximum dose experiment refers to the administration of permissive maximum dose tested substances under reasonable concentration and administration volume, and observation of reactions by animals.

Other Methods

Other methods for acute toxicity testing include fixed-dose procedure, approximate lethal dose procedure, dose detection test, expand test, and dose accumulation test.

Acute toxicity study of Chinese herbs usually references the *Guidelines on Acute Toxicity Test for Chinese Medicines and Natural Drugs* (2005, 2nd ed., NO. [Z] GPT2-1) issued from the *State Food and Drug Administration* (SFDA) of China.

7.2.7 General Observations in Acute Toxicity Study

According to the *Guidelines on Acute Toxicity Test for Chinese Medicines and Natural Drugs* from SFDA, observations in acute toxicity studies should cover the following systems. The toxic targets can be speculated based on the indications given in Tables 7.5–7.8.

Table 7.5 Observation on Respiratory System

Indication	Possible related organ, tissue, or system
Dyspnea: dyspnea or being strenuous, pant, general lower breath frequency	
Abdominal breathing: barrier diaphragm breathing, abdominal region slants obviously when respiration	Apneustic center of CNS, costa myoparalysis, cholinergic nerve palsy
Pant: obvious air-breathing sound when deep inspiration	Apneustic center of CNS, edema of lung, accumulation of respiratory passage excreta, increased function of bilineurine
Apnea: brief respiratory arrest appear when forced breathing	Apneustic center of CNS, insufficiency of lung and heart function
Cyanosis: Blue emerge in the tail assembly, mouth, and foot pad	Insufficiency of lung and heart function, edema of lung
Polypnea: Breath is quick and shallow	Apneustic center stimulus, insufficiency of lung and heart function
Nasal discharge: red or achromasy	Edema of lung, bleeding

Table 7.6 Observation on Locomotor System A: Motor Function

Indication	Possible related organ, tissue, or system
Increased or decreased spontaneous activity, spy, hair scratching, or motion	Body movement, CNS
Drowsiness: deep slumber appears, but easy to be awaken and recover normal activity.	Sleep center of CNS
Normal reflect and righting reflex disappear	CNS, sense organ, nerve and muscle
Paralysis: normal reflect and nociceptive reflex disappear	CNS, sense organ
Anochlesia; carus cataleptica	CNS, sense organ, nerve and muscle, autonomic nerve
Incoordination	CNS, sense organ, autonomic nerve
Abnormal activity: spasm, toe gait, pedal, etc	CNS, sense organ, nerve and muscle
Pronation	CNS, sense organ, nerve and muscle
Trepidation	Nerve and muscle, CNS
Fasciculation	Nerve and muscle, CNS, autonomic nerve

Table 7.6 Observation on Locomotor System B: Twitch (Eclampsia)

Indication	Possible related organ, tissue, or system
Clonicity twitches	CNS, respiratory failure, nerve and muscle, autonomic nerve
Rigidity twitches	CNS, respiratory failure, nerve and muscle, autonomic nerve
Clonicity twitches and rigidity twitches	CNS, respiratory failure, nerve and muscle, autonomic nerve
Fainting twiches	CNS, respiratory failure, nerve and muscle, autonomic nerve
Opisthotonos	CNS, respiratory failure, nerve and muscle, autonomic nerve

Table 7.6 Observation on Locomotor System C: Reflex

Indication	Possible related organ, tissue, or system
Cornea catacleisis	Sense organ, nerve and muscle
Basic reflex	Sense organ, nerve and muscle
Righting reflex	CNS, Sense organ, nerve and muscle
Liddell-Sherrington reflex	Sense organ, nerve and muscle
Light reflex	Sense organ, nerve and muscle, autonomic nerve
Startle reflex	Sense organ, nerve and muscle

Table 7.7 Observation Results on Cardiovascular System

Indication	Possible related organ, tissue, or system
Bradycardia	Autonomic nerve, lung and heart dysfunction
Pyknocardia	Autonomic nerve, lung and heart dysfunction
Anapetia	Autonomic nerve, CNS, cardiac output increase, high environmental temperature.
Vasoconstriction	Autonomic nerve, CNS, cardiac output decrease, low environmental temperature.
Anisorhythmia	CNS, autonomic nerve, lung heart function degrade, heart muscle ischemia

Table 7.8 Observation on other Systems

Observation	Indication	Possible related organ, tissue, or system
Palpebra indication	Dacryorrhea: hyperdacryosis	Autonomic nerve
	Myosis	Autonomic nerve
	Mydriasis	Autonomic nerve
	Exophthalmos	Autonomic nerve
	Lapsus palpebrae superioris	Autonomic nerve
	Bloody tears	Autonomic nerve, bleeding, infection
	Superior palpebrarum	Autonomic nerve
	Conjunctiva muddy	Eyes stimulus
Salivary secretion	Polysialia	Autonomic nerve
Piloerection	Arrector muscles of hair in folliculus pili contract	Autonomic nerve
Anodynia	Reactions to pain stimulus degrade	Sense organ, CNS
Muscular tension	Tensile force decrease	Autonomic nerve
	Tensile force increase	Autonomic nerve
Defecation	Solid, arescent, few amounts	Autonomic nerve, constipation, stomach intestine dynamia
	Flowability degrade, watery stool	Autonomic nerve, dysentery, stomach intestine dynamia
Vomit	Vomit or nausea	Sense organ, CNS, autonomic nerve
Hyperdiuresis	Red urine	Kidney damage
	Anischuria	Independence sence organ
Skin	Cutaneous	Stimulus, kidney nonfiction, tissue damage, long-term immobility
	Erythema	Stimulus, inflammation, sensibilization

- Respiratory system: Breath block of apertura narium, changes of breath frequency and depth, changes of body surface color (see Table 7.5)
- Locomotor system: A. Motor function: changes of motion frequency and characteristic; B. Twitch (eclampsia): obvious voluntary muscle contracts with unconsciousness or eclampsia; C. Reflex (see Table 7.5).
- Cardiovascular system: (see Table 7.7)
- Other systems: (see Table 7.8).

7.3 CHRONIC TOXICITY STUDY

Chronic toxicity testing includes sub-chronic toxicity and chronic toxicity. The difference between them is the exposing period and group size of animals. Chronic toxicity testing is relatively expensive and time-consuming. Therefore, great care must be taken in the design, execution, and interpretation of results from such studies.

7.3.1 Sub-Chronic Toxicity Study

The 90-day sub-chronic toxicity study provides information on the possible health hazards likely to arise from repeated administration of a sample over a period of "rapid growth" of the animals into young adulthood, mostly 3–4-month-old mice or rats. These studies give insights into the major toxic effects of a sample and indicate the target organs of toxicity and the possibility of drug accumulation. The results also provide an estimate of a no-observed-adverse-effect level (NOAEL) of exposure, which can be used in selecting dose levels for chronic studies and for establishing safety criteria for human administration. The 3-month toxicity studies precede the 6–12 months chronic toxicity studies.

7.3.2 Chronic Toxicity Study

The objective of 6–12 month chronic toxicity tests is to characterize the profile of a pharmaceutical or chemical substance in a mammalian species following a prolonged and repeated administration period. These studies should generate data to correlate the majority of chronic effects with the test article, and to determine dose-response relationships with those toxicities observed. Ideally, the design and conduction of these tests should allow for the detection of general toxicity including neurological, physiological, biochemical, and hematological effects, and exposure-related morphological (pathology) effects.

The duration of these chronic toxicity studies is usually 6, 9, or 12 months duration in mice or rats by most routes. The primary objective of chronic toxicity studies is to examine the toxic effects and the NOAEL of the test sample by repeated administration for a long period. The examination includes observing symptoms during the administration, and damage and recovery of tissues and organs after administration. These data may then be used in the establishment of a safe daily dosage for human beings.

Different countries have different requirements for chronic toxicity testing, including duration. To follow a single development plan for chronic toxicity testing of new medicinal products and avoid duplication, ICH issued the *Guidance for Duration of Chronic Toxicity Testing in Animals (Rodent and Nonrodent Toxicity Testing)* (S4A) in 1999 on the basis of the review of the testing of chronic toxicity in three regions (the European Union, Japan, and the United States). The following studies are considered acceptable for submission in the three regions: (1) Rodents: A study of 6 months' duration; (2) Non-rodents: A study of 9 months' duration.

7.3.3 Animals and Groups

To obtain more information, two kinds of animals, usually rats for rodent and canines for non-rodent, are preferred for a chronic toxicity study. According to the *Technique Requirements of New Chinese Medicine*, purebred dogs are demanded for a new drug application if it is a preparation of active ingredients from TCM or natural medicine in China.

Ages for the experimental animals are 5–8 weeks and 4–12 months, respectively. The sexual requirement is half male and half female. Other species (rodent or non-rodent) may be used depending on the samples or results from similarly conducted studies.

Laboratory strains of young healthy animals are usually employed and test samples are given as soon as possible after weaning. At the commencement of the study, variation of body weights of the animals should not exceed ±20% of the mean value. The animals should be kept under the experimental housing and feeding conditions for at least 5 days prior to the test. Before the test, healthy young animals are randomized and assigned to the control and treated groups. The number of animals in each group depends on the administration period of the sample. To ensure the efficacy of the result statistics, the minimal number of rats in each group is 20, and of canines 6.

7.3.4 Route, Frequency, Duration, and Dosage of Administration

The designs of route, frequency, duration, and dosage of administration are very important for the success of the chronic toxicity study.

Route of Administration

The choice of the route of administration depends on the physical and chemical characteristics of the test samples. It should be the same as for humans in clinic, if possible. If taken orally in clinic, the sample is administered by gavage for animals, rather than mixed with food. If the sample is taken through intravenous injection in clinic but proves difficult to accomplish for animals, intraperitoneal injection (*i.p.*) is allowed. Some factors, such as pH, irritation to animals, and osmotic pressure should be considered when preparing the solution. If it is still difficult or impossible to perform these two administrations, other routes of administration are allowed.

Frequency and Duration of Administration

Frequency of administration is normally daily. Sometimes, the sample is added into drinking water or mixed into the diet.

The duration of administration depends on the intended administration period of humans in clinic. As mentioned above, a study of 6 months' duration for rodents and a study of 9 months' duration for non-rodents are considered acceptable for the submission of a new drug application in the European Union, Japan, and the United States, according to ICH guidance S4A.

FDA considers 9-month studies in non-rodents acceptable for most drug development programs; shorter studies may be equally acceptable in some circumstances and longer studies may be more appropriate in others, as follows:

- Six-month studies may be acceptable for indications of chronic conditions associated with short-term, intermittent drug exposure, such as bacterial infections, migraine, erectile dysfunction, and herpes.
- Six-month studies may be acceptable for drugs intended for indications for life-threatening diseases for which substantial long-term human clinical data are available, such as cancer chemotherapy in advanced disease, or in adjuvant use.
- Twelve-month studies may be more appropriate for chronically used drugs to be approved on the basis of short-term clinical trials employing efficacy surrogate markers where safety data from humans are limited to short-term exposure, such as some acquired immunodeficiency syndrome (AIDS) therapies.
- Twelve-month studies may be more appropriate for new molecular entities acting at new molecular targets where post-marketing experience is not available for the pharmacological class. Thus, the therapeutic is the first in a pharmacological class for which there is limited human or animal experience on its long-term toxic potential.

For chronic toxicity study of Chinese herbs in China, generally, if the administration of sample in clinic is only 1–3 days, chronic toxicity may not be necessary; if it is less than 1 week, the duration for chronic study should be 2 weeks; if it is less than 2 weeks, the duration should be 4 weeks; if it is more than 4 weeks, then the duration should be doubled. For some medicines that treat chronic diseases, the test must follow the longest experimental period, that is, 6 months chronic toxicity test observation for rats, and 9 months for canine.[23]

Doses of Administration

In general, three different dose levels (high, middle, and low) should be designed. For bigger animals, two levels may be applied. It is better to perform a preliminary test with a few animals to find the right range of each dose level. Animals treated with the highest dosage should display significant toxic reaction, and a few (<20%) of them may die. A dosage of 1/4–1/10 of the LD_{50} may be used as a start. Those treated with median dosage should display slight or moderate toxic reaction. A dosage of 1/10–1/30 of the LD_{50} may be used as a start. Those treated with the lowest dosage should display no toxic reaction. The dosage should be slightly higher than the dosage at which the animals will display significant efficacy or that used for clinical study.[23]

7.3.5 Observation and Evaluation of Results

All animals should be carefully observed and examined at least once each day to detect the onset and progression of any toxic reaction. Any abnormal sign, including neurological and ocular changes as well as mortality, should be recorded for all animals. Time of onset and progression of toxic conditions, including suspected tumors, should be documented.

Appropriate actions should be taken for weak or moribund animals by isolation or sacrifice to minimize loss of animals due to disease, autolysis, or cannibalism. The dead animals should be refrigerated for later necropsy.

Observations are usually conducted by both vision and under microscope. The items include general conditions, death rate, and changes in body weights and organs. Blood and urinary tests are also performed.

General Conditions

General conditions should be observed daily. Any abnormalities in outlook, energy level, behavior, drinking and eating, sleeping habits; or times, shape, and color of stool and urine should be recorded. Body weight should be recorded once a week. All deaths should be recorded with reasons, and the death rate should be calculated.

Blood and Urine Tests

1. *Hematologic test*

Hematologic tests are usually carried out for counts of red blood cell, white blood cell, white blood cell percentage, reticulocyte, hemoglobin, hematocrit, platelet count, and so on. In addition, any other item that relates to the toxicity of the test sample should be performed, and it is recommended that the internationally adopted measurement methods and units be used.

2. *Blood biochemistry test*

This test is generally performed for determination of total protein, A/G ratio, blood sugar, triglyceride, phospholipid, total cholesterol, urea nitrogen, creatinine, uric acid, sodium, potassium, chlorine, calcium, phosphorus, glutamate oxaloacetate transaminase (GOT), glutamate pyruvate transaminase (GPT), lipotropic hormone (LPH), alkaline phosphatase, creatine phosphokinase, γ-guanosine triphosphate (GTP), ornithine decarboxylase, and so on. Just as in the hematologic test, any other item that relates to toxicity of the test sample should be performed, and the internationally adopted measurement methods and units are recommended.

3. *Urine test*

This test generally includes urine volume, photogrammetry, occult blood, total protein, sugar, ketone body, urobilinogen, and bilirubin.

Pathology Exams

Autopsies should be performed on all animals in the experiment (including those that have died or been put down during the experiment). The main organs should be weighed after autopsies. The performance usually starts on the maximum dosage group and control group. If the differences between the maximum dosage group and control group are significant, then the performance should also be carried out on the lower dosage groups. All internal organs and tissues should be carefully observed, including heart, lungs, liver, spleen, kidneys, brain, stomach, duodenum, ileum, colon, pituitary, spinal cord, lymph nodes, urinary bladder, optic nerve, uterus, ovary, gonads, testis, thymus, adrenals, thyroid, prostate, and local tissues which were given medicine. Other organ tissues should be checked, if necessary.

Evaluation of Results

Evaluation and analysis of the results are necessary for chronic toxicity studies. The report should give an objective reflection of the original data and information during the whole experiment, including the occurrence and recovery period of toxic reaction, analysis of the possibility of cumulative toxicity, and the details of toxic symptoms, anatomy inspection, and pathology tests. To discover the target organ of toxicity, synthetic analysis of the related index must be performed. Data from the tests should be scientifically calculated and statistically evaluated. Only results analyzed by statistical method showing significant differences between the sample and control groups are considered toxic or causing pathological change. It is necessary to combine the statistical results with practical clinical meaning.

Much more attention should be paid to the analysis of poisoning or dead animals and the detailed description of toxic reaction, recovery, and death time. The cornerstone of experimental toxicology is pathological examination. Pathological exams usually include visible inflammation, inflammatory cell infiltration, focal hemorrhage, congestion, and suspected cell degeneration. They may be caused by the sample, or not. Comparison between the sample and control groups must be described truthfully. Without evidence, it cannot be said that pathological changes have no relevance to the sample. Results from the pathological exam should have quantitative and semiquantitative standards and use the right statistical methods.

As mentioned before, two kinds of animals are selected for the chronic toxicity studies according to *Technique Requirements of Research of New Chinese Herb*. If pathological, hematological, and biochemical changes are observed from one species of animal but not another, one should not draw a negative conclusion.

Meanwhile, it is not uncommon in chronic toxicity study to find pathological or other changes that occur in low incidence and that are not dose-related but occur only in treated animals. Such reactions may be idiosyncratic in nature or may be due to the hypersensitivity of certain animals. Nevertheless, they deserve special attention since they may be indicative of a hitherto unsuspected toxic effect. The clinical history and other data from such animals should be reviewed with great care and an attempt should be made to determine the reason for the observed changes. Toxic effects that occur in extremely low incidence present special problems in interpretation. There is no substitute for experience in this respect, and the prudent investigator should consult the knowledgeable experts in this field.

7.4 SPECIAL TOXICITY STUDY

The purpose of a special toxicity study is to observe and evaluate whether herbal medicine can cause some special toxicities. To evaluate the safety of new drugs, special toxicity study mainly focuses on mutagenicity, reproductive toxicology, and carcinogenicity. But it also includes drug dependency test, hypersensitive test, skin irritation test, immunotoxicity test, photosensitization test, eye irritation test, and ear irritation test. Most herbal medicines have a long history of use; therefore, their toxicities, including those mentioned in this chapter, have been well discovered through clinic. Thus, it becomes unnecessary to perform specific toxicity test to all new herbal medicines or products. In China, only new drugs in the first category, which include new artificial products of Chinese herbal materials, newly discovered herbs and their products, and compounds isolated from Chinese herbs, are required to undergo specific toxicity tests.

7.4.1 Genotoxicology Study

ICH *Guidances on Genotoxicity Testing and Data Interpretation for Pharmaceuticals Intended for Human Use* (S2[R1]), *Specific Aspects of Regulatory Genotoxicity Tests for Pharmaceuticals* (S2A), and *Genotoxicity: A Standard Battery for Genotoxicity Testing of Pharmaceuticals* (S2B) should be used as references when carcinogenicity study is performed for herbal extracts and compounds isolated from herbs for new drug application.

Genotoxicology study uses cell cultures, usually mammalian, to determine gene mutations, changes in chromosome structure and number, and other gene toxicities caused by medical devices, materials, or their extracts, including herbal extracts. These that exert their adverse effect through interaction with the genetic material (DNA) of cells are known as genotoxicity. Since genotoxicity tests mainly determine gene mutations, these tests are also called mutagenicity tests in some books or web sites. Mutagenicity refers to the capacity of substances to induce mutations and cause permanent genetic alterations. Substances able to induce mutation are called mutagens. By comparison, genotoxicity addresses the adverse effect of mutagenicity.

Based on the damage level of the chemicals to DNA, the type of mutation can be divided into gene mutations and chromosomal aberrations. Gene mutation, also called point mutation, refers to minor DNA injury that involves a single or a few DNA building blocks (DNA base). Chromosome aberrations refer to greater injury such that the structure or number in one or several chromosomes have been changed and can be observed under optical microscope. There is no essential difference between gene mutations and chromosomes aberrations in reference to their mutagenic character. They only differ at the damage level.

The result of chemical mutagenesis depends on the target cell. If the target is a somatic cell, the mutation will only impact the individual him or herself, but not be carried on to the offspring. The most concerning consequence is carcinogenicity. Reports also show that somatic mutations may relate to arteriosclerosis. If the target is an embryonic cell, it will lead to teratogenicity, also known as congenital malformations. There are many *in vitro* and *in vivo* mutagenesis experimental methods using different biological detection systems, but not many are well recognized worldwide.

Genotoxicity tests are performed on drugs, including herbal medicines, to evaluate whether they can induce damage of genetic material and alteration of heritage and to judge what kind of genetic damage or mutation they can cause.

Methods for genotoxicology study can be classified into three subtypes based on the genetic end points: tests to examine gene mutations, tests to exam chromatin mutations, and tests to examine DNA effects (see Tables 7.9–7.11). If any gene in a chromosome is mutated, it can cause the alteration of gene structure and genetic information. Mutations can also generate a new gene, but it is unable to change the location and number of genes in a chromosome. Chromatin mutation examinations examine aberrations of chromosome structure and chromosome number. DNA effect tests examine the alteration of DNA.

Most of the countries have specific guidelines for testing of pharmaceuticals for genotoxicity, and there are regional variations in requirements (methods and procedures). Genotoxicity test guidelines of the ICH and the Organization for Economic Cooperation and Development (OECD) have been used as references in most countries for drafting. The ICH guidelines are more important because they contain pharmaceutically related recommendations. Review on practices and strategies of genotoxicity tests for pharmaceuticals, the limitations of different test systems, and the integration of new test methods are available from literature.[24–26]

Residing in the country where herbal medicine is most widely used, Chinese scientists have performed mutagenesis studies on many herbs. By evaluating results

Table 7.9 Tests for Examination of Mutagenic and Carcinogenic Agents

Test level	Name of test
Prokaryotic microorganism	*Salmonella typhimurium* reverse mutation assay
	Escherichia coli reverse mutation assay
Eukaryotic microorganisms	*Streptomyces gilvosporeus* gene mutation test
	Aspergillus nidulans gene mutation test
	Brewers Yeast gene mutation test
In vitro	Mammalian cells gene mutation test
In vivo	*Drosophila melanogaster* sex-linked recessive lethal test
	Mouse spot test
	Mouse visible specific locus test (to examine the effect of germinal cells)

Table 7.10 Tests for Examination of Chromatin Mutation

Name of test	Purpose
Mammalian cultured cells genetic test	To screen mutagenic and carcinogenic agent
Mammalian bone marrow chromosome aberration test	To screen mutagenic and carcinogenic agent
Mammalian bone marrow micronucleus test	To screen the agent that can interfere mitosis
Drosophila melanogaster heritable translocation test	To examine the agent that can induce germ cells mutation
Rodent dominant lethal test	To evaluate germ cells effect and genetic risk
Mammalian germ cells cytogenetic test	To evaluate germ cells effect and genetic risk
Mouse heritable translocation test	To evaluate germ cells effect and genetic risk

Table 7.11 Tests for Examination of DNA Effect

Name of test	Purpose
Mammalian cultured cells genetic test	To screen mutagenic and carcinogenic agent
Brewers yeast mitotic recombination test	To screen mutagenic and carcinogenic agent
Unscheduled DNA synthesis test	To screen mutagenic and carcinogenic agent
Mammalian cultured cells sister chromatid exchange test	To screen mutagenic and carcinogenic agent

of many mutagenesis methods, four tests are selected in the Examination and Approval of New Drugs for toxicology study of drugs, including Chinese herbs. They are the microbial reverse mutation test, chromosomal aberrations test in mammalian culture cells, rodent micronuclear test, and rodent dominant lethal test. Only these four tests will be briefly introduced in this section. Detailed requirements and procedures for each assay are available in guidelines from organizations such as ICH and OECD from their official web sites or in literature; therefore, they will not be repeated here. Descriptions of more methods are also available from many toxicological books, journals, and web sites.

1. *Bacterial reverse mutation assay (Ames test)*

This test generally uses amino acid requiring strains of *Salmonella typhimurium* or *Escherichia coli* to detect point mutations, which involve substitution, addition, or deletion of one or a few DNA base pairs. In the early 1970s, Prof. Ames from California University recommended five strains of f *S. typhimurium* (TA 1535, TA 1537, TA1358, TA98, and TA100) for reverse mutation test.[27,28] Later in the 1980s, Maron and Ames used TA97 and TA102 to replace the first three strains for better sensitivity. Since then, TA97, TA98, TA100, and TA102 have been used as standard testing strains. The bacterial reverse mutation test is often called the Ames test.[29] The Ames test has been used worldwide as an initial screening to determine the mutagenic potential of new chemicals and drugs, as well as herbal products.[23]

2. *Chromosome aberrations in mammalian cells*

The purpose of the chromosomal aberration test is to verify whether the test samples can cause abnormal changes in structure or number of chromosome. There are two types of aberrations: chromosome and chromatid. A chromatid is one of the two identical copies of DNA making up a replicated chromosome. The majority of chemical-induced aberrations are of the chromatid type, but chromosome type aberrations also may occur.

Four tests are commonly used for chromosomal aberration study: *in vivo* tests on bone marrow cells of rodent animals or mouse spermatogonia, and *in vitro* tests on cultural cells and peripheral blood lymphocytes of mammalian cells. The bone marrow cells *in vivo* test is mostly applied for herbal study in China. For the *in vitro* cultural cell test, the Chinese hamster lung cell (CHL) is found to be more sensitive for chromosomal aberration by screening. Results obtained from the chromosomal aberration using CHL basically parallel those from the Ames test.

3. *The rodent micronucleus test*

The micronucleus is mainly caused by chromosome breakage, which is also the root cause of chromosome aberration. The sensitivity of the micronucleus test is basically consistent with that of the chromosome aberration test. In comparison, the micronucleus test is relatively simple in the technical operation, but some test samples may give false negative results. The NIH mouse is preferred for this test. The bone marrow cells are often used.

4. *Dominant lethal test*

A dominant lethal mutation is one occurring in a germ cell that does not cause dysfunction of the gamete, but which is lethal to the fertilized egg or developing embryo. Dominant lethal test is an *in vivo* test, in which the male animals are first exposed to the test substance, then are mated with nonexposed female animals. Embryonic or fetal death is observed to evaluate the mutagenicity. A positive result indicates that the substance has affected the germ tissue of the test species. Dominant lethals are generally accepted to be the result of chromosomal damage (structural and numerical anomalies), but gene mutations and toxic effects cannot be excluded.

7.4.2 Reproductive Toxicity

The ICH *Guidelines for Industry—Detection of Toxicity to Reproduction for Medicinal Products and Toxicity to Male Fertility* (S5A) and its addendum (S5B) should be used as references when carcinogenicity study is performed for herbal extracts and compounds isolated from herbs for new drug application.

The main purpose of reproductive toxicity is to evaluate the potential effect of the test substances on reproductive function, embryo growth development, and growth development of prenatal and birth initial stage.

Broadly, reproductive toxicity study includes the teratogenicity tests. Deformity could be found in both the reproductive toxicity tests and the teratogenicity tests. But the natures of the occurrences are different. In the general reproductive toxicity tests, the test substances are given to animals before mating, cause the germ cell mutation, and lead to a deformity that is inherited. In the teratogenicity tests, test substances are given to animals during the sensitive period and cause deformity of somatic cell in the period of embryonic organogenesis. This deformity cannot be inherited.

The Chinese Ministry of Health, Japan's Ministry of Health and Welfare, and the U.S. Food and Drug Administration employ three-segment reproduction tests using whole-animal models for reproductive toxicity study, that is, general reproductive toxicity test, teratogenicity, or embryotoxicity test, and peri-postnatal toxicity test. This is different from the multigeneration studies used for food additives (three-generation) in food industry or pesticide contaminant (two- or three-generation) for environmental protection. The three segment reproduction tests include three different tests; basically, only one generation of animal is observed. But the three-generation test is one test in which three generations of animals are observed.

Three segments reproduction toxicity test generally includes general reproductive test (*segment 1*), teratogenicity or embryotoxicity test (*segment 2*), and perinatal toxicity test (*segment 3*). In the first segment, the test medicine is given to animals before copulation. The purpose is to evaluate effect of the medicine on fertilization, reproductive system, and filial generation of germ cells. In the second segment, the medicine is given to animals in organogenetic period, and the purpose is to evaluate the potential fetal toxicity and teratogenecity of medicine. In the third segment, the medicine is given around the perinatal and lactation periods. The purpose is to evaluate the effect of medicine on growth and development of offspring.[30] Among the three tests, the teratogenic test is the most important and widely applied in terms of evaluation of drug safety.

The *general reproductive test* is usually performed on mouse or rat with a minimal test subject number of 20 males and females. The males are treated for 60–80 days before mating, and the females are treated for 14 days before mating; after mating, the treatment continues for the mated females.

The *teratogenicity or embryotoxicity test* is performed on 15–20 pregnant rats or mice, or 8–12 pregnant rabbits. Pregnant animals are treated during the period of organgenesis (days 7–17 for rat, days 6–15 for mice, and days 6–18 for rabbits).

The *peri-postnatal toxicity test* is performed on 15–20 pregnant rats or mice, or 8–12 pregnant rabbits. Pregnant animals are treated from the last gestation through parturition till weaning.

Times of starting and ending treatment may be different in protocols from different countries. There should be two to three test drug groups with different dosages and negative and positive control groups. The route should be as similar to clinical application as possible. Drug administered orally is given by gavage. Observation and details of each test are available from many toxicology books and papers.

The reproductive toxicology tests are required when developing new drugs. But for herbal products, the requirement differs between countries. In China, it is only required for new artificial products of Chinese herbal materials. Again, it is better to follow the protocols from your own country when performing the reproductive toxicology tests for new drug development.[23]

7.4.3 Carcinogenicity Study

ICH guidelines, *Need for Carcinogenicity Studies of Pharmaceuticals* (S1A), *Testing for Carcinogenicity of Pharmaceuticals* (S1B), and *Dose Selection for Carcinogenicity Studies of Pharmaceuticals* (S1C) should be used as references when carcinogenicity study is performed for herbal extracts and compounds isolated from herbs for new drug application.

Cancer is any malignant growth or tumor caused by abnormal and uncontrolled cell division. About 80–85% of human tumors are caused by chemicals. Chemicals that trigger animal and human tumors or increase tumor incidence and mortality are called carcinogens. Identification and test of carcinogens is an important part of toxicology study. Carcinogenicity studies are important for the evaluation of a drug's safety. Since there is no specific method for carcinogenicity study of herbal medicine thus far, methods for Western drugs are mainly referred.[23]

At present, carcinogenicity studies can be classified into three categories, including short-term study, long-term animal carcinogenicity study, and human epidemiology survey.

The *in vitro* short-term carcinogenicity study includes mutagenesis tests and transformation tests. The *in vivo* short-term carcinogenicity study includes mice cutaneous tumor provocation test, mice lung tumor provocation test, rat liver transformation focus provocation test, female rat breast cancer provocation test, and carcinogenic promoting agent test.

Long-term animal carcinogenicity test is usually costly and requires about 2-3 years including pretest. Thus, one should be cautious in making the decision and careful in designing the experiment. For drug development, the performance is only necessary when (1) the structure of the new drug or its metabolite is similar or related to the known carcinogens; (2) cytotoxicity or abnormality of organs or tissues is found in chronic toxicology study of the new drug; or (3) the mutagenicity test of the new drug shows positive results.

For herbal medicine, different countries have different requirements for carcinogenicity studies of herbal products. For example, carcinogenic studies are only required in the following situations in China: (1) the structure of compounds in an herbal medicine is similar to known carcinogen, metabolites, or related to known carcinogens; (2) in long-term toxicity tests, the herbal medicine or its product shows cytotoxic effect or has abnormal promoting effects on some organs or histocytes; (3) for some antitumor herbal medicine, carcinogenicity tests are necessary; (4) the artificial products of herbal medicine (if necessary); (5) newly discovered herbal medicines and their preparations (if necessary); (6) the active single or group of components isolated from herbal medicine and their preparations (if necessary).

7.5 EXAMPLES OF *IN VIVO* TOXICOLOGICAL EXPERIMENTS FOR COMPOUNDS OR EXTRACTS FROM HERBAL MEDICINES

7.5.1 Acute Toxicity of Aristolochic Acid (AA) in Rodents

Introduction

Of the 35,000 medicinal plants used worldwide, the Aristolochia family is one of the oldest species used in herbal medicines. AA is the active component of the extract derived from *Aristolochia* species, for example, *Aristolochia manshuriensis* and *Aristolochia fangchi*, and has been used to treat arthritis, gout, rheumatism, and snake bites. AA has been reported to be responsible for the toxic effects of herbs from *Aristolochia* species, thus attracted considerable attention. Experiments have shown AA possess carcinogenic,[31] mutagenic,[32] nephrotoxic,[33] cytostatic,[34] and anti-fertility effects.[35] Study of its acute toxic effects in rodents[36] is given here as an example.

Materials

Animals The experimental animals were NMRI mice and Wistar rats of both sexes, having body weights of 20 and 200 g, respectively. The animals were housed in plastic cages (Makrolon type III) under conventional laboratory conditions at a room temperature of 22 ± 1°C and a relative humidity of 50 ± 10%. They were given standardized pellet food and drinking water as required. Food was withdrawn for approximately 16 h before oral administration. The acclimatization time was 7 days.

Apparatus Microscopy; balance; surgical instruments

Chemicals AA; hematoxylin; eosin; scarlet red

Methods

1. *Treatment:* AA was given in a single dose as the sodium salt (mixture of 77.24% AAI and 21.18% AAII %) through a rigid gastric tube or intravenously into a tail vein. The dose ranges tested are set out in Table 7.12. The drug was dissolved in distilled water or physiological saline. The volumes in which it was administered were 10 or 20 mL/kg, depending on mode of administration and species. To ascertain dosage in each individual the animals were weighed immediately before treatment.
2. *Observations:* All the rats and mice were kept under observation for the whole day after administration of the drug and were then inspected at least once daily for up to 21 days thereafter. Body weight was measured from Monday to Friday per week for 3 weeks.

Postmortem examinations: Necropsies were carried out as soon as possible after death on all rats and mice that died during the course of the trial. After the macroscopic findings had been recorded, the heart, lungs, thymus, spleen, liver, stomach, duodenum, kidneys, adrenals, testes, epididymis, and ovaries were fixed in formalin for histological processing. Paraffin and frozen sections were stained with hematoxylin and eosin or scarlet red, respectively.

Table 7.12 LD_{50} Values in Rats and Mice after Oral and Intravenous Administration of AA

Species	Sex	n[a]	Mode of administration	Dose range (mg/kg)	LD_{50} (mg/kg)	Survival time (days)
Rats	♂	10	*p.o.*	120–295	203.4	4–12
	♀	10	*i.v.*	150–300	183.9	5–10
	♂	10		62–110	82.5	1–10
	♀	10		38–86	74	4–8
Mice	♂	10	*p.o.*	Oct-70	55.9	1–15
	♀	10	*i.v.*	60–120	106.1	1–15
	♂	10		17–102	38.4	1–13
	♀	10		40–125	70.1	1–8

[a] Number of animals per dose group.

3. *Statistics:* LD_{50} values were calculated from the mortality data from each dose group by means of probit analysis.

Results

1. *Clinical signs:* Administration of AA in large single doses by the intragastric or intravenous routes was followed within a few days by dose-dependent reactions in rats and mice of both sexes. These consisted of sedation, pilo-erection, abnormalities of coordination, dyspnea, and kyphotic posture. In the terminal stages the animals adopted the prone position and were totally apathetic. The animals died within 15 days, the interval depending on the dose. Clinically, the behavior of the survivors had reverted almost entirely to normal by this time. After treatment with AA, the rats and mice lost up to 22% of their body weight within 15 days. In most cases they did not regain their original weight by the end of the experiment.
2. *Mortality rates:* The LD_{50} values for rats and mice are set out in Table 7.12. The figures suggest that AA is slightly more toxic to mice than to rats. The highest LD_{50} after intragastric administration was 203 mg/kg for male rats, while the lowest figure was 56 mg/kg for male mice. The corresponding LD_{50} values for intravenous injection were 83 mg/kg for male rats and 38 mg/kg for male mice. The LD_{50} values for male mice were lower by a factor of approximately 2 than those for females. There were no differences between male and female rats.
3. *Postmortem examinations:* The macroscopic and microscopic changes were largely independent of the mode of administration or species, and are hence described collectively below.
4. *Macroscopical findings:* The predominant necropsy finding among rats and mice dying up to 15 days after treatment was atrophy of lymphoid organs. The kidneys were slightly enlarged and were conspicuously pale, as was the liver surface. The cut surfaces of the kidneys showed a yellowish cortex of soft consistency. Among those animals that died within 5 days after oral administration of AA, the fore-stomach showed severe inflammation and the entire mucosa had separated. The stomach as well as the small intestine contained massive amounts of blood clots. Those animals that died at a later stage showed small focal erosions in the forestomach, together with definite thickening of its wall.
5. *Histological findings:* The organs of the lymphoid system—spleen, thymus, and any lymph nodes fortuitously included in the histological preparations—showed dose-dependent lymphocyte depletion due to cell destruction (Fig. 7.1). In the kidneys there was extensive tubular necrosis throughout the cortex, involving practically every nephron included in the sections (Fig. 7.2). Masses of cell debris and protein casts were present within the renal tubules, more especially in the collecting tubes of the medulla. The changes in the adrenals consisted of loss of the normal sudanophilia of the cortex,

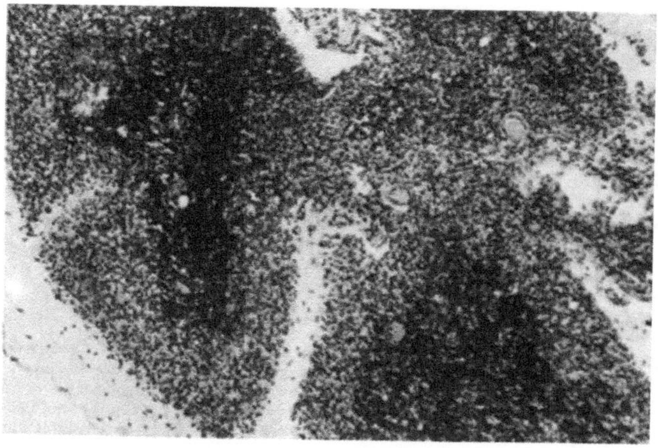

Figure 7.1 Moderate lymphocyte depletion in the cortex of the rat thymus 6 days after oral treatment with 200 mg/kg AA. Magnification: 50×, H.E.

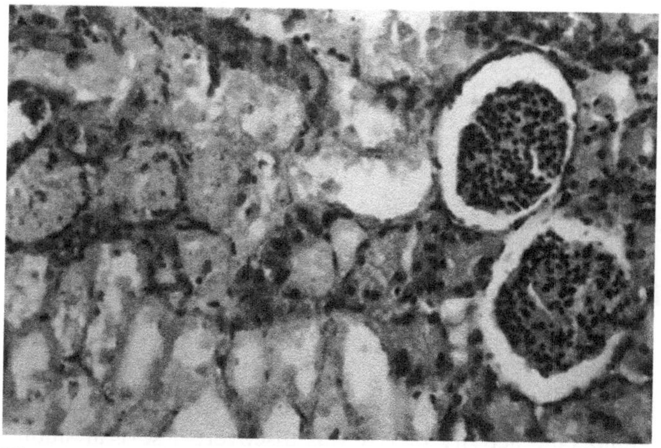

Figure 7.2 Severe tubular necrosis of the rat kidney 6 days after oral treatment with 200 mg/kg AA. Magnification: 250×, H.E.

together with pyknosis of some nuclei and single-cell necrosis. Regressive changes were also found in the liver and in the duodenum. In the testes, spermiogenesis was severely curtailed. Regressive changes in the germinal epithelium were also noted in a few instances. The tubules of the epididymes contained degeneration products of epithelial cells and only a few sperms.

After intragastric administration of AA, the forestomach showed large areas of ulceration with almost total loss of the squamous epithelium and dense infiltration by granulocytes. Among those animals that died 10 days after poisoning or later, the lesions were healed, and the main changes were hyperplasia and hyperkeratosis of the squamous epithelium (Fig. 7.3). In a few animals, the glandular mucosa of the

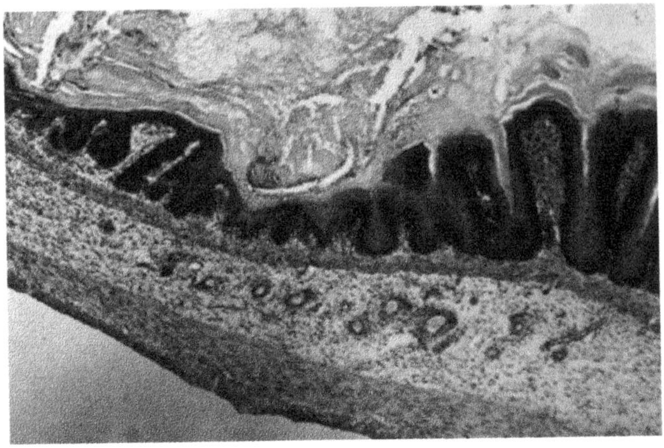

Figure 7.3 Hyperplasia and hyperkeratosis of the rat forestomach epithelium with mononuclear cell infiltration and edema in the submucosa 14 days after oral treatment with 200 mg/kg AA. Magnification: 50×, H. E.

stomach showed superficial gastritis of minor degree. In animals that had received the drug intravenously, there was distinct atrophy of the mucosa of the forestomach, and in some cases of the glandular mucosa as well. Severe necrotic lesions of the hepatic parenchyma were noted after intravenous administration of AA, particularly in mice.

The lungs, heart, and ovaries showed no changes.

7.5.2 Mutagenicity of AA in Bacteria

Introduction

In the early 1980s, two independent studies showed that AA was a potent carcinogen in rodents. After the carcinogenic activity of AA was discovered, several studies were conducted to investigate the mutagenicity and other genotoxic activities of AA. Mutagenicity study of AA in bacteria using modified Ames test[32] is given here as an example.

Materials

Chemicals, Cells, and S9 Activation System AA, benzo[*a*]pyrene, 2-aminoanthracene, 2-nitrofluorene, sodium azide, and 5-trifluorothymidine (TFT); liver microsomal enzymes (S9 homogenate) (the homogenate was prepared from male Sprague–Dawley rats that had been injected [*i.p.*] with Aroclor™ 1254 [200 mg/mL in corn oil] at 500 mg/kg). *S. typhimurium* histidine auxotrophs TA98 and TA100. The tester strains were checked for genetic integrity at the time of use.

Apparatus Colony counter system; incubator

Methods

The bacterial mutation screening test was conducted according to the procedure described by Ames and his colleagues, but only two tester strains of *Salmonella* (TA98 and TA100) were used, and duplicate plates were used for each treatment level. Briefly, the tester strains were exposed to AA via the plate incorporation methodology with and without an exogenous metabolic activation system (S9 mix). AA, tester strain, and S9 mix (when required) were added to molten top agar supplemented with histidine and biotin. After the required components had been added, the mixture was vortexed and overlaid onto the surface of 25 mL of minimal bottom agar contained in a 15 × 100 mm petri dish. After the overlay solidified, the plates were inverted and incubated for 52 ± 4 h at 37 ± 2°C. After incubation, the revertant colonies were counted. For an agent to be considered mutagenic, it had to produce at least a twofold concentration-dependent increase in the mean revertants per plate of at least one of these tester strains over the mean revertants per plate of the appropriate vehicle control.

Results

The mutagenicity of AA in bacteria was evaluated in *Salmonella* tester strains TA98 and TA100 up to a maximal concentration of 1000 µg/plate (Table 7.13). The results show that AA was detected as mutagenic in both TA98 and TA100 tester strains, particularly when tested without S9 mix. The screening version included only two strains of *S. typhimurium*, one that detects base-pair substitution mutations and one that detects frameshift mutations.

7.5.3 Genotoxicity of Crotonaldehyde in Spermatocyte Chromosomal Aberration Assay

Introduction

Some chemical mutagents choose germ cell as targets and cause them to mutate. The mutation of germ cells cannot only affect the parent generation, but also transmit to further generations. Crotonaldehyde is a toxic compound that occurs naturally in meat, fish, in many fruits, some vegetables, herbs, and food products, which has been listed as an "extremely hazardous substance," as defined by the U.S. Emergency Planning and Community Right-to-Know Act. Anand and his colleagues investigated the genotoxicity of crotonaldehyde in the bone marrow and germ cells of laboratory mice by employing bone marrow and spermatocyte chromosomal aberration and dominant lethal mutation assays in Swiss albino mice.[37] Only spermatogonium chromosomal aberration assay for crotonaldehyde is given here as an example of examining its potential genetic effect.

Materials

Animal Male laboratory bred Swiss albino mice (*Mus musculus*)

Table 7.13 Mutagenic Effect of AA in Bacteria (*Salmonella typhimurium*)

Treatment	Dose (μg/plate)	Tester strain	
		TA98 revertant/plate	TA100 revertant/plate
S9 Activation			
Vehicle control		27 ± 1	129 ± 1
Positive control[a]		311 ± 4[b]	1352 ± 108[b]
AA	1.00	21 ± 7	132 ± 6
	3.33	31 ± 9	128 ± 4
	10.0	26 ± 1	136 ± 4
	33.3	28 ± 3	215 ± 5
	100	32 ± 6	405 ± 27[b]
	333	49 ± 6	35 ± 5
	1000	78 ± 9[b]	0 ± 0
Nonactivation			
Vehicle control		16 ± 7	100 ± 7
Positive control[c]		237 ± 12	974 ± 59
AA	1.00	15 ± 8	102 ± 4
	3.33	15 ± 0	95 ± 12
	10.0	17 ± 1	114 ± 5
	33.3	10 ± 1	161 ± 19
	100	25 ± 4	297 ± 13[b]
	333	45 ± 4[b]	7 ± 4
	1000	105 ± 25[b]	0 ± 4

[a] For TA98: benzo[*a*]pyrene at 2.5 μg/plate; for TA100: 2-aminoanthracene at 2.5 μg/plate.

[b] Positive increase (twofold) compared with concurrent vehicle control.

[c] For TA98: 2-nitrofluorene at 1.0 μg/plate; for TA100: sodium azide at 2.0 μg/plate.

Note: No reduction of background lawn was observed at any concentration level tested.

Drugs Crotonaldehyde, mitomycin C, and cyclophosphamide

Methods

1. *Selection of dose:* The doses of crotonaldehyde were selected on the basis of toxicity study. Selected for treatment was 8, 16 and 32 μL/kg b.w. of crotonaldehyde. Chosen for positive control group animals were 1.5 mg/kg b.w. of mitomycin C and 40 mg/kg b.w. of cyclophosphamide.
2. *Procedure:* Male mice weighing approximately 25–30 g (5–6 weeks) were treated with single *i.p.* of 8, 16, and 32 μL/kg b.w. crotonaldehyde solution in olive oil for 24 h. The negative and positive control groups were treated

Table 7.14 Chromosomal Aberrations Induced by Crotonaldehyde in Spermatocytes of Swiss Albino Mice

Treatment	Dose (per kg b.w.)	Number of metaphase with[a]					%Abn. cells (mean ± S,E.)
		AU	SU	AU + SU	Fg	MV	
Olive oil	0.2 mL	1	0	0	6	0	1.04 ± 0.65
CP	25 mg	29	33	8	12	6	17.60 ± 1.45***
Crotonaldehyde	8 μL	2	5	0	2	0	1.80 ± 0.54
	16 μL	6	9	5	9	0	5.80 ± 0.85**
	32 μL	14	15	8	16	6	11.60 ± 1.20***

Olive oil (negative control); CP, cyclophosphamide (positive control); AU, autosomal univalents; SU:,sex chromosomal (X–Y) univalents; Fg, fragments; MV, multivalents having a chain of four chromosomes. *, ** and ***, significantly differ from control at $p < 0.05$, <0.01, and <0.001, respectively, in Student's *t*-test.

[a] 100 diakinesis-metaphase I cells analyzed per animal and 500 per dose.

with 0.2 mL olive oil and 25 mg/kg b.w. cyclophosphamide with the same method as above, respectively. Four hours before sacrifice, the mice were injected with 0.4 mL of freshly prepared colchicine solution (4 mg/kg b.w.). Meiotic chromosome preparations were made according to the air-drying technique. Slides were stained with 7% Giemsa in phosphate buffer (pH 6.8). A group of five mice was used for each treatment and 100 well-spread diakinesis-metaphase I cells per animal were analyzed for chromosome aberrations. The types of aberrations recorded in diakinesis-metaphase I cells include sex-chromosomal univalents (X–Y), autosomal univalents, fragments or breaks, and translocations, that is, multivalent formation.

3. *Data analysis:* For statistical analysis of data on primary spermatocytes, the difference between the control and experimental groups was analyzed by one tailed Student's *t*-test.

Results

In the negative control group of animals (i.e., animals treated with 0.02 ml/kg b.w. olive oil), a very low frequency of chromosomal aberrations (1.04 ± 0.65) was recorded. The percentage of the induced chromosome aberrations in diakinesis-metaphase-I cells showed dose-dependent increase and reached 11.60 ± 1.20 ($p < 0.001$) after *i.p.* with the highest tested dose, that is, 32 μL/kg b.w. Chromosome aberrations observed were autosomal univalents, X–Y univalents, fragments or breaks, and multivalents (i.e., diakinesis-metaphase-I cells with chain of IV) (see Table 7.14).

7.5.4 Reproductive Toxicity of *Cassia occidentalis*

Introduction

C. occidentalis L. (Leguminosae) has long been used as natural medicine in rain forests and other tropical regions for the treatment of inflammation, fever, liver disorders, constipation, worms, fungal infections, ulcers, respiratory infections, snakebite, and as a potent abortifacient. The study example given here investigated the effects of oral subacute administration of *C. occidentalis* on the reproductive process of Wistar rats.[38]

Materials

Animal Pregnant Wistar rats (Rattus norvegicus) aged 4 months and weighing 250–300 g

Drugs The hydroalcoholic extract of the aerial parts (only stems and leaves) of *C. occidentalis*

Methods

The animals were randomly divided into three groups ($n = 7$–10 per group). Observation of the presence of sperm in the vaginal smear was used to establish the first day of pregnancy.

1. *Administration:* Group 1 (control group) received water (1 mL/100 g body weight; *p.o.*), groups 2 and 3 received the extract of *C. occidentalis* at doses of 250 and 500 mg/kg. The administration was given to rats during the preimplantation period (first to sixth day) and organogenic period (seventh to fourteenth day).
2. *Observation:* During pregnancy, the rats were evaluated for survival, altered appearance, and any clinical signs of toxicity, such as changes in food and water intake, piloerection, diarrhea, change of locomotor activity, and vaginal bleeding. The maternal weight was also recorded on days 1, 7, 14, and 20.
3. *Autopsy:* On the twentieth day of pregnancy, the rats were subjected to euthanasia by cervical dislocation, laparotomized, and their uterine horns removed. The number of implants, resorption, and live and dead fetuses was then recorded. Ovaries were weighed and the corpora lutea were counted. The fetuses and placentas were weighed and examined. Any visible and macroscopic fetal abnormalities were recorded.
4. *Data analysis:* The implantation index (total number of implantation sites/total number of corpora lutea × 100), the number of preimplantation (number

of corpora lutea—number of implantations/number of corpora lutea × 100) and the postimplantation loss rate (number of implantations—number of live fetuses/number of implantations × 100) were calculated. All data were expressed as mean ± S.E.M. Statistical analysis was performed using a one-way or two-way repeated measure analysis of variance (ANOVA) followed by the Tukey test. *p*-Values less than 0.05 were considered statistically significant.

Results

The results revealed no statistically significant differences between the control and treated groups in terms of offspring/dam relationship; fetuses, placenta and ovaries weights; number of implantation and resorption sites; number of corpora lutea in the ovaries and pre- and postimplantation loss rates. However, dead fetuses were encountered after treatment, during the periods of preimplantation and organogenesis, with the doses of 500 and 250 mg/kg *C. occidentalis*, respectively (see Tables 7.15 and 7.16).

Table 7.15 Reproductive Parameters of Female Rats Treated with Extract of *Cassia occidentali* (CO) (*p.o.*) from first to sixth Day (Preimplantation Period) of Pregnancy

Reproductive parameters	Control	CO (250 mg/kg)	CO (500 mg/kg)
Number of pregnant rats	8	10	9
Mass gain in the preimplantation period (g)[a]	21.04 ± 1.89	17.70 ± 2.65	16.89 ± 2.63
Number of live fetuses	84	93	91
Number of dead fetuses	0	0	4
Offspring/dam relationship[a]	10.50 ± 1.49	9.30 ± 0.82	10.11 ± 0.87
Fetus mass (g)[a]	2.25 ± 0.10	2.11 ± 0.04	2.44 ± 0.30
Placentae mass (g)[a]	0.46 ± 0.02	0.42 ± 0.02	0.44 ± 0.03
Ovary mass (mg/100 g)[a]	17.89 ± 0.55	17.96 ± 1.17	16.89 ± 1.09
Number of corpora lutea	112	131	110
Corpora lutea/dam relationship[a]	14.00 ± 0.76	13.10 ± 0.59	12.22 ± 0.60
Number of implantation sites	91	106	104
Implantation index (%)[b]	100	87.86	100
Number of resorption sites	7	13	9
Preimplantation loss (%)[b]	0	12.14	0
Postimplantation loss (%)[b]	3.57	7.73	11.11

[a] The values are expressed as mean ± S.E.M. (statistical analysis: ANOVA, $p > 0.05$).

[b] The values are expressed as median (statistical analysis: Kruskal–Wallis test, $p > 0.05$).

Table 7.16 Reproductive Parameters of Female Rats Treated with *Cassia occidentalis* (CO) (*p.o.*) from seventh to fourteenth Day (Organogenic Period) of Pregnancy

Reproductive parameters	Control	CO (250 mg/kg)	CO (500 mg/kg)
Number of pregnant rats	7	8	9
Mass gain in the preimplantation period (g)[a]	19.14 ± 1.55	17.63 ± 2.38	23.11 ± 2.95
Number of live fetuses	70	88	105
Number of dead fetuses	0	2	0
Offspring/dam relationship[a]	10.00 ± 0.87	11.00 ± 0.78	11.67 ± 1.11
Fetus mass (g)[a]	2.23 ± 0.06	2.24 ± 0.08	2.02 ± 0.21
Placentae mass (g)[a]	0.44 ± 0.02	0.45 ± 0.02	0.40 ± 0.02
Ovary mass (mg/100 g)[a]	18.92 ± 1.73	19.13 ± 0.88	18.02 ± 1.33
Number of corpora lutea	79	112	123
Corpora lutea/dam relationship[a]	11.29 ± 0.57	15.38 ± 2.35	13.67 ± 0.76
Number of implantation sites	70	91	106
Implantation index (%)[b]	100	82.95	100
Number of resorption sites	0	3	1
Preimplantation loss (%)[b]	0	17.05	0
Postimplantation loss (%)[b]	0	0	0

[a] The values are expressed as mean ± S.E.M. (statistical analysis: ANOVA, $p > 0.05$).

[b] The values are expressed as median (statistical analysis: Kruskal–Wallis test, $p > 0.05$).

REFERENCES

1. Dai, N., et al. (2007) Gynura root induces hepatic veno-occlusive disease: a case report and review of the literature. *World Journal of Gastroenterology* 13(10):1628–1631.
2. Colson, C.R. and De Broe, M.E. (2005) Kidney injury from alternative medicines. *Advances in Chronic Kidney Disease* 12(3):261–275.
3. Saad, B., et al. (2006) Safety of traditional Arab herbal medicine. *Evidence-based Complementary and Alternative Medicine* 3(4):433–439.
4. Dunnick, J.K., et al. (2007) Cardiotoxicity of ma huang/caffeine or ephedrine/caffeine in a rodent model system. *Toxicologic Pathology* 35(5):657–664.
5. Demma, J., et al. (2009) Potential genotoxicity of plant extracts used in Ethiopian traditional medicine. *Journal of Ethnopharmacology* 122(1):136–142.
6. He, J., et al. (2009) Dual effects of Ginkgo biloba leaf extract on human red blood cells. *Basic & Clinical Pharmacology &Toxicology* 104(2):138–144.
7. Yuan, X., et al. (2009) The Analysis of heavy metals in Chinese Herbal Medicine by flow injection-mercury hydride system and raphite furnace atomic absorption spectrometry. *Phytochemical Anaysis* 20(4):293–297.
8. Wang, L., et al. (2003) Prevention of oxidative injury in PC12 cells by a traditional Chinese medicine, shengmai san, as a model of an antioxidant-based composite formula. *Biological & Pharmaceutical Bulletin* 26(7):1000–1004.
9. Claude, J.R. and Claude, N. (2004) Safety pharmacology: an essential interface of pharmacology and toxicology in the non-clinical assessment of new pharmaceuticals. *Toxicology Letters* 151(1):25–28.

10. VALENTIN, J.P. and HAMMOND, T.J. (2008) Safety and secondary pharmacology: successes, threats, challenges and opportunities. *Journal of Pharmacological and Toxicological Methods* 58(2):77–87.
11. SHAYNE, C. (2003) *Safety Pharmacology in Pharmaceutical Development*. Boca Raton, FL, CRC Press.
12. SULLIVAN, A.T. and KINTER, L.B. (1995) Status of safety pharmacology in the pharmaceutical industry. *Drug Development Research* 35(3):166–172.
13. PORSOLT, R.D., et al. (2002) New perspectives in CNS safety pharmacology. *Fundamental & Clinical Pharmacology* 16(3):197–207.
14. ANDERSON, H. and SPLIID, H. (2000) Statistical analysis of time to event data from preclinical safety pharmacology studies. *Toxicology Mechanisms and Methods* 10(2):111–125.
15. CHAMPEROUX, P., et al. (2009) Calculation of QT shift in non clinical safety pharmacology Studies. *Journal of Pharmacological and Toxicological Methods* 59(2):73–85.
16. LACROIX, P. and PROVOST, D. (2000) Basic safety pharmacology: the cardiovascular system. *Therapie* 55(1):63–69.
17. HAUSER, D.S., et al. (2005) Cardiovascular parameters in anaesthetized guinea pigs: a safety pharmacology screening model. *Journal of Pharmacological and Toxicological Methods* 52(1):106–114.
18. PORSOLT, R.D., et al. (2002) New perspectives in CNS safety pharmacology. *Fundamental & Clinical Pharmacology* 16(3):197–207.
19. MURPHY, D.J. (1994) Safety pharmacology of the respiratory system: techniques and study design. *Drug Development Research* 32(4):237–246.
20. MURPHY, D.J. (2002) Assessment of respiratory function in safety pharmacology. *Fundamental & Clinical Pharmacology* 16(3):183–196.
21. AKHILA, J.S., et al. (2007) Acute toxicity studies and determination of median lethal dose. *Current Science* 93(7):917–920.
22. ROSIELLO, A.P., et al. (1977) Rapid and accurate determination of the median lethal dose (LD_{50}) and its error with a small computer. *Journal of Toxicology and Environmental Health* 3(5–6):797–809.
23. CHEN, Q. (2006) *Research Methodology of Chinese Herbal Pharmacology* (2nd ed.), Beijing, People's Health Publishing House.
24. PURVES, D., et al. (1995) Genotoxicity testing: current practices and strategies used by the pharmaceutical industry. *Mutagenesis* 10(4):297–312.
25. JENA, G.B., et al. (2002) Genotoxicity testing, a regulatory requirement for drug discovery and development: impact of ICH guidelines. *Indian Journal of Pharmacology* 34:86–99.
26. LORGE, E., et al. (2007) Genetic toxicity assessment: employing the best science for human safety evaluation Part IV: a strategy in genotoxicity testing in drug development: some examples. *Toxicological Sciences* 98(1):39–42.
27. AMES, B.N., et al. (1972) Carcinogens as frameshift mutagens: metabolites and derivatives of 2-acetylaminofluorene and other aromatic amine carcinogens. *Proceedings of the National Academy of Sciences of the United States of America* 69(11):3128–3132.
28. AMES, B.N., et al. (1973) An improved bacterial test system for the detection and classification of mutagens and carcinogens. *Proceedings of the National Academy of Sciences of the United States of America* 70(3):782–786.
29. MORTELMANS, K. and ZEIGER, E. (2000) The Ames Salmonella/microsome mutagenicity assay. *Mutation Research* 455(1–2):29–60.
30. DUFFUS, J.H. and WORTH, H.G.J. (1996) *Fundamental Toxicology for Chemists*. Cambridge, Royal Society of Chemistry.
31. MENGS, U. (1982) The carcinogenic action of aristolochic acid in rats. *Archives of Toxicology.* 51:107–119.
32. ZHANG, H., et al. (2004) Application of simplified in vitro screening tests to detect genotoxicity of aristolochic acid. *Food and Chemical Toxicology* 42:2021–2028.
33. MENGS, U., et al. (1993) Renal toxicity of aristolochic acid in rats as an example of nephrotoxicity testing in routine toxicology. *Archives of Toxicology.* 67:307–311.
34. PEZZUTO, J.M., et al. (1988) Evaluation of the mutagenic and cytostatic potential of aristolochic acid (3,4-methylenedioxy-8-methoxy-10-nitrophenanthrene-1-carboxylic acid) and several of its derivatives. *Mutation Research* 206:447–454.

35. Che, C.T., et al. (1984) Studies on Aristolochia III. Isolation and biological evaluation of constituents of Aristolochia indica roots for fertility-regulating activity. *Journal of Natural Products* 47: 331–341.
36. Mengs, U. (1987) Acute toxicity of aristolochic acid in rodents. *Archives of Toxicology.* 59: 328–331.
37. Anand, M., et al. (2007) Genotoxicity of crotonaldehyde in the bone marrow and germ cells of laboratory mice. *Mutation Research* 632:69–77.
38. Aragão, T.P., et al. (2009) Toxicological reproductive study of *Cassia occidentalis* L. in female Wistar rats. *Journal of Ethnopharmacology* 123(1):163–166.

Chapter 8

Clinical Study of Traditional Herbal Medicine

Hong Yuan, Guo-ping Yang, and Zhi-jun Huang

8.1 INTRODUCTION TO CLINICAL TRIALS AND CHALLENGE OF CLINICAL TRIALS ON HERBAL MEDICINES

8.1.1 Significance of Clinical Trial

In health care, clinical trials are conducted to allow safety and efficacy data to be collected for test drugs or devices. More specific to medicine, including herbal medicine, clinical trial is a systematic study of medicines in human subjects (patients or healthy volunteers), intended to validate or reveal their action, adverse reaction, and/or absorption, distribution, metabolism, and excretion, and to identify their efficacy and safety.

In terms of evaluation of efficacy and safety, clinical trial is more significant than preclinical experimental investigations in both modern and traditional herbal medicines. This is because the bioactivity, pharmacodynamic, and toxic results obtained from *in vitro* bioassays and *in vivo* animal studies might be different from the efficacy and toxicity on human bodies, due to the absorption and metabolism as well as species differences between animals and human beings.

According to statistics, it takes on average about 10 years to develop a new drug, from basic investigation to being approved as a new drug for marketing. The average research and development costs for each new drug is about US$300–500 million, of which 70% of the total budget and time is spent on clinical trials.

8.1.2 History Review of Clinical Trial

As early as 1938, the U.S. Congress passed the *Federal Food, Drug, and Cosmetics Act* (FDCA), which stipulated that a clinical trial must be carried out to validate

Traditional Herbal Medicine Research Methods, Edited by Willow J.H. Liu

the safety of drugs, and that safety trial results should be submitted through "new drug approval" procedures prior to entering the market. However, there was no specific and definite regulation on the efficacy evaluation or the clinical study method in this act.

In the 1960s, the thalidomide tragedy shocked the world. At the time, there were no strict regulations and management of clinical trials in European countries. Thus, thalidomide was widely used in Europe and some other countries without any clinical trial. No attention was drawn on this fact until it was observed that thousands of pregnant women taking this drug gave birth to the same type of fetus deformity, resulting in the birth of hundreds of thousands of fetus deformities in 20 or more countries. Fortunately for the United States, this drug was still at the stage of clinical trial (not approved) under the supervision and management of the Food and Drug Administration (FDA); thus only nine fetus deformities were reported. This tragedy elucidated to many governments the importance of passing legislation to better monitor drug development. It was agreed that each drug, prior to marketing approval, must go through safety and efficacy evaluation via clinical trials. Moreover, drug supervision and administration departments were granted the right of approving new drugs and performing mandatory supervisions and inspections.

In the 1970s, questions in the process of clinical trials were gradually discovered in some developed countries, such as the scientificity of the clinical trial method, the reliability of clinical data, and ethical issues. These problems drew the attention of the Council for International Organizations of Medical Sciences (CIOMS) and led to the passing of the "Declaration of Helsinki," a recommended guidance for doctors performing medical research, on the 18th World Medical Assemble (WMA) in Helsinki, Finland, in July 1964. Since then, the declaration has been revised several times, during which principles with detailed provisions for human trials have been provided and revised. These principles laid a solid foundation for the core contents of current Good Clinical Practices (GCP), emphasizing the benefits to subjects/patients, strictly supervising quality control of the whole process of clinical trials, and ensuring that the rights of subjects/patients are protected. In the United States, a series of laws and regulations, such as the Responsibilities of Sponsor and Inspector (1977), the Responsibilities of Researcher (1978), and Protection of Subjects' Interests (1981), were passed, one after another.

Some other countries, like South Korea (1987), Norway (1989), Japan (1989), Canada (1989), Australia (1991), and China (1998) also formulated and promulgated their own GCP. Although the principles of these regulations are the same, the specific details vary. The promulgation and implementation of these regulations have surfaced as one of the most important duties of drug supervision and management departments in many governments all over the world, and are playing increasingly important roles in the protection of human health.

In the early 1990s, based on the GCP issued from each country, the World Health Organization (WHO) formulated a set of "GCP Guiding Principles," coming into effect in 1993. It is applicable to all its member states. Moreover, its six members, the FDA in the United States, Pharmaceutical Research and Manufacturers of America (PRMA), European Union (EU), European Federation of Pharmaceutical

Industries Associations (EFPIA), Ministry of Health, Labor and Welfare (MHLW), and Japan Pharmaceutical Manufacturers Association (JPMA), initiated the International Conference on Harmonization (ICH) of Technical Requirements for Registration of Pharmaceuticals for Human Use in 1991, in Brussels. Since then, the conference has been held every 2 years to discuss the formulation and revision of international standards of GCP and to formulate and perfect the standards and guiding principles of each aspect of the registration of pharmaceuticals for human use (including ICH GCP, Definition and Standard of Rapid Report, and the Content and Format of Clinical Trial Report). Currently, all clinical trials around the world, especially the multicenter of clinical trials in many countries, have adopted the GCP Guiding Principles established by WHO and ICH as the reference standard. All of these have promoted a new age of international standardization of the administration of drug clinical trials. The ICH guidelines about quality, safety, and efficacy of pharmaceuticals are available from the ICH web site: http://www.ich.org/cache/compo/276-254-1.html.

8.1.3 Challenge of Clinical Trial of Herbal Medicines

According to statistics offered by WHO, about 4 billion people in the world have been treated with botanical drugs up to 2007. Traditional herbal medicine or natural product is not only an important source of modern drug development, but is also becoming an important competitive developing direction in the health field worldwide. In June 2004, *Guidance for Industry—Botanical Drug Products* was published by the FDA. This guidance aims to provide suggestions on the regulation of botanical drug products, especially the requirements related to its clinical research and suggestions on the application of phase I, II, or III clinical trials. Moreover, it also specifies the application procedure of registering for botanical drug product and offers suggestions on how a product with complicated chemical composition can meet the FDA's strict requirement for new drug application (NDA).

According to the FDA's definition, botanical drug is the drug that contains only the extracts of plants, algae, or fungi. Different from the majority of conventional Western drugs that have only one chemical structure, botanical drugs are usually complex combinations of many natural chemical compounds. According to FDA regulations, botanical drug is a modernized traditional herbal medicine that has a detailed pharmacological description for its bioactive components.

In 2006, the FDA's approval of Veregen™, a topic ointment produced by MediGen, a German biotechnology company, became the first approved botanical drug for prescription in the United States. This marked a breakthrough of modernization of traditional herbal medicines. MediGene AG, the owner of Veregen, was the first German biotechnology company to have listed products. Although the company was very capable in the development of tea extracts, food, beverage, and health foods, it transferred the patent and technology of Veregen to Epitome, a Canadian pharmaceutical company, due to its lack of experience and resources in the realm of clinical trial of new drugs. In 1997, Epitome purchased the global development right of the tea extract of Veregen through Japan's Mitsui Co., Ltd. In 1999, Epitome transferred the patent of a genital wart treatment with Veregen in dermatology back to MediGene AG. On November 5, 2005, MediGene AG successfully waived for phase I and II clinical trials just 1 year after the implementation of *Guidance for Industry for Botanical Drugs* and directly applied for phase III clinical trial of Veregen for the treatment of *Condyloma acuminatum* in the United States and other countries. The FDA accepted this conventional NDA, and approved the drug to be listed on October 31, 2006. Now, other indications of this drug have entered into phase II clinical trials, and the preliminary clinical data are encouraging. It is the first time that the FDA has approved a multimolecule botanical drug for almost half a century. In the product press conference held by the American Botanical Committee, Freddie Ann Hoffman, a doctor of medicine, botanical drug expert, and former FDA officer, said, "This approval shows that the FDA considered the plant not only as a food or dietary supplement but also as an approved drug."

It has been widely misinterpreted that herbal medicines must be safe and nontoxic because they are natural and have mild action; many countries regulate herbal products as dietary supplements. The reality is that certain types of herbal medicines are not as safe as many people expect, and an improper use of herbal medicine without guidance by experts may easily cause adverse or toxic reaction, or an interaction with other therapeutic drugs.

In fact, Chinese herbs are so diverse that in the first Chinese herbal book, *Shen Nong Ben Cao Jiang* (*The Divine Farmer's Materia Medica Classic*), written 2000 years ago, 365 different kinds of traditional Chinese medicines (TCM) were divided into three groups based on their therapeutic efficacy, safety, and toxicity. Herbs in the first group are generally harmless to humans, with "tonifying or stimulating properties." Herbs in the second group are intended to treat the sick, but with more or less adverse or toxic reaction, depending on the dosage and period of administration. Herbs in the third group have a violent action on physiological functions and are usually poisonous; the principle of utilizing such herbs is the use of toxic substances to attack poison accumulated within the body.

There is still a lack of recognition of the use of herbal medicine. Recently, due to fast growth of consumption of traditional herbal medicine, the toxicity and safety of traditional herbs have become a hot discussion. The absence of strict administration, slipshod and incomplete research, and erroneous marketing guidance are undoubtedly responsible for the concern over the safety of patients who are less than knowledgeable about herbal medicine. WHO has noticed this and begun to pay great attention to the proper use of traditional herbal medicines, especially with regard to their safety.

Tragedy caused by improper use of ephedra is an example that appeals for more strict regulation on dietary supplement from the government. Ephedra has long been used for inducing perspiration and treating respiratory tract congestion in TCM. However, Western countries marketed its extract for weight loss and as a stimulant, leading many people to overuse it, resulting in heart attack and apoplexy (stroke), and even a few deaths. For this reason, ephedra was prohibited in the United States and many European countries. To be fair, the herb ephedra in itself is innocent because it has been safely used under the guidance of TCM for thousands of years. The real culprits are its abused application, misleading marketing, and improper regulation.

It is estimated that nearly 48% of clinical reports on the efficacy of herbs are not reliable due to defect in design or analysis. Recently, most developed countries have placed greater emphasis on the rationality of clinical trial design and data processing in the research of traditional herbal medicine. Some countries have established extremely detailed and rigorous regulations on the products associated with herbal medicine. A series of relevant laws, administrative regulations, and technical guidelines have been formulated recently to repeatedly assure the safety, efficacy, and quality control of herbal products. The following aspects of these issues have been pointed out by the WHO:

1. lacking in methodology of research
2. insufficient in scientific evidence
3. lacking in international or national standard to ensure safety, efficacy, and quality control
4. lacking in appropriate management rules and registration system
5. insufficient in the support of research
6. registration problem of practitioner of traditional medicine.

In the *Guidance for Industry for Botanical Drugs* published by the FDA, emphasis on two aspects in the application of botanical drugs remains unchanged: the requirement of phase III clinical trial (expanded clinical trial) as in pharmaceutical chemicals, and the basic principles of randomized, double-blind, and placebo-controlled (dose-effect relationship research) experiments for the clinical trials. This indicates that the FDA pays great attention to the perfection of clinical development of botanical drugs. Fortunately, the FDA gives lenient standards for preliminary clinical trials for botanical drugs, that is, relatively lax requirements for phase I and II studies. The reason is that botanical drugs have already had a long history of folk use and clinical practices. However, the requirements of safety and efficacy standards for final approval of botanical drugs are the same as regular pharmaceutical chemicals with the same indications.

In the *Guidance*, a large portion of the content is used to elucidate the key points of the scientific clinical trial methods for botanical drugs. This *Guidance* might be a textbook worthy of thorough study and exploration by researchers. Regulations on botanical drugs have become a special focus among administrative laws and regulations, and have formed a special field, which can be called "Worldwide Administrative Regulations on Botanical Drugs."

Similar to *in vivo* pharmacological study of herbal medicine, the challenge of clinical trials of herbal medicine also comes from the complicated chemical composition in herbs, which makes its quality control and standardization for clinical trial difficult to handle. The chemical composition and concentrations of biological chemical compounds of the same species may vary not only between materials collected in different times and from different regions, but also among extracts from different manufacturers. Therefore, clinical trials of herbal medicine, from beginning to end, should use the same batch of product whose chemical fingerprints from high-performance liquid chromatography (HPLC) or thin layer chromatography (TLC), as well as concentrations of its major or bioactive compounds, have been well analyzed and documented.

An additional challenge for clinical trials of herbal medicine also comes from the design and performance of the trial. The population, size, and criteria of the subjects selected; the course and route of administration of the test sample; the method used to observe or monitor the result; and the statistic methods of analysis of the data will all impact the result. Another big difference between modern drugs and botanical drugs is that the latter are complex mixtures. Most of the bioactive compounds in herbs have milder effects than modern drugs and are less concentrated in mixtures of herbal extracts; thus, it usually takes longer to see the effect in animals or human beings. Therefore, the treatment course of the test herbal medicine in trials should be based on its clinical practice, rather than determined by that of the positive control drug.

It is quite often that clinical trials for the same herbal medicine performed by different research groups give different results. For example, products of St. John's wort (*Hypericum perforatum*) have been widely marketed as an antidepressant dietary supplement. A very recent review showed that many clinical trial studies supported this application.[1] However, a National Institutes of Health (NIH)-funded study shows that an extract of St. John's wort was no more effective for treating major depression of moderate severity than a placebo.[2] Such heterogeneous results may have resulted from differences of material collection, methods of extraction, as well as designs (sizes, controls, courses, observation index, etc.) of clinical trials, and so on.

8.2 ESSENTIAL ELEMENTS AND PRINCIPLES OF CLINICAL TRIAL DESIGN

Clinical trial is very important from the perspectives of safety, efficacy, and capital investment. The successful development of both modern and botanic drugs ultimately relies upon tests on human beings. Clinical trials should be cautiously designed not only to ensure that the efficacy and safety are being fairly evaluated, but also to prevent missing the occurrence of serious adverse events or toxicity in future clinical applications.

A scientific, thoughtful, and detailed plan must be established for a successful clinical trial according to the purpose of the research, before any stages of clinical trial, so as to guide the trials to move smoothly and to obtain accurate and reliable results with less manpower, material, and time. This plan is called clinical trial design.

First, the purpose of the clinical trial should be clearly determined. The clinical trial of a botanical drug should be designed to discover and validate its therapeutic efficacy, safety, or toxicity. The goal of the trial should be set up based on data from preclinical study; it should be careful to avoid being too broad or incomplete.

Variation of results from clinical trials of one drug may be caused by many factors. Therefore, when designing a clinical trial, the following questions must be answered:

1. Which population will be selected as study subjects?
2. What is the proper sample size in the study?
3. Which drug is used as a suitable comparator?
4. What kinds of factors of the study drug will be investigated?
5. How will the experimental effect be determined?
6. How will the group study subjects be scientifically grouped?
7. How will the proper type of trial design be determined?

To answer these questions, researchers must clearly know the basic concepts related to trial design, which include three elements of trial design (study subject, experimental factors, and experimental effect), four principles of trial design (randomization, control, replication, and homogeneity), and types of trial design. This section mainly introduces the three elements and four principles. Detailed introductions about design or research of clinical trials are available in the literature[3–6] and on web sites. The web sites mentioned here are the pages of clinical trials from the U.S. NIH (http://clinicaltrials.gov). Readers can find much important information about clinical trials, such as definitions used in clinical study and investigator instructions.

8.2.1 Three Elements of Clinical Trial Design

The "study subject, experimental factor, and experimental effect" are the three elements of experimental design. For example, when observing the therapeutic effects of an antihypertensive agent on hypertension patients, the test agent is an experimental factor; the patients with high blood pressure are the study subjects; and the value change of blood pressure before and after medication is the experimental effect. Together, they constitute the three elements of clinical trial design. None of them can be dismissed.

Experimental Factors

The experimental conditions that researchers emphasize for investigation are called *experimental factors*. Take the trial for antihypertension medicine as an example. The value changes of blood pressure of hypertensive patients before and after treatment with different doses of an antihypertensive drug are investigated. In this study, five different test dosages of the antihypertensive drug are called experimental factors, while the specific five doses are called the five different levels. In this trial, the difference in body weight and gender of patients might exert strong influence on study results. Therefore, in order to achieve more accurate assessment of the test agent, even patients with different body weights and genders should be grouped together to eliminate the impact of weight and gender in statistical analysis. Influential factors like the weight and gender of hypertensive patients are not the investigated experimental factors, they are called *nonexperimental factors*.

Generally, many factors can influence experimental results. The investigators are unable and do not need to study all of the influential factors. To maximally reduce the self-condition influence on experimental results from the subjects, it is very important to researchers, with professional knowledge and experimental conditions, to find out and control the nonexperimental factors during trial design. These factors are also called block factors. If gender is chosen as a block factor, then the different genders are different levels. The difference in arrangement and control of experimental factors and block factors gives rise to different types of trial design.

Pay attention to the features of the disease for treatment, such as its onset time and natural course, the characteristics of the positive drug, and its therapeutic effect. If the design of observation time or course of treatment in the clinical trial is improper, features of the studied drug will fail to be fully displayed.

Design of Observation Time Points If the indication of a test drug is for a disease characterized with acute onset and a short course, it is generally required that patients be enrolled and given the positive drug within 24 h. If the enrollment of patients and administration of the test drug is within 72 h in the clinical trial design, then, features of the test drug are difficult to evaluate because the delay of patient enrollment and observation time points lead to the differences of onset time and action intensity between the test drug and positive drug.

Design of Course of Treatment The course of treatment should be designed rationally in accordance with the study purpose and based on a specific disease. For an example, first-degree burns can automatically recover in 2 or 3 days. If the treatment course of a test drug for the treatment of first degree burn is designed for 10 days, then obviously, this design is not reasonable. In another case, although some clinical research reports revealed that the therapeutic effects of drugs for Alzheimer's disease could be observed in 3 months or so, the treatment course of a test drug for this disease designed for 3 months seems not long enough in a trial for efficacy assessment.

Mistakes as mentioned above are commonly seen in clinical trial designs of botanical drugs. Sometimes, several defects and drawbacks may be found in one design, making the efficacy and safety of the test drug unable to be assessed correctly; consequently, this leads to not only a huge waste of manpower, time, and medical and financial resources, but also a wrong conclusion.

Study Subjects

In clinical trials, the study subjects refer to people who receive the treatment. Sometimes, the subject may be the *ex vivo* specimens or cells of the human body.

For instance, both the patients of laryngeal cancer and the sliced specimens from their cancer tissue are the study subjects. But usually the patient of laryngeal cancer is considered as the study unit rather than the tissue slice. In ophthalmologic research, if one patient has disease in one eye, then the study subject is either the patient or the diseased eye, but if both eyes have problems, then it is more proper to take each diseased eye as a study subject.

Different studies require different type of subjects. The inclusion and exclusion criteria of study subjects should be established before the study so as to assure homogeneity. In clinical studies, in addition to type of disease, pathology and course of disease, race, region, sex, weight, age, health condition and family history, social factors (profession, living condition, and financial status), and psychological factors of the study subjects should also be taken into consideration.

The total number of study subjects in a complete trial design is called "sample size." Determination of the sample size depends on the specific type of clinical trial. A sample that is too small will easily lead to an incorrect conclusion due to statistical bias. A sample that is too large not only costs more, but also makes the data processing more difficult due to increased nonexperimental factors.

Experimental Effect

Experimental effect refers to the effect of the test drug or herbal medicine on the study subjects, which is reflected by *observation index*. Based on the attribute, the observation index can be divided into objective index and subjective index.

In a clinical trial of a new antilipidemic drug for the treatment of hyperlipidemia patients, changes of triglycerol and cholesterol levels before and after medication are quantitatively measured. The triglycerol and cholesterol levels are considered as the *objective index*.

In another clinical trial of a new psychotropic drug, the patients' feelings, such as nervousness and anxiety before and after medication, are recorded to evaluate the drug's therapeutic effect. The patients' feelings are considered as the *subjective index*.

Some indexes are objective, but their judgment is easily influenced by the subjective feelings of the observer, such as the diagnosis through reading of the same X-ray films by different doctors. The X-ray reading is considered as the *semi-objective index*.

The selection of observation index directly determines the success of the study. The precondition as an observation index is that the index employed in the study must have essential connections with the study purpose and be able to faithfully and accurately reflect the effect of experimental factors. For example, to correctly evaluate the efficacy of an anti-hepatitis B virus (HBV) drug in the clinical trial, the alteration of HBV DNA level of patients rather than the change of transaminase concentration should be measured as the observed index, because the transaminase is nonspecific to type B hepatitis.

The observation index should be of a certain sensitivity and specificity. Relatively speaking, sensitivity refers to the capability of detecting the true positive, while specificity is the ability to identify the true negative. High sensitivity and specificity embodies the availability of the index. For example, although the electrocardiogram (ECG) and coronary angiography (CAG) are both used for diagnosing coronary heart disease in clinic, the ECG can only be taken as an auxiliary index due to its decreased sensitivity and specificity as compared with CAG.

If available, the objective index is preferably selected as the observation index. The measurement of the index requires exactness, that is, accuracy and precision. Accuracy refers to the proximal degree of measured value to the true value, and reflects the systematic error. Precision refers to the proximal degree of measured value to mean value, and is expressed by coefficient of variability, standard deviation, and variance in statistics.

8.2.2 Four Basic Principles of Clinical Trial Design

In order to accurately reveal experimental effects on the study objects produced by experimental factors and the change patterns of the observation index reflected from different subjects, a concrete and scientific trial design is necessary to achieve the goal of the study. To make the clinical trial design more scientific, in addition to considering the three factors mentioned above, the following four basic principles must be strictly complied with: control, randomization, replication, and homogeneity.[7]

Controls

The establishment of a control group is the basis of comparison. Control studies are intended to discriminate whether there is a significant statistical difference between the test drug and the positive control in terms of the therapeutic effects. Therapeutic effects through clinical treatment may be achieved by the test drug or some other factor, such as psychological effect. Therefore, control studies are adopted to minimize the influence of other factors to the utmost degree.

By comparing the data from test sample groups with the data from the control group, the influence of experimental factors and nonexperimental factors on experimental effects, that is, the effect of the test drug, can be identified and isolated. There are often many factors influencing experimental results in a single study, but usually only one or a few experimental factors are investigated. Establishment of the control group can help make a reliable estimation on the experimental factors when nonexperimental factors are the same in both the experimental group and control group. Many diseases, such as upper respiratory tract infection and dental ulcer, could actually be relieved naturally. In this case, whether the effects of the treatment in a trial are from the experimental factors or natural turnover can only be judged by comparison.

In the ICH *Harmonized Tripartite Guideline—Choice of Control Group and Related Issue in Clinical Trial* (E10) (2000), five types of controls are introduced: the placebo control, no treatment control, dose-response control, active (positive) control, and external control (including historical control). In the following introduced controls, all except the external (including historical) control are concurrent, meaning that the control group and test groups are chosen from the same population and treated concurrently.

1. *Placebo control*

A placebo is a substance or procedure a patient accepts as medicine or therapy, but which has no verifiable therapeutic activity. About 30% of the human population is susceptible to the placebo effect. In placebo controlled trials, some participants take a placebo while others receive the therapy being investigated. The use of a placebo (fake treatment) allows the researchers to isolate the effect of the study treatment. Placebo control studies can reliably justify the therapeutic effects of the test drug and evaluate its "absolute" efficacy and safety, minimize the effect of expectations of subjects and investigators, and thus make the comparison more accurate.

Although placebo control studies can check the "absolute" efficacy and safety of the studied drug, they are unable to provide valuable information obtained by comparison of therapeutic effects. If there is a positive drug available, it is better to add it in the trial. Otherwise, it is difficult to evaluate the efficacy of the test drug merely using placebo control.

Provided there is no serious harm, the setting of placebo is noncoercive, and patients are fully informed about possible consequences of the delay of treatment; it is generally considered ethical to ask patients to participate in a placebo-controlled trial, even if they may experience discomfort as a result. Sometimes, however, a delay may give rise to ethical issues, except for when a new treatment is tested for a disease that no effective treatment is available. In order to protect the interests of study subjects, sometimes baseline treatment is added on to placebo-controlled trials. Superiority design is employed in placebo-controlled studies.

2. *Blank control (no-treatment concurrent control)*

Blank control refers to a control group that receives no treatment. It is called *no-treatment concurrent control* in ICH-E10. This kind of study reflects the self-changing nature of the study object during the experiment. It is frequently used in animal experiments and clinical trials for chronic diseases.

3. *Standard control (active or positive control)*

In order to compare the efficacy of a test drug, the existing widely accepted drug of the same type is often chosen as a standard control for comparison. It is called active (positive) concurrent control in ICH-E10.

The selection of the positive drug is critical in design of clinical trials. An improper choice of positive drug will bring unfavorable influences on the evaluation of efficacy and safety of the studied drug. The selection of the positive drug should refer to the previously reported clinical trials using the same drug as a positive control. The function and indication of the selected positive comparator drug should be the same as that of the studied drug, as well as its function and classification.

The first choice for active control drug selection should be a drug with an identical mechanism of action as the test one. If it is not available, then a drug with similar mechanism of action as the test one should be chosen. If neither of these is available, a drug for treatment of the same indication should be selected as active control drug. The positive drug should be an officially approved, marketed drug that has been commonly used in clinic with definite therapeutic effect. It must be used for the approved indication at the specified dosage.

The active control study has advantages in ethical and practical concerns. It can be easily performed in a trial with a larger sample size and it provides more safety information. But positive control study is unable to directly measure the absolute efficacy of the test drug and it is difficult to quantify the safety profile. The trial may need a larger sample size to carefully estimate the therapeutic effects of the comparator drug. Non-inferiority or superiority design may be employed in positive control studies, depending on specific cases.

A scientific clinical trial also means that the population with targeted indication in the study is comparable with that of intend-to-perform studies. Under certain circumstances, therapeutic effects of a positive drug may not always be observed as expected or be replicated. This is frequently seen when the indication(s) of the disease state has greater variability; when placebo contributes a substantial improvement or change of disease state; or when the diseases lack sufficiently objective therapeutic indexes, such as depression, anxiety, dementia, and angina. In these cases, sensitivity of the observing efficacy and detection of the index are very critical to the success of a clinical trial. It is recommended to employ a three-arm-study design, which includes study drug, positive drug, and placebo control groups.

4. *Dose-response control*

Subjects may either be placed on their fixed dose initially or be raised to that dose gradually. But the intended comparison is between groups on their final doses.

5. *Self control*

Each subject is treated as both a control and a study subject. For instance, to evaluate the efficacy of a drug treating tinea on feet and hands, the tested drug can be used on one arm and the conventional drug as a positive control in the other arm, if both arms of a patient have it. This allows the study to compare the therapeutic effects of these two drugs. In clinic, the comparison of therapeutic effect before and after medication is also considered as self-control.

6. *Experimental control*

Sometimes experimental factors are mingled with important nonexperimental factors. In such cases, blank control alone is unable to well assess the result, and an experimental control group that exclusively includes the nonexperimental factor should be added. For example, in order to observe the nutritive value of bread containing diaminocaproic acid, in addition to giving a certain amount of bread containing diaminocaproic acid to people in the experimental group in the morning and afternoon, the same amount of bread containing no diaminocaproic acid is given to people in the control group. The latter is called an experimental control.

7. *External control, historical control, and baseline control*

According to ICH-E10, an *external control* trial compares treatment results between a group of subjects receiving the test treatment and a group of patients who are from another population with a different treatment, which can be a group of patients treated at an earlier time (historical control) or a group during the same time period but in another setting.

Historical control study uses previous experimental results of the same kind for comparison with the current experimental results. It is usually not suitable to be adopted. For example, in a report, 34 children patients (22 male and 12 female) of combined severe pulmonary hypertension congenital heart disease received a consecutive operational treatment from November 1996 to March 2000. Patients received treatment from November 1996 to August 1998 were used as a historical control group, while those who received treatment from August 1998 to March 2000 were used as lung protection group, according to operation time. The percentage of blood vessel endothelium eNOS, tracheal epithelium iNOS, and macrophagocyte iNOS were observed. Obviously, results from such study are not reliable.

Baseline-control study is a kind of clinical study performed using the baseline values as control, in which the status of subjects (e.g., blood pressure or tumor size) in therapy is compared with the status before treatment. The baseline has several meanings, depending on the application: (1) information gathered at the beginning of a study from which variations found in the study are measured; (2) a known value or quantity with which an unknown is compared when measured or assessed; (3) the initial time point in a clinical trial, just before a participant starts to receive the experimental treatment that is being tested. At this reference point, measurable values such as CD_4 count are recorded. Safety and efficacy of a drug are often determined by monitoring changes from the baseline values.

A proper choice of control should be made in accordance with its phase of study and study purpose. In clinical trial design, multiple control groups, that is, more than one type of control, are often selected to remove the impact of some important nonexperimental factors. For example, in order to investigate the changes of blood pressure value of hypertensive patients before and after treatment with different doses of hypotensive drug, four groups are established in the experimental design: blank control group, positive control group of conventional drug, treatment group of dose 1 and treatment group of dose 2. Hypertensive patients are equally affected by such important nonexperimental factors as weight, gender, and age. The setup of the blank control group is to reflect the state of blood pressure of patients without any medication; the positive control group is to reflect the decrease of blood pressure after medication of a known hypotensive drug. Treatment groups of two doses form a mutual control to each other. Without positive control, even though changes of blood pressure are statistically significant between treatment groups of different doses, the results can only indicate that the test drug has different impacts on blood pressure with different doses, but these data are insufficient to justify whether there is a statistically significant difference of therapeutic effects between the test drug and the known hypotensive drug. Without the blank control, it is unable to explain how the factor "medication or not" influences experimental results.

The selection of controls for clinical trials of botanical drugs is suggested to refer to three-arm-study, that is, the positive drug group, placebo group, and test drug group. It is a design with many superiorities, especially suitable for diseases that are susceptible to psychological factors, such as depressive disorder, senile dementia, and other diseases. The positive control drugs and placebo are also commonly used in clinical trials for TCM.

It is important to mention that a control group often refers to one level group among several levels in a single experimental factor design. If it is a multifactor study design, there should be more than one control group, forming mutual controls to each other.

Replication

Due to the existence of influential factors, such as individual difference, the same treatment on different subjects will result in different effects. As a consequence, the values of some specific indexes may be varied. Only under the condition that a large number of experiments are repeated can the actual effects of the treatment be revealed. Therefore, the replication of experiments must be performed in studies.

"Replication" usually has three meanings in a study: replication of experiments, replication of sampling, and replication of measurements.

Replication of experiments means that a certain number of samples (study subjects) are observed repeatedly under the same experimental condition, in order to reduce the possible experimental errors caused by individual differences.

Replication of sampling refers to obtaining many specimens from the same subject or sample at the same time, with the purpose of understanding the distribution of a quantitative observation index in subjects or samples.

Replication of measurements means that the same sample (or study subject) is measured under partly or totally different experimental conditions to observe the change of results. The purpose of such replication is to learn an overall reflection of the dynamic alteration of the observation indexes, thus excluding the influence of experimental condition on the observation results.

It should be noticed that an extra-large sample size would cause a waste of manpower and material resources, and create difficulty in controlling nonexperimental factors. Meanwhile, a sample size that is too small would easily result in the wrong conclusion. Therefore, it is better to generate a reasonable estimation based on the experimental and statistic requirements in specific cases.

Randomization

A randomized controlled trial means each study subject is randomly assigned to receive either the treatment or a placebo. In a randomized trial, patients in two groups are evenly allocated without the subjects' input. Randomly assigning a single subject into a treatment or control group will assure the populations are similar in the two groups. Such design can provide the most compelling evidence that the treatment produces the expected effect on human health.

Randomization is often employed in statistics to keep the distribution of nonexperimental factors even in both the experimental and control groups. The randomization should be implemented in runs throughout the whole trial. It is mainly embodied in the following three aspects.

- Each individual has the same probability to be selected as a sample when sampling or grouping, so as to ensure the representativeness of samples.
- Each individual has the same probability of being assigned into any group, so as to ensure the homogeneity of samples among groups and the comparability of each group.
- Each individual has the same probability to be treated, so as to eliminate the influence caused by order of treatment during study.

Randomization can reduce the influence of systematic errors, thus avoiding systematic differences between groups. There are many ways to achieve randomization, like lot drawing, checking random digital tables or random arrangement tables, and the pseudo-random number generated by computer or the specialized software Randal O. Simply randomly assigning only eliminates error of allocation, rather than subjective and objective biases. Thus, the implementation of randomized, controlled, and double-blinded studies can make the reliability of studies conspicuously higher.

Randomized, *double-blind*, and *placebo*-control is performed in some phase II and most phase III drug trials. "Blind" in clinical trial means that the subjects involved in the study do not know which treatment they receive. Blinding is intended to minimize the potential biases resulting from management, treatment, assessment, or interpretation of results from subjects or investigators knowing the assigned treatment, thus ensuring that objective assessments and evaluations are not affected by knowledge of treatment assignment.

In a *single-blind trial*, only the subjects do not know whether they receive test treatment or placebo. In a *double-blind trial*, neither the subjects nor the researchers involved know which subject is given the treatment. If a physician knew which patient is getting the treatment and which is getting the placebo, he/she might be tempted to give the (presumably helpful) test drug to a patient who could more easily benefit from it, or give extra care to the patients who receive the placebos to compensate for their ineffectiveness. A form of double-blind study called a "double-dummy" design allows additional insurance against bias or placebo effect. In this kind of study, all patients are given both placebo and active doses in alternating periods of time during the study.

Homogeneity

Homogeneity requires balance of nonexperimental factors in both the experimental groups and control groups. A good homogeneity between groups must be maintained during the study so that the comparison can be achievable. For example, in a study observing the vimentin expression in epithelium lentis cells of senile cataract patients, investigators choose the crystalline lens of diseased patients (age range between 65 and 83 years old) as an experimental group, and the crystalline lens of young people (age ranges between 25 and 35 years old) who died of no disease but accidents as a control group. Comparison of the expressed contents of vimentin between crystalline lens of the two groups leads to the conclusion that the contents of vimentin in senile cataract crystalline lens of old patients are lower than that of young people. The design of this study ignored the fact that the fiber of crystalline lens gradually ages when people get older; thus, the vimentin expression of epithelial cell will change whether the subject is a cataractous patient or not. The comparison here is improper because people in one group are older than those in another, but age is an important nonexperimental factor in this study. This is an obviously a nonhomogeneous example. It illustrates that the effect of an experimental factor on observation index can be truly revealed only when the nonhomogeneity influence of other experimental factors or some important nonexperimental factors are excluded in a clinical study.

8.3 DESIGN OF CLINICAL TRIALS

The result of a clinical trial depends on many factors in its design, such as different number of subjects. As an example, if 3 kinds of adverse reactions of a test drug occur in a small-scale phase I human tolerance test, 6 kinds may appear in the following large-scale phase II human therapeutic effect test; 11 kinds may be in a larger-scale phase III human clinical trial; and even more possible adverse effects (AEs) may emerge in phase IV trial after being widely marketed. Therefore, the scientific design of a clinical trial is very important from both the perspective of safety and efficacy and wise capital investment. The clinical trial should be conducted with caution based on thoughtful protocol to prevent the occurrence of serious adverse events and the emergence of ineffective or even harmful drugs. Similar to those of the modern drug, the purpose of clinical trials for herbal medicine is to scientifically confirm therapeutic effects and safety even though many of them have been long used in practice.

A successful clinical trial for a drug or an herbal medicine needs a detailed, considerate, and scientific protocol that is carefully planned before the trial starts. The design should not only reference the ICH guidelines related to clinical trial, such as *Structure and Content of Clinical Study Reports* (ICH-E3), *Good Clinical Practice: Consolidated Guidance* (ICH-E6), *Statistical Principles for Clinical Trials* (ICH-E9), and *Choice of Control Group and Related Issues in Clinical Trials* (ICH-E10), but also follow the requirements or standards of the relevant government departments or agents.

8.3.1 Clinical Trial Protocol

A clinical trial protocol describes the scientific rationale, objective(s), designs of the procedures and methods, statistical analysis, and organization, including the administrators and the team of researchers. It is supposed to contain a precise study plan with details about study population, control group, treatment procedure (dose, course, and measurement), statistic methods, and so on, for executing the clinical trial. It is important not only to assure the safety and health of the trial subjects, but also to provide an exact template for trial implementations that all involved investigators perform the study in exactly the same way. The format and content of clinical trial protocols sponsored by pharmaceutical, biotechnology, or medical device companies, or universities in the United States, EU, or Japan has been standardized to follow ICH *Guideline for Good Clinical Practice* (E6) issued in 2002.

8.3.2 Ethical Issues

When a new treatment is tested, ethical problems might arise, especially when a placebo is used. Even though most herbal medicine products have been practiced for hundreds or even thousands of years and their safety and toxicities are well known, the FDA in the United States still requires clinical trials if the product is submitted for approval as a botanical drug. Just as in the application of other new treatments, ethical issues have to be considered. A signed "informed consent" and approval of clinical trial protocol are procedures that must be performed before starting the clinical trial.

Recruiting study subjects is an essential part of the process of initiating a clinical trial. A document called "informed consent" is required to be signed by the subjects. This is a legally defined process wherein a person is told about key facts about being involved in a clinical trial before deciding whether or not to participate. The document should include trial details, such as trial purpose, duration, required procedures, risks, potential benefits, and key contacts. Participants decide whether or not to sign the document in agreement after fully acknowledging the possible benefits and risks. They have the right to withdraw at any time.

In the United States, the clinical trial protocol and "informed consent" must be approved by an institutional review board (IRB), also known as an independent ethics committee (IEC) or ethical review board (ERB). An IRB performs critical oversight functions to ensure that research conducted on human subjects is scientific, ethical, and regulatory. It is a committee whose task is to ensure that the rights and welfare of human subjects are protected in all medical, behavioral, and social sciences research. The FDA, Department of Health and Human Services (HHS), and Office for Human Research Protections (OHRP) have empowered IRBs to approve, require modifications (to secure approval) for, or disapprove research. An IRB must review and approve research involving human subjects prior to the initiation of the research. It is the responsibility of the IRB to determine whether proposed research exposes subjects to unreasonable or unnecessary risks, to review "informed consent" forms and process, and to monitor the progress of research. In its deliberations, the IRB will use the ethical principles as detailed in the Belmont Report (1979) to make its determination.

8.3.3 Statistics in Clinical Trials

The goal of a clinical trial is mostly to obtain a statistically significant results showing a significant difference in outcome between the control group(s) and treatment groups. According to the *Guidance for Industry—Statistical Principles for Clinical Trials* (E9) issued jointly by the FDA, HHS, and ICH, the results of a trial depend on the sample size (number of participating patients or volunteers), types of controls, statistic data analysis, and so on.

The number of patients depends on the individual trial, the test drug, the disease the drug treats for, and the number and type of control groups selected. It should have enough participants to give a statistically significant result.

The number of patients enrolled in a study has a large bearing on the ability of the study to reliably detect the size of the effect of the study intervention. This is described as the "power" of the trial. The larger the sample size or number of participants in the trial, the greater the statistical power. However, when designing a clinical trial, this consideration must be balanced with the fact that more patients make for a more expensive trial. The power of a trial is not a single, unique value; it estimates the ability of a trial to detect a difference of a particular size (or larger) between the treated (tested drug/device) and control (placebo or standard treatment) groups. For example, a trial of a lipid-lowering drug versus placebo with 100 patients in each group might have a power of 0.90 to detect a difference between patients receiving study drug and patients receiving placebo of 10 mg/dL or more, but only have a power of 0.70 to detect a difference of 5 mg/dL.

8.3.4 Group Design

An appropriate grouping method should be chosen in a clinical trial based on study purposes. Examples of study design include parallel group, crossover, and factorial designs. In *parallel group design*, the treatment and the control (or the two treatments) are applied simultaneously to *two separate groups of subjects*. It is the most commonly used method of design for confirmatory clinical trials. *Crossover design* is a self-control method, during which each subject is randomly arranged in the treatment orders of two or more therapeutic regimen, fro example, getting the treatment and then the control, or the control and then the treatment, in sequence. An adequately long wash-out period should be given to avoid the retarding effects of any previous therapeutic regimen. *Factorial design* is used to concurrently evaluate two or more therapeutic drugs through different treatment combinations; it is chiefly used to assess the interactions of different combinational factors, such as compound preparation.

8.3.5 Superiority and Non-Inferiority Design

The clinical trial is an important investigation to explore and confirm the safety and efficacy of the study drug in a population with targeted indication. Based on study purposes, the design of clinical trials can be chiefly classified into two categories: superiority design and non-inferiority design.

A clinical study of superiority design is intended to demonstrate that one treatment is more effective than another, that is, to assess whether the efficacy and safety of the test drug is superior to any control treatment (placebo, no treatment, low dose of test drug, or active control). In comparison with a clinical trial of non-inferiority design, the design, implementation, and result analysis of clinical trial of superiority design are comparatively simple and easy to operate, and this design is commonly applied in the exploratory and confirmatory studies of new drugs.

A clinical study of non-inferiority design is intended to demonstrate that a treatment is at least not appreciably worse than another, that is, to assess the relative efficacy, safety, and benefits/risks of two therapeutic drugs. If efficacy of the test drug is not lower than the preestablished margin (delta) when compared with the active control drug with known efficacy, then it is claimed that the test drug is non-inferior to the active control drug, and its efficacy is equivalent to the known efficacy of positive control drug. A non-inferiority design is often applied in the following situations.

1. The marketed drugs have favorable efficacy that the test drug is less likely to surpass. For example, non-inferiority design is often employed in clinical trials of new antibiotics.
2. The test drugs are therapeutically equivalent to the active control drug, but they may have some other features, such as better safety, or may bring additional benefits to patients besides therapeutic effects, such as convenience or better patient compliance. For example, a test lipid-lowering drug cannot only reduce the level of low-density lipoprotein (LDL), but also increase the level of high-density lipoprotein (HDL), having an overall lipid-regulatory function.
3. Ethical considerations: to the population with life-threatening diseases, if there are safe and effective drugs or treatments available as active controls, then only placebo-controlled trials would not be a good choice. Active control drug or standard treatment should be used to confirm that efficacy of the test drug is non-inferior to active control treatment. Meanwhile, the test drug can bring additional benefits to patients.

Different countries hold different opinions on the acceptance of this kind of non-inferiority design. Some argue that, to the population with life-threatening diseases, the therapeutic effects of the investigational drug are remarkably important, and its efficacy should be superior to the marketed products; thus, non-inferiority design is unacceptable.

4. There are other conditions where non-inferiority design is applicable. An example is the comparison of therapeutic effects between a compound drug and its individual high-concentrated single agent.

Non-inferiority trials are based on the important hypothesis that therapeutic effects of a study drug are non-inferior to the active control drug or nondrug standard treatment. (In view of the fact that only in rare occasions is a study drug compared with nondrug standard treatment in clinical studies, the following discussion will not take it into consideration.) This hypothesis stems from the basis that the efficacy of the active drug is clearly shown in clinical trial.

As a matter of fact, many factors will impact the design, implementation, and result analysis in non-inferior clinical study. The following key points should be addressed when designing the trial.

1. *Selection of active control drug:* The drug used as an active control should be chosen carefully, as the purpose of non-inferiority trials is to validate the efficacy of the study drug is non-inferior to the active control drug.
2. *Establishment of non-inferiority margin (delta):* The margin is the degree of inferiority of the study drug to the active control drug the trial will attempt to exclude statistically. If the confidence interval of the difference between the study drug and the positive control drug excludes a degree of inferiority of the study drug equivalent to or larger than the margin, then the study drug can be declared non-inferior; if the confidence interval includes a difference equivalent to the margin, then the study drug could not be declared non-inferior, thus is not considered effective. As the determination of the margin is based upon the superiority of the design of the active drug in historical control, it is critical to select an appropriate active control drug with supportive historical study results, which is to say, collection and analysis of the historical study results are indispensable. If no information is available from historical evidence for determining the margin, special caution should be given when applying non-inferiority design in clinical trials. This is because the design may be unable to perform. It is strictly forbidden to make up a margin without the support of sufficient research data.
3. *Assay sensitivity and consistency:* According to ICH-E10, assay sensitivity of a clinical trial is defined as the ability to distinguish an effective treatment from a less effective treatment. Assay sensitivity is important in any trial but has different implications for trials intended to show differences between treatments (superiority trials) and trials intended to show non-inferiority (ICH Steering Committee, 2000). The assay sensitivity in a non-inferiority trial is required, as the actual or absolute efficacy of the active control drug is not measured during study, thus there is no internal standard to measure or assure assay sensitivity. As a result, the assay sensitivity of a study is based upon the inferences and assumptions from the historically controlled studies as well as the evidence of appropriate conduct of trials.

Details about design and issues of non-inferiority trials can reference the article.[8]

8.3.6 Phases in Clinical Trial

In the United States, generally before human studies can begin, an Investigational New Drug (IND) must be submitted to the FDA, containing information on any anticipated risks based on the results of pharmacologic and toxicological data collected during studies of the drug in animals. The studies are designed to permit the selection of a safe starting dose for humans, to gain an understanding of which organs may be the targets of toxicity, to estimate the margin of safety between a clinical and a toxic dose, and to predict pharmacokinetic and pharmacodynamic parameters.

According to the *Exploratory IND Studies, Guidance for Industry, Investigators, and Reviewers* (a draft in 2006) from the FDA, preclinical and clinical approaches, as well as chemistry, manufacturing, and controls information should be considered when planning exploratory studies in humans.

Preclinical studies for new drug development and functional evaluation of herbal medicines involve *in vitro* experiments generally using enzymes, receptors, genes, cells, study subjects, and *in situ* or *in vivo* experiments using animals to obtain preliminary efficacy, toxicity, and pharmacokinetic data of drug candidates or herbal extracts. Pharmaceutical companies use these research results to decide whether a drug candidate has scientific potential for further development. For herbal medicines, researchers evaluate their mechanisms of functions and toxicity, and if there is significant efficacy, they develop botanical drugs based on the results from the preclinical research. Drug development is a logical, stepwise procedure in which information from small early studies is used to support and plan larger, more definitive studies.

According to the United States National Library of Medicine (NLM) of the NIH, clinical trials are conducted in a series of steps, called phases; each phase is designed to answer a separate research question. There are usually four phases.

1. *Phase I:* Researchers test a new drug or treatment in a small group of people for the first time to evaluate its safety, determine a safe dosage range, and identify side effects.
2. *Phase II:* The drug or treatment is given to a larger group of people to see if it is effective and to further evaluate its safety.
3. *Phase III:* The drug or treatment is given to large groups of people to confirm its effectiveness, monitor side effects, compare it to commonly used treatments, and collect information that will allow the drug or treatment to be used safely.
4. *Phase IV:* Studies are done after the drug or treatment has been marketed to gather information on the drug's effect in various populations and any side effects associated with long-term use.

Since the goals are different in each phase study, the sample sizes are also different. Specific requirements for NDA vary in different countries. In China, population recruited for phase I study is 20–30, phase II 100, phase III 300, and phase IV 2000.

For traditional herbal medicines that have long been used by folks or commonly prescribed to patients by practitioners, different countries have different requirements for their application as drugs. According to the *Guidance for Industry for*

Botanical Drugs published by the FDA (2004), clinical trial for the application of a botanical drug registration can directly start on phase III in the United States. Therefore, the term "clinical trials" as applied to herbal medicine is most commonly associated with the large, randomized studies typical of phase III. Even so, all the phases of the clinical trial are introduced here.

Phase I Trials

Phase I trial is the initial study to evaluate the clinical pharmacology and safety in human subjects. Normally a small (20–80) group of healthy volunteers is selected. It focuses on assessments of the preliminary safety and tolerability, and gives instructions to dosage of administration and dosage regimen for later studies. Tolerance test of phase I trial is to evaluate the preliminary safety of the test drug in the human body. Pharmacokinetics experiments are performed in phase I for the same purpose.

Through phase I trials, the following information can be determined: the tolerable dose range of the test drug in the human body; the basis for dose selection in phase II trials; the administration frequency and administration time in phase II trials; and expectation of adverse reactions and dose ranges in phase II trials.

Assessment of activity and potential therapeutic effects is not the major objective of phase I trials, because they will be further investigated in following studies. However, if drug activity is readily measurable with a short duration of drug exposure in patients at this early stage, it is proper to conduct this type of study in phase I trial.

Phase I trial is usually open and baseline-controlled. Sometimes, it may also be randomized, blinded, or placebo-controlled in order to improve observation results.

Subjects Phase I trial is usually conducted in healthy volunteers, the nontherapeutic objectives. However, for genotoxic potential drugs, patients are used as study subjects. Age difference of subjects should be less than 10 years, half male and half female. Physical examinations are performed on subjects, and no serious functional impairment should be found in the heart, liver, kidney, and blood system. Pregnant women and children should be excluded, unless special pediatric requirements apply.

Doses Initial dose of the study drug shall be determined by experienced clinical pharmacological researchers and clinicians in accordance with animal pharmacodynamic and toxicity study data or the dose of similar products. However, this principle should not be applied mechanically, for there is an overt species differences between human beings and animals. The maximum dose should be preestablished and based upon animal toxicity study results or the dose of similar products, and it should be at least equivalent or slightly higher than the upper limit of clinical dose. Intergroup dose difference depends upon drug toxicity and investigator's experience. Usually, there are 3–5 dose groups, with 5–6 subjects in each group, and 8–10 subjects for dose groups in the vicinity of clinical dose.

Route of Administration It should be determined by the dosage form of investigated drug, medication purpose, and the route of administration applied in preclinical pharmacological and toxicity studies. It should be consistent with the preestablished route of administration in clinic. Herbal products should reference the clinical application of these herbs that the product is made from.

Clinical Observation Indexes In addition to conventional clinical symptoms, physical signs and laboratory examinations, toxic target organs in animal studies, target organs potentially or already appearing with toxicity, or toxic target organs caused by similar products should also be added as the observation index.

Duration If no adverse reactions are exhibited at maximum dose, then the study can be terminated. If serious adverse events show up during dose escalation, the study should be suspended even if specified maximum tolerance dose is not reached. In multiple dose tolerance studies, duration of a study drug with a short half-life is usually around 1 week, and that with a long half-life is often 2 weeks.

Single Ascending Dose (SAD) Study SAD studies are those in which small groups of people are given a single dose of the drug, and are observed and tested for a period of time. If they do not exhibit any adverse side effects, and the pharmacokinetic data are within the safe levels, the dose is escalated and a new group of people is then given a higher dose. This is continued until pre-calculated pharmacokinetic safety levels are reached, or intolerable side effects start showing up.

Multiple Ascending Dose (MAD) Study In these studies, a group of people receives multiple low doses of the drug. Samples of blood and other fluids are collected at various time points and analyzed to understand how the drug is processed within the body. The dose is subsequently escalated for further groups, up to a predetermined level.

Food Effect A short trial of food effect is designed to investigate any differences in absorption caused by eating pre-dose, and its effect on the pharmacokinetic profile. These studies are usually run as a crossover study, with volunteers given two identical doses of the drug on different occasions: one while fasting, and one after being fed.

Tolerance Test In the tolerance test, the observation should start from low dose group and gradually increase to next dose group in the condition that the previous dose is safe and well tolerated. One subject can only receive a single dose. It is not allowed to conduct a single-dose tolerance test in parallel with MAD study on the same subject. Single-dose crossover tolerance study is absolutely prohibited. The tolerance test shall be conducted under the guidance of experienced clinical physicians in a medical facility with conditions for careful observation and timely emergency rescue. Maximum tolerated dose (MTD) refers to the point at which intolerable side effects start showing up.

In phase I trials, pharmacokinetic study is intended to understand the kinetic features of the absorption, distribution, and elimination of the tested drug in the human body, so as to provide instructions to design a rational dosage regimen and to provide a theoretical basis for safe use of the tested drug in phase II trials. Usually, three dose groups of single dose and one group of multiple dose pharmacokinetic study are carried out. For oral dosage form, a trial shall be designed to investigate food effect on drug absorption.

Phase II Trials

Phase II trials are designed for the preliminary assessment of efficacy of the test drug. The objective is to initially evaluate the therapeutic actions and safety in patients of the target disease, to determine potential study end points, therapeutic regimens, and target populations (e.g., comparison of mild with severe disease), and to provide rationale for the design of phase III trials and dosage regimen. The number of subjects for phase II trial is 100–300.

Briefly, phase II trials have two characteristics: (1) exploration: exploration of therapeutic effects and indication; (2) stages: at the early stage (IIA), initial assessment of dose-response relationship through dose escalation design, which is called dose-ranging study; at a later stage (IIB), confirmation of dose-response relationship between study drug and indication through parallel dose-response design. In this phase, subjects are selected with narrow criteria, and the therapeutic regimen may not be established.

In phase II trials, when therapeutic effect and/or safety of study drug are compared with placebo, the absolute effective rate of the study drug is obtained; when compared with an active comparator, the relative effective rate is obtained, which is to say, phase II trials can be varied in design according to specific study purposes, such as randomized, blinded, or controlled clinical studies.

Subjects Phase II trials include the exploratory research of therapeutic effects and safety of study drugs to specified indications, as well as the probe study of indications (such as type of tumors). Typically, subjects in phase II trials are enrolled in accordance to comparatively narrow inclusion criteria. They represent a homogeneous population and should be under close monitoring.

Doses During a dose exploratory study, a randomized and parallel dose-response study with three or more doses is a widely recognized and successful, but not exclusive, method to obtain data of the average dose-response relationship of a population. With this method, the relationship between clinical benefits or untoward effects and dosage or concentration of the study drug can be established if appropriate test doses are selected. For a more comprehensive dose-response relationship study, additional exploratory study should be performed to evaluate the influence of demographic differences (age, gender, race, etc.) among subsets of data on dose-response relationship. To reach the goal, it is crucial to understand whether there are pharmacokinetic (PK) differences (such as differences of metabolism, body shape, and body composition) among groups. The marketing authorization of new drug requires results from dose-response relationship studies with rational design.

Dose exploratory study should provide the following benefits: establishment of initial therapeutic dose; establishment of rational and therapeutic effect-guided dosage regulation procedures and intervals, and proper adjustment according to characteristics of patients; and establishment of a therapeutic dose, above which no more benefits will be observed or adverse reaction will be increased.

The approval of a study drug depends upon the overall review of related documents. Although dose-response relationship information should be practicable, based on the types and degrees of efficacy of the study drug, incomplete data are acceptable under the condition that further studies will be performed after approval. Just like the information of therapeutic effects of drugs on special populations, long-term administration, potential drug-drug interaction, or drug-disease information, dose-response relationship information is expected; the timeline to submit dose-response relationship data may be allowed to be prolonged if significant therapeutic benefits have been received, or extreme low toxicity has been ensured.

Phase III Trials

Phase III trials are designed for a confirmatory assessment of the therapeutic efficacy of the test drug or an additional new function of a current drug, or functional evaluation of an herbal medicine. The purpose is to further validate that the study drug is safe and effective for use in the intended indication and recipient population. Because of the different purpose, as a comparison, phase II can be called an exploratory study and phase III can be called a confirmatory study.

Phase III clinical studies should be performed with sufficient number of subjects (1000–3000) depending upon the disease to be investigated, the objective of the study, and the study end points. The most accepted phase III trial is the randomized, controlled, double-blinded study with parallel group design. Therefore, phase III is the most expensive, time-consuming, and difficult trial to design and run, especially when the test drugs or herbs are for chronic diseases.

Based upon phase II clinical trials, phase III studies are conducted to further confirm the therapeutic effects and safety profile, finalize a clear indication, and determine a stable therapeutic regimen. These studies carried out in phase III will complete the last piece of information needed to support the package insert of the drug and provide an adequate basis for marketing approval. The following aspects should be considered during trial design.

Route of Administration It is noted that when route of administration is pre-established, it shall be consistent with that in phase II trials. Dosage and course of treatment are not required to be fixed in exploratory studies, but are usually established in confirmatory studies.

Indication Sufficient evidence should be provided to establish indication of the study. Indication of a new test drug should be based on preclinical animal studies. For herbal medicine, it should base on the clinical application of the test herbs or formula.

Subject Selecting Criteria Inclusion criteria should consider patient population, special diagnostic standards, and special disease requirements. These criteria should be widely recognized in clinic. Exclusion criteria should be of specificity and rationale. Predetermined reasons to exclude patients during treatment or analysis of observational outcomes should be explicitly listed. Although it is very important to establish subject inclusion criteria, more emphasis should be put on the implementation of these standards in clinical study operations.

Statistical Assessments Statistical assessments should be performed with sufficient samples (patients) based on the expected magnitude of the treatment effect, the variability of the data, the specified probability of error, and the desire for information on subsets of the population or secondary end points. In some circumstances, a larger database may be needed to establish the safety of a drug. It should be noted that the number of subjects in a clinical trial should always be large enough to provide a reliable answer to the questions addressed.

Efficacy Observation Index Efficacy observation index should be defined in advance. The index that reflects status changes of patients must be selected. Specific observation methods and quantitative methods should be available, and the observation method should be objective and timely. There are primary and secondary clinical observation indexes. The result of the primary index should reflect clinical effects and be selected in accordance with the major purpose of investigation, while the result of the secondary index is used to assess other actions of the study drug, which may be correlated or irrelevant to the primary one.

In addition, routine and special examinations should be comprehensively performed by following a fixed schedule. Specific efficacy judgment criteria and time (such as clinic recognition of antineoplastic agent) should be clearly established. Meanwhile, lab test indexes should also be covered in the clinical study protocol. Course of medication; test time points before, during, and after course of treatment; and time for suspension, stop, and termination of clinical study should be clearly defined. Personnel training is required to secure observation of subjective efficacy indexes, while test reagents are critical to ensure determination of the lab test indexes.

Safety Observation Index Besides the general items to be observed, the safety observation index should be tightly focused on the toxic effects and special target organs suggested in animal studies, as well as the toxic target organs of similar products.

Multi-center Clinical Trials A multicenter research trial is a clinical trial conducted at more than one medical center or clinic. Most large clinical trials, particularly phase III, are conducted at several clinical research centers. Through multicenter trials with a larger number of participants at different geographic locations, the inclusion of a wider range of population groups becomes easier. The ability to compare results among centers increases the generalizability of the study.

In a multicenter trial, a large number of study subjects are enrolled and study drugs are tested in a broader range of clinical settings; thus the result is more representative for future application. The involvement of a great number of investigators also gives the potential for a wider range of clinical judgment concerning the value of the investigational drug. If the trial is conducted in a number of different countries, it will facilitate generalizability even further. To some special diseases, multicenter trials are a practical option to collect sufficient subjects within limited time. Multicenter study may also establish a better foundation for the popularization and application of the study results, thus serving as an effective way to evaluate new investigational medicinal products.

A multicenter trial requires the design of a unified protocol and a standardized method in all study centers. Training of personnel should be performed in advance of the trial. A sound protocol should maximally minimize the degree of heterogeneity in all study centers. Subject assignment in each study site should be as identical as possible, including its numbers.

In many cases, efficacy may vary significantly between population groups with "demographic" factors, such as different genetic, environmental, and ethnic or cultural backgrounds. Normally only geographically dispersed trials can properly evaluate this.

Phase IV Trials

Phase IV trials are performed by sponsors after new drug approval. Studies in phase IV are intended to assess therapeutic effects and adverse reactions of study drugs in a wider population, to evaluate the benefits and risks in general or specific populations, and to optimize dosage of administration.

Phase IV studies mainly focus on the approved indication, but side effects are also investigated. In fact, some side effects may become noticeable after several years' clinical application; the recent withdrawal of Vioxx is an example. It is important for optimizing the drug's application. The investigation may take any form, but should have reasonable scientific objectives, such as drug-drug interaction, dose-response, safety studies, or other studies that support application under the approved indication, mortality/morbidity studies, and epidemiological studies.

8.4 EXAMPLES OF CLINICAL TRIALS

8.4.1 Phase I Tolerance Test of Morusin Capsule

Morusin Capsule is manufactured from the extract of Cortex Mori, a Chinese herbal medicine from the inner layer of the root bark of *Morus alba* L. The major components in the extract are azacyclo-carbohydrate type of compounds, in which the concentration of 1-deoxynojirimycin exceeds 50% of the extract. Pharmacodynamic studies have shown that this extract exhibited significant therapeutic effects on rats of type II diabetes induced by streptozotocin and high-calorie feed. The phase I tolerance test reported here was conducted for new drug registration in China.

The purpose of this study was to complete a tolerance test of Morusin Capsule in the human body and to provide references for the safety and effective dosage regimen for phase II trials through the observational outcomes of the single-dose and multiple-dose tolerance studies of Morusin Capsule in healthy volunteers.

Trial Design

Single Dose Tolerance Study A randomized, blinded, and placebo-controlled single dose escalation study is performed in a study center. Thirty-three healthy volunteers, half male and half female, are screened and assigned into five dose groups with 6, 8, 8, 8, and 8 subjects in each group, respectively. Two subjects in each group are randomly selected to take placeboes, and the rest are administered with study drug. *Dose escalation plan*: the initial dose, low dose, medium dose, high dose, and maximum dose are, respectively, 25 mg, 50 mg, 100 mg, 150 mg, and 200 mg of extract of Cortex Mori (42.5 mg of total alkaloids in 50 mg of extract). Escalation to next dose is only permitted on the condition that the previous dose is observed to be safe and well tolerated. One subject only receives a single dose.

Multiple-Dose Tolerance Study This study is a randomized, blinded, and placebo-controlled study conducted in a study center. Twenty-four healthy volunteers, half male and half female, are screened and intended to be allocated into two dose groups with 7 days' successive administration, three times per day. The initial dose in multiple-dose tolerance study is the dose second to maximum tolerance dose in single dose tolerance study (the dose without occurrence of obvious moderate or serious adverse reactions). If moderate or serious adverse reactions appear after the first dose, then the dose third to "maximum tolerance dose" in single-dose tolerance study is used as the second dose in the multiple-dose tolerance study. If no conspicuous moderate or serious adverse reactions appear, then "maximum tolerance dose" is taken as the second dose in the multiple-dose tolerance study.

Observation Indexes Vital sign, ECG, routine blood (including blood platelets count) and urine tests, hepatic function, renal function, blood clotting function, blood glucose, blood lipid, adverse events.

Data Processing and Statistical Analysis Baseline information of selected subjects was statistically analyzed. Physical and laboratory examination indexes before, between, and after drug administration were also measured and analyzed. The significant difference in the average value and standard deviation of test indexes, as well as the *t*-test results of test indexes before and after administration in each dose group were calculated. Tolerance of subjects in different dose groups was studied and the relationship between administration dose and frequency as well as severity of adverse reactions was evaluated.

Result

All of the 38 subjects in 5 dose groups finished the single-dose tolerance study according to study protocol, and all of the 24 subjects enrolled in the multiple-dose tolerance study completed the study, being given 150 mg and 200 mg of study drug in accordance with study protocol.

Adverse Events Among all healthy volunteers, 14 of them (8 male and 6 female) experienced mild adverse events characterized by dizziness, abdominal distension, diarrhea, nausea, bitter taste of mouth, debilitation, mild premature pulse, and hypolekocytosis. Ten of the 14 subjects were administered with study drug, and 4 of them were given a placebo. One of them was in the maximum dose group of single dose tolerance study, 5 of them were in high-dose group of multiple-dose tolerance study, and 8 of them from maximum-dose group of multiple-dose tolerance study.

Analysis of outcome shows that the relationship between adverse events and dose of study drug is as follows: abdominal distension, diarrhea, and nausea are probably relevant to study drug; bitter taste of mouth, debilitation, and mild premature pulse may not be relevant to the study drug; dizziness and hypolekocytosis are absolutely irrelevant to the study drug.

Vital Signs Study results indicated that the study drug has no obvious influence on blood pressure, heart rate, body temperature, and respiration.

Laboratory Tests One subject (administered with placebo) in maximum-dose group of multiple-dose tolerance study showed significant abnormal results in lab test. However, the rest of subjects were normal or abnormal without clinical significance.

Conclusion In the clinical trial of Morusin Capsule, the designed dose range is well tolerated by study subjects, so administration of 100 mg per time, two to three times per day is recommended in phase II trials.

8.4.2 Phase II/III Clinical Trial of Polyphenon E

This is an example of a phase II and III clinical trial obtained from literature.[9]

Polyphenon® E is a defined extract of catechins of green tea leaves of *Camellia sinensis*, a species of the Theaceae family. It contains more than 80% of tea polyphenols/catechins accountable for the major biological properties, such as immunomodulatory, protein binding, particularly to enzymes, as well as strong antioxidant activities. These potential properties are supportive of its use in wart treatment together with the observed anti-inflammatory activities. MediGene has developed two new formulations, a 10% cream and a 15% ointment, with enhanced penetration properties to further improve the efficacy of patient-applied Polyphenon® E in the treatment of external genital warts.

The objective of this study was to investigate the clinical efficacy and safety of Polyphenon® E 10% cream and Polyphenon® E 15% ointment in comparison with placebo in the treatment of external genital warts in male and female patients.

Trial Design

The study was a randomized, double-blind, placebo-controlled, four-arm parallel-group phase II/III, multicenter performed trial. Two hundred forty-two outpatients with 2–30 warts (total wart area, 12–600 mm^2) were randomly assigned into groups of Polyphenon® E 10% cream ($n = 79$, 41 males), Polyphenon® E 15% ointment ($n = 80$, 42 males), or placebo ($n = 83$, 42 males), stratified by gender. Patients were instructed to apply the medication topically to all external genital warts three times per day, regardless of being baseline or new warts. Dosing was anticipated to be <250 mg of study medication per application. The maximum duration of treatment was 12 weeks or until complete clearance of all baseline warts, whichever came first, followed by a 12-week treatment-free follow-up phase for complete responders.

Main Observation Indexes The total wart area, local reactions, and adverse events.

Statistical Analysis Descriptive statistics were performed to analyze the secondary efficacy involving complete/partial clearance, time to complete/partial clearance, and recurrence rates as well as safety parameters involving local signs/symptoms, AEs, physical examination, vital signs, or laboratory parameters. Complete clearance of all (baseline and new warts occurring during treatment) warts was additionally analyzed as requested for the subsequent pivotal phase III trials by the U.S. FDA. Baseline comparisons were done using Wilcoxon's two-sample test (two-sided, $a = 0.05$). In addition, an influence analysis (logistic regression model) was carried out on wart location and area, age and numbers of warts, age of patients, usage of cream and ointment, and previous treatments to investigate contributing prognostic factors.

Result

For 15% ointment, statistically significant differences to placebo were achieved regarding complete clearance of all baseline external genital warts (61.0% vs. 40.5%

in males, 56.8% vs. 34.1% in females; combined gender: $p = 0.0066$) and 75% to 100% clearance (80.8% vs. 51.8%; $p = 0.0001$) in both the intent-to-treat and per-protocol populations. For 10% cream, 53.8% males and 39.5% females achieved complete clearance, no significant difference to placebo. Recurrence rates after 12 weeks treatment were 10.6%, 11.8%, and 10.3% for 15% ointment, 10% cream, and placebo, respectively. Adverse events were observed in only 7.9% of patients, with no serious adverse events or deaths reported. Local skin reactions were generally mild to moderate and resolved with continued treatment without sequelae.

Conclusions Polyphenon® E 15% ointment, composed of a defined green tea extract, proved to be efficacious and safe for both gender in the treatment of external genital warts.

8.4.3 Phase III Clinical Trial of Nanoparticle Albumin-Bound Paclitaxel

This is an example of phase III clinical trial obtained from literature.[10]

Paclitaxel, extracted from bark and needle of *Taxus brevifolia*, is a new treatment for ovarian cancer, breast cancer, and lung cancer. However, intravenous administration of paclitaxel is hindered by poor water solubility of the drug. Currently, paclitaxel is dissolved in a mixture of ethanol and Cremophor EL. But this formulation is associated with significant side effects, which are thought to be related to the pharmaceutical vehicle. ABI-007 is a novel, biologically interactive, nanometer sized albumin-bound paclitaxel particle with shorter infusion schedules (30 min vs. 3 h, respectively) and no need for premedication. This preparation was initially developed to avoid the toxicities associated with polyethylated castor oil. It allows the safe infusion of significantly higher doses of paclitaxel than the doses used in standard paclitaxel therapy. Thus it is a better treatment for patient with ovarian cancer, breast cancer, and lung cancer.

ABI-007, the first biologically interactive albumin-bound paclitaxel in a nanometer particle, free of solvents, was compared with polyethylated castor oil–based standard paclitaxel in patients with metastatic breast cancer (MBC). This phase III study was performed to confirm preclinical studies demonstrating superior efficacy and reduced toxicity of ABI-007 compared with standard paclitaxel.

Trial Design

The study was an international multicenter, randomized, open-label, parallel-controlled, phase III trial. Four hundred fifty-four women with MBC were randomly assigned into either treatment group or active control group (1:1), in which patients received treatment with ABI-007 (260 mg/m^2 intravenously over 30 min without corticosteroid or antihistamine premedication or special infusion sets) and standard paclitaxel (175 mg/m^2 intravenously over 3 h with premedication and special infusion sets as indicated in the prescribing information) every 3 weeks. Up to two dose reductions (40 mg/m^2 each) were allowed from the initial 260 mg/m^2 dose of

ABI-007. Reductions in standard paclitaxel doses were allowed according to the package insert for each country.

Observation Indexes Complete response was the disappearance of all clinical evidence of visible tumor; partial response was a ≥50% decrease in product of largest perpendicular diameters of measurable lesions (i.e., "decrease of perpendicular diameter of the largest tumor"). Responses were assessed according to *Response Evaluation Criteria in Solid Tumor* guidelines. Objective response rate and time to progression (TTP) were measured as the primary and secondary efficacy indexes, respectively. The overall survival was counted. Quality-of-life assessment data (The Eastern Cooperative Oncology Group, performance status, scores from the European Organisation for Research and Treatment of Cancer Quality of Life Questionnaire C30, and body weight) were also collected.

AEs were coded by using the Medical Dictionary for Drug Regulatory Affairs and further classified according to National Cancer Institute Common Toxicity Criteria.

Statistical Analysis All statistical analyses were performed with SAS Version 8.2 (SAS Institute Inc, Cary, NC). The primary efficacy analysis consisted of three nested tests: non-inferiority, with all patients; superiority, with all patients; and superiority, with patients receiving study drug as first-line therapy. They were conducted sequentially and contingent on the prior successful test(s). Treatment differences in TTP and survival were analyzed using the Kaplan-Meier method.

Results

ABI-007 demonstrated significantly higher response rates compared with standard paclitaxel (33% v 19%, respectively; $p = 0.001$) and significantly longer time to tumor progression (23.0 vs. 16.9 weeks, respectively; hazard ratio = 0.75; $p = 0.006$). The incidence of grade 4 neutropenia was significantly lower in the ABI-007 treatment group compared with the standard paclitaxel one (9% vs. 22%, respectively; $p < 0.001$) despite a 49% higher paclitaxel dose. The incidence of febrile neutropenia was lower (<2%) with no difference between the two study arms. Grade 3 sensory neuropathy was more common in the ABI-007 arm than in the standard paclitaxel one (10% vs. 2%, respectively; $p < 0.001$), but was easily managed for rapid improvement (median, 22 days). No hypersensitivity reactions occurred with ABI-007 despite the absence of premedication and shorter administration time.

Conclusion ABI-007 demonstrated greater efficacy and a favorable safety profile compared with standard paclitaxel in this trial. The improved therapeutic index and elimination of corticosteroid premedication that required for solvent-based taxanes make the novel albumin bound paclitaxel ABI-007 become an important advanced treatment for MBC.

REFERENCES

1. Linda, K., et al. (2008) St John's wort for major depression. *Cochrance Database Systematic Reviews* (Online) 8(4):CD000448.
2. Hypericum Depression Trial Study Group (2002) Effect of Hypericum perforatum (St John's wort) in major depressive disorder: a randomized controlled trial. *The Journal of the American Medical Association* 287(14):1807–1814.
3. Pocock, S.J. (1991) *Clinical Trials: A Practical Approach* (2nd ed.). New York, John Wiley & Sons.
4. Friedman, L.M. (1998) *Fundamentals of Clinical Trials* (3rd ed.). New York, Springer.
5. Chow, S.C. and Liu, J.P. (2004) *Design and Analysis of Clinical Trials—Concepts and Methodology* (2nd ed.). New York, John Wiley & Sons.
6. Gallin, J.I. (2007) *Principles and Practice of Clinical Research* (2nd ed.). London, Academic Press.
7. Montgomery, D.C. (2001) *Design and Analysis of Experiments* (5th ed.). New York, John Wiley & Sons.
8. D'Agostino, R.B., et al. (2003) Non-inferiority trials: design concepts and issues—the encounters of academic consultants in statistics. *Statistics in Medicine* 22(2):169–186.
9. Gross, G., et al. (2007) A randomized, double-blind, four-arm parallel-group, placebo-controlled Phase II/III study to investigate the clinical efficacy of two galenic formulations of polyphenon E in the treatment of external genital warts. *Journal of the European Academy of Dermatology and Venereology* 21(10):1404–1412.
10. Gradishar, W.J., et al. (2005) Phase III trial of nanoparticle albumin-bound paclitaxel compared with polyethylated castor oil-based paclitaxel in women with breast cancer. *Journal of Clinical Oncology* 23(31):7794–7803.

Chapter 9

Standardization and Quality Control of Herbal Extracts and Products

Jian-guo Zeng, Man-liang Tan, Xuan Peng, and Qi Luo

There are widespread resources of various medicinal herbs with a long historical use all over the world. Although the efficacy of herbs has been acknowledged by more and more people, the quality of herbal products is a main setback to their growing acceptance. This is because the quality not only determines the efficacy of an herbal product but, more importantly, the safety of consumers.

Because herbs are composed of complex compounds, the quality of herbal products is decided by more factors in comparison with synthetic drugs. These factors can be divided into three aspects: the selection of herbal materials, the design of extraction and processing procedures, and the hygiene condition of the environment for manufacturing and storage. Also, because of their complex components, the quality measurement of herbal products is more difficult than that of synthetic drugs. Quality control (QC) should include both quantitative analysis for major characteristic compounds and qualitative analysis for the profile of whole composition. It is not only important for ensuring the safety of the herbal products to consumers, but also for correctly evaluating the efficacy of herbal medicine because the quality of samples for animal experiments and clinical trials directly impact the accuracy of the study result.

To ensure the quality and safety of herbal products, quality assurance (QA), QC, good manufacturing practice (GMP), and standard operating procedures (SOP) are being widely applied to the field of medicinal herbs in many countries. This chapter will give a brief introduction to QA, QC, GMP, and SOP, as well as equipment and methods that are generally applied to qualitative and quantitative analysis

Traditional Herbal Medicine Research Methods, Edited by Willow J.H. Liu

for herbal medicines or products. In addition, product quality review (PQR), quality risk management (QRM), and quality unit (QU) will also be mentioned.

9.1 INTRODUCTION OF QA, QC, AND GMP

The core of GMP characterizes modern quality management systems for drugs. The World Health Organization (WHO), the International Conference on Harmonization of Technical Requirements for Registration of Pharmaceuticals for Human (ICH), the U.S. Food and Drug Administration (FDA), the European Medicines Agency (EMEA), and official organizations in many other countries have established laws or regulations to protect the quality management of drugs. The drug quality system has been developed from the initial "End-point Testing" control to today's Comprehensive Quality Systems (CQS) or Total Quality Systems (TQS). It has been popularly accepted that quality is built into a product (meaning that quality is involved in the whole procedure), rather than resting solely on its final products. GMP requires not only that the final product meet the standards, but that the products be made with the same procedures under the same conditions for every batch.

The basic concepts of GMP, QA, QC, QRM, and QU are interrelated. They are described here in order to emphasize their relationships and their fundamental importance to the production and control of medicinal products including herbal products.

9.1.1 Principles

Authorized drug manufacturers must guarantee that their medicinal products are not only fit for their intended use, but that they also comply with the requirements of the authorized marketing field. Any products with risk due to inadequate safety, quality, or efficacy assurance are banned from the market. To achieve the reliable quality objective, there must be a comprehensively designed and correctly implemented system of QA incorporating GMP, QC, and QRM. The system must be fully documented and its effectiveness should be monitored. Every aspect of the QA system should be adequately resourced with competent personnel, suitable and sufficient premises, equipment, and facilities. In addition, the authorized manufacturer and QA personnel are responsible for liability.

9.1.2 Good Manufacturing Practice (GMP) for Medicinal Products

Brief Introduction to GMP

GMP is a term that is recognized worldwide for the management and control of manufacturing and quality testing of foods, pharmaceutical products, and medical devices. It is also referred to as current GMP (cGMP), which appears first in 501(B) of the 1938 U.S. Food, Drug, and Cosmetic Act (21USC351). The word "current" reminds manufacturers that they must apply up-to-date technologies and systems in order to comply with the regulation.

GMPs came out of the realization that end-point quality testing was insufficient to assure the quality of the individual medication unit (such as the tablet, the capsule, and the vial) dispensed to patients. Manufacturers need to ensure that every step of the manufacturing process is well set and each product unit meets its quality specifications.

GMP is part of QA, which ensures that products are consistently produced and controlled with the quality standards appropriate to their intended use as required by the marketing authorization or product specification. GMP covers all aspects of the manufacturing process: designing manufacturing procedures; validating critical manufacturing steps; ensuring suitable premises, storage, and transport; having qualified trained personnel for production and QC; possessing adequate laboratory facilities; following the approved written procedures and instructions; keeping records showing that all steps are being taken in accordance with defined procedures; fully tracing a product through batch records and distribution records; and establishing systems for recall and investigation of complaints. GMP involves both production and QC.

History of GMP

In 1905, a book titled *The Jungle*, written by Upton Sinclair, reported that the remains of poisoned rats and even unfortunate workers who fell into the machinery were often found in the meat produced in Chicago. This report rightly caused a huge public outrage at the time and led to Congress passing the Pure Food and Drug Act in 1906.

In 1937, the Elixir of Sulfanilamide disaster alerted the United States authorities that the safety of drugs needed to be proven. In this case, elixir of sulfanilamide, the first of a new generation of "wonder drugs" popularly used to treat strep throat and gonorrhea, was marketed for use in children. This liquid formulation, however, contained polyethylene glycol, a toxic ingredient used as antifreeze, and killed 107 people, most of them children. In response, Congress passed the Federal Food, Drug and Cosmetic (FD&C) Act of 1938. For the first time, companies were required to prove that their products were safe before marketing.

From 1957 to 1962, thalidomide was sold in about 100 countries under different names. About 12,000 mothers who took it in their first trimester pregnancy to alleviate morning sickness gave birth to children with severely deformed appendages. Fortunately, the impact in the United States was minimized because Frances Oldham Kelsey, an FDA officer, had rejected the application from Richardson Merrell to market it, due to the belief that this drug needed more study. In 1962, President Kennedy awarded Kelsey the President's Distinguished Federal Civilian Service Award, the highest honor a government employee may earn as a civilian. Since then, the FDA has had the authority to regulate advertising of prescription drugs. And in 1963, the first GMPs for finished pharmaceuticals were enacted.

At the twenty-second WHO conference, in 1969, the organization proposed that members apply GMP to pharmaceutical production to ensure the drug quality and to join in "the Certification Scheme on the Quality of Pharmaceutical Products Moving in International Commerce." In November 1975, WHO formally issued its GMP and recommended it again to its members at the twenty-eight WHO conference, in 1977. Since then, the GMP has been established as one of the WHO regulations and has been enrolled in the Official Records of the WHO (XII, appendix 226). The GMP of WHO is an important part of the entire quality management of pharmaceutical production. It is a necessary and reliable approach to ensure drug quality and minimize errors, accidents, confusion, and contamination of drugs.

The Japan Pharmaceutical Manufacturers Association (JPMA) established its own GMP in 1973. The Japanese government promulgated its GMP and guidance for implementation in 1974, and then decided to implement it formally in 1980. The United States issued its new modified GMP in 1978. Thereafter, the United Kingdom and most other European countries started to propose, study, and draft their own GMP. By 1980, 63 countries had issued their own GMPs. Today, GMPs have been implemented in more than 100 countries.

Along with the social development and scientific and technological progress, each country has modified and improved its GMP constantly with a variety of detailed regulations and guidelines. As examples, the United Kingdom had already published its fourth version of GMP guidelines describing details of specific provisions by 1985; Japan also modified its GMP continuously to provide more details, including the self-inspection of drug manufacturers. The Japanese government also established GMP on active pharmaceutical ingredients (APIs) in 1988 and implemented it formally in 1990. In the United States, about 70–80% or more of the APIs used for making products in manufacturers were imported from other countries, where manufacturing standards may not be stringent enough. For this reason, both the European Union (EU) and the United States recently published draft guidance for manufacturers of APIs. In 2001, the ICH published Q7A *Good Manufacturing Practice Guidance for Active Pharmaceutical Ingredients*. This document has been considered the *de facto* standard for manufacturing APIs.

Since China is the most populous country and serves as the origin of Chinese herbal medicine, it is necessary to give a brief introduction about the performance of GMP in China. GMP was initially applied in part to products in some enterprises at the end of the 1970s along with the reform and opening policy of drug export in China. The first tentative version of GMP was drafted by the China National Pharmaceutical Industry Corporation Limited (CNPIC) in 1982, then modified and issued by former State Medicine and Drugs Administration in 1985. At the end of the same year, CNPIC prepared and issued the *Guidelines for Implementation of GMP* (version 1985). Two other versions of GMP were promulgated by the Ministry of Health (MOH) of the People's Republic of China separately in 1988 and 1992. The *Guidelines for Implementation of GMP* was modified by CNPIC in 1992, and a newly modified GMP (version 1998) was issued in June 1999.

Since GMP certification is an advanced scientific management approach to ensure the stability, safety, and effectiveness of drugs, and has been implemented and developed worldwide, it becomes vital to the international trade and administration of drugs. The Chinese MOH issued decree no. 53 about performing GMP certification for drugs in 1995. In the same year, the China Certification Committee for Drugs (CCCD) was established. In 1998, the State Drug Administration (SDA), the precursor of the State Food and Drug Administration (SFDA), was established. Meanwhile, the SDA certification center was founded. Since then, the United States has begun to supervise drug manufacturing enterprises and products via inspecting the implementation status on GMPs. Only well-implemented GMPs can get GMP certification. If the manufacturers of APIs or finished products failed to pass the inspection of GMP implementation, and no GMP certificate was issued from SFDA since July 1, 2004, they were then forbidden to make products. On January 1, 2008, a more scientific and stricter Evaluation Criterion of Inspection on Drug GMP Compliance was implemented.

Basic Requirements of GMP

Although many countries have their own GMP, their basic requirements are similar. According to the EU *Guidelines to Good Manufacturing Practice, Medicinal Products for Human and Veterinary Use* (European Communities, 1998), the basic requirements of GMP include the following.

1. All manufacturing processes are clearly defined, systematically reviewed in the light of experience, and shown to be capable of consistently manufacturing medicinal products of the required quality and complying with their specifications.
2. Critical steps of manufacturing processes and significant changes to the process are validated.
3. All necessary facilities for GMP are provided, including appropriately qualified and trained personnel; adequate premises and space; suitable equipment and services; correct materials, containers, and labels; approved procedures and instructions; and suitable storage and transport.
4. Instructions and procedures are written in an instructional form in clear and unambiguous language, specifically applicable to the facilities provided.
5. Operators are trained to carry out procedures correctly.
6. Records are made, manually and/or by recording instruments, during manufacture, which demonstrate that all the steps required by the defined procedures and instructions were in fact taken and that the quantity and quality of the product was as expected. Any significant deviations are fully recorded and investigated.
7. Records of manufacture, including distribution, which enables the complete history of a batch to be traced, are retained in a comprehensible and accessible form.
8. The distribution (wholesaling) of the products minimizes any risk to their quality.

9. A system is available to recall any batch of product, from sale or supply.
10. Complaints about marketed products are examined, the causes of quality defects investigated, and appropriate measures are taken in respect of the defective products and to prevent reoccurrence.

Ten Principles of GMP

The following guidelines are from the web site of the GMP Institute, A Division of International Association for Pharmaceutical Engineering (ISPE) (http://www.gmp1st.com).

1. Writing detailed step-by-step procedures that provide a road map for controlled and consistent performance.
2. Carefully following written procedures to prevent contamination, mix-ups, and errors.
3. Promptly and accurately documenting work for compliance and traceability.
4. Proving that systems do what they are designed to do by validating work.
5. Integrating productivity, product quality, and employee safety into the design and construction of facilities and equipment.
6. Properly maintaining facilities and equipment.
7. Clearly defining, developing, and demonstrating job competence.
8. Protecting products against contamination by making cleanliness a daily habit.
9. Building quality into products by systematically controlling our components and product-related processes such as manufacturing, packaging and labeling, testing, distribution, and marketing.
10. Conducting planned and periodic audits for compliance and performance.

9.1.3 Quality Assurance (QA)

In a document issued from the International Organization for Standardization (ISO) titled *The Quality Management and Quality Assurance—Vocabulary* (ISO8402, 1995), QA is defined as, "All those planned and systematic actions necessary to provide adequate confidence that a product or service will satisfy given requirements for quality." *QA personnel* refer to people who are in charge of works on QA.

In the pharmaceutical industry, QA is a wide-ranging concept that covers all aspects that influence the quality of a product. It is the whole of all measures that ensure medicinal products meet the quality required for their intended use. QA, therefore, incorporates not only with GMP, but also with other factors that are beyond the scope of this chapter.

According to the EU *Guidelines to Good Manufacturing Practice, Medicinal Products for Human and Veterinary Use* (European Communities, 1998), the basic requirements of QA include the following.

1. Medicinal products are designed and developed in a way that takes account of the requirements of GMP.
2. Production and control operations are clearly specified and GMP adopted.
3. Managerial responsibilities are clearly specified.
4. Arrangements are made for the manufacture, supply, and use of the correct starting and packaging materials.
5. All necessary controls on intermediate products, and any other in-process controls and validations are carried out.
6. The finished product is correctly processed and checked, according to the defined procedures.
7. Medicinal products are not sold or supplied before a qualified person has certified that each production batch has been produced and controlled in accordance with the requirements of the marketing authorization and any other regulations relevant to the production, control, and release of medicinal products.
8. Satisfactory arrangements exist to ensure, as far as possible, that the medicinal products are stored, distributed, and subsequently handled so that quality is maintained throughout their shelf life.
9. There is a procedure for self-inspection and/or quality audit, which regularly appraises the effectiveness and applicability of the QA system.

The following are functions of QA, as summarized by Dr. Jean Huxsoll, the former director of regulatory Affairs and Quality Assurance in Bayer: (1) testing; (2) documentation; (3) labeling; (4) vendor audits; (5) vendor approval; (6) raw material receipt; (7) product release; (8) product specifications; (9) training; (10) validation; (11) participating in planning for long- and short-term goals of the company; (12) involvement of business aspect; (13) communicating with outside world, such as regulatory agencies and other companies having similar product lines, and attending conferences in order to upgrade the industry and regulatory standards; (14) risk analysis/decision making.

9.1.4 Quality Control (QC)

In ISO8402, QC is defined as, "The operational techniques and activities that are used to fulfill requirements for quality. People in charge of works on QC are named QA personnel."

In the pharmaceutical industry, QC is that part of GMP which is concerned with sampling, specifications, and testing, and with the organization, documentation, and release procedures which ensure that the necessary and relevant tests are actually carried out and that materials are not released for use, nor products released for sale or supply, until their quality has been judged to be satisfactory.

According to the EU *Guidelines to Good Manufacturing Practice, Medicinal Products for Human and Veterinary Use in the Rules Governing Medicinal Products in the European Union* (vol. 4), the basic requirements of QC include the following.

1. Adequate facilities, trained personnel, and approved procedures are available for sampling, inspecting, and testing starting materials, packaging materials, intermediate, bulk, and finished products, and, where appropriate, for monitoring environmental conditions for GMP purposes.
2. Samples of starting materials, packaging materials, intermediate products, bulk products, and finished products are taken by personnel and by methods approved by QC.
3. Test methods are validated.
4. Records are made, manually and/or by recording instruments, which demonstrate that all the required sampling, inspecting, and testing procedures were actually carried out. Any deviations are fully recorded and investigated.
5. The finished products contain active ingredients complying with the qualitative and quantitative composition of the marketing authorization, are of the purity required, and are enclosed within their proper containers and correctly labeled.
6. Records are made of the results of inspection and testing of materials, intermediate, bulk, and finished products is formally assessed against specification. Product assessment includes a review and evaluation of relevant production documentation and an assessment of deviations from specified procedures.
7. No batch of product is released for sale or supply prior to certification by a qualified person that it is in accordance with the requirements of the relevant authorizations.
8. Sufficient reference samples of starting materials and products are retained to permit future examination of the product if necessary and the product is retained in its final pack unless exceptionally large packs are produced.

According to the *Guidance for Industry—Quality Systems Approach to Pharmaceutical CGMP Regulations*, issued from the U.S. Department of Health and Human Services and Food and Drug Administration (2006), the functions of QC include the following.

1. Assessing the suitability of incoming components, containers, closures, labeling, in-process materials, and the finished product.
2. Evaluating the performance of the manufacturing process to ensure adherence to proper specifications and limits.
3. Determining the acceptability of each batch for release.

9.1.5 Product Quality Review (PQR)

According to the EU *Guidelines to Good Manufacturing Practice, Medicinal Products for Human and Veterinary Use in the Rules Governing Medicinal Products in the European Union* (vol. 4), regular periodic or rolling quality reviews of all licensed medicinal products, including export-only products, should be conducted with the objective of verifying the consistency of the existing process, the appropriateness of current specifications for both starting materials and finished product to highlight any trends, and to identify product and process improvements. Such reviews should normally be conducted and documented annually, taking into account previous reviews.

PQR should include the following.

1. A review of starting materials, including packaging materials used for the product, especially those from new suppliers.
2. A review of critical in-process controls and finished product results.
3. A review of all batches that failed to meet the established specification(s) and their investigation.
4. A review of all significant deviations or nonconformances, their related investigations, and the effectiveness of resultant corrective and preventative action taken.
5. A review of all changes carried out to the processing or analytical methods.
6. A review of Marketing Authorization variations submitted/granted/refused, including those for third country (for export only) dossiers.
7. A review of the results of the stability monitoring program and any adverse trends.
8. A review of all quality-related returns, complaints, and recalls, and the investigations performed at that time.
9. A review of adequacy of any other previous product process or equipment correction actions.
10. For new marketing authorizations and variations to marketing authorizations, a review of post-marketing commitments.
11. A review of the qualification status of relevant equipment and utilities, for example, heating, ventilating, and air conditioning (HVAC), water, and compressed gases.
12. A review of any contractual arrangements to ensure that they are up-to-date.

The manufacturer and marketing authorization holder should evaluate the results of these reviews to decide whether corrective and preventative actions or any revalidation need to be taken. If so, reasons for these actions should be documented. Agreed corrective and preventative actions should be timely and effectively performed. The ongoing management activity should be performed according to the setup procedures. The effectiveness of these actions and procedures should be validated during self-inspection. Quality reviews may be scientifically carried out according to product type, for example, oral solid preparation, oral liquid preparation, and sterile products.

If the manufacturer is not the marketing authorization holder, there should be a technical agreement between different parties that defines their respective responsibilities in producing the quality review. The QA person who issues certification should work together with the marketing authorization holder to ensure that the quality review is performed in a timely and accurate manner.

9.1.6 Quality Risk Management (QRM)

In 2006, ICH issued a new *Guidance for Industry: Quality Risk Management* (Q9). QRM is a valuable component of an effective quality systems framework. QRM can, for example, help guide the process parameters for drug manufacturing, assess and mitigate the risk of changing a process or specification, and determine the extent of discrepancy investigations and corrective actions.

QRM is a systematic process for the assessment, control, communication, and review of risks to the quality of the medicinal products. It can be applied both proactively and retrospectively.

According to the EU *Guidelines to Good Manufacturing Practice, Medicinal Products for Human and Veterinary Use in the Rules Governing Medicinal Products in the European Union* (vol. 4), the QRM system should ensure that

1. the evaluation of the risk to quality should be based on scientific knowledge and experience with the process and ultimately links to the protection of the patients.
2. The level of effort, formality, and documentation of the QRM process is commensurate with the level of risk.

9.1.7 The Quality Unit (QU)

The QU was first used in the FDA *Guidance for Industry: Quality Systems Approach to Pharmaceutical CGMP Regulations* (2006). The guidance states, "The cGMP regulations specifically assign the QU the authority to create, monitor, and implement a quality system." Such activities do not substitute for, or preclude, the daily responsibility of manufacturing personnel to build quality into the product. The QU should not take on the responsibilities of other units of a manufacturer's organization, such as the responsibilities handled by manufacturing personnel, engineers, and development scientists. Manufacturing personnel and the QU are both critical in fulfilling the manufacturer's responsibility to produce quality products.

The responsibilities and procedures applied in the QU should be documented in writing. The QU should have adequate laboratory testing equipments in order to implement the following duties.

1. Accept or reject all materials, containers, closures, intermediate products, packaging materials, labeling, and finished products made by its own company or drug products that manufactured, processed, packed, or held by other companies under contract.
2. Review production records to assure there are no errors, or if errors occurred, they should be fully investigated for any unexplained discrepancies.
3. Approve or reject all procedures or specifications involving identification, quantity, quality, and purity of drug products.

The above are consistent with the approaches of a modern quality system, with the purpose of ensuring that QCs are implemented and completed satisfactorily during manufacturing operations and the developed procedures and specifications are appropriate for implementation, including those used by contractors.

Under a quality system, the product and process development unit, the manufacturing unit, and the QU are generally independent. In rare circumstances, a single individual can perform both production and quality functions. This person is responsible for implementing all the QCs and reviewing the results of manufacture to ensure that products have met quality standards. Under such circumstances, it is recommended that another qualified individual who is not involved in the production operation conduct an additional, periodic review of QU activities.

9.1.8 QA, QC, GMP, and Herbal Extracts

Traditional medicine, also called complementary or alternative medicine, natural medicine, nonconventional medicine, or holistic medicine, has always maintained its popularity worldwide. Over the last decade, we have seen its increasing use in many developed and developing countries. Currently, as a main part of traditional medicine, herbal medicine can be prescribed as drugs by doctors in some countries, such as China, India, and Germany, but only be used as dietary supplements in other countries, such as the United States.

The safety and efficacy of herbal medicine, as well as QC, have become important concerns for both health authorities and the public. Health-care professionals, providers, and consumers are calling for regulations of products of herbal medicine. Some countries, such as Japan and China, have applied GMP management to herbal manufactories. In the United States, cGMP requirements will also be fully applied to all manufacturers of dietary supplements by June 2010. This will not only ensure the safety of herbal products, but also promote recognition of traditional medicine and their products, and further define their role in modern health-care systems.

Despite the use of herbal medicine over many centuries, only a relatively small number of medicinal plant species have been studied for their medical applications. Safety and efficacy data are available for an even smaller number of medicinal herbs or their extracts, active ingredients, and preparations.

A single herbal extract may be defined as a food, a functional food, a dietary supplement, or an herbal medicine in different countries, depending on the regulations applying to foods and medicines in each country. Therefore, there are no international standards or systems to carry out the QA and QC on herbal extracts. Currently, most herbal extracts are sold as materials for preparations in the international market while a small amount of them are listed as drugs in some countries. Although standards of chemical analysis for herbal extracts cannot be compared with synthetic drugs because of the different characteristics of their components, herbal products listed as drugs are regulated with standards similar to the conventional drugs, that is, products are not allowed to enter into the market if they are made by manufacturers without GMP certification. However, many extracts sold as materials for further processing are not well regulated. Their quality standards are commonly self-established by manufacturers where management and QC vary due to the different levels of knowledge and skills of the management staff.

Member countries of WHO have strengthened the management on herbal extracts under the coordination of WHO in the past decades. Impressively, Xijun Yan, the president of Tianjin Tasly Group Co. Ltd., in China, first proposed the conception of Good Extracting Practice (GEP) in 2001, arousing extensive attention of experts worldwide. It includes performance of extraction, concentration, chromatography, crystallization, filtration, and dryness. GEP provides not only solutions to isolate the toxic compounds from herbs but also solutions against heavy metals and pesticide residues in herbs. Now, GEP has become a good practice in China to ensure the quality of herbal extracts. GEP shows promise to serve the international administration principle on herbal extracts.

9.2 STANDARDIZATIONS AND SOP OF HERBAL EXTRACTS

Two factors hinder herbal extracts from fully spreading in the international market. One is that there is no good connection between the description of traditional functions and the modern mechanisms of actions and pharmacology. The other is a lack of quality control of raw materials and the production process of herbal extracts, resulting in inconsistent quality of different batches, as well as poor control against toxic contaminants.

For herbs whose components are totally unclear, the only control for extraction is to select raw materials with relatively consistent features of appearance and normative processes (see Chapter 2); otherwise, even the color and odor of the extract may be inconsistent, let alone the uniformity of quality. In such situations, it is necessary to do some chemical investigations on the herb first, so that the quality can be controlled based on the major components.

To control and standardize the production of herbal extracts effectively, the specifications of raw materials and herbal extracts, and the SOP of raw materials collection and process, extraction, and analysis should be applied as a guideline of the technical route. This is simply called "two specifications and three SOPs" (see Fig. 9.1).

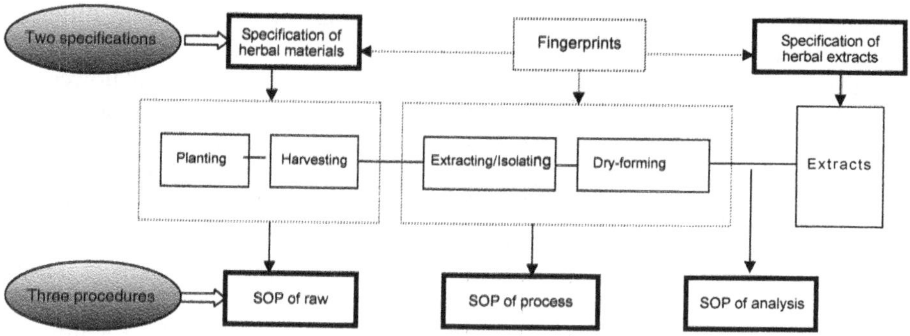

Figure 9.1 Two specifications and three procedures.

The biggest difference between a synthetic drug and an herbal medicine is that the former is made of a single compound, while the latter is composed of complex compounds. Although main compounds in most commonly used herbal medicines have been investigated, many of the major peaks in high-performance liquid chromatography (HPLC) chromatograms have not been identified, meaning these components are still unclear.

Since one compound may commonly exist in different herbal species, especially those in the same genus, quantification of such compounds is unable to reveal whether the materials used for extraction are the right species. For this reason, specifications should include not only quantification of a few main known compounds (used as index compounds for quantification), but also the HPLC or thin layer chromatography (TLC) fingerprint for comparing the profile of components in the sample. The following sections give reasons why the "two specifications and three SOPs" should be applied for QC of herbal products.

9.2.1 Reasons for Establishing Specifications of Raw Herbal Materials

Sometimes, different herbal materials are traditionally used as the same herbal medicine. The quality is different not only between materials of different species used as the same herbal medicine, but also between materials of the same species growing from different areas. Here are two examples.

Immature fruit of *Citrus aurantium* L is officially recorded as *C. aurantium* in *Chinese Pharmacopoeia.* But commercially, immature fruits of mandarin orange, tangerine, and orange are all used as *C. aurantium.* Synephrine, one of the components in *C. aurantium,* has been used as the characteristic component of extracts of *C. aurantium.* Thus, the requirement of the extract factories for the raw material of *C. aurantium* is usually 1% of synephrine. In this case, materials such as immature fruits of mandarin orange, tangerine, and orange could all be purchased as the raw material of *C. aurantium* for extraction. However, these different fruits contain other compounds; therefore, it is obvious that the qualities of extracts from different sources would be quite different.

Red clover is the inflorescence or the branches and leaves with flowers of *Trifolium pratense* L. There are mainly four isoflavones in this herb, formononetin, biochanin A, daidzein, and genistein. Among them, daidzein and genistein are lower concentrated while formononetin and biochanin A are higher. Therefore, the total amount of formononetin and biochanin A are often used as index compounds in tests to indicate the level of isoflavones in red clover. In general, the total isoflavones in red clover extract is required to be more than 20%. If only materials with a total level formononetin and biochanin A more than 0.5% being extracted with a standard procedure can meet this requirement (total isoflavones is more than 20%), a different extraction process has to be designed for the material that contains only a total level of 0.1% of formononetin and biochanin A to obtain the required extract. It needs to be mentioned that genistein has actually been reported as the most active phytoestrogen in this herb. This means that the total level formononetin and biochanin A in the extract may not absolutely indicate the phytoestrogenic activity of red clover. Fingerprints of these isoflavones in HPLC chromatograms will help in this case.

In addition, the concentration of pesticide residues, heavy metals, and other exogenous contaminants in raw materials should be limited. Therefore, appropriate specifications of raw herbal materials should be established corresponding to the specifications of related extracts.

9.2.2 Reasons for SOP of Raw Herbal Materials

The harvest season, parts of plants to be collected, processing methods (cutting and drying), and storage conditions of raw herbal materials are all directly related to the quality of the final extracts. Therefore, it is feasible and practical to establish handy SOPs for collecting and processing methods and storage conditions.

Hypericum is the dry aerial part of *Hyperium perforatum* (Guttiferae). HPLC analysis shows that the contents of hypericin and hypeforin in the raw material are different between those collected from different parts (see Table 9.1) and with different processing methods (see Table 9.2). The variation was significant even when the materials were from the same origin and extracted with the same method.

Table 9.1 Contents of Hypericin and Hypeforin in Different Parts of *Hypericum perforatum*

	Hypericin (%)	Hypeforin (%)
Flower	0.085	0.198
Leaf	0.040	0.190
Culm	0.005	0.067
Root	0.008	0.063
Whole grass	0.034	0.135

Table 9.2 Contents of Hypericin and Hypeforin in *Hypericum perforatum* Processed with Different methods

	Hypericin (%)	Hypeforin (%)
Dried naturally in the shade for 1 week	0.075	0.178
Dried under the sun for 2 days	0.040	0.089
Dried over heat (80°C) for 6h	0.035	0.067
Fresh material (within 6h after harvest)	0.098	0.183

Because the chemical composition of the same herbal material collected at different times or with different processing methods is different, SOPs of the production should be strictly specified in order to control the product quality and minimize variations between different product batches.

9.2.3 Reasons for SOP of Production Process

Quantitative analysis for index components in specification is unable to represent all of the components in an herbal extract, so QC standards should not be considered the only way to control the extract quality. SOPs should be adopted into extraction procedures of the same material. Otherwise, the composition of the product, which can be indicated in the HPLC fingerprints, will be varied because of different processes, even if all the index components comply with the specifications.

Take the extract of red clover as an example, again. Even if red clover was collected from the same origin and same part of the plant, processed with the same method, and the contents of formononetin are uniform between batches, the quality of these extracts will be varied if different extraction methods were used, meaning levels of other components, such as genistein, in these extracts will be different depending on the solvents or other materials used for isolation. In other words, even if all the index components comply with the specifications, it is not possible to ensure that other bioactive components have been extracted out. Therefore, standardized extracting processes should be applied in order to get extracts with consistent components. It should be mentioned that the recognition of bioactive compounds in an herbal medicine would become wider if simultaneous investigations were carried out by scientists over the world with time. So, the SOPs for extraction should be upgraded in a timely manner when new bioactive compounds are reported.

9.2.4 Reasons for SOPs of Analysis

Quality analysis results of an herbal extract may vary when different analytical methods, apparatuses, and sample preparation methods are used. An improper pretreatment method may lead to distortion of test results. Therefore, SOPs of analysis should be established to ensure the uniformity of quality evaluation for products.

For the same extract, different sample preparation methods will result in different quality evaluations about the product. Therefore, different pretreatment methods for analysis should be compared for method evaluation. Use red clover as an example again. Formononetin was much more highly concentrated in the extract that was pretreated with recirculation than that pretreated with ultrasonic extraction.

The selected quantitative analysis should make sure that the sample preparation method will completely extract the index compounds and the analytical method is sensitive enough. Therefore, SOPs of analysis should be carefully established and strictly specified when evaluating the quality of an extract.

9.2.5 Reason for Establishing Specifications of Herbal Extracts

To ensure the therapeutic efficacy and safety of herbal products and their acceptance in the international market, QC of herbal extracts or products should be specified using international standards, especially the quantitative analysis of index components, qualitative analysis by means of HPLC or TLC fingerprint, content limitations of pesticide residues, heavy metals, microbials, and other exogenous contaminants.

9.3 EQUIPMENT FOR QUALITY CONTROL OF HERBAL EXTRACTS AND PRODUCTS

Selection of equipment for analysis of herbal extracts or products depends on the purpose of the analysis and the properties of compounds in the herbal extracts to be analyzed. The chromatography tools are mainly used for isolation (e.g., HPLC, gas chromatography [GC], capillary electrophoresis [CE], and supercritical fluid extraction [SFE]), while the spectroscopic tools are mainly used for detection and identification (e.g., ultraviolet-visible [UV-Vis], infrared [IR]/near-IR [NIR], mass spectrometry [MS], and atomic absorption spectrometry [AAS]/atomic fluorescence spectrometry [AFS]). They are often coupled together to carry out a variety of tasks. Some of the combinations are suitable for analysis of polar compounds (e.g., HPLC with reversed phase column and electrospray ionization [ESI]-MS), ionic compounds (CE and ESI-MS), or macromolecular compounds (CE and ESI-MS for protein analysis), while others are more appropriate for small lipid molecules (e.g., GC and CI-MS). Devices with good reproducibility are suitable for quantitative analysis, while quantification of trace active ingredients also requires equipment with high resolution and sensitivity. Some equipment is more suitable for qualitative than quantitative analysis, such as ion trap LC-MS. Therefore, it is best to know the capabilities and features of commonly used equipment. Abbreviations for the names of the equipment are given below.

HPLC and GC are the most widely applied analytical equipment for herbal study. When coupling with different types of detectors, they are not only used for identification and quantification of compounds in herbal extracts, but also for the comparison of chromatogram fingerprint of herbal products, and the analysis of *in vivo* metabolism of herbal compounds for pharmacokinetics study.

9.3.1 High Performance Liquid Chromatography (HPLC)

HPLC has been widely used for analysis of pharmaceuticals, food, human metabolites, and biomacromolecules. It is currently the most commonly used tool for quantitative and qualitative analysis of chemicals in herbal extracts or products by means of efficient separation and automatic operation. It is characterized by high-pressure pumps ($150–350 \times 10^5$ Pa), high performance of stationary phase (>30,000 plates/m), and high sensitivity of detector(s) ($10^{-9}–10^{-11}$ g). The typical structural system of HPLC is shown in Figure 9.2.

The components of a standard HPLC system generally include liquid reservoir(s), (two) high-pressure pumps, a gradient elution device, an injector, chromatographic columns, a thermostat, a detector, and a recorder. As shown in Figure 9.2, carrier liquid (usually needs to be degassed) stored in the reservoir is transported to the inlet of the column by pumps with high pressure. A dual-pump system is usually required for transportation when gradient elution is used. Test sample is injected into the carrier fluid system through the injector and then transported to the column for separation. Isolated compounds are detected by the detector and output signals are sent to the recorder or the data-processing device.

Types of HPLC column for analysis include liquid-solid, liquid-liquid, ion exchange, and ion pair chromatography. Reversed phase chromatography is most commonly used for analysis of herbal extracts, especially hydrophilic samples. A few samples, such as fatty acids, may be more effectively isolated by normal phase columns. Ion exchange chromatography is used when the sample contains ionic compounds that may be dissolved in an acidic or alkaline aqueous solution.

HPLC is traditionally applied for separation of test samples with a relative molecular mass between 200 and 2000. But protein separation has also been reported very recently.[1] Detectors of HPLC could be UV photometric detector (UVD), photodiode array detector, fluorescence detector, differential refractive index detector,

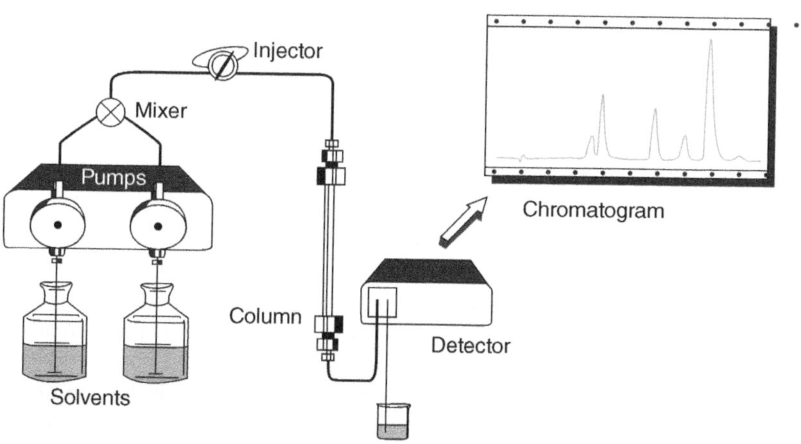

Figure 9.2 System of high performance liquid chromatography (HPLC).

electrical conductivity detector, mass spectroscopy detector, and so on. Recently, nuclear magnetic resonance (NMR) has also been coupled to HPLC for online structural elucidation.

9.3.2 Ultra Performance Liquid Chromatography (UPLC)

UPLC is developed based on the theory and principles of HPLC. It takes advantage of innovative technologies, such as small particle fillers and rapid detection, with increased analytical capacity and sensitivity. Compared with conventional HPLC, its most significant characteristic is the application of smaller particles in column, for example, size for HPLC scale is 5 μm, but for UPLC scale it is 1.7 μm. This creates higher column efficiency and system pressure (commonly up to 10,000 psi) for separation. The application of special materials and fillers, as well as its unique design, enables UPLC to tolerate high pressure. In addition, the detection cell is optimally designed with smaller cell volume, resulting in reduced diffusion of samples and higher detection sensitivity. Therefore, the UPLC system provides higher efficacy and sensitivity in comparison with common HPLC. Furthermore, without loss in column efficacy, UPLC could greatly reduce the duration of analysis by using shorter columns and higher eluting velocity. The analytical velocity, sensitivity, and resolution of UPLC are, respectively, 9-, 3-, and 1.7-folds of the traditional HPLC. Thus, it is widely used in fields of pharmaceuticals and foods. Additionally, the use of UPLC in conjunction with a mass spectrometer has greatly accelerated research of herbal medicine.

Currently, commercialized UPLC systems have been launched by several companies. Among them, ACQUITY-UPLC, manufactured by Waters (Milford, MA), is better known.[2] The ACQUITY-UPLC column from Waters is designed specifically for UPLC system with a very low diffusion volume. The innovative design of ACQUITY-UPLC significantly enhances resolution, sample capacity, and detection sensitivity.

QC of herbal products is a great challenge due to their complex composition. UPLC is becoming popular for QC of herb products due to its superior separation capability. The combined application of UPLC, photo-diode array detector, and mass spectrometer makes the quantitative analysis more accurate and reliable, and provides a powerful tool for analysis of the complex medicinal herbs.

9.3.3 Gas Chromatography (GC)

GC is another widely used method for quantitative and qualitative analysis. It can be applied in analysis of the gas, fluid, or solid samples. This includes food, water, and herbal medicine, in which the analytes are volatile or able to be converted to volatile with good thermal stability, such as volatile oils as well as pesticide residues in herbs or herbal products.

GC is chromatography using gas as the mobile phase, in which the sample is carried by the carrier gas through the stationary phase in column for separation; the separated compounds are then carried to the detector for identification. Figure 9.3 is a simple flow chart of the procedure.

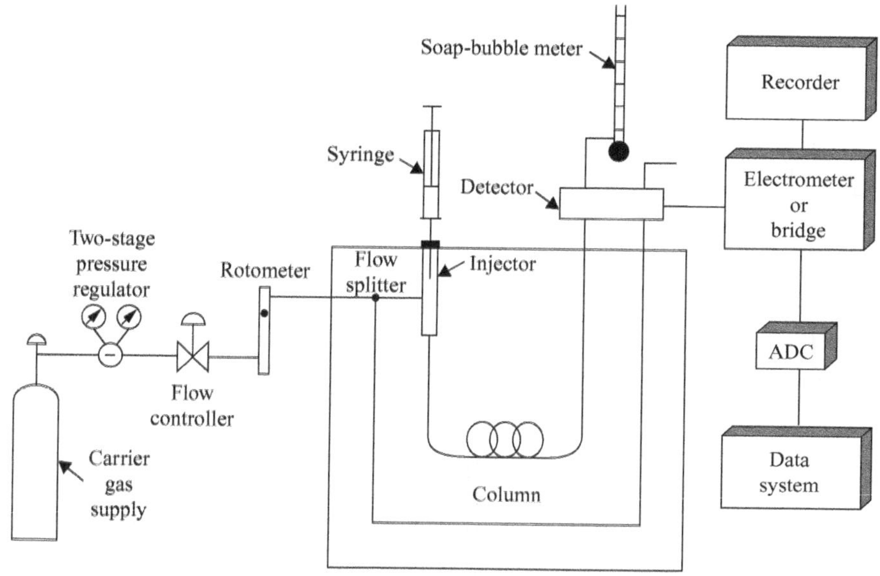

Figure 9.3 System of gas chromatography (GC).

The carrier gas supplied from a high-pressure steel cylinder enters a purifying and drying tube for removal of water after decompression by a pressure-reducing valve. The flow and pressure of the carrier gas are controlled by a needle valve. It then passes through the injector (including the gasification chamber), where the test sample is injected. If the test sample is liquid, it will be gasified instantaneously during passage through the gasification chamber and carried into the column by a constant flow of carrier gas. Compounds in the sample are separated by column, and then successively enter into the detector. Signals from the detector are recorded by a recorder and then a chromatogram is obtained.

GC displays high performance, high selectivity, and high sensitivity, and is easily operated. By GC with a highly sensitive detector, substances can be detected to weights as low as 10^{-11}–10^{-13} g. Therefore, it is a powerful tool for trace analysis, such as impurities with a mass fraction from 10^{-6} to even 10^{-10} orders of magnitude.

9.3.4 Capillary Electrophoresis (CE)

CE was designed to separate ionic compounds including proteins, peptides, DNA, and organic acids on the basis of their size-to-charge ratio in the interior of a small capillary filled with an electrolyte. Based on differences in the capillary and separation systems, CE can be categorized into several modes of separation: capillary zone electrophoresis (CZE), isotachophoresis, micellar electrokinetic capillary chromatography (MECC), capillary isoelecatric focusing (CIEF), as well

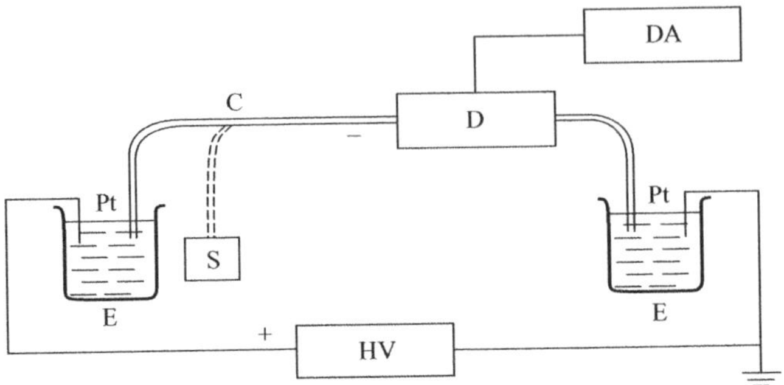

Figure 9.4 High-performance capillary electrophoresis. HV, high-voltage power (0-30kV); C, capillary tube; E, electrode reservoir; Pt, platinum electrode; D, on-column detector; S, sample; DA, data acquisition and processing system.

as capillary electro-chromatography, a very important mode combining capillary and chromatography.

As shown in Figure 9.4, instruments of CE mainly include a high-voltage power supply, a capillary tube, a detector, and two reservoirs (electrode reservoirs). Two ends of the capillary tube are immersed in both electrode reservoirs containing the same electrolyte solution (buffer solution), respectively. The capillary tube is also filled with this buffer. One end is the inlet for the sample and the other end is connected to the detector. After the sample is introduced into the inlet of the capillary tube by an injector, an electric field is applied between both ends of the capillary tube for electrophoretic separation and analysis.

Compared with other separation systems such as HPLC and common electrophoresis, CE has the following advantages for analysis of ionic compounds.

1. *High resolution.* According to theoretical assumptions, the theoretical plate height in CE is proportional to the diffusion coefficient of solute. Its theoretical plates per meter could be several hundred thousands or even up to millions or 10 million, while that of HPLC is generally several thousands to tens of thousands. Therefore, for most molecules, the column efficiency of CE is much higher than that of HPLC.
2. *High sensitivity.* Due to its ultrahigh column efficiency, separation by CE results in sharp and narrow chromatographic peaks. The detection limit of coupled UV detector is greatly lower than that of HPLC. A much lower detection limit could be obtained by using the combined CE-MS technique.
3. *High speed.* CE analysis is carried out under a condition of a high electric field, which costs less time than HPLC. The analytical duration of CE is usually less than 30 min while the shortest is less than 60 s.

CE has become one of the instruments for separation and analysis that has developed the most rapidly in recent years. The high-separation efficiency of CE leads to its more extensive use for analysis of complex components in plants.[3] CE has increasingly become one of the more important approaches in research and control of quality of herbal extracts or products.[4]

9.3.5 Supercritical Fluid Chromatography (SFC)

SFC is a chromatographic technique that utilizes supercritical fluid, typically carbon dioxide (CO_2), as the mobile phase. Compounds are dissolved and separated by means of controlling pressure and adjusting density of the mobile phase to regulate their solubility. The supercritical fluid usually has great solubility, good mobility, and faster mass transfer rate.

SFC integrates the advantages of HPLC and GC. It is particularly suitable to separate and analyze thermo-sensitive and low-volatile compounds; thus, it has a broader scope of application than GC. SFC is superior to HPLC with faster separation speed and easier pretreatment of samples because of the high diffusibility and low viscosity of supercritical fluid. In addition, SFC can be used in conjunction with various large analytical apparatuses such as a mass spectrometer, Fourier transform infrared spectrometer, and nuclear magnetic resonance analyzer, making it an effective approach for analysis of complex samples.

SFC usually consists of chromatographic columns, a column oven, an automated flow-injection apparatus, a supercritical fluid control system (including a supercritical fluid pump, a modifier pump, and a current limiter), a detector, and a data process system. Supercritical fluid refers to a state of substance at whose temperature and pressure above its critical point with characteristics of both gas and fluid. CO_2 is mostly applied as the mobile phase in SFC. Therefore, the entire chromatographic flow path must be pressurized. The stationary phase of SFC contains either solid adsorbents, such as silica gel, or polymers bonded to a carrier or the wall of the capillary tube. There are packed column and open tubular column for SFC.

Application of HPLC columns loaded with high-performance stationary phase fillers to SFC greatly reduces analytical duration and increases separation efficiency. Silica gel without inactivated modification is suitable for analysis of nonpolar compounds. The utilization of packed columns with chemically bonded fillers significantly extends the range of SFC application. Currently, detectors most commonly used in SFC are UVD and flame ionization detector (FID). Both are characterized with high sensitivity and good selectivity.

The application of SFC has gradually become extensive with the development of chromatographic techniques. SFC can be used to analyze many types of natural products, such as thermally labile esters, steroids, polyunsaturated fatty acids, pigments, amino acids, and sugars. It is particularly suitable for heat-labile compounds. Preparative SFC has many advantages in separation and purification of compounds in herbal extracts because of its high speed and efficiency. Extracts of natural products, food additives, and chiral compounds can all be separated favorably by SFC.

9.3.6 Ultraviolet-Visible Spectrometry (UV-Vis)

UV-Vis spectrometry is an analytical method based on molecular absorption at the UV and visible light regions. Test samples using this method need to have strong absorptions between wavelengths of 200–800 nm. In other words, compounds measured by UV-Vis spectrometry usually have one or more chromophores and conjugation systems in their structures.

When a specific intensity of incident beam of light passes through the medium containing a homogeneous test sample in a cell, the light will be partly absorbed. The unabsorbed light will permeate the test sample solution and be lost through scattering and reflection on the liquid and vessel surfaces. Sometimes the loss may be up to 10%. These influences must be deducted by using a reference solution in a reference cell when measuring the samples. When the wavelength of the incident light is in a certain range, the absorbance and concentration of the test solution is proportional to its concentration and thickness of the solution (Lambert-Beer Law). Now UV-Vis is mostly coupled with high- or medium-pressure liquid chromatography as a detector for quantitative or qualitative analysis of the column separated compounds.

9.3.7 Near Infrared Spectrometry (NIRS)

IR radiation is electromagnetic radiation whose wavelength is between the visible and microwave regions. IR can be divided into three regions: far-IR (400–10 cm^{-1} or 1000–30 μm), mid-IR (4000–400 cm^{-1} or 30–2.5 μm), and NIR (14,000–4000 cm^{-1} or 2.5–0.8 μm). According to *ASTM International*, originally known as the *American Society for Testing and Materials*, the precise range of NIR is from 780 nm to 2526 nm. Relatively, NIR has higher energy and potent penetration capability. It produces less heat due to its small absorption coefficient.

IR radiation is absorbed by organic molecules and converted into energy of molecular vibration, either stretching or bending. Different functional groups absorb different wavelengths of IR radiation due to their different types of bonds. The general IR spectroscopy is utilized in chemistry to record absorption of organic compounds in the mid-IR to provide spectra of organic compounds. It is mainly used for structural identification and elucidation of organic compounds, including natural compounds isolated from herbs (see Chapter 4 for detailed introduction).

NIR spectrum falls into the category of doubling and dominant frequency absorption spectrum of the molecule vibration spectrum, which is produced by the transition from ground to high-energy state due to harmonic molecular vibration. NIR is primarily absorbed by doubling and combining frequency of hydrogen-containing groups X-H (X=C, N, O), which contain information of most organic compounds and molecular structures.[5] Different organic compounds will show different NIR absorptions, and absorptions will be altered in different physical and chemical environments.

Under NIR light, if the frequency of the wavelength of the light is the same as the vibration frequency of hydrogen-containing compounds, resonance phenomenon will happen. The energy of the light can be transmitted to molecules while molecular dipole moment changes. But if the frequency of the light differs from the vibrating frequency of the samples, there will be no absorption. Therefore, when a sample is detected with IR lights of continuous wave length spectra, an absorption spectrum will be obtained, and the sample can be determined through analysis of the wavelength and intensity of the absorbance.

Different from other analytical methods, NIR is an indirect analysis method, in which a calibration model between the attribute value and IR spectrum data is constructed by statistical methods. Therefore, before analyzing a real sample, a group of known samples (called calibration samples) needs to be collected to acquire the NIR spectral data, and analyzed with another analytical method (called reference method) for comparison or reference.

Dubbed as an analytical "giant," NIR spectrometry is a state-of-the-art method in analytical chemistry that has been developed rapidly in recent years. It is drawing more and more attention from analytical chemists and experts. NIRS can be used for both qualitative and quantitative analyses. Its advantages include: (1) requires no pretreatment for analysis; therefore, no reagents are consumed and the sample is not processed; (2) analysis of several parameters can be completed by one sampling within several minutes; (3) has good repeatability and the cost of investment and performance is low.

However, NIR also has disadvantages: (1) its sensitivity is low and the relative error is large; (2) because it needs a reference method, it requires a certain number of sample data; thus the accuracy of this method is unable to reach that of the reference method. Meanwhile, the construction of the model requires not only expertise in chemometrics but also expense in money and time; (3) the scope of NIR detection is limited to components containing hydrogen groups or other relevant substances, and these components should be more than 0.1% of the content. For these reasons, this technique is effective and convenient for QCs of routinely tested samples, for example, batch samples of products in pharmaceutical manufactories, including herbal products, but it is not suitable for occasional sample analysis or scattered samples.

Recently, the combination of NIR spectroscopic technique with computers and the fiber-optic techniques allows for direct analysis of opaque samples in forms of solid granule, powder, or paste without the application of chemical reagents or treatment of samples, using optical detection methods such as transmission, scattering, and diffuse reflectance. This provides a convenient and rapid analytical technique for real-time online analysis and process control of drug production.

NIRS has been widely used in QC and evaluation of drugs, including quality evaluation of drug substances and excipients, determination of physicochemical properties (crystal size and form) of drug, QC of formulated granules, and quantitative analyses of drugs. It has also been applied toward qualitative and quantitative analyses of herbal medicines and identification of herbal materials.[6,7]

9.3.8 Mass Spectrometry (MS) and Its Combination with Chromatography

MS is an analytical technique for the determination of the elemental composition of a compound or sample. It has been increasingly used in qualification and quantification of samples, such as pharmaceuticals, foods, natural medicines, and biochemical products.

A mass spectrometer instrument primarily consists of a vacuum system, an injection system, an ion source, a mass analyzer, an ion detector, and a recorder. Types of ion sources commonly used in MS system include electron ionization (EI), chemical ionization (CI), matrix-assisted laser desorption/ionization (MALDI), fast atom bombardment (FAB), field ionization and field desorption (FI, FD), ESI, atmospheric pressure CI (APCI), and inductively coupled plasma (ICP). Types of mass analyzers include ion trap, quadruple (single and triple quadruple), quadruple ion trap, time-of-flight (TOF), and a Fourier transform ion cyclotron resonance (FTICR). A detailed introduction to MS principles and prescriptions of these ion sources and analyzers was given in Chapter 4.

The selection of the MS instrument is very important for successful analysis of a sample. If a lab decides to buy a new MS instrument, the decision for the type of MS should depend on the properties of the most frequently analyzed samples and the purposes of the analysis. If a lab already has MS equipment, researchers should be aware of the characteristics of each instrument and know whether the available MS is suitable for analysis of the samples. The ion source and analyzer of the MS instrument are the two important factors determining the capability of MS analysis. In general, EI and CI are usually used for gases and evaporated samples, while ESI, APCI, FAB, and MALDI are often used with liquid and solid biological samples. EI and CI are commonly used in GC-MS systems, while ESI and APCI are choices in LC-MS systems. Relatively, ion trap analyzers are the choice for qualitative identification of compounds, while quadruple analyzers are better for quantification. TOF is the fastest MS analyzer, so might be favored as the choice for high throughput. Characteristics of the commonly used ion sources and analyzers are briefly summarized here.

The coupling of MS with chromatography technique combines the advantage of GC or HPLC in separating a mixture and the high resolution of MS in identifying a compound in a mixture. Therefore, it is a very powerful tool for qualitatively and quantitatively analyzing complex compositions of herbal extracts. HPLC-MS and GC-MS have become the most essential and reliable weapons for identification and quantification of compounds in herbal extracts, comparisons of the chromatogram fingerprint of herbal products, and analysis of *in vivo* metabolism of herbal compounds.

Characteristics of Ion Sources and Analyzers

1. *Ion sources*

EI is the classical ionization technique in MS. It is the one used to produce "classical" compound spectra that are library searchable and/or interpretable, therefore, useful for compound identification and structure elucidation. This is a comparatively rugged and sensitive ionization technique. The analytes for EI need to be vaporized first. The evaporated sample is bombarded with 70 eV electrons in the ion source with low pressure. There is no ion-molecule reactions, for example, no $[M + H]^+$ signal due to no observed proton transfer. The application of EI is restricted to thermally stable samples with low molecular masses (<ca. 2000 Da).

CI is an ionization technique similar to the classical EI. In CI source, the analyte is ionized by chemical ion molecule reactions during collisions in the source. In comparison with EI, it provides better molecular weight information. Quantification by CI usually needs internal standards.

FAB is an ion source with high-energy beams of neutral atoms. The FAB spectrum contains only a few fragments and a signal for the pseudo-molecular ion, for example, $[M + H]^+$, $[M + Na]^+$ adducts, which makes FAB useful for molecular weight determination. FAB provides an efficient means of analyzing polar, ionic, thermally labile, or large molecular compounds that are not easy to analyze or are incapable of being analyzed by EI or CI; therefore, it has been used for the analysis of polar biomolecules, detergents, surfactants, and natural products.

MALDI is a laser ionization method that is suitable for analyzing large, polar, and ionic compounds such as proteins, DNA fragments, polymers, or metal complexes. It is a soft ionization method that can provide molecular weight information, but less fragmentation. The sensitivity depends strongly upon the structure of the analyte. It can measure molecular weight up to 500 kDa, commonly 5–100 kDa, depending on the analyzer. It can be coupled with a TOF analyzer or a Fourier-transform mass spectrometer.

FI produces mass spectra with little or no fragmentation. They are dominated by molecular radical cations M^+ and less often, protonated molecules $[M + H]^+$.

API is used in conjunction with LC-MS techniques. The ions are formed at atmospheric pressure. ESI and APCI are the two common types of atmospheric pressure ionization.

ESI is a soft ionization method that provides molecular weight information and less fragmentation. The sensitivity depends strongly upon the structure of analyte. It not only enables LC-MS coupling for detecting most small molecular compounds, but is also suitable for analyzing polar and ionic compounds (e.g., metal complexes) and large bio- or synthetic polymers.

APCI can provide molecular weight information as well. In comparison with ESI, it is more suitable for analyzing less polar compounds with increased fragmentation. Its sensitivity also depends strongly upon the structure of analyte. When coupling MS with LC, the flow rate could be up to 1 mL/min.

ICP-MS utilizes a plasma, a gas that contains a sufficient concentration of ions and electrons to conduct electricity. ICP-MS is highly sensitive and capable of determining a range of metal and several nonmetals at concentrations below one part in 10^{12}.

2. *Analyzers*

Magnetic sector MS has high resolution, high sensitivity, high dynamic range, very high reproducibility, and the best quantitative performance of all mass spectrometer analyzers. However, its larger size and higher cost limit its popularity.

Quadruple provides classical mass spectra with good reproducibility. There are single quadruple and triple quadruples. The triple quadruple can produce low-energy collision-induced dissociation (CID) MS/MS spectra. The relatively small and low-cost systems make it the majority in benchtop GC-MS and LC-MS systems.

TOF is the fastest MS analyzer with high ion transmission and highest practical mass range of all MS analyzers. It requires pulsed ionization method or ion beam switching, so it is frequently coupled with MALDI. However, there is limited precursor-ion selectivity for most MS/MS experiments and limited dynamic range due to the fast digitizers used in TOF. It provides a very fast GC-MS system. But relatively, the resolution and accuracy of the spectra are low.

There are two principal trapped-ion mass analyzers: three-dimensional quadruple ion traps and ion cyclotron resonance mass spectrometers.

Ion cyclotron resonance, also called "static" traps, has the highest recorded mass resolution of all mass spectrometers, powerful capabilities for ion chemistry and MS/MS experiments, nondestructive ion detection, and stable mass calibration in superconducting magnet FTICR systems. It is well-suited for high-resolution MALDI and ESI for high-mass analytes, ion chemistry, and laser desorption for materials and surface characterization.

Quadruple ion trap, also called "dynamic" traps, has high sensitivity, multistage mass spectrometry, but very poor dynamic range. It is commonly seen in benchtop GC-MS, LC-MS, and MS-MS systems, and is suitable for qualitative identification of compounds, including ionized compounds in natural products, but not a good choice for quantification.

Fourier transform mass spectrometry (FTMS) is relatively expensive, difficult to handle, has low dynamical range, but is very accurate.

GC-MS

Similar to HPLC, the advantage of GC-MS is that it can perform separation, identification, and quantification for a sample by one injection, greatly promoting the analysis of herbal samples. The GC-MS detection limit can reach up to 10^{-9}–10^{-12} g. Generally, GC-MS can be used to qualitatively analyze and quantitatively test most herbal samples. The wide application of GC-MS also covers analysis of pesticide residues and solvent residue in herbal extracts or related products. However, GC-MS is not an appropriate choice for analyzing highly polar, heat-labile, and nonvolatile macromolecular organic compounds.

For MS in the GC-MS system, EI and CI are the most commonly used ion sources, and the quadruple is the most commonly used analyzer. For quantification, triple quadruple is more selective than single quadruple.

LC-MS

In contrast to GC-MS, LC-MS is more suitable for separation and analysis of heat-sensitive and relatively polar samples. Breakthroughs in LC-MS development have been accompanied by the advent of techniques including APCI, ESI, and particle beam (PB) interfaces. In liquid chromatographic technique, the application of smaller (3 μm) particle size stationary phase and narrow-bore columns increases the column efficiency while reducing the flow of the mobile phase. LC-MS has now become one of the most important tools in the fields of pharmaceutical, clinical medicine, chemistry, and chemical engineering. It is especially suitable for structural identification and quantitative analysis of herbal extracts and products.

CE-MS

CE-MS is a newly hyphenated technique developed in the end of the 1990s. It integrates the advantage of the high resolution of CE and that of MS. Although there are some similarities, CE-MS differs primarily from LC-MS in that the flow of background electrolyte of CE is much smaller than that of the mobile phase of HPLC. Therefore, the interfaces are different between them. In general, a mass spectrometer that can be used in LC-MS may also be suitable for CE-MS when the interface is changed. The same mass spectrometer may be jointly used by CE and HPLC by installing an interface. Despite its recent advent in this century, CE-MS has become an effective tool for analysis of complex mixtures. At present, CE-MS is mainly used for qualitative identification and composition analysis of biomacromolecules, such as proteins and peptides in herbal medicines, and their products.[8–10]

9.3.9 Atomic Absorption Spectrometry (AAS) and Atomic Fluorescence Spectrometry (AFS)

Both AAS and AFS are atomic optical spectroscopes. It is known that electrons exist in energy levels within an atom. They can move between different levels by absorbing or emitting an amount of energy that equals the energy difference between levels. When energy is absorbed, the electron moves to a higher level; when energy is emitted, the electron moves to a lower level. Different atoms need different energies for transition. Because the wavelength of light is related to its energy, the wavelength of light is usually measured for quantification.

There are two types of AAS instruments available: single beam and double beam. In addition, multiple-channel AAS can determine multiple atoms. Detailed introduction about the instrument is available from books[11] and web sites.

AAS is an analytical method for determining the content of metal elements in a sample.[12] It is a preferred method for quantitative analyses of metal ions, especially heavy metal ions, in samples of herbal extracts or related products. It is a quantitative analytical method with good efficiency, accuracy, and sensitivity. AAS has been increasingly used due to the simplicity of the instruments and ease in operation.

Under experimental conditions of AAS, the electrons of the atoms are rapidly excited to higher orbits by absorbing energy emitted from light of a given wavelength. The amount of energy is specific to a particular electron transition in a particular element. In general, each wavelength corresponds to only one element. The calculation of the energy for transition is based on the Beer-Lambert law. The intensity of a signal is proportional to the concentration of the element being measured.

AFS has been widely used in analysis of inorganic elements and quantitative analysis of metals including heavy metal residues in biological samples and agricultural samples, including herbal medicines.

When energy stored in the atoms is released as light, this is known as fluorescence. AFS measures this emitted light. The intensity of the fluoresced light is directly proportional to the concentration of atoms. The wavelength reveals the identity of the atoms.

The advantages of AFS include low detection limit, high sensitivity, simple spectral line, less interference, wide linear range (up to 3–5 orders of magnitude), and simultaneous measurement of multiple elements. The main advantage of fluorescence detection compared with absorption measurements is the greater sensitivity achievable because the fluorescence signal has a very low background. For example, the detection of Cd and Zn could reach as low as 10^{-12} g/mL and 10^{-11} g/mL, respectively.

9.4 QUALITATIVE ANALYSIS OF HERBAL EXTRACTS AND PRODUCTS

The qualitative analysis of herb extracts and related products include identifying the bioactive, major, or characteristic components in herbal extracts; distinguishing the authenticity of species; and evaluating their qualities. The analysis can be carried out from a simple description of raw materials, physical properties, and chemical reactions to advanced chromatography and spectroscopy analyses, depending on the analytical purposes. Identification of natural compounds or herbal extracts usually requires a combination of several methods, as introduced below. In general, it is necessary to preliminarily judge the color, odor, taste, solubility, and so on, of the target compounds or herbal extracts, and then to identify and recognize them by means of chemical reaction, chromatography, spectroscopy, fingerprint, and so on.

9.4.1 Preliminary Identification

Preliminary identification refers to the identification and determination of the characters, solubilities, physical constants, and so on, of herbal extracts; these observations will help to judge the species varieties and quality of herbs.

Character Identification

The characters of herbal extracts, including appearance, particle size, color, odor, and taste, can be easily identified by simple identification methods such as visualization, touch, smell, and taste. These traditional identification methods are quick and easy, so they are usually the first step of identification.

The color, odor, and taste all come from the components in the raw material. Generally, the color of an herbal extract is related to the parts of the plant from which it originates. For example, a red or purple colored fruit extract is relatively colored due to the existence of anthocyanins, proanthocyanins, or other colored compounds; stem and leaf extracts usually show green or yellow colors because of the chlorophyll or flavonoids in it; a dark colored root material will produce a dark extract, such as the root of *Rehmannia glutinosa,* while a white root material will give a light colored extract, such as *Dioscorea opposita.*

The odor comes from the essential oil (e.g., cinnamon and star anise), coumarins (angelica), sulfur (garlic), or other compounds. These special odors can be used for preliminary identification.

The tastes vary also depending on the components of the material. For example, the extracts of dark plum and hawthorn should taste acidic because of the organic acids in the fruits; extracts of coptis and phellodendron should taste bitter because of the alkaloids; and extracts of glycyrrhiza should taste sweet because of the glycyrrhizin. If the taste of an herb extract becomes abnormal, the species or quality of the raw material should be confirmed.

Solubility

Solubility is one physical property of herbal extracts. The solubility of an herbal extract in different solvents is usually described as "very easily soluble," "easily soluble," "slightly soluble," "almost insoluble," or "insoluble."

Physical Constants

Physical properties are important indexes to identify the quality of herbal extracts or products. The detected physical constants usually include relative density, optical rotation, refractive index, hardness, viscosity, boiling point, freezing point, and melting point. The physical constants play important roles in identification of the authenticity and purity of herbal extracts, especially of volatile oils.

9.4.2 Chemical Reaction

Chemical reaction for identification of herbal extracts is based on chemical constituents in the extracts. The purpose is to judge whether herbal extracts contain the types of compounds that have been reported in the raw materials of the right species. These chemical reactions are relatively simple and fast, usually easily observable through production of color, gas, fluorescence, or precipitation under suitable conditions. Characteristic chemical reactions and methods for identification of each type of compound have been introduced in Chapter 4. A summary of the most common methods for chemical identification is given in Table 9.3. The following are only given as examples.

Table 9.3 Chemical Identification of Compounds in Herbal Extracts

Compounds	Preparation	Test method	Result
Phenols	Acidic alcohol extract	Ferric chloride test	
		Diazotization test	Turns yellow and slightly red
Tannins	Hot water extract	Ferric chloride test	Turns purplish red
		Sodium chloride gelatin test	Turns turbid
		Alkaloid reaction	Turns turbid when dropping
Alkaloids	Acidic alcohol extract	Bismuth potassium iodide test	Precipitations generated
		Mercuric potassium iodide test	Precipitations generated
		Silicotungstic acid test	Precipitations generated
		Iodized potassium iodine test	Precipitations generated
Carbohydrates, polysaccharide, glycosides	Hot water extract	Alkaline tartaric acidic test	Brick red precipitations generated
		α-Naphthol test	Purple rings generated
Saponins	Hot water extract	Foam test	Foam generated and last for a long time
		Acetic anhydride—conc. sulfuric acid test	Purple rings generated
		Chloroform—conc. sulfuric acid test	The chloroform layer turns dark red; the acid layer has fluorescence
Steroids, terpenes	Alcohol extract	Glacial acetic acid—conc. sulfuric acid test	A series of changes in color
		Chloroform—conc. sulfuric acid test	The chloroform layer turns dark red; the acid layer has green fluorescence
Flavonoids	Alcohol extract	Hydrochloride—magnesium power reaction	Turns dark red
		Lead acetate precipitation reaction	Yellow precipitations generated
		Magnesium acetate test	Obvious color changes
Anthraquinone	Methanol extract	Alkali test	Obvious color changes
Cardiac glycosides	Methanol extract	3,5-Dinitrobenzoic acid test	Turns red
		Acid picric acid test	Turns orange
		Sodium nitroprusside test	Turns red
Lactone, coumarins	Methanol extract	Ferric hydroxamic acid test	Turns purple
		Diazotization test	Turns red

Table 9.3 (*Continued*)

Compounds	Preparation	Test method	Result
Amino acids, polypeptides, proteins	Cold water extract	Precipitation test	Turns turbid
		Biuret test (for peptide bonds)	Turns violet
		Ninhydrin test (for NH_3 or NH_2- and –NH-)	Turns blue or purple
Volatile oil, fats	Petroleum ether extract	Vanillin-conc. sulfuric acid reaction on TLC for volatile oil	The spots of volatile oil on developed TLC turn to bright color
		Filter paper test for volatile oil and fats	The volatile oil on the paper will present as an oily spot first after solvent volatilized, then slowly disappear, but the fat spot will remain.

Color Reaction

Some chemical constituents in herbal extracts can react with certain reagents to produce special color reactions under certain conditions. For examples, ferric chloride can react with compounds containing phenolic hydroxyl groups (e.g., flavonoids and anthroquinone) or other substitutuent groups that can be hydrolyzed into phenolic hydroxyl groups to generate different colors; ferric hydroxamic acid can react with compounds containing aromatic acids and their esters, amides, and so on., such as coumarin and terpene lactone; ninhydrin can react with compounds containing aliphatic amine, such as amino acid, polypeptide, and protein in their structure.

Precipitation Reaction

Some kinds of compounds in herbal extracts can react with certain reagents to produce differently colored precipitations under certain conditions. For example, some alkaloids can react with bismuth potassium iodide, ammonium reineckate, and metallic salts (e.g., lead salt, barium salt, or copper salt) to produce different colored precipitates; Therefore, for identification, an extract of Vietnamese Sophora Root (*Sophora tonkinensis*) dissolved in hydrochloride can react with mercuric potassium iodide solution to form a light yellow precipitate; extract of grape seed can react with copper acetate to produce a brown-yellow precipitate; extract of Decumbent Corydalis Tuber (*Corydalis decumbentis*) can react with bismuth potassium iodide to generate a red precipitate; and extract of celandine (*Chelidonium majus*) can react with ammonium reineckate to generate a black precipitate.

Fluorescence Reaction

Some compounds in herbal extracts can generate different colored fluorescence under UV or natural light. For example, an extract of cortex periplocae shows a purple fluorescence, while a water extract of ash bark shows an azure fluorescence. Some herbal extracts have no fluorescence under regular light, but can produce fluorescence under a UV lamp after being treated with acid, alkali, or other chemical methods. For instance, when a water extract of aloe is heated with borax, aloin in the extract will react with borax and generate a yellowish green fluorescence; when extract of bitter orange is dropped on a filter paper and sprayed with a solution of 0.5% magnesium acetate-methanol then dried, it will show a light blue fluorescence.

Foam Reaction and Hemolysis Index

An aqueous solution of herbal extract containing saponins can produce persistent foams and hemolytic phenomena.

Microsublimation

A few compounds in some herbal extracts can sublimate at a certain temperature. The shape and color of the crystal obtained from the sublimates can be used as identification characteristics. For examples, the sublimate of mint extract is a cluster of colorless needle crystals. The crystals will change from yellow to orange yellow with the addition of two drops of concentrated sulfuric acid and a small amount of vanillin crystal, and to purplish red after adding one drop of distilled water. The sublimate of rhubarb extract displays as yellow needle crystals at low temperature and dendritic and pinnate crystals at high temperature; the crystals will change into red when added to alkali solution.

Microchemistry

Microchemistry is a method in which a small amount of herbal extract solution is dropped onto a glass slide, then reacted with chemical reagents to form precipitates or crystals that can be observed under a microscope. For example, after adding alcohol first and then water onto pepper extract, piperine will precipitate as needle crystals along the margin of the cover glass. By adding a few drops of chloroform, then a few drops of 2% phenylhydrazine hydrochloride, to cinnamon extract, yellow needle or rod crystals can be observed.

9.4.3 Chromatographic Methods

Chromatography has become an indispensable routine and effective method in identification of herbal extracts or products. It has the dual advantages of separation and qualitative analysis, especially for herbs or their products with complex composition. The common chromatographic methods used to identify the authenticity of herbal extracts include TLC, paper chromatography (PC), GC, HPLC, and CE. Identification of a compound by a chromatography method is mainly based on its retention factor (R*f*) in TLC and PC or retention time (Rt) in HPLC, GC, or CE. It needs a known standard compound as a reference. When the chromatography instruments are connected with the spectrometers, the identification becomes more reliable.

TLC

At present, TLC is still the most commonly used, cheapest, and easiest method in identifying compounds in herbal extracts for many phytochemists. It can be used for identification of most minimally polar and small molecular natural compounds. A detailed introduction of TLC and its application for identification has been introduced in Chapters 3 and 4.

When TLC is used to identify the authenticity of herbal extracts, standard compounds of the known major or characteristic components in the herb are normally used as references for comparison. If the positions (R*f*) and colors of major compounds in the test extract are the same as those of the reference standards on three plates that were respectively developed with different solvent systems, the species of material for extraction can be confirmed. For example, there are several species used as radix curcumae (*Curcuma wenyujin, Curcuma longa, Curcuma kwangsiensis, Curcuma phaeocaulis*). In order to identify the varieties, TLC can be used to compare the samples with standards of the known principal components including curcumin, germacrone, curdione, and curcumol that have been reported in radix curcumae.

Introduced here is a detailed method for TLC identification of extracts between Cortex Phellodendri (*Phellodendron chinense* and *Phellodendron amurense*) and aspen bark (*Populus davidiana*), its counterfeit species. Take 5 mg extracts of aspen bark and phellodendron, add 10 mL of methanol, heat on a water bath and reflux for 10 min, then filter to obtain the solutions of test samples. Dissolve a standard of berberine hydrochloride into methanol as reference solution. Drop 1 μL of the three solutions, respectively, on a silica gel G TLC plate, develop the plate with a reagent of benzene: ethyl acetate: methanol: isopropanol: ammonia (6:3:1.5:1.5:0.5) in an ammonia vaporsaturated TLC tank. Take out the developed TLC plate, air-dry it, and observe the result under a UV lamp at a wavelength of 365 nm. There should be two obvious yellow spots for phellodendron, one of which matches the yellow spot of berberine on TLC. For aspen bark, there should be only one bright blue spot (see Fig. 9.5).

PC

PC is a chromatographic method using the principle of partition. Although most secondary natural products are analyzed by TLC, PC is still used for separation and analysis of amino acids, sugars, and other kinds of water-soluble compounds in

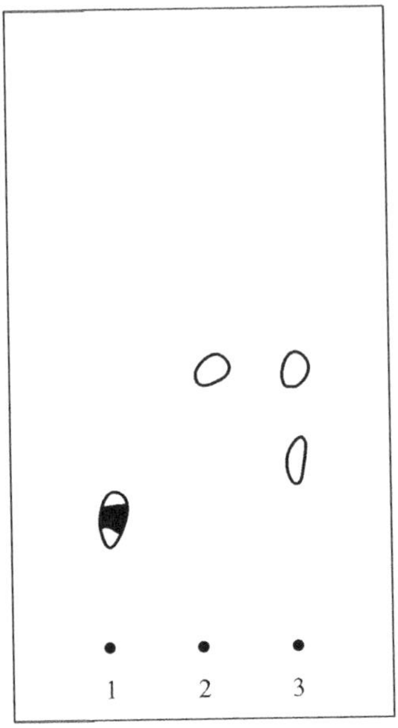

Figure 9.5 Thin layer chromatogram of aspen bark extract and phellodendron extract. 1, Aspen bark extract; 2, berberine hydrochloride; 3, phellodendron extract.

herbal extracts. The detailed introduction of PC and its application for identification have been introduced in Chapter 3.

GC

In comparison with HPLC, GC has the advantages of high sensitivity and high-separation efficiency. It is particularly suitable for analysis of volatile oils or other volatile lipophilic compounds in herbal extracts. Compounds containing functional groups with active hydrogens (e.g., -COOH, -OH, -NH, and -SH) can be made into less polar derivatives by alkylation, acylation, and silylation for GC analysis.[13] The most common one is the trimethylsyl (TMS) derivatives. However, these compounds are now mostly analyzed directly by HPLC. Nonvolatile components can also be identified by pyrolysis GC.[14] MS is the most common detector that couples to GC.

HPLC

In comparison with GC, HPLC is more widely applied for analysis of natural compounds, including those with low volatility and poor thermal stability that GC is incapable of analyzing. HPLC can couple with different types of detectors such as refractive index (RI), UV, fluorescent, radiochemical, electrochemical, NIR, MS, NMR, and light scattering (LS). HPLC is applied to identify compounds based on Rt and spectrum (e.g., UV spectrum in diode array or MS spectrum in MS detector) of each chromatographic peak and distinguish different herbal extracts based on fingerprints of chromatogram.

Alkaloids, terpenoids, flavonoids, steroids, amino acids, proteins, and other types of natural compounds can all be qualitatively analyzed by HPLC. Each type of compound needs appropriate columns and detectors. For example, most flavonoids, anthroquinones, and coumarins have strong and characteristic UV absorptions. Therefore, UV is a good detector for these compounds. But UV may not a good choice for saponins and other compounds that have no UV absorptions due to the lack of conjugated double bonds. MS is currently the detector that can detect the highest variety of compounds. Relatively, the ion trap analyzer is more suitable for structural elucidation of unknown compounds than the quadruple analyzer because of its capability of MS/MS.

Many HPLC methods for the analysis of herbal extracts or related products are available from the literature.

9.4.4 Spectroscopic Methods

Spectral identification of a compound is more reliable than chromatography methods. It is not only able to identify a known compound, but also capable of elucidating the structure of an isolated unknown compound (see Chapter 4). When the spectrometers are coupled to chromatography instruments, online identification of a compound in a mixture makes the qualitative analysis of an herbal extract or product simple, quick, and reliable. Different types of spectrometry instruments have different characteristics and different ranges of application.

UV-Vis Spectrometry

Most phenolic compounds, such as flavonoids, anthroquinones, coumarins, anthocyanins, and other compounds containing conjugated double bond(s) with chromophore(s) in herbs have strong UV-Vis absorptions.

The qualitative analysis of natural organic compounds by UV-Vis spectrometry is based on their spectral characteristics, including the shape, peak number, wavelength, and intensity of each peak, and value of absorption coefficient, which depend on the chromophores, auxochromes, and conjugation status in the structure of each compound.

When UV-Vis spectrum is used for identification of a compound or an herbal extract, the spectrum of a standard sample is generally used for comparison with the test sample under the same condition. If both spectra are exactly the same, it can be said that the test sample has the same conjugated system(s) and chromophore(s) as the standard, but their structures may not be exactly the same. If there is no standard sample, a standard spectrum can be used for comparison. The UV-Vis spectra of an herbal extract is the superposition of absorption spectra of each component in the extract. Different batches of an herbal extract should have the same UV-Vis absorption spectrum under the same condition. However, on the other hand, if two different extracts have the same spectra, it can only be said that the structures of the components in the two extracts have the same chromophores and conjugate system, but not that they have the exact same structures.

Derivative Spectrometry

Derivative spectrum is also called differential spectrum. It is a technique to eliminate the interference in UV spectrum. It has the characteristics of sharp signal and high resolution, and can therefore effectively separate the overlapped spectra. It can also distinguish between small absorption peaks and shoulder peaks in the UV spectrum to provide more information. It has a greater practical value to herbal extracts that cannot be easily discriminated by UV spectrum.

Derivative spectrometry provides an effective tool for discriminating different species of herbal extracts and the extracts obtained from the same plant but collected from different regions or extracted with different extraction procedures.

For example, essential oil samples extracted from Fructus Alpiniae oxyphyllae were analyzed by first-order and second-order derivative spectrum, and compared for the difference between samples originated from Hainan and Guangzhou.[15] The result showed that the two samples have difference absorption spectrogram as well as first-order and second-order derivative spectrograms when measured by first derivative UV-Vis spectrometers. Therefore, this method can be used as a reference standard for qualitative identification of essential oil extracted from fruits of *Alpinia Oxyphylla* Miq.

Near Infrared Spectrometry (NIRS)

NIRS is recently becoming popular for chemical analysis. Its applications for discrimination of varieties of paddy, tea, apple, and Chinese herbs have been reported in Chinese journals. However, only a few English references for NIRS applications on herbal medicine are available.[16] The advantages and disadvantages of NIRS for chemical analysis have been mentioned in Section 9.3.7.

AAS

AAS is particularly suitable for analyzing metal elements and some other elements, such as Pb, As, Hg, Sb, Se, and Te. It is a commonly used method for detecting heavy metals and trace elements in herbal extracts and related products. AAS is characterized by specificity, high sensitivity, rapid determination, simple spectral line, less interference, wide linear range (up to 3–5 orders of magnitude), and detecting multiple elements spontaneously. One of the important applications of AAS in herbal medicine is to monitor the contamination by heavy metals of herbal plants.

9.4.5 Application of Fingerprint in Qualitative Analysis of Herbal Extracts

Herbal medicine is composed of complex chemical constituents. Simple analysis of one or two of the major compounds in an herbal extract is not specific enough for QC because this compound may exist in many plants, especially plants in the same genus or family, and most important, the bioactivity of an herbal medicine is an integral contribution of several types of compounds in the plant. For example, saponins, flavonoids, polysaccharide, and several other types of compounds in ginseng have all been reported to show different bioactivities by working on different targets. Thus, in addition to quantitative determination of major compounds, the whole pattern of pharmacologically active and chemically characteristic constituents need to be presented to distinguish and identify herbal extracts in order to accurately demonstrate the similarities and differences between various samples.

Fingerprint method has been accepted worldwide as an effective method for QC of herbal extracts or products. It can not only effectively identify species for authentication of samples, but also quantify, based on peak area or proportion of the main characteristic peaks in a chromatographic fingerprint for QC.

Fingerprint analysis for herbal medicine utilizes modern separation and detection approaches to obtain profiles of herbal medicine. It includes DNA fingerprints of raw materials and chemical chromatographic fingerprints of extracts or products. The DNA fingerprint methods have been introduced in Chapter 2. Fingerprint in this chapter mainly refers to chromatographic fingerprint of chemical composition. TLC and HPLC fingerprints are the two most commonly used methods.

Chromatographic fingerprint is an analytical method with characteristics of integrity very suitable for QC of medicinal herbs that are composed of complex compounds. To obtain whole patterns of compounds in an herbal extract with better isolation, the running time for a chromatographic fingerprint is usually longer than analysis for only one or two major compounds. TLC fingerprint analysis is an easy operating and time-saving method with low cost. Except TLC fingerprint, chromatographic fingerprint obtained from the hyphenated chromatography and spectroscopy (e.g., HPLC-diode array detector [DAD], HPLC-MS, HPLC-NMR, GC-MS, CE-DVD, and CE-MS) provides not only chromatograms, but also spectra, depending on the detector used. Therefore, in addition to the chromatogram pattern (Rt and areas of all peaks), identification will be further confirmed based on the spectrum of each peak. For HPLC and GC, a gradient solvent system and increase of temperature is generally applied, respectively. Application of fingerprint for quality evaluation by HPLC-DAD,[17] HPLC-MS,[18] and hyphenation of HPLC with other detectors[19] are available in the literature.

An example of the TLC and HPLC-DAD fingerprints of extracts of Du Huo (Radix Angelicae pubescentis) and its substitutes is given in Figures 9.6–9.8. Du huo is a Chinese herb with many substitutes. But the root of *Angelica pubescens* Maxim *f. biserrata* is the only official species listed as Du Huo in the *Chinese Pharmacopoeas*. TLC and HPLC fingerprints provide easy and accurate methods for distinguishing Radix Angelicae pubescentis from other substitutes.[17]

9.5 QUANTITATIVE ANALYSIS OF HERBAL EXTRACTS AND PRODUCTS

With the development of modern analytical technologies and improvement of human health consciousness, many countries demand stricter QC on herbal products; requirements for quantitative analysis in particular are increasing. In fact,

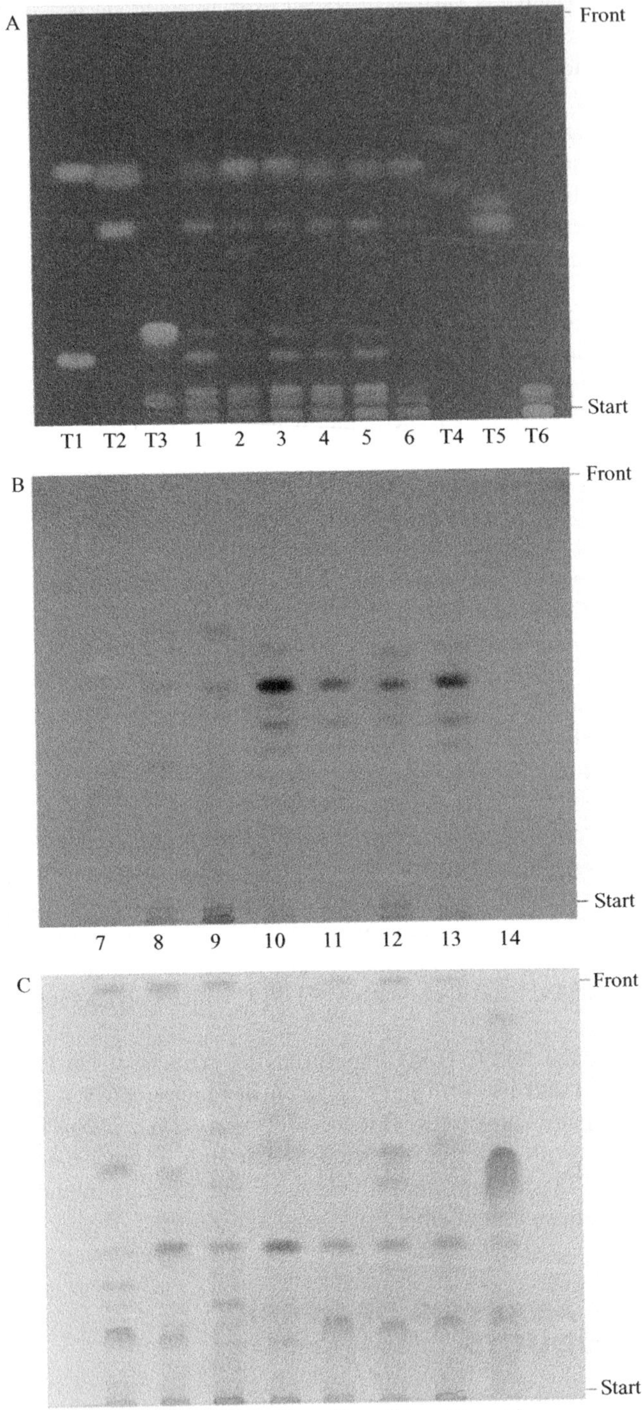

Figure 9.6 TLC fingerprints of extracts of *Angelica pubescens f. biserrata* and its substitutes. (A) TLC under UV 365 nm; (B) TLC under UV 254 nm; (C) TLC after spraying with anisaldehyde sulphuric acid regent (VIS).

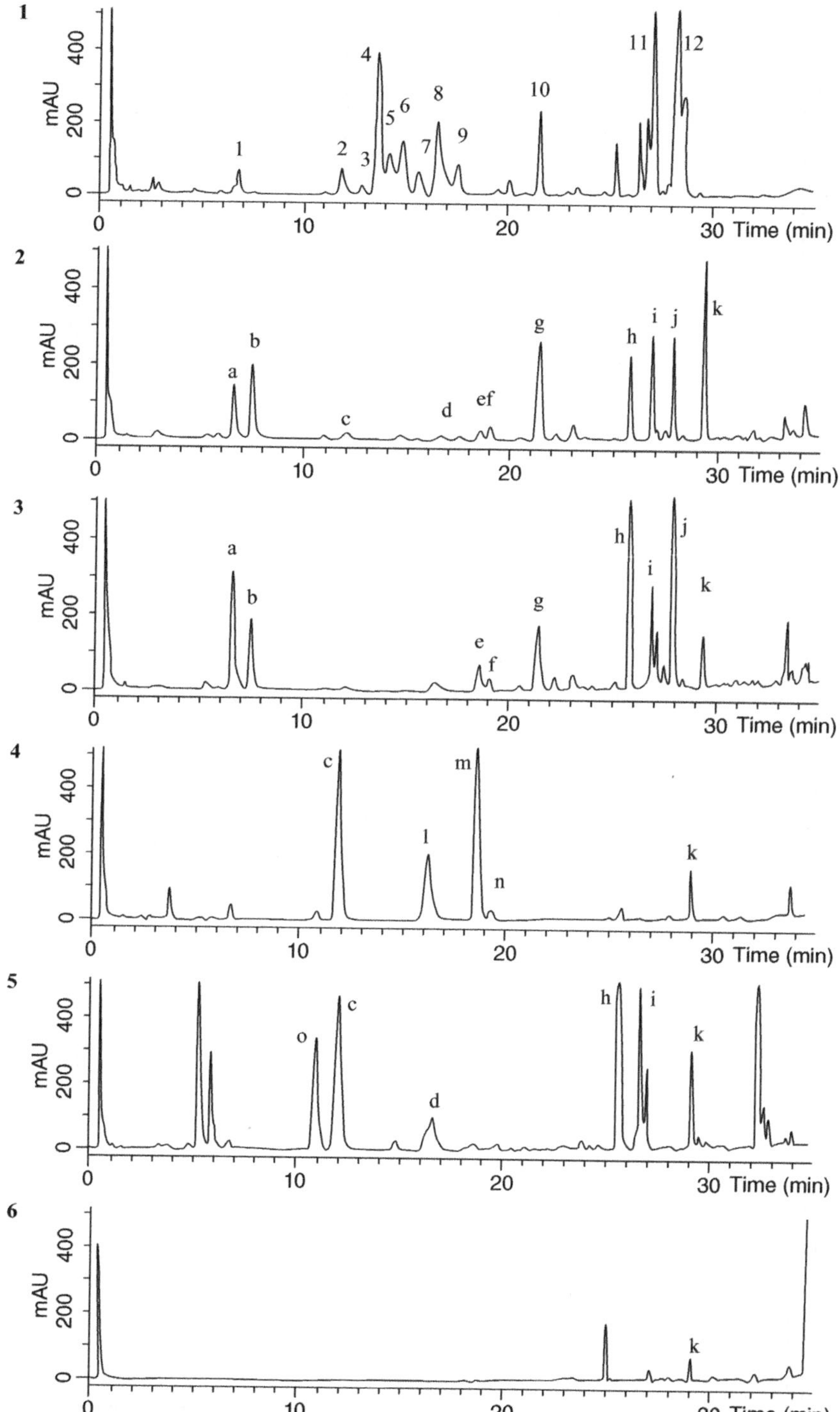

Figure 9.7 HPLC fingerprint analysis of extracts from *Angelica pubescens f. biserrata* and some substitutes using DAD detector. 1, *Angelica pubescens f. biserrata*; 2, *Angelica dahurica*; 3, *Angelica apaensis*; 4, *Heracleum moellendorffii*; 5, *Heracleum candicans*; 6, *Aralia cordata*.

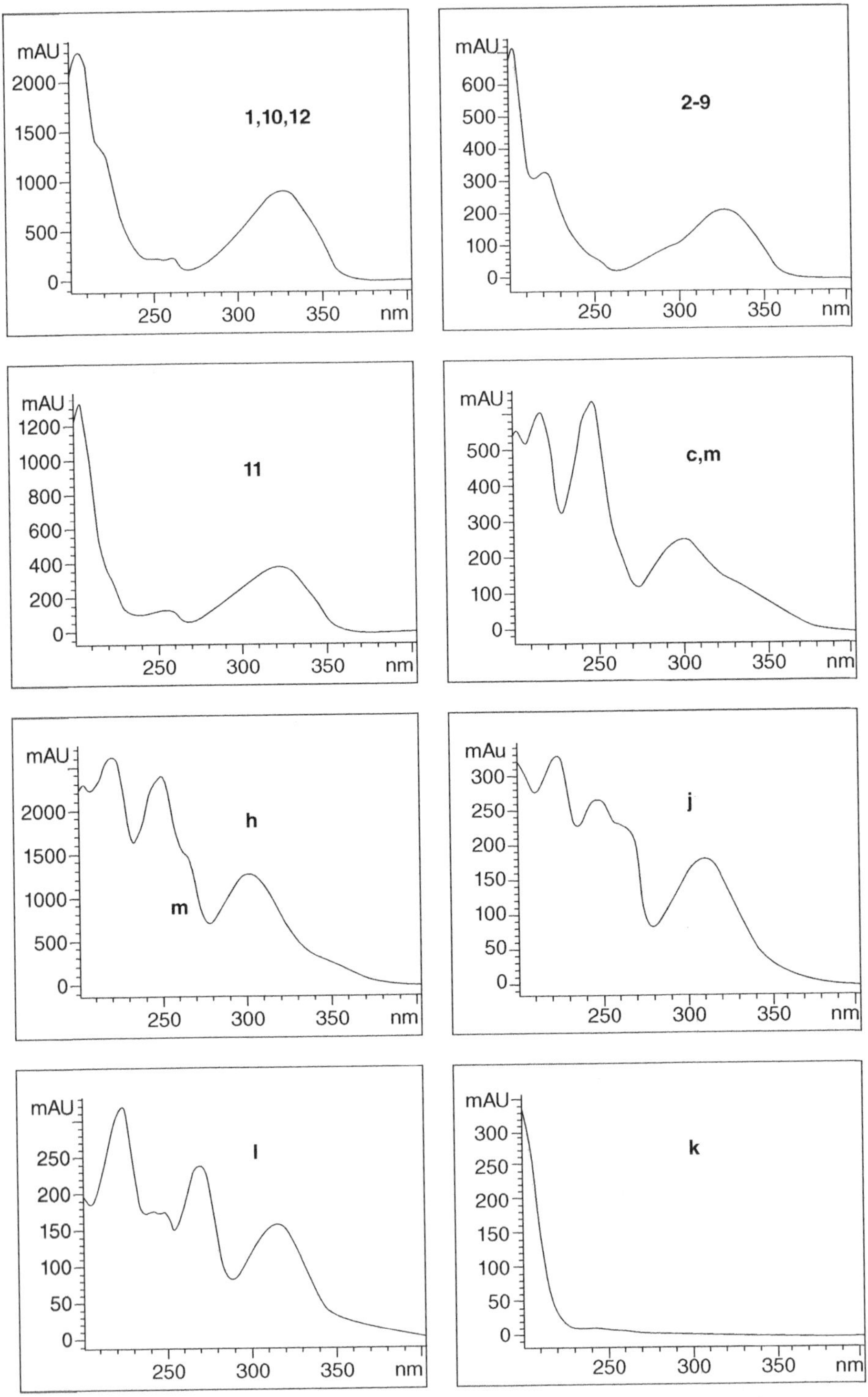

Figure 9.8 UV spectra of the major compounds in *Angelica pubescens f. biserrata* and some substitutes (the numbers and letters in Fig. 9.8 match the HPLC peaks in Fig. 9.7).

quantitative analysis is applied to herbs with many reasons: to compare the contents of major characteristic or bioactive compounds in a species growing in different areas or collected at different times; to measure the contamination of pesticides, heavy metals, and other toxic chemicals on herbs and organic solvents in extracts for safety consideration; to validate extraction methods or processing procedures; and to control the quality of the products.

The key to quantitatively analyzing chemical constituents in herbs or herbal extracts is to establish an appropriate evaluation system, including selection of characteristic components to be quantified, establishment of an appropriate sample preparation method, and an accurate and precise quantitative approach, as well as elimination of interferences of other components on characteristic compounds to be quantified. For quantitative analysis of compounds in complex herbal extracts, make sure the sensitivity and reproducibility of the detection, as well as resolution of the peak of analytes, are good enough.

HPLC-UV-Vis, HPLC-MS, and GC-MS have been widely applied for quantitative analysis of herbal extracts due to their specificity and accuracy. The analysis for extracts or products from manufacturers should be performed in a GMP environment. To accurately quantify compounds in an extract with instruments such as HPLC and GC, the machine needs to be validated for its stability and accuracy. The accuracy and precision of each method should also be validated.

Standard curves should be made for all methods of quantitative analysis to ensure the analysis results are within a straight line of the curve. Ideally, no matter what analytical method will be used, five to seven different concentrated solutions of standard compound should be prepared for making the calibration curve when establishing a quantitative analysis. Meanwhile, one of the standard solutions with a concentration within the linear range of the calibration curve is repeatedly injected seven to nine times to evaluate the reproducibility of the system. Solutions of samples should be prepared in which the concentration of the quantified compound will fall into the linear range of the calibration curve.

When developing a new analytical method, it is necessary to evaluate the accuracy, precision, linearity, specificity, limit of detection (LOD), and limits of quantitation (LOQ) of the method and to compare the method to available alternative methods. Statistics have to be applied in the analysis of results.

Definitions related to quantitative analysis (e.g., accuracy, precision, errors, mean, and standard deviation), principles, and methods of quantitative analysis should be well known by analysts performing the quantification analysis. Many books are available for references, such as *Analytical Chemistry in a GMP Environment—A Practice Guide*.[20]

In addition to components in herbs, contamination of pesticides, heavy metals, and microorganisms in herbal extracts and products should also be analyzed quantitatively.

9.5.1 Sample Preparation

A successful quantitative analysis depends on many aspects, starting from collection, processing, transportation, and storage of materials to sample preparation, then analytical methods. Commonly used sample preparation methods for raw herbal materials include solvent immersion, solvent partition, reflux extraction, ultrasonic extraction, and SFE. Details of these methods and their principles have been introduced in Chapter 3 of this book.

If the test herbal sample is an extract or a product, it should be completely dissolved in a proper solvent without any particles. Small size solid phase (SPE) cartridges are commonly applied in sample preparation for both filtration and separation purposes. The selection of sample preparation methods is mainly based on the solubility and stability of the target compounds to be analyzed. It should maximally extract the target compounds without changing their structures and exclude other compounds to minimize their interferences for detection. Safety, cost, and time are also factors to be considered. These are very important for a reliable and accurate quantitative analysis.

9.5.2 Quantitative Analysis Methods

The quantitative analysis methods that are suitable and commonly used for herbal extracts or products by chromatography and spectroscopy include internal standard method, external standard method, calibration curve method, standard addition method, and normalization method. Introduced here is only brief information for a few commonly used methods; detailed introductions about each method are available in many books,[21] articles in journals, and web sites.

1. *The internal standard method*

This is the most suitable method for quantification of compounds in a complex herbal extract. In this method, an appropriate substance needs to be selected as an internal standard substance and added quantitatively into the test samples. The calculation of the contents is based on the ratio of peak area of components to be analyzed and standard substance. This method is particularly useful when only several component contents need to be quantitatively analyzed. The internal standard method is ideal for quantification of characteristic or bioactive components in herbal extracts.

The selected internal standard should meet the following requirements: it should not be a compound in the sample; its peak should appear near or between the peaks of the test substances but with complete separation; the addition amount in the sample should be about the amount of the test component.

The application of internal standard can offset errors caused by instrument instability, inaccurate injection, or other factors. To offset the loss of test compound during the pretreatment procedure, it is better to add the internal standard before sample preparation. Because of the utilization of internal standard in this method, the chromatographic conditions do not influence the results too much, so the accuracy and precision are relatively higher. However, one disadvantage is difficulty in choosing an ideal internal standard.

2. *The external standard method*

The external standard is usually the purified compound to be quantitatively analyzed in the test sample. In this case, no response factors will be required for calculation. Most of the standards are commercially available; if they are not, then they need to be isolated. The external standard solution should be prepared at a concentration close to the one in test samples in order to eliminate errors from the instrument. The external standard and test sample are run through the HPLC or GC column separately but under exactly the same conditions. When analyzing a series of samples, the standard solution is better repeatedly injected before and after, sometimes even between several samples, to reduce the effect caused by unstable changes of instrument during analysis. To minimize error caused by loss of the analyzed compound during sample preparation, it is better to process the standard with the same procedure as for test samples. The peak areas of the analyzed compound in test sample and external standard are measured for calculation.

A slight difference of injection volume may cause significant difference of result in this method. It is better to inject the sample through a quantitative ring when the external standard method is applied. Application of an auto-sampler usually gives more accurate injection.

3. *The calibration curve method*

It is also called standard curve method. In this method, a series of standard solutions with different concentrations (generally 5 to 7) are prepared and measured. When establishing the method, the selection of these standard concentrations should be first extensively distributed in order to know the linearity range, as well as the detection limit of the method or system. For quantitative analysis of samples, a set of standard solutions should be prepared to give the linear curve where sample results will fall between. These concentrations of the standards should not be intensively distributed at either the higher or lower part of the linear range. A calibration curve can be drawn based on the points from the measured data and a formula ($y = mx + b$, y: signal, x: concentration) can be obtained with the Excel software. The regression of the curve should be above 0.99 ($R^2 > 0.99$). The determination of concentration of the target compound in a test sample is then calculated using this formula. Some instruments may provide the software for calculation. This method can be used not only for analysis by chromatography, but also spectroscopy. When absorption (A) is measured, it should be in the range of 0.2–0.7.

The standard solutions should be run together with samples in each series of analysis, in which the operation condition should remain constant during the whole analysis procedure. The structure of the standard should be the same or, if not possible, similar to the measured compound sample as much as possible. The sample should be prepared at a concentration in which the content of the measured compounds will fall within the linear range of the calibration curve.

For HPLC or GC, to minimize the errors caused by loss of samples during the preparation procedure, the same amount of internal standard can be added into all standards and samples. The curve is drawn with the ratio of the peak area of the internal standard to the test substance against that of the standard solutions, and the ratio of the peak area of the internal standard to that of the test compounds is used for calculation.

4. *Standard addition method*

Similar to the calibration curve method, this method is used in instrumental analysis to determine the concentration of a substance in a sample by comparing one sample to another, in which a certain known concentration of standard of this substance is added. The difference between the standard addition and calibration curve methods is that the standard is added (spiked) into the sample rather than in a pure solution as in the standard addition method. The concentration of the substance in a sample is calculated based on the difference of the measurement between the spiked sample and unspiked sample. It can be applied to most chromatographic and spectroscopic analysis, such as GC, AAS, and ICP, particularly in the case that the analysis is easily interfered with by other components (matrix) in the complex sample that would cause an incorrect determination.

9.5.3 Analysis with HPLC

HPLC can couple with many types of detectors, including UV detector, DAD, electrical conductivity detector, fluorescence detector, refractive index detector, and the more advanced mass spectrometry detector. Among these, different types of UV detectors are most widely applied. Such systems generally have the characteristics of effective separation, good selectivity, and high sensitivity. Because many types of columns are now available for HPLC, HPLC has been used for analysis of a wide range of compounds from nonpolar to polar or ionic compounds, from heat sensitive to nonvolatile compounds, from small to larger size molecules. It has become the most widely applied effective separation and analysis tool for herbal products including their pesticide contamination.

The common methods for preparation of herbal samples for HPLC analysis include reflux extraction, Soxhlet extraction, ultrasonic extraction, microwave-assisted extraction, SFE, accelerated solvent extraction, solid phase extraction, liquid membrane extraction, and liquid phase micro-extraction. Selection of methods depends on the properties of analytes.

HPLC-UV-Vis or DAD

UV-Vis detector is the most common detector linked with HPLC. HPLC-UV-Vis is particularly suitable for quantification of phenolic compounds such as flavonoids, coumarins, anthroquinones, and other compounds with conjugated double bonds. The sensitivity and selection depend on the type of UV detectors, which include the fixed wavelength UV detector, the multiwavelength detector, and DAD.

Many compounds absorb light in the wavelength range between 200 and 800 nm. When the elute from column passes through the small cylindrical cell in the detector having a capacity between 1 and 10 mL, the UV-Vis light passes through the cell and falls on a photoelectric cell or array. The output from the photocell is sent to a modifying amplifier and then to a recorder or data acquisition system. The relationship between the intensity of UV light transmitted through the cell and the concentration of solute in it follows the Beer's Law.

The fixed wavelength UV detector uses the light of a single wavelength produced by a specific type of discharge lamp. The most popular wavelength applied is 254 nm, generated from a low pressure mercury vapor lamp. Quantitative analysis using this detector is limited to only those compounds that have good absorption at the available wavelength.

The multiwavelength detector employs a light source that emits light over a wide range of wavelengths. With this detector, light of a specific wavelength can be selected for detection. An absorption maximum is usually chosen to obtain the maximum sensitivity. Meanwhile, the absorption spectra of the isolated compound in the elute could also be obtained for identification purposes by scanning over a range of wavelengths. There are two basic types of multiwavelength detectors, the dispersion detector and the DAD, with the latter being the more popular.

DAD utilizes a deuterium or xenon lamp that emits light over the UV spectrum range. Light from the lamp is focused by means of an achromatic lens through the sample cell and onto a holographic grating. The dispersed light from the grating is arranged to fall on a linear diode array.

The selection of the wavelength for quantitative determination is usually based on the UV spectrum of the quantified compound, which is determined by its structure. Choosing a right wavelength is very important for measurement because it will significantly affect the selectivity and sensitivity of the quantification. The principle is to maximize the sensitivity of compounds to be quantified for better detection and minimize the sensitivity of other compounds for least interference. Keep in mind that the solvents will cause a shift of the maximum absorbance of a compound.

Quantification of different compounds in an extract may need to be measured at different a wavelength. For example, hyperricin and amentoflavone are two characteristic compounds in St. John's wort. The UV wavelength for quantification could be set around 590–600 nm for hyperricin and 260 nm for amentoflavone based on their UV spectra if an HPLC solvent system of water (containing 0.3% H_3PO_4) and acetonitrile (CH_3CN) is used.

HPLC-MS and UPLC-MS

LC-MS has been widely applied in many fields such as analysis of drugs, foods, and environmental samples. It combines the advantages of chromatography and mass spectrometry, that is, the effective separation capacity of chromatography and the high selectivity and good sensitivity of MS. In comparison to LC-UV-Vis, LC-MS is relatively more expensive. The sensitivity and selectivity of the analysis depend on the selection of the MS analyzer and scan mode. Particularly when determining trace components in a complex mixture, be sure to choose the proper analyzer and mode.

For quantification, MS with quadruple analyzer gives better selectivity and resolution than that with an ion trap.

Full scan has lower sensitivity due to the poor selectivity; thus the total ion chromatogram is usually used to collect spectra for structural identification or a library search, but is not a good choice for quantification.

Selected ion monitoring (SIM) provides relatively higher sensitivity and selectivity, and thus is often used for quantitative applications. In this mode, the mass filter is set to pass a selected *m/z*.

Multiple-reaction monitoring (MRM) in MS detectors that are capable for tandem MS/MS such as triple-quadruple provides even greater selectivity than SIM mode. In this mode, the most abundant ion is usually selected as the precursor ion and allowed to pass through the first mass selector and is fragmented by CID to get product ions (also called daughter ions), among which one characteristic ion is selected and allowed to pass the next analyzing stage to be detected. The combination of such two ions provides greater selectivity and sensitivity due to the signal-to-noise.

Analysis of trace compounds in a complex mixture, for example. determination of pesticide residues and other harmful components in herb extracts, usually requires higher sensitivity of the detector. LC-MS/MS is particularly suitable for such analysis. When two compounds have the same molecular weights and differ only in positions of substituent groups, they may not be well separated by HPLC and detected separately by MS in the SIM mode. In this case, application of the MS/MS mode in triple-quadruple MS is ideal to minimize interference from one another.

Sample preparation methods for HPLC-MS include solvent extraction, solid phase extraction, ultra-filtration, and many others depending on the properties of the quantified compounds. Protein in herbal extract can be precipitated by methanol or acetonitrile, or hydrolyzed by acid or enzyme derivative method.

To obtain good and reproducible results, it is important to maintain high sensitivity of the instrument by preventing instrument contamination, decreasing analysis background, and eliminating the interference. Regular maintenance of MS, such as cleaning of the ion source, is very important.

UPLC-MS provides faster analysis with higher sensitivity and resolution, with the utilization of the smaller particles. Methods for sample preparation and analytical analysis of UPLC are basically the same with that of traditional HPLC.

9.5.4 Analysis with GC

The GC method is particularly suitable for quantitative analysis of natural volatile components, and lipophilic compounds with low boiling point and good thermal stability in medicinal herbs. It is also used to determine the residues of pesticide herbs and organic solvents in herbal extracts.

Although high boiling point and heat-sensitive compounds are now mostly analyzed by HPLC, some scientists still like to use GC to quantify them by making them into less polar derivatives, because GC provides better resolution due to its longer column than HPLC, and less interference by the mobile phase. For nonvolatile big molecular compounds, decomposition reaction can be applied to make them into small volatile molecular compounds for GC analysis. For polar or thermally sensitive compounds, derivatization can be applied to make them into less polar or stable compounds. For examples, esterification is commonly used for analysis of long-chain fatty acids, and silane is often used for analysis of compounds with high boiling point or poor thermal stability that contain hydroxyl, carboxyl, and amino groups.

MS is the most common detector coupled with GC. GC-MS is very sensitive and has high resolution. It has been widely applied in research labs for study and drug manufacturing and for QC. It is an indispensable tool in the aspects of identification of volatile components and determination of residues of pesticides in foods and herbs.

An herbal sample for GC-MS can be prepared with methods such as solvent extraction, distillation, solid phase extraction, gas extraction, membrane separation, thermal desorption, and derivative technique. Different from HPLC analysis, a sample solution for GC analysis should not contain any water. Similar to HPLC-MS analysis, the sensitivity and selectivity of the analytical method depend on selection of the type of MS, particularly the analyzer and scan mode.

9.5.5 Analysis with UV-Vis Spectrometry

This method is particularly suitable for analysis of compounds with UV-Vis absorption between 200 and 800 nm. Some other compounds, such as sugars, may also be measured after color reaction with reagents.

UV-Vis is a simple and sensitive detector for quantitative analysis of certain types of compounds in herbal medicine. One of its important traits is additivity of UV spectrum when a mixture is analyzed without separation. The absorption of each compound contributes to the total absorption of this mixture under the same conditions.

Several decades ago, before HPLC became popular, UV-Vis spectrometry was a main tool used for quantitative analysis of compounds isolated from herbs because of its high sensitivity. It is easy, thus still a commonly used method for measurement of total content of certain types of compounds in mixtures, such as total flavonoids and total anthroquinones in extracts.

However, the selectivity of UV spectrometry is much less than that of HPLC-UV-Vis, in which compounds are determined after separation by HPLC. This method is unable to determine one or a group of compounds at the wavelength where other compounds in the extract also have absorption. Isolation must be performed to exclude the interfering compounds before the analysis in this case. Because it is cheap and easy to operate, this method is still used in many manufactories for QC of herbal extracts.

The quantification methods for a single compound mainly include the contrast method, absorption coefficient method, and calibration curve method. Methods for quantification of multicomponents include dual wavelength spectrometry, derivative spectrometry, and convolution spectrometry. Absorption of a sample closely depends on the determination conditions. The shape, wavelength, and intensity of absorption peaks in a spectrum may be altered by factors such as temperature, solvent, and pH value.

1. *Standard control method* is to measure, respectively, the absorbance of sample solution and control solution at selected wavelengths, and then to calculate the sample contents with the equation of $Cx = (C_S \times A_X)/A_S$ (C: concentration; A: absorbance; X: sample; S: standard).
2. *Standard curve method* is to make a group of different concentration solutions with a standard sample, then to measure, respectively, their absorbance and draw the absorbance-solution (A-S) standard curve. The concentration of the sample solution is then determined by the standard curve. In general, the concentration of a sample solution is calculated with the measured absorbance using the regression equation.
3. *Absorption coefficient method* is based on Beer Law ($C = A/\varepsilon \times L$). If the thickness of the absorption cell (L) and absorption coefficient (ε) is known, the density can be calculated according to the absorbance (A) of the tested sample.
4. *Dual wavelength spectrometric method* uses two different wavelength of homogeneous light λ_1 (as a reference wavelength) and λ_2 (as the determining wavelength) to irradiate on the sample solution alternately. The difference of absorbance (ΔA) between the two wavelengths is then detected. The concentration of the sample is calculated with the equation of $\Delta A = A_2 - A_1 = (\varepsilon_2 - \varepsilon_1) \times C \times L$.

9.5.6 Analysis with TLC Scanning

TLC scanning had been widely applied in separation, identification, and quantitative analysis of compounds in herbal extracts or their products before HPLC became popular. In this method, a beam of a certain wavelength and intensity is used to irradiate the spots on the developed TLC plate. The intensity of the beam pre- and postirradiation is measured to determine the quantity of each spot. The advantage of this method is that several samples can be simultaneously analyzed on one TLC plate.

Although this method can provide quantitative results, the procedure is more cumbersome than other chromatographic methods. It is cheap to prepare and run the TLC, but the accuracy of the result cannot be compared with that obtained by HPLC or GC. Scanning equipment for better results has now become available but is expensive, virtually eliminating the cost advantage of the technique.

In addition to direct scanning on TLC plate, quantitative analysis can also be carried out by scraping the spot out of the plate first, eluting the isolated compound with solvent from the adsorbent, then condensing the eluent to a certain concentration for determination with an appropriate method, depending on its properties, for example, quantification of a flavonoid compound by UV spectroscopy.

According to the Bruker's web site, Bruker has recently introduced a direct coupling of TLC with MALDI-TOF mass spectrometry (TLC-MALDI) and an adapter target for TLC plates and an extension of the standard flex software that turns the MALDI into a molecular TLC readout platform. This system has been successfully applied for lipid analysis in the company (Bruker: Application note # MT-94).

9.5.7 Analysis with AAS

AAS is mainly used to measure the residues of heavy metals in herbs. This method has the characteristics of high accuracy, high sensitivity, and good selectivity.

The test sample must be prepared into a solution, either by dissolving in a solvent or melting at high temperature.

Standard and sample solution should be prepared at a concentration of which the absorbance of the test sample will be between 0.2 and 0.7.

Quantification by AAS may be impacted by the following interferences.

1. *Ionization interference:* It refers to the interference cause by atom ionization and can be effectively suppressed and eliminated by adding a deionizing agent.
2. *Physical interference:* It refers to the decrease of sample absorbance due to the change of the sample's physical characteristics during transfer, evaporation, and atomization. The most common method used to eliminate the interference is to have a reference substance that has composition similar to the test sample, or that uses the standard addition method.
3. *Optical interference:* It mainly includes interferences on spectral line and nonabsorption line. Methods for elimination include adjusting zero absorption of the instrument and calibrating adjacent nonresonant lines and background of the continuous light sources as well as Zeeman efficacy.
4. *Chemical interference:* It refers to the interference due to the formation of nonvolatized or non-dissociated compounds by chemical reaction between the measured element(s) and other components in the sample solution or gas phase. The common effective methods for elimination include adding releasing protecting reagents and appropriately increasing flame temperature.

9.5.8 Analysis with Inductively Coupled Plasma Mass Spectrometry (ICP-MS)

ICP-MS can also be used for measurement of heavy metals. In comparison with AAS, which can only measure a single element at a time, ICP-MS has the capability to scan for all elements simultaneously.

In ICP-MS method, the sample is ionized by ICP and separated and detected according to the charge-to-mass ratio.[22] The intensity of the chromatographic peak of each ion is measured for quantification. It is mainly used in the analysis of heavy metals, trace elements, and toxic elements in medicinal herbs and their extracts.

The sample is generally prepared as a solution. But herbal samples containing organic compounds should be digested first to get rid of organic compounds before preparing the solution. The microwave digestion method is generally applied, in which a reagent is added to digest the sample into inorganic elements in a digestion tank in a microwave.

The common quantitative methods include the external standard method, internal standard method, standard addition method, and isotope dilution method. In the isotope dilution method, a known quantity of radioisotope or a rare stable isotope is added (spiked) into the sample.

REFERENCES

1. Dauly, C., et al. (2006) Protein separation and characterization by np-RP-HPLC followed by intact MALDI-TOF mass spectrometry and peptide mass mapping analyses. *Journal of Proteome Research* 5(7):1688–1700.
2. Swartz, M.E. (2005) Ultra performance liquid chromatography (UPLC): an introduction. *Separation Science Redefined*. http://chromatographyonline.findanalytichem.com/lcgc/data/articlestandard//lcgc/242005/164646/article.pdf.
3. Frazier, R.A. and Papadopoulou, A. (2003) Recent advances in the application of capillary electrophoresis for food analysis. *Electrophoresis* 24(22–23):4095–4105.
4. Fung, Y.S. and Tung, H.S. (2001) Application of capillary electrophoresis for organic acid analysis in herbal studies. *Electrophoresis* 22(11):2242–2250.
5. Osborne, B.G. (2007) Principles and practice of near infra-red (NIR) reflectance analysis. *International Journal of Food Science & Technology* 16(1):13–19.
6. Woo, Y.A., et al. (1999) Identification of herbal medicine using pattern recognition techniques with near-infrared reflectance spectra. *Microchemical Journal* 63(1):61–70.
7. Mao, J. and Xu, J. (2006) Discrimination of herbal medicines by molecular spectroscopy and chemical pattern recognition. *Spectrochim Acta A: Mol Biomol Spectrosc* 65(2):497–500.
8. Tessier, B., et al. (2005) Prediction of the amino acid composition of small peptides contained in a plant protein hydrolysate by LC–MS and CE–MS. *Food Research International* 38:577–584.
9. Nesbitt, C.A., et al. (2008) Recent applications of capillary electrophoresis–mass spectrometry (CE–MS): CE performing functions beyond separation. *Analytica Ahimica Acta* 627(1):3–24.
10. Sawalha, S.M.S., et al. (2009) Quantification of main phenolic compounds in sweet and bitter orange peel using CE–MS/MS. *Food Chemistry* 116(2):67–574.
11. Robinson, J.W. and Frame, G.M. (2005) *Undergraduate Instrument Analysis* (6th ed.). Boca Raton, FL, CRC Press.
12. Sperling, M.B. and Welz, B. (1999) *Atomic Absorption Spectrometry*. Weinheim, Wiley-VCH.
13. Knapp, D.R. (1979) *Handbook of Analytical Derivatization Reactions*. New York, John Wiley & Sons.
14. Wampler, T.P. (1999) Introduction to Pyrolysis-capillary Gas Chromatography (Review). *Journal of Chromatography A* 842(1–2):207–220.
15. Wang, N.S. (1990) Application of derivative spectra method in qualitative identification of volatile oil from fructus alpiniae oxyphyllae. *Journal of Guangzhou University of Traditional Chinese Medicine* 7(4):225–226.
16. Halgerson, J.L., et al. (2004) Near-infrared reflectance spectroscopy prediction of leaf and mineral concentrations in alfalfa. *Agronomy Journal* 96:344–351.

17. Liu, J.H., et al. (1998) Comparison of Radix Angelicae pubescentis and substitutes—constituents and inhibitory effect in 5-lipoxygenase and cyclooxygenase. *Pharmaceutical Biology* 36(3): 207–216.
18. Zhang, L., et al. (2007) Development of the fingerprints for the quality evaluation of Scutellariae Radix by HPLC-DAD and LC-MS-MS. *Chromatographia* 66(1–2):13–20.
19. Shi, X.M., et al. (2008) Fingerprint analysis of Lingzhi (*Ganoderma*) strains by high-performance liquid chromatography coupled with chemometric methods. *World Journal of Microbiology and Biotechnology* 24(11):2443–2450.
20. Miller, J.M. and Crowther, J.B. (2000) *Analytical Chemistry in A GMP Environment—A Practice Guide*. New York, John Wiley & Sons.
21. Harris, D.C. (2003) *Quantitative Chemical Analysis* (6th ed.). New York, W.H. Freeman.
22. Montaser, A. (1998) *Inductively Coupled Plasma Mass Spectrometry*. Weinheim, Wiley-VCH.

Chapter 10

Understanding Traditional Chinese Medicine and Chinese Herbs

Willow J.H. Liu

Traditional Chinese Medicine (TCM) has been practiced for more than 2000 years. Its theoretical system was partly derived from the philosophy that influences Taoist and Buddhist thought and is complemented through practice. The basic theories include yin and yang, the five elements, Zang-Fu theory, the theory of *qi* (pronounced "chee"), blood, and body fluid, the theory of meridians and collaterals, and the theory of etiology and pathogenesis. These theories have been applied to diagnosis of diseases and guide treatment with herbs, acupuncture, diet, and other methods. Among all these treatment methods, Chinese herbs are most widely applied.

TCM has survived over several thousand years. Even though conventional medicine has dominated, TCM is not only still popular in China, but has also extended to many countries all over the world. This is because it has therapeutic effects on some diseases that modern medicine cannot treat, with fewer side effects. In China, even in modern hospitals, TCM treatment has been combined with the methods of modern medicine for some diseases, such as rehabilitation of stroke sequelae and treatment of diabetes complications. In the United States and many European countries, Chinese doctors are welcomed by more and more people. However, TCM is opposed by most Western medical doctors because they know little about it. I would say that the main reason for TCM's lack of recognition among Western doctors is due to the difficulty of understanding its theories and terminologies. To scientists who are interested in research on Chinese herbs and trained by modern science but without sufficient knowledge and clinical experience of TCM, the study is mostly copying methods for drug development. Therefore, they are often disappointed when the research shows that the bioactivities of individual compounds isolated from herbs are less effective than modern drugs. Some may gradually lose

Traditional Herbal Medicine Research Methods, Edited by Willow J.H. Liu

confidence in Chinese herbal medicine. Others may wonder why a Chinese formula works better than modern drugs, when no compounds with better bioactivity were found in these herbs. To understand this, one should first understand the TCM theories that guide treatment.

10.1 UNDERSTANDING TCM THEORIES WITH MODERN MEDICAL TERMINOLOGY

To learn or understand TCM, *The Yellow Emperor's Inner Classic*, the earliest available TCM text, written 2000 years ago, is a must-read because it is still very valuable today (English translation is available[1]). This famous book describes most of the structures and organs in the human body and their physiological functions, as well as the etiology and pathology of diseases, and is unbelievably advanced for its time, although with terminologies different from those of modern medicine. Compared with modern medicine, knowledge, in particular, anatomical knowledge, was limited in some respects at that time, because a corpse was not allowed to be cut open for observation due to the ethics of the time.

Therefore, even though most of the structures, organs, or tissues illustrated in TCM 2000 years ago can be matched with those in modern medicine by direct language translation, a few were missed; for example, neither a word for the pancreas nor one for nerves is found in TCM. The author would say that the function of the pancreas was covered with that of the spleen because the spleen plays a very important role in TCM, while the function of the nerves is part of the function of *qi*. It has to be admitted that some illustrations are inaccurate due to the limited knowledge of physiological science, and a small number of views seem a bit superstitious. But, just like a few slight defects in a precious jade, these flaws do not decrease its value.

Languages and terminologies used in TCM are significantly different from those in modern medicine. Concepts such as *qi*, yin and yang, and meridians give people the impression that TCM is mysterious. But, with careful comparison, it is not difficult to discover that TCM actually covers much of the knowledge presented in modern medicine. For example, *qi* and meridians are the most characteristic aspects of TCM, but were not able to be translated directly to modern medicinal terminologies. Some think *qi* and meridians are mysterious or unexplainable; others may even consider them superstitious and unscientific. But with the combined knowledge of TCM and modern medicine, the author would agree that the functions of *qi* actually cover functions of all organs and systems, while the functions of *qi* in meridians directly relate to that of the nerve, blood, and lymphatic circulatory systems, as well as immune and endocrine systems. This is because their functions are carried out via the circulatory system. Another example is that TCM uses evil *qi* (cold, hot, fire, wind, dampness, etc.) to describe the exogenous pathogens that are named bacteria, viruses, fungi, and so on, in modern medicine. Meanwhile, the defensive *qi* is used to describe the functions of the immune system in fighting these foreign pathogens—the evil *qi*.

TCM has stood the test of thousands of years of clinical trials. According to the laws of nature, it would have disappeared if it did not prove its effect. The task of scientists who are interested in TCM, therefore, is to try to learn and understand the theories of TCM first and then apply them to the evaluation of theories and preventive and therapeutic efficacies of TCM with the knowledge of modern science. The following points are a few of the author's insights on and understanding of TCM. It is by no means a complete representation or summary of TCM, but will hopefully serve as a good start to help people lift the veil of TCM mystery.

10.1.1 Key Differences between Chinese and Modern Medicines

Needless to say, Chinese and modern medicines are different in almost every respect, such as in their length of history, regions of origin, language and terminology applied, theories, diagnosis, treatment, and etiology and pathology. But the author believes that the following three critical differences are the key points that make TCM treatment for some diseases superior than conventional treatment.

1. *TCM emphasizes prevention, while modern medicine focuses on treatment.*

For example, modern medicine has the advantage of surgery for coronary heart disease (CHD) and stroke. However, surgery only fixes the blocked part of the vessel, ignoring the cause of the blockage. Therefore, some patients may receive a second or third surgery later when other parts of the vessel become blocked. Former president Bill Clinton received two surgeries for his CHD within five years. One of the American professors I worked with received three surgeries in succession, all because of his CHD. TCM is good at preventing the onset and progress of such diseases. As is said in *The Yellow Emperor's Inner Classic*, "the superior doctors are good at treating diseases that have not been developed." Some Chinese herbs, such as hawthorn and salvia, are able to significantly prevent CHD *by improving blood circulation of the coronary arteries.*

2. *TCM emphasizes impacts of environment, diet, and emotion on health, as well as interactions between internal organs, while Western medicine focuses on the problematic point.*

TCM is an integral or "holistic" medical system. The holism of TCM is embodied through theories of yin and yang, the five elements, *qi* and blood, meridians, and etiology and pathology. It is applied in both diagnosis and treatment.

Although modern medicine also acknowledges that the environment, diet, and emotion have impacts on health, not many doctors emphasize the importance of diet and emotion to their patients. The task, for most doctors except psychiatrists, is focused on prescribing medicines to their patients, seldom giving advice on diet and emotion control, which are thought to be the business of dietitians and psychiatrists, respectively. But TCM believes that abnormal emotional change and poor diet are the cause of many diseases.

For example, stress from working or being laid off from the job can impact one's emotions, which is called *qi* disorder in TCM. Consequently, the emotional change can trigger physiological reactions to cause sadness, anger, or depression, which can further lead to poor digestion, insomnia, palpitation, and so on, and gradually affect functions of the nervous, circulatory, endocrine, and immune systems that control or involve the functions of internal organs, which are called "emotion induced illness" by Dr. John A. Schindler in his book, *How to Live 365 Days a Year*. If a person eats too much or too little, eats only a few types of foods, or has absorption problems, the function of the body will gradually deteriorate because body cells, especially nerve cells, cannot obtain sufficient nutrients.

Emotional changes, especially depression and sadness, may easily impact the function of the digestive system, then slowly other systems because of the insufficient nutrient supply through the blood. The five elements theory in TCM tells us that a problem in one region is probably related to disorders of other organs. It is absolutely true because the blood circulates through liver, heart, lung, spleen, and kidney. In addition, blockage of circulation in any organ will also impact the function of other organs. Therefore, diet and excretion are always inquired about when TCM doctors begin to diagnose patients.

3. *TCM emphasizes treating the causal roots, the fundamental causes that result in symptoms, while modern medicine focuses on relieving symptoms or removing the final, unwanted effect.*

Unlike Western doctors who treat only the head for headache—as goes a saying in Chinese—TCM treatment is based on the whole body, rather than simply in response to a patient's complaints about certain spots. In addition to the environment, emotion, diet, and trauma, TCM believes that blockage of *qi* and blood circulation may cause hundreds of types of diseases. When treating chronic disease, TCM also emphasizes the functional regulation of digestion and absorption, which is called spleen and stomach *qi*, the source of *qi* in the human body. Therefore, in the formula for treatment of many diseases, there are often herbs that invigorate blood circulation and herbs that improve the digestive function. For a headache patient, TCM doctors may treat with a formula that contains herbs that not only relieve pain, but also herbs that improve the brain circulation, or with acupuncture on the foot, neck, head, and other places of the body depending on the differentiation of diagnosis. For this reason, rather than a singular herb, Chinese doctors often prescribe formulas that are usually composed of several or even more than 10 herbs having different actions.

As mentioned in Chapter 1, Section 1.4, uterine fibroids are formed due to abnormal hormone release, and gynecologists usually remove them by surgery when they cause heavy abnormal bleeding or become too big. The uterus is often removed together with the fibroids to prevent the regrowth, though women who want to have children are reluctant to accept such treatment. Take one of my patients as an example. When this woman came to seek treatment for her heavy uterine bleeding caused by fibroids, she said that her gynecologist wanted to remove the fibroids with her uterus because the medicine could not stop her bleeding. She had been married for two years, but she would not be able to become pregnant. She tried the Chinese herbs which were said to stop her bleeding so that she would be able to have children later. By regulating her abnormal *qi* and blood that caused both her fibroid and infertility, the herbal formula stopped the bleeding within 10 days. Soon after, she became pregnant and gave birth to a daughter.

Take insomnia as another example. Western doctors usually treat insomnia using drugs like benzodiazepines and non-benzodiazepines that selectively or unselectively bind to $GABA_A$ receptors as agonists of $GABA_A$ to make up the deficiency of γ-aminobutyric acid (GABA) or other chemicals regulating sleep. Such medicines are only able to temporarily help sleep rather than cure insomnia. Mostly, patients have to depend on them, and quite often need to increase dosages gradually because the medicine will produce dependence. Chinese doctors believe that insomnia could be caused by many reasons. Thus they use different herbal formulas to treat insomnia patients, depending on particular diagnoses. Through experience from my personal clinical practice and combining TCM theory and knowledge of biochemistry and pharmacology, I believe that the causes of chronic insomnia include long-term poor digestion, poor blood circulation in the brain, and poor blood supply to the brain. Of course, psychological reasons, such as stress from work or study and sadness due to loss of a family member, should not be overlooked.

From the view of chemical composition in blood, all the nutrients related to regulating the function of nerve cells, such as cofactors including vitamins, coenzymes, precursors of neurotransmitters or receptor ligands, are supplied through blood. Long-term poor digestion and absorption can lead to an insufficient supply of the nutrients that are involved in reactions to form chemicals that participate in synthesis of neurotransmitters or receptor ligands within the brain. Weak functioning of the heart will also lead to insufficient blood supply to the brain; muscle tightness on the neck, or cervical hyperplasia, pressing upon the blood vessel can decrease the blood supply to the brain; and narrowed or blocked capillaries in special areas within the brain may cause poor blood supply to local brain cells. In other words, abnormal quantity or quality of blood supply to the brain is one of the causes of chronic insomnia. Depending on the diagnosis, the Chinese formula for insomnia is usually composed of herbs that directly work on the central nervous system, with those improving digestion, blood circulation, or heart function to quantitatively or qualitatively improve blood supply to brain. I have successfully treated many patients suffering from insomnia with other symptoms, using Chinese herbal formulas based on differentiation of diagnosis.

It is necessary to mention here that TCM believes that the obstruction of circulation is one of the important inner causes of many diseases and aging. Details will be introduced in Section 10.1.7.

10.1.2 Yin and Yang Theory in TCM Emphasizes Balance

As mentioned in Chapter 1, the relationship of yin and yang is opposing, interdependent, inter-transforming (in a state of constant change), and balanced. This theory is applied in TCM for diagnosis and as principles of treatment. The imbalance and fluctuation of yin and yang are considered the basic causative factors of disease occurrence and development. The goal of clinical treatment is to restore yin-yang balance of patients.

Yin and yang can be applied to the two opposite aspects of anything. Overregulation of yin or yang to extremes is not advisable because it may make a disease worse. Such examples are seen in both herbal and modern medicines. For example, drugs for hypertension and diabetes may cause lower blood pressure and lower blood sugar; long-term administration of yin-tonifying herbs may cause yang deficiency, while long-term administration of yang-tonifying herbs may lead to yin deficiency. It was reported a few years ago that a famous man living in Hollywood was given warfarin for a blockage in his blood vessel, but one day when he fell down accidently on a sharp stone in his yard, he kept bleeding because of the warfarin he was taking; he died on the way to the hospital. Of course, such unfortunate cases happen to both Chinese and Western doctors, but good Chinese doctors are able to correct the side effects of one herb with a proper combination of others in a formula to correct the excessiveness of each other.

10.1.3 Five Elements and Zang-Fu Theories in TCM Emphasize Dependence and Interaction between Organs and Tissues

As mentioned in Chapter 1, the theory of the five elements is about relationship of the five elements: wood, fire, earth, gold, and water. They are inter-promoting, interacting, counteracting, and mutually related to each other. The five elements theory can be applied to many things, such as colors, tastes, and emotions. Specifically, wood, fire, earth, gold, and water are, respectively, linked to the liver, heart, spleen, lungs, and kidney of the body; anger, joy, sadness, grief, and fear of emotions; windy, cold, damp, dry, and cold of climates; spring, summer, late summer, autumn, and winter of the seasons; sour, bitter, sweet, pungent, and salty of tastes; green (or blue), red, yellow, white, and black of colors; and so on.

Now we know that the tastes and colors of the plants are decided by the structures of chemical compounds, and bioactivities of compounds are decided by their structures. Although ancient Chinese did not know the structures of compounds, they found out that the tastes and colors of foods and herbs do have a relationship with their physiological and therapeutic functions.

It is already known that the colors of flowers and fruits, and tastes of fruits, change along with seasons due to chemical reactions in the plants. Although these changes may not follow the law of the five elements exactly, they are dependent and they interact with each other, at least. When the five elements are applied in TCM, it can be used to explain the physiological and pathological interrelationship among Zang and Fu organs, and guide diagnosis and treatment of diseases. Figures about the relationships of the five elements and their corresponding Zang and Fu are available in many web sites and Chinese TCM books.

The liver, heart, spleen, lung, and kidney are known as the five Zang organs; they correspond to wood, fire, earth, gold, and water, respectively. The gallbladder, small intestine, stomach, large intestine, and bladder are the five Fu organs that correspond in order to the above Zang organs with an internal-external relationship. In addition, the pericardium and triple energizer forms another internal-external pair of Zang-Fu (see the explanation in Chapter 1, Section 1.5.1). Meridians connect not only between but also within Zang and Fu organs, forming an internal cross-network. Without such connection, each Zang or Fu organ would become an isolated and static organ, failing to perform its functional activities. On the other hand, the mutual interconnections between the meridians, Zang, and Fu organs mean that when a disease occurs on one particular meridian due to invasion of pathogenic factors, there may be a transmission of pathological changes to other meridians, particularly the externally-internally related one. This is why, when treating a disease in one organ, Chinese doctors usually add herbs either to protect possible damage to other organs or to enhance the effect via improving the functions of related organs.

In fact, function impacts between organs are also known in modern medicine. When the lung has problems, such as pleural effusion, it will affect the function of the heart. When the liver fails to produce bile, lipids will not be decomposed and can deposit in blood vessels, including the coronary arteries, and cause heart attack. It is also known that heart disease is a frequent problem among people with kidney disease. One of the reasons is that kidneys produce renin, an enzyme that regulates the blood pressure. If the kidney becomes dysfunctional, more renin enzyme will be produced, leading to higher blood pressure, which has the potential to damage the heart and lead to heart attacks. In addition, the new practice of neuroendocrine-immunology has convinced many that nerve, endocrine, and immune systems interact as a network.

Several patients in my clinic with post-surgical symptoms mentioned that their doctors told them the removal of the gallbladder or appendix would have no harm on the body. Of course, they have to be removed if keeping them would threaten patients' lives. However, the message that these doctors gave was inaccurate.

10.1.4 Scientifically Understanding the *Qi* in TCM

There is no English word or modern scientific term with similar meaning to that of the *qi*. Some say that *qi* is a magnetic field within the body. But according to *The Yellow Emperor's Inner Classic*, *qi* can be interpreted as covering all substances necessary for growth, development, and maintenance of normal functions of human body. *Qi* also refers to all life-sustaining functions. As a summary, *qi* in TCM has two meanings. One refers to the vital substances comprising the human body and maintaining its life activities, such as oxygen, water, nutrients from digested foods, and substances that are made of blood. Another refers to physiological functions of the body, such as functions of Zang, Fu, and meridians.

From the following introduction about the formation, functions, and categorization of *qi*, you will find *qi* covers almost every aspect of physiological functions of the human body. Regulation of *qi* by acupuncture and herbs is also briefly introduced.

1. *Formation of qi*

According to *The Yellow Emperor's Inner Classic*, *qi* has two sources. One is the innate vital substances inherited from parents before birth, carried by kidney. Another is the food essence and fresh air a person receives after birth, which have to be processed and transformed by internal organs, particularly the stomach and spleen, before becoming the *qi* of the human body.

It should be noted that the kidney mentioned in TCM usually refers to the kidney meridian, which involves the function of reproductive and urinary systems, rather than the kidney organ alone. In this case, it is easy to accept that TCM states the innate vital substance is related to kidney. Since there is no name found in old TCM books to match the pancreas organ, the author would like to add that the functions of spleen meridians in TCM include those of the pancreas, which is physiologically located just between the spleen and the stomach. One possible reason for this is the limited amount of physiological knowledge due to restrictions in access to human anatomy 2000 years ago.

The fundamental substances of *qi* include genetic traits from parents and the nutrients ingested from foods and oxygen breathed from air. TCM believes that *qi* formation involves the spleen, stomach, lung, and kidney. This is not in conflict with the knowledge of modern medicine. It is not difficult to associate the innate vital substances inherited from parents with genes and health conditions when born. It is also known that ingested foods need to be first digested, absorbed, decomposed, and distributed, then transformed into energy with the help of enzymes, or stored in other forms for later use. The foods are first chemically and physically digested by the stomach, then further by bile that is produced by the liver and stored in the gallbladder and by digestive enzymes released from the pancreas. After that, the digested nutrients are absorbed mainly in the small intestine and transported through portal veins to the liver where plasma protein synthesis, metabolism, and detoxification are carried out. The synthesized proteins and metabolized or detoxified substances in liver, that is, the food essence, are then transported to the heart, from which the blood carries nutrients and distributes them to all tissues, organs, and cells throughout the body for nourishment. The lung brings breathed oxygen to the heart; the spleen stores blood, disintegrates old blood cells, and filters foreign substances from the blood; the kidney filters the blood. In fact, all these Zang organs participate in the transportation and transformation of *qi*.

2. *Classification of* qi

Qi in TCM can be classified in different ways. If based on its origination sources and functions, there are inborn *qi*, pectoral *qi*, nourishing *qi*, and defending *qi*; if based on its effective organs, there are liver *qi*, heart *qi*, spleen *qi*, lung *qi*, kidney *qi*, stomach *qi*, and so on. There are also Jing *qi* (meridian *qi*) and luo *qi* (collateral *qi*), righteous *qi*, and evil *qi*. Acupuncture treatment mainly regulates Jing *qi* through adjusting the conducting induction along meridians.

1 *Inborn* qi (元气)

Inborn *qi* is also called primary *qi*. It is derived from congenital essence and is the most important and fundamental *qi* of all. It comes mainly from the innate essence stored in the kidney and spreads to the entire body. It needs to be supplemented and nourished by the *qi* obtained after birth from food essence. Inborn qi mainly stimulates and promotes the functional activities of Zang-Fu organs and the associated tissues.

Sufficient inborn *qi* guarantees normal growth and the development of healthy and vigorous activities of all functions of the body. On the other hand, insufficient inborn *qi* or deficiency due to a prolonged illness may lead to late and slow growth and development, and possibly reduce nearly all physiological functions, indicated by lassitude or general debility, and susceptibility to diseases.

In other words, the substances of inborn *qi* are mainly congenital and inherited from parents; therefore, their functions are involved in promoting growth and development and functions of all organs.

2 *Pectoral* qi (宗气)

Pectoral *qi* is the *qi* stored in the chest. It is a combination of the fresh air inhaled by the lung and the food essence produced by the spleen and stomach from water and grains. The pectoral *qi* flows through the respiratory tract to promote the respiratory movement of the lung, functions of the heart, and circulation of blood, and, as a consequence, enhances vital energy. Pectoral *qi* thus exerts an influence on the warmth and motor activities of limbs, loudness or softness of voice, and so on. The manifestations of a deficiency of pectoral *qi* include weakened functions of the heart and lung, such as shallow breathing, soft voice, abnormal heartbeat, slow blood flow, cold limbs, lassitude, and difficulty with movement.

In other words, the substances of pectoral *qi* come mainly from the air breathed in and foods digested and absorbed.

3 *Nourishing* qi (营气)

Nourishing *qi* refers to the *qi* circulating within the blood vessels that have a nourishing function. It comes mainly from the food essence transformed and transported by the spleen and stomach. The primary function of nourishing *qi* is to produce blood to nourish all organs and tissues via circulation throughout the whole body. It goes up and down along the channels, circulating around the body and thus providing nutrients for the physiological activities of all the viscera and bowels, meridians and collaterals, tissues, and other organs.

In other words, the substances of nourishing *qi* are necessary substances that travel through blood vessels to all tissues and organs to provide them nutrients and support the functions of cells.

4 *Defending* qi (卫气)

Defending *qi* is also derived from the food essence, but it travels between skin and flesh, and moves outside the vessels. Even so, it still leans against the channels when moving. The protective function of defending *qi* displays in three respects: guarding the surface of the body; keeping a relatively constant body temperature by controlling the opening and closing of the pores and adjusting the excretion of sweat; and nourishing the viscera, bowels, muscles, skin and hair. In addition, the circulation of defending *qi* is associated with sleep. One goes to sleep when defending *qi* circulates in the interior of the body, and wakes up when it circulates on the surface of the body. When defending *qi* is insufficient, the defending function of the human body is weakened, the exogenous pathogens invade the body more easily, and the disease is hard to cure. Abnormal circulation of defending *qi* may cause sleep disorders. From this point, defending *qi* covers the function of the immune system.

In other words, the substances of defending *qi* are the necessary substances outside of the blood vessels that are involved in regulating immune functions, temperature, sweating, and sleeping.

Based on the above definitions, it is not difficult to distinguish the difference of various *qi*. With the exception of the inborn *qi*, the pectoral *qi*, nourishing *qi*, and defending *qi* all depend on the food essence, that is, nutrients from foods. Therefore, TCM puts great emphasis on proper diets. Skilled Chinese doctors always add herbs that adjust for spleen and stomach in the formula to improve the function of digestive system when treating chronic diseases. The pectoral *qi* is stored in the chest and is derived from the combination of food essences and air. The nourishing *qi* circulates within the blood vessels and belongs to yin, and the defending *qi* moves outside of blood vessels and belongs to yang. When the nourishing *qi* and defending *qi* fail to coordinate with each other because of attack by exogenous pathogens, symptoms such as aversion to wind, fever, and sweating will occur.

5 *Other kinds of* qi

The *qi* can also be named depending on their functional locations. It is common to hear Chinese doctors mention liver *qi*, heart *qi*, spleen *qi*, lung *qi*, kidney *qi*, stomach *qi*, and so on. Jing *qi* means the *qi* circulates within the meridian *qi*, while luo *qi* is the *qi* circulated along the collaterals. Righteous *qi* refers to normal functioning, but evil *qi* refers to *qi* that causes diseases, such as the six exogenous pathological factors. It is said that when the righteous *qi* is stronger than the evil *qi*, the person is in health, but when the evil *qi* is stronger than the righteous *qi*, the person will be sick.

3. *Functions of* qi

The functions of *qi* in TCM cover different aspects of modern physiological activities of the human body.

1. *Promoting action: Qi* can help activate growth and development of the human body, promote functions of Zang and Fu organs, meridians and collateral, and speed up the formation and circulation of blood and metabolism of body fluid as well.
2. *Warming action: Qi* is the main source of heat needed by the body. The body keeps its constant temperature mainly through the warming action of its *qi*.

This action reflects the regulation of body temperature by the nervous system. Cold extremities usually indicates poor blood circulation.

3. *Defending action: Qi* not only guards the surface of the skin against foreign pathogens, but also combats the invading pathogens. When the defending function of *qi* is normal, it is difficult for pathogens to invade the body. Even when they manage to enter the body and cause diseases, the defending *qi* will inhibit or kill them so the disease is cured. This action actually reflects functions of the immune system in modern medicine. As mentioned before, TCM uses different terminologies to describe similar phenomena found in modern medicine.
4. *Consolidating and governing action: Qi* has the ability to command, control, and consolidate substances and organs in the body. This is done by keeping blood flowing within, not exploding out of the vessels; controlling and adjusting the secretion and excretion of sweat, urine, and saliva, and preventing bodily fluids from escaping; consolidating and storing sperm and preventing emission and premature ejaculation; and consolidating the organs so as to prevent them from descending. This action mainly reflects functions of the nervous and endocrine systems.
5. *Promoting metabolism and transformation: Qi* performs its functions through the process by which substances in the body are metabolized, and substances and energy are transformed. In detail, the ingested food is changed into food essence, and food essence is, in turn, transformed into vital energy, blood, or bodily fluid, which can then be changed into any one of the others according to the physiological need of the body. Although such explanation is more superficial, this action reflects the function of enzymes and other chemical reactions involving chemical exchanges between proteins, carbohydrates, and lipids, as well as energy transformation such as the tricarboxylic acid cycle (TCA cycle, Krebs cycle).

From the above functions of *qi*, it can be summarized that *qi* promotes growth, development, functioning of organs, blood formation, metabolism and energy transformation of human body; regulates body temperature; enhances functions of the immune, nervous, and endocrine systems; controls blood flow, secretion and excretion of sweat, urine, and saliva, and sexual function; and prevents organs from prolapsing. In brief, the functions of *qi* cover the functions of all organs and systems such as nervous, circulatory, immune, endocrine, and reproductive systems.

A decrease in the function of *qi* is called *qi* deficiency in TCM. A deficiency of *qi* can cause lowered body temperature, intolerance to cold and cold limbs, easy invasion by pathogens, hemorrhage, spontaneous perspiration, polyuria, salivation, spermatorrhea, premature ejaculation, prolapse of the stomach, kidney, and uterus, and many other problems. Dysfunction of *qi* also affects digestion, absorption, metabolism, and transportation of food necessary for formation, movement, and transformation of vital energy, blood, and bodily fluid, as a result, causing various symptoms associated with abnormal metabolism.

4. *Movement of* qi

Qi can move with an ascending, descending, outward, or inward direction. The functional activities of the Zang-Fu organs and meridians, and the relationship between the Zang-Fu organs, meridians, *qi*, blood, yin, and yang are maintained by the ascending, descending, outward, and inward movements of *qi*.

Abnormal ascending and descending refer to pathological states of the Zang-Fu organs, meridians, yin and yang, *qi*, and blood in which they fail to maintain their normal state of governing the ascending and descending of *qi*, leading to a variety of pathological changes. For example, the function of lung *qi* is descending and dispersing; if the lung *qi* moves toward the wrong direction, it causes cough or asthma. The function of spleen *qi* is sending up clear essence of food; if the spleen *qi* moves opposite, it causes loose stool and diarrhea. The function of stomach *qi* is sending down partially digested food. If the stomach *qi* moves up, then it causes belching, acid reflex, and vomit.

5. *Understanding regulation of* qi *by acupuncture and Chinese herbs*

Both herbs and acupuncture can treat patients by regulating the abnormal *qi*. Although this book is about herbal medicine, acupuncture is mentioned here to help in the understanding of the functions and regulation of *qi*.

TCM believes that Jing *qi* (meridian *qi*) can be stimulated by acupuncture, and the effect of acupuncture is to induce the conduction of meridians and collaterals, which will be introduced in the next section (Section 10.1.5). Many scholars who have received training from both traditional and modern medicines believe that the function of acupuncture is related to stimulation of the nervous, endocrine, immune, and circulatory systems. In addition to pain symptoms, acupuncture can actually treat many diseases or disorders, including heart diseases, digestion and metabolizing disorders, insomnia, stress, ear ringing, and paralysis caused by stroke. With knowledge of both acupuncture and physiology, it is easy to find that most acupuncture points are actually the places where nerves pass by, which encourages scientists to connect the results of acupuncture treatment with regulation of the central nervous system, autonomic nervous system, or motor systems.

In my clinic, I saved the life of a 90-year-old woman with heart failure simply through acupuncture in the spring of 2007. This woman had severe coronary heart disease; 75% of her major coronary arteries had been blocked, but her health condition did not allow her to receive surgery. The Western doctors had given up treatment except for giving her oxygen for life sustenance.

Before I gave her treatment, this woman was too weak to open her eyes, and her face was as pale as a sheet of paper. Unexpectedly, right after the first acupuncture treatment, this woman took off the oxygen tube from her nose with her own hand and said to me with smile, "I am fine now." Witnessing the miracle taking place in front of me, I was convinced that acupuncture effectively worked through adjusting her heart function via stimulation of the autonomic nerves. In the past 3 years, I have rescued her three times with acupuncture. This woman is still alive, with the help of Chinese herbs.

In comparison, Chinese herbs work with more complex mechanisms. There are two groups of Chinese herbs that mainly work on *qi*; one group is called *qi*-regulating herbs and the other group is known as *qi*-tonifying herbs. Modern pharmacological studies show that *qi*-regulating herbs could regulate gastrointestinal (GI) movement and improve secretion of digestive liquid. Some of them, such as hawthorn berry, can increase myocardial contractility by directly working on α- or β-adrenergic receptor and decreasing heart blood pressure. It has been demonstrated that most *qi*-tonifying herbs can increase functions of immune, cardiovascular, and digestive systems. Many of them can enhance functions of the endocrine system including the activity of hypothalamus-pituitary-adrenal axis and hypothalamus-pituitary-gonadal axis, promote metabolism and hematopoiesis, and eliminate free radicals.[2]

6. *Scientific basis of relationship between* qi *and blood*

Qi and blood are closely related. It is a saying that *qi* generates blood, promotes blood flow, and keeps blood moving within the vessels. Meanwhile, blood is the carrier of *qi*, *qi* moves along with blood flow, and is unable to exist without blood. And now we know that blood carries not only nutrients from food, but also endocrine hormones and immune cells and factors.

Using modern scientific methods, researchers from Hunan University of Traditional Chinese Medicine, China, recently confirmed that electronic acupuncture at Zu-san-li and some other acupunctural points on the stomach meridians that are commonly applied to treat digestive disorders were able to adjust the gastric mucosal blood flow, gastric motility, and secretion of the brain-gut peptide.[3] The brain-gut peptide works not only as a hormone but also as a neurotransmitter, and plays a very important role in the regulation of visceral sense, movement, and secretion. It has been known that this peptide is related to the cause of irritable bowel syndrome (IBS).

Using acupuncture or herbal formulas mainly composed of herbs for regulating and tonifying *qi* and invigorating blood circulation, I have successfully treated many patients with complaints of abdominal fullness and other digestive problems that Western doctors have failed to treat in my clinic. Besides these examples, regulation of *qi* and blood is effective in treating many other diseases.

In summary, the fundamental substances of *qi* are composed of the food essence and neurotransmitters, hormones, and all other chemicals necessary for functions of the body, and the functions of *qi* cover the functions of all of the organs and physiological systems.

10.1.5 Meridians and Collaterals in TCM Involve the Neuroendocrine-Immune Network and Circulatory System

The meridians and collaterals are pathways in which the *qi* and blood are circulated within the human body. They pertain to the Zang-Fu organs interiorly and extend over the body exteriorly, forming a network and linking the tissues and organs into an organic whole. As a result, they carry out functions of transporting *qi* and blood, balancing yin and yang, preventing diseases, reflecting symptoms of diseases, adjusting excess and deficiency, and so on.

It is known that the nervous and circulatory systems form networks in our body. The circulatory system can be further divided into cardiovascular and lymphatic systems. Therefore, by learning both anatomy and TCM theories, it is easy to associate the main nerve bundles, arteries and veins, and lymphatic vessels with meridians, and the nerve branches, and blood and lymphatic capillaries to collaterals.

It is known that all the functions in our body are controlled by the nervous system, including the pumping of blood from the heart and the contraction of blood vessels. On the other hand, survival and functionality of nerve cells relies on blood supply via the cardiovascular system. The function of the cardiovascular system is collecting and distributing oxygen, nutrients, hormones, and other necessary substances such as enzymes, receptor ligands, vitamins, antibodies, some cells, and many other substances to the tissues of the entire body to nourish all cells and regulate their functions, including nerve cells; meanwhile, the lymphatic system is collecting and removing waste products left behind in the tissues, including cell metabolites, dead blood cells, pathogens, and cancer cells. It is also known that immune and endocrine systems interact with the nervous system, forming the neuroendocrine-immune (NEI) network. In addition, the two systems depend on the circulatory system (including the cardiovascular and lymphatic systems) because they function via the delivery of substances through blood and lymphatic vessels. Besides, all the internal organs are connected through blood vessels.

If nutrients supplied from blood are reduced or stopped somewhere due to poor circulation or obstruction of blood flow, the local nerve cells will become dysfunctional or dead, resulting in numbness, pain, and many other problems. If metabolized wastes from cells or dead cells are not carried out right away and accumulate locally due to poor circulation, they would impact functions of cells around them, resulting in swelling or blockage, and gradually form cysts or fibrosis in organs such as liver, lung, or kidney, thus affecting the functions of the tissues or organs. If obstruction happens to capillaries within the brain, the symptoms may be ear ringing, insomnia, headache, and so on. If inflammation is involved in the obstruction, it may also cause pain, fever, or edema. The symptoms and degrees of the risk to life depend on where the circulatory obstruction happens.

10.1.6 Understanding Etiology and Pathology in TCM

TCM believes that there are four main causes of diseases. These are (1) the exogenous *qi*; (2) the emotions or passions; (3) improper diet and overstrain or lack of physical exercise; and (4) stagnant blood and phlegm fluid. The onset of disease can be generalized as being due to disharmony of yin and yang and conflict between righteous *qi* (anti-pathogenic *qi*) and evil *qi* (pathogenic *qi*).

There are six exogenous factors and seven emotional factors. The six exogenous factors are wind, cold, summer heat, dampness, dryness, and fire. They could be considered as evil *qi*. The emotional factors are joy, anger, melancholy, worry, grief, fear, and fright.

The names of these exogenous factors are derived from the symptoms they cause. For example, if a patient has high fever, it is said the disease is caused by heat or fire; if a patient is adverse to cold, it is said to be caused by wind cold; if a patient gets a stomach flu in the summer, it is said to be caused by summer dampness; if a patient feels pain worsen in joints or limbs as the weather cools, it is said to be caused by wind dampness. The author believes that the exogenous factors actually refer to extreme cold or hot weather, dry or damp environments, fire, infections by bacteria, viruses, fungi, and other natural microorganisms that come from nature.

In regard to passions or emotions, *The Yellow Emperor's Inner Classic* says, "anger injures liver, joy injures heart, thinking injures spleen, sorrow injures lung, fear injures kidney." Modern studies have shown that emotion changes can alter the release of neurotransmitters, hormones, and immune factors, thus impacting functions of nervous, endocrine, and immune systems, consequently affecting functioning of various internal organs. It is also known that inappropriate diet may cause imbalance of nutrients in the body, while overwork and insufficient exercise will impact the physiological functions of the body, thus gradually resulting in illness.

It is easy to understand and accept the above causes. Thus, there is little need for further explanation. However, I would like to further explain that all of them can result in another cause—phlegm and stagnant blood, which may further damage the internal organs, obstruct meridians, impact skin, muscles, bones or tendons, and cause many diseases.

The concept of phlegm in TCM is not as simple as the word in English. It is an abnormal product that can be divided into visible and invisible. The visible phlegm is the same as phlegm or sputum in English; the invisible phlegm stays within the body and may cause diseases. TCM believes that formation of phlegm results from abnormal accumulation of body fluid due to dysfunction of the lung, spleen, and kidney, as well as impairment of water metabolism. Such dysfunction may be caused by infection, inflammation, deficient nutrients, or accumulation of fat due to improper diets, lack of exercise, stress, and emotional changes. Clinical manifestations characterized by invisible phlegm accumulation include dizziness, blurred eyes, abdominal fullness, and overweight, depending on its retention area. Phlegm in TCM could be interpreted as the accumulation of any abnormal metabolized liquid and solid substances within any part of the body. It may be edema, a cyst, enlarged tissue, a tumor, or some syndromes that modern medicine still does not know the cause of yet.

Stagnant blood is mainly caused by impaired circulation. TCM believes it is formed due to either coldness, deficiency, or stagnation of *qi*. Coldness in TCM does not simply mean lower temperature, but frequently refers to pathogens that cause pain, which is commonly related to either inflammation or noninflammatory edema, both of which might locally impact circulation. Deficiency of *qi* may indicate either lack of nutrients in the blood due to poor digestion, improper diet, excessive bleeding, or functional decrease of organs and systems. Stagnation of *qi* may indicate dysfunctions of nerve regulation.

Clinical manifestations of stagnant blood vary according to the affected areas. If stagnation localizes within the brain, it may be stroke, ear ringing, headache, insomnia, or degenerated symptoms such as dementia; if it happens in internal organs, it may be coronary heart disease, liver or lung fibrosis, failure of kidney, abdominal pain or discomfort; if happens in the limbs, it may be pain, numbness, varicose veins, and so on.

TCM believes that pathogens, emotions, improper diets, or lack of exercise may all lead to poor circulation of cardiovascular and lymphatic systems. If it happens within the blood vessels and capillaries, it causes stagnant blood; if it happens between lymphatic vessels and capillaries, it causes phlegm. No matter where the obstruction of circulation is, the function of cells around the blocked area will be impacted because the blockage can affect the transportation of nutrients and elimination of metabolized wastes through circulation. If nerve cells are affected, then their function of transporting signals to the central nervous system will be impacted and they will fail to perform the regulatory function of nerve cells on the local tissue or organ. As a consequence, further dysfunctions of the involved organs or tissues are generated.

10.1.7 Why Obstruction of Circulation Is the Root Cause of Many Diseases and Aging

As mentioned in Section 10.1.6, blood stasis is believed to cause many diseases. Since learning this truth through several years of personal practice in my clinic, I have been utilizing it as a powerful weapon in the treatment of diseases. Looking back to traditional books, I have found that most of the formulas for treating chronic diseases include blood-invigorating herbs. Therefore, I added this section to discuss more about blood stasis.

Blood supplies nutrients to nourish the whole body and cleans metabolized wastes with the combination of lymph. It circulates internally within the vessels and capillaries to all the organs, and externally to skin, muscles, bones, and tendons to maintain their physiological functions. Once circulation is obstructed, their normal functions will be affected due to deficient nutrients to the cells and accumulation of wastes. Obstructed circulation may not only cause many diseases, but also accelerate aging.[4]

QingRen Wang, a Chinese doctor in the Qing dynasty, is the most famous representative who emphasized the importance of anatomy. He addressed in particular the role of stagnant blood in causing diseases. As mentioned before, in ancient China, cadavers were buried by the families and the dead bodies were not allowed to be cut open. Thus, Dr. Wang could only study the bodies of those who either died from plague or death penalty and who had no family members willing to assist in their burial. By using bodies that had either decomposed or been torn apart by wild animals, he was able to find several mistakes about the structural illustration of the human body as documented by his ancestors, but at the same time, made some new mistakes of his own. Nonetheless, his theory of stagnant blood serving as the cause of diseases has been considered a great contribution to TCM, and his many formulas treating chronic diseases with blood-invigorating herbs as major ingredients have been accepted and widely applied by following generations. He listed more than fifty diseases that related to dysfunction of *qi* and blood circulation in his book *Corrections of Errors among Physicians*. Diseases and symptoms related to stagnant blood include palpitation, chest pain, stroke, paralysis, cough, asthma, dizziness, insomnia, hair loss or whitening, sight or hearing problems, dry skin, senile plaques, purpura, fatigue, edema, frequent urination, decreasing sexuality, numbness, and pain. Female disorders such as amenorrhea, dysmenorrheal, irregular period, constant bleeding and fibrosis, as well as liver, lung, and kidney dysfunctions and tumors are also related to stagnant circulation.

Many of these correlations have been confirmed by modern medicine. Blood stagnation may be related to many chronic diseases that Western doctors have so far failed to cure because their causes are still unknown in modern medicine.

In addition to using blood-invigorating herbs to treat diseases, Chinese doctors like to combine tonifying or other herbs for antiaging. The author's explanation is that after several decades of continual work and strain on the body, blood vessels and capillaries, just like all pipes in life, will become more or less narrowed, fragile, or blocked somewhere. Once the circulation is blocked, the cells around the blockage area will not only lose nutrient supply, but also accumulate metabolized wastes, resulting first in dysfunction of cells and later in cell death. The author believes that the cause of degenerative diseases, such as Alzheimer and Parkinson's, more or less relates to capillaries' blockage. Many scientists are focusing on the mutant genes of these diseases, hoping to find the relationship of mutant genes to the diseases. But what causes the mutation? Is it possible that it is involved with or is accelerated by blocked microcirculation?

When I was at Columbia University, I joined the research on the pathological study of age-related macular degenerative disease (AMD). AMD is a disease causing vision loss in many people over age 50. The autofluorescent lipofuscin, particularly the major lipofuscin fluorophore, A2E, which accumulates with age in retinal pigment epithelial cells, was presumed to contribute to an age-related decline in cell function. Using mass spectra analysis, the biosynthesis of A2E within the epithelial cell was confirmed to be the result of the reaction of all-trans-retinal and phosphatidylethanolamine (PE) in the photoreceptor outer segment membrane. These two compounds are actually normal substances within the body, but once cell metabolism degenerates, A2E will be formed and accumulated there, impacting cell function.[5]

The view that the author wants to present with this example is that while aging is the law of nature, many factors may accelerate its progress; one of them being blockage of circulation which might result in accumulation of certain substances and impact or kill surrounding cells. What scientists can then do is to find the cause of such acceleration, so that a method to delay aging may be developed.

10.2 CHEMICALS ARE FUNDAMENTAL SUBSTANCES REFLECTING FUNCTIONS OF CHINESE HERBS

Chinese herbs have been used for several thousand years. However, modern pharmacological study on them just started in the 1920s, when K.K. Chen, a Chinese pharmacologist, first discovered the adrenergic effect of ephedrine isolated from the Chinese herb ephedra with the collaboration of chemists. Since then, more and more chemicals from Chinese herbs and their pharmacological activities have been studied. First, the studies simply copied the methods of Western drug development. However, with extensive studies, scientists gradually recognized that modern studies of Chinese herbs should be based on TCM theories. In 1980, the committee of Chinese Herbal Pharmacology was established in China, and the first corresponding textbook and journal were also published.

To date, research on Chinese herbs with modern chemistry, bioassay, pharmacology, and clinical trials have achieved great success. The main and active biological components in most commonly used Chinese herbs have been well known, and their pharmacological activities have been widely studied with different animal models. Scientists are gaining deeper understanding about the TCM theories that guide diagnosis and treatment with herbs.

The difference between modern drugs and traditional herbs is that the former contains only one single compound, while the latter is composed of many different types of compounds with different biological activities. The complexity and "mysteriousness" of composition makes herbal medicine difficult to accept by consumers and difficult to research for researchers.

In TCM, one disease can be treated with different formulas and thus have different results depending on the doctor's skill, and his or her understanding and application of TCM theories. Such different results then lead to different comments or conclusions about efficacy of TCM. This is why some believe TCM and testify to its "miraculous" nature, while others are left in doubt due to poor results. My clinical practice tells me that TCM can indeed do miracles to save one's life or help desperate people whom Western medicine have let down.

As a scientist having been trained by both traditional and modern science and medicine, I believe that fundamental substances of the efficacy of herbal medicine result from its complicated composition. The information about chemical compounds in herbs and their extraction and isolation is available in Chapter 3. However, if studies on herbs only focus on the efficacy of chemicals in herbs using methods for modern drug development, the "miraculous" nature of herbs cannot be easily explained. Researchers who are willing to commit themselves to the study of traditional medicine should understand the system first. Thus, the study of Chinese herbs must be combined with TCM theory and properties of herbs.

In the following sections, I will try to share my understanding of Chinese medicines with the knowledge of modern science by providing an explanation of medical terminology.

10.3 BRIEF INTRODUCTION TO THE PROPERTIES OF CHINESE HERBS

In most Chinese textbooks, Chinese herbs are categorized based on their functions. Each Chinese herb has its own properties: five tastes, four natures (temperatures), directions, and meridian attributions. Dating back to 2000 years ago, chemical structures and biology were unknown. But TCM believes that tastes and colors of herbs are related to their function (see Table 10.1). Different terminologies were used to describe the properties and corresponding biological activities of herbs in TCM.

It is known that different tastes of materials are the result of different chemical compounds in the materials, and different compounds have different biological functions. Different tastes of herbs are actually direct reflections of the properties of the compounds in them as sensed by the taste organ, as explained in detail below. Thus, based on the tastes of herbs, their functions could be roughly speculated. Temperature, direction, and meridian attribution are in fact indications of herbal functions or efficacies from different aspects. They were summarized based on accumulated experience of treatment results.

Properties of some herbs can be changed by herbal processing. Processing is applied to some herbs for the purpose of changing or weakening the properties, such as nature and direction. The processing methods include heating with sand, frying with vinegar, steaming, and boiling. Honey or salts are also used with some herbs. Processing has been demonstrated to change the structures of compounds in herbs, thus their bioactivities. This further confirms that chemicals in herbs are the fundamental substances that decide their properties.

10.3.1 Five Tastes

The five tastes of Chinese herbs refer to acrid or pungent, sweet or bland, sour or astringent, bitter, and salty. Chemical studies have shown the following relationships between tastes and chemical composition. Herbs with similar actions usually have same or similar tastes. The details of actions are available from many Chinese herbology books in English, so an explanation of herbal actions is not given here.

Table 10.1 Relationship of Tastes and Colors of Herbs with the Body Meridians

Taste of herbs	Color of herbs	Body meridian
Sour	Green or blue	Liver
Bitter	Red	Heart
Sweet	Yellow	Spleen
Pungent	White	Lungs
Salty	Black	Kidney

Acrid or pungent herbs generally contain more essential oils, followed by glycosides or alkaloids. TCM believes that herbs with acrid or pungent tastes are able to generate sweat, relieve cold, regulate *qi*, remove wind, and stop pain.

Sweet or bland herbs usually contain more nutrient substances that are necessary for body metabolism such as sugars, polysaccharides, proteins, amino acids, vitamins, or glycosides. TCM believes that sweet or bland herbs are able to tonify or harmonize functional systems in the body, particularly regulate the immune system, and enhance disease resistance. Some sweet-bland herbs can also drain dampness through diuresis.

Sour or astringent herbs contain more organic acids or tannins. These compounds usually have astringent or consolidating effects.

Bitter herbs generally contain more alkaloids or glycosides. TCM believes that bitter herbs are able to dispel heat, purge the bowels, and remove dampness by drying them out. Alkaloids or glycosides compounds have antibiotic, anti-inflammatory activities, and many other activities.

Salty herbs usually contain more inorganic salts or iodine. Most salty Chinese medicines are of an animal or mineral source. They can soften hard masses such as enlarged thyroid as well as purge and open the bowels.

10.3.2 Four Natures or Temperatures

The four natures of Chinese herbs refer to cold, cool, warm, and hot properties of herbs. Some books translate them as temperatures of herbs. Neutral is also used sometimes as description. As introduced in Chapter 1, TCM believes that diseases can be caused by external cold and heat pathogens. Correspondingly, symptoms can also be divided into cold and hot, described as cold *zheng* and hot *zheng* (see the explanation of *zheng* in Section 6.2.5), respectively. In general, patients with cold *zheng* should be treated with hot or warm herbs, while those with hot *zheng* should be treated with cold or cool herbs. The natures of herbs actually roughly indicate the herbal functions.

The hot *zheng* usually include fever or any hot feeling and burning sensation, while cold *zheng* include aversion to cold, cold extremities, or any cold feeling such as cold stomach. Cold or cool herbs are those able to reduce fever or hot feeling, while hot or warm herbs are those able to relieve cold feeling and warm the body. As mentioned in Section 10.1.4, defensive *qi* is thought to regulate body temperature. Modern pharmacological studies have demonstrated that Chinese herbs can regulate the body temperature through regulating functions of nerve, endocrine, or circulatory systems, and energy metabolism. For example, some alkaloids with structures similar to catechol ammonia were found in several hot or warm herbs such as radix Aconiti, Fructus Evodiae, Rhizoma Alpiniae. Studies showed that they worked as agonists of β-receptor, while extracts of some cold or cool herbs, such as Radix Anemarrhenae and Cortex Lycii could inhibit the activity of Na^+, K^+-ATPase, thus suppressing heat generation.

10.3.3 Directions

The directions of Chinese herbs include ascending or upwards and descending or floating. This is another property that indicates the function of Chinese herbs. TCM believes that herbs are able to correct wrong dynamic movement in diseases. For example, lung *qi* and stomach *qi* are expected to move downward; symptoms like coughing and vomiting are thought to be caused by reversed, upward effects of lung *qi* and stomach *qi*, so descending herbs are used to correct the wrong direction of the qi. While spleen *qi* is expected to move upward, diarrhea, dysentery, and prolapse of rectum or stomach are thought to be caused by reversed, downwards effects of spleen *qi*, so ascending herbs are used to redirect the spleen *qi*.

The directions of herbs are actually related to other properties. For example, upwards or ascending herbs are usually acrid or pungent and warm or hot, while downwards or descending herbs are generally sour, bitter, or salty, and cold or cool. From this point, the direction of and herb is another illustration of the herbal functions.

10.3.4 Meridians Attribution

The meridians attribution tells which meridian or organ a Chinese herb acts upon. TCM believes that there are 12 classic meridians in the human body that go along either the arms or the legs to the body trunk. They form six internal-external pairs: lung and large intestine, spleen and stomach, heart and small intestine, kidney and urinary bladder, pericardium and triple energizer (also translated as triple heater or triple burner), and liver and gallbladder. Each meridian has its own biological functions, not simply referring to the structural organs. TCM believes that symptoms are the reflection of disorders of organs or their related meridians; thus, functions of herbs are attributed to meridians based on their efficacy on the symptoms generated from the dysfunction of organs or meridians. For example, ephedra is used to stop cough and relieve the symptoms of asthma, thus it is said to enter the Lung meridian; rhubarb has purgative functions to cause diarrhea, so is said to enter the large intestine meridian.

The meridian attribution of each herb is actually determined based on the collected information of treatment results. It has been scientifically confirmed that the meridian attribution does have very close relationship with Chinese pharmacological activities.

10.4 MODERN PHARMACOLOGY OF CHINESE HERBAL MEDICINE

Currently in China, several thousand herbs are used by Chinese doctors, but of these only 300–500 species are used on a regular basis. Many others are only used in some regions or by some minorities. Most of the frequently used herbs have been well studied for their chemical compositions and pharmacological activities. There are several well edited English books for Chinese herbology or Chinese pharmacology[6,7] that cover the detailed properties and functions of each herb.

Different terminologies are not only applied in TCM to describe symptoms, causes, and diagnosis of diseases, but also the functions of Chinese herbs. As mentioned before, Chinese herbs are composed of complex chemical compounds. Therefore, modern study of Chinese pharmacology is far more difficult than modern pharmacology study. First, researchers need to match the symptoms described in TCM with modern medicinal terminology, such as the meaning of *Bi* (obstruction) syndrome; then they must find the right animal models for experiments. Sometimes, models for modern pharmacological study may not exactly match the illustration of symptoms in TCM, such as deficient or stagnant symptoms. Thus, researchers need to build up a new model based on TCM etiology or pathology.[8] In addition, pharmacologists must become familiar with the causes and symptoms of diseases in TCM. For this purpose, TCM theories should also be well known.

The following are very brief introductions of modern pharmacological studies on Chinese herbs with references to *Chinese Pharmacology*[2] and other books. Complete systematic pharmacological study results of individual herbs are available in the recent published *Zhong Hua Ben Cao*.[9] Very recent research results are partly available in English journals such as *Plant Medica*, *Phytochemistry*, *Phytomedicine*, *Journal of Agricultural and Food Chemistry*, and *Journal* of *Natural Products*.

10.4.1 Exterior-Releasing Herbs

Exterior-releasing herbs include herbs that release wind-cold and wind-heat. These herbs can dispel pathogenic factors from the external part of the body; for example, the beginning of the flu or other infections. The pathogenic factors treated by this group of herbs mainly include cold, heat, or wind. Herbs with acrid and warm properties are used to release wind-cold, while those with acrid, sweet or bitter, and cold or cool properties are used to release wind-heat. Modern pharmacological studies reveal that most of them have analgesic and sedative activities; many of them have antipyretic, anti-inflammatory, antibacteria, and antivirus activities; and few of the wind-cold-releasing herbs such as ephedra, cinnamon twigs, and ginger have diaphoretic effects. A few herbs also have antiallergic activities.

10.4.2 Heat-Clearing Herbs

Herbs in this group are bitter or sweet and cold or cool, and thus can regulate imbalances caused by excessive heat at various stages of pathological penetration by clearing heat, purging fire, or cooling organs or related tissues injured by heat. Modern pharmacological studies have shown that most of them have antibacterial activities; many have anti-inflammatory and antipyretic effects; some have antivirus, antitoxin (endotoxin and exotoxin) activity, or anti-neoplastic activities, while others enhance the immune system.

10.4.3 Downward Draining Herbs

These herbs are bitter and cold. They can eliminate pathogenic factors and conditions via purgative, laxative, or cathartic means. Modern pharmacological studies indicate that most herbs in this group have laxative and antibacterial activities; some have anti-neoplastic activities; and the subgroup of cathartics have diuretic effects.

10.4.4 Wind-Damp Dispelling Herbs

Most herbs in this group are acrid and warm, but a few are sweet and cool or cold. They can dispel wind and eliminate dampness from muscles, joints, tendons, and bones to relieve pain and obstruction associated with painful obstruction syndrome. They are commonly used for all kinds of arthritis. Both wind and dampness can cause pain, but pain caused by wind is movable rather than fixed, that is, patients may complain of leg pain the first day and shoulder pain the second day. But pain caused by dampness is accompanied by a heaving feeling. Modern pharmacological studies demonstrate that most of herbs in this group have anti-inflammatory and analgesic activities. Some of them, such as Cortex Acanthopanacis, Radix Angelicae pubescens, Herba Siegesbeckiae, and Caulis Sinomenii, can significantly suppress immune function.

10.4.5 Aromatic Damp-Dissolving Herbs

These herbs are aromatic, fragrant, and warm due to the rich essential oils in them. Thus, they can enhance the function of the spleen to dissolve, dry, or disperse dampness, transform and transport substances that would otherwise accumulate to cause damp imbalance or obstruction.

In comparison with the organ in modern medicine, the spleen in TCM covers much wider functions. It is an organ disliking dampness. If dampness accumulates in the spleen, it will be disabled from carrying out its transporting and transforming functions, leading to digestive disorder. Dampness in TCM does not simply mean water accumulation like edema. The symptoms of summer dampness are actually similar to those of the stomach flu.

As is known, digestion is involved in the functioning of stomach, gallbladder, liver, pancreas, and intestines. Spleen and stomach are in the relationship of Zang-Fu. Same with liver and gallbladder, meaning they are internal-external linked as well as in yin and yang relationship. There is no word for "pancreas" in TCM. Personally, I believe the spleen functions in TCM to cover the functions of pancreas, which not only releases a variety of digestive enzymes, but also is the largest gland in the body.

Modern pharmacological studies reveal that most of the damp-dissolving herbs can stimulate movement of GI tract; some inhibit acid secretion or prevent ulcer formation; and others inhibit growth of bacteria or virus.

10.4.6 Water-Regulating and Damp-Resolving Herbs

Most of these herbs have sweet and cool or cold properties. They can regulate water and resolve dampness to normalize water circulation, eliminate water accumulation, and drain dampness. The symptoms of dampness treated by this group of herbs include edema, dysuria, anuria, and phlegm. Pharmacological studies exhibit that most of them have diuretic and anti-infectious activities; some improve liver and gallbladder functions; and others enhance immune function and inhibit the growth of tumors. For example, both poria and poluporus have diuretic, anti-infectious, liver-protecting, and anti-neoplastic activities.

10.4.7 Interior-Warming Herbs

All of these herbs are acrid and warm or hot. Therefore, they can warm the interior of the body and dispel cold because of their warm and acrid properties; they especially warm the spleen, stomach, heart, and kidney. Modern pharmacological studies demonstrate that most of them can dilate blood vessels, enhance heart function, and improve digestion. Some of them have analgesic activities or regulate sympathetic systems.

10.4.8 *Qi*-Regulating Herbs

Most of them are acrid, many of them are bitter and warm; they especially warm the spleen and stomach. They can regulate *qi* function to promote normal circulation of *qi*, correct reversed flow of *qi*, and eliminate *qi* stagnation. Modern pharmacological studies show that these herbs can regulate GI movement and improve secretion of digestive liquid. Some can significantly increase heart blood pressure and myocardial contractility by directly work on α- or β-adrenergic receptor. Others have cholagogic activities or dilate smooth muscles of bronchi or uterus.

10.4.9 Digestive Herbs

Most of these herbs have sweet and neutral properties. They can promote digestion function to strengthen the spleen, enhance the appetite, promote digestion, and relieve food stagnation. Some of these herbs contain digestive enzymes and vitamins. Modern pharmacological studies demonstrate that they can stimulate GI movement and secretion of digestive liquid, and expel abdominal gas.

10.4.10 Stop-Bleeding Herbs

Most herbs in this group are bitter and cool or cold; some are acrid and neutral. They can stop both internal and external bleeding. Modern pharmacological studies confirm that they can all stop bleeding via either stimulating coagulation, anti-fibrinolysis, or accelerating platelet aggregation. Some can also improve blood circulation and function of blood vessel.

10.4.11 Blood-Invigorating and Stasis-Removing Herbs

Most of these herbs are acrid and/or bitter and warm; some are neutral; and a few are cool. They can activate blood circulation and eliminate blood stasis. Modern pharmacological studies indicate that they can increase cardiac output, dilate vessels, inhibit platelet aggregation, and improve circulation. Products made from some of these herbs such as salvia root and cnidium are commonly used in China for patients with coronary heart diseases and stroke.

10.4.12 Phlegm-Resolving and Coughing- and Wheezing-Relieving Herbs

These herbs can be subdivided into three groups: (1) herbs that resolve phlegm caused by heat; (2) herbs that resolve phlegm caused by cold; (3) herbs that stop cough and wheezing. Herbs in the first group are bitter or sweet and cold; those in the second and third groups are acrid and warm. Most herbs in this category have the ability to resolve phlegm, arrest coughing, and relieve wheezing and dyspnea. Modern pharmacological studies show that they have expectorant, antitussive, and antiasthmatic activities.

10.4.13 Shen-Calming Herbs

Shen in TCM means spirit and emotion. Some of the medicines in this group are from minerals and animals. The plant-sourced herbs in this group are sweet and neutral. They can tranquilize and calm the minds. Modern pharmacological studies demonstrate that they have sedative, hypnosis, and anticonvulsion activities.

10.4.14 Liver-Calming and Wind-Extinguishing Herbs

Liver wind and excessive liver yang are disorders with symptoms like spontaneous movement or seizure, mood disturbance, and problems of the head, eyes, and face. Modern pharmacological studies show that they have sedative, anticonvulsion, and antihypertension activities. Primary hypertension is believed to be directly related to excessive liver yang, which may originally be caused by kidney deficiency.

10.4.15 Astringent Herbs

Most herbs in this group are sour or sweet and astringent, and warm or neutral. They can bind, retain, restrain, and prevent the loss of precious body fluids and substances.

Modern pharmacological studies show that they can inhibit secretion of glands, stop diarrhea, and have astringent and antibacterial activities.

10.4.16 Tonic Herbs

These herbs can be further divided into four subgroups: *qi*-tonifying herbs, blood-tonified herbs,yin-tonified herbs, and yang-tonified herbs. Most of them are sweet and neutral, while yin-tonified herbs are cool or cold and yang-tonified herbs are warm and hot. They can separately strengthen *qi*, tonify blood, nourish yin, and build yang by improving function of internal organs, strengthen bodily constitution and improve overall health. Modern pharmacological studies demonstrate that most of them can increase functions of the immune system and the cardiovascular system. Many of them promote metabolism and hematopoiesis, and eliminate free radicals. Many *qi*-tonifying herbs can enhance the functions of the endocrine system, including the activity of hypothalamus-pituitary-adrenal axis and the hypothalamus-pituitary-gonadal axis, as well as improve functions of the digestive system. Most yang-tonifying herbs also increase the activity of hypothalamus-pituitary-gonadal axis.

10.5 CHINESE HERBAL FORMULAS

A Chinese herbal formula is a manifestation of the concrete application of TCM integrity theory in treatment. A formula is usually composed of 2–15 herbs with different indications or functions; sometimes the number will be extended to 20 or 30, depending on the complexity of the treated disease. It is necessary to acknowledge that a Chinese herbal formula is not the simple addition of several single herbs with similar functions, but the synergetic effect by working on different symptoms with different functions. TCM theories should not be overlooked when performing formula studies, for the reasons mentioned below.

Many traditional formulas have been practiced in clinic for hundreds or thousands of years. Some of them came from the two most famous books, *Shang Han Lun* and *Jin Gui Yao Lue*, which were written 2000 years ago. Experienced Chinese doctors seldom simply use the formulas left from their ancestors. They either modify these formulas from herbal books by adding or subtracting, and increasing or decreasing the quantity of specific herbs to adjust the therapeutic actions, or create their own, based on the symptoms and diagnosis. The strategy of a combination of herbs in a formula is mainly to enhance or extend therapeutic strengths and reduce the side effects. Many contemporary TCM doctors as myself apply their knowledge of modern Chinese herbal pharmacological study results to the formulation.

The herbs in a formula can be categorized into several groups (chief, deputy, assistant, envoy) based on the roles they play. The chief ingredient refers to the one that is indispensible to the formula and works directly against the principal pattern or disease. The deputy ingredient aids the chief herb in treating the principal pattern or disease and serves as the main ingredient directed against the principal pattern or disease. The assistant herb reinforces the effect of the chief or deputy ingredients, and directly treats a less important aspect of the pattern or disease, moderates or eliminates the toxicity or harsh properties of the chief or deputy ingredients, or has an effect against that of the chief ingredient in a complex case. A formula does not require all of these four types of ingredients. Detailed introduction about Chinese herbal formula is available in the English book *Formula & Strategies*.[7]

The study of a formula is either performed with the whole formula, or with each individual or pairs of herbs in the formula. Chinese scientists have performed many studies on Chinese herbal formulas. However, there are not many available systematic research results about formulas because study of a formula is much more difficult than that on individual herbs, because of their more complex chemical composition.

According to the online e! Science News on August 18, 2009, the study results from the University of Texas Health Science Center at Houston suggests that ancient Chinese herbal formulas used primarily for cardiovascular indications including heart disease may produce large amounts of artery-widening nitric oxide. The results from this center revealed that these formulas "have profound nitric oxide bioactivity primarily through the enhancement of nitric oxide in the inner walls of blood vessels, but also through their ability to convert nitrite and nitrate into nitric oxide."

Chinese doctors have known the interaction between herbs for a long time. Combinations of a few herbs may create undesirable side effects and adverse reactions. There are 18 incompatibles (shi ba fan) and 19 counteractions (shi jiu wei) recorded in most herbal textbooks.[2,6] Combinations of the listed herbs may likely lead to adverse effects and/or toxic reactions. An example is that Gan Cao (Radix Glycyrrhizae) is incompatible with Gan Sui (Radix Euphorbiae kansui), Da Ji (Radix Euphorbiae Pekinensis or Radix Knoxiae), Yuan Hua (Flos Genkwa), and Hai Zao (Sargassum). Modern studies have scientifically showed that during the boiling process, saponins in Gan Cao could form complexes with the toxic steroid compounds in Gan Sui; as a result, the dissolution rate of the latter is increased, that is, the concentration of the steroid compounds is increased, leading to a raised toxicity.[2]

Although Chinese herbal formulas are the main focus in TCM clinical application, there are not many reports about systematic mechanism study and clinical trials (not including case studies) on Chinese herbal formulas so far. As mentioned before, one main reason is that the more complex composition of chemicals in the formulas makes it difficult. But the author believes that with efforts of the scientists, the secrets of these traditional formulas will be soon explored. And discovery of the secrets in these formulas may likely help scientists to understand diseases from a broader perspective of modern medicine and, thus, to find more effective and safe treatment methods for diseases that currently are not well controlled.

REFERENCES

1. Veith, I. (Translation) (2002) *The Yellow Emperor's Classic of Internal Medicine (in English)*. Berkeley and Los Angeles, CA, University of California Press.

2. SHEN, Y.J. and CHEN, C.X. (2008) *Traditional Chinese Pharmacology* (5th ed.), Shanghai, Shanghai Science and Technology Publisher.
3. LIN, Y.P., et al. (2007) Effect of acupuncture at Foot-Yangming Meridian on gastric ucosal blood flow, gastric motility and brain-gut peptide. *World Journal of Gastroenterology* 13(15):2229–2233.
4. YAN, D.X. (1995) *Aging & Blood Stasis—A New TCM Approach to Geriatrics*. Boulder, CO, Blue Poppy Press.
5. LIU, J.H., et al. (2000) The biosynthesis of A2E, a fluorophore of aging retina, involves the formation of the precursor, A2-PE, in the photoreceptor outer segment membrane. *Journal of Biological Chemistry* 275(38):29354–29360.
6. CHEN, J.K. and CHEN, T.T. (2001) *Chinese Medical Herbology & Pharmacology*. City of Industry, CA, Art of Medicine Press.
7. BENSKY, D. and BAROLET, R. (1990) *Chinese Herbal Medicine: Formula & Strategies*. Seattle, WA, Eastland Press.
8. CHEN, Q. (2006) *Pharmacological Research Methodology of Chinese Medicine* (2nd ed.). Beijing, People's Health Publishing House.
9. STATE ADMINISTRATION OF TRADITIONAL CHINESE MEDICINE (1999) *Zhong Hua Ben Cao*. Shanghai, Shanghai Science and Technology.

Index

Traditional Herbal Medicine Research Methods, Edited by Willow J.H. Liu

Printed and bound by CPI Group (UK) Ltd, Croydon, CR0 4YY
16/09/2026
14963458-0008